Anuran Communication

ADMIRANDA TI
LEVIUM SPECTACULA
RERUM

ANURAN COMMUNICATION

EDITED BY MICHAEL J. RYAN

SMITHSONIAN INSTITUTION PRESS
Washington and London

TO AUSTIN STANLEY RAND
FOR HIS CONTRIBUTIONS TO THE SCIENCE AND SCIENTISTS OF THE TROPICS

Frontispiece from *Historian Naturalis Ranarum* by August Roesel von Rosenhof, printed 1758.

Grateful acknowledgment is made for permission to reprint figures from the following works: Figure 6.2 from I. B. Stiebler and P. M. Narins, Temperature-dependence of auditory nerve response properties in the frog, Hearing Research 46:63–82, Figure 6, © 1990, Elsevier Science; Figure 6.3 from M. S. Smotherman and P. M. Narins, The effect of temperature on electrical resonance in leopard frog saccular hair cells, Journal of Neurophysiology 79:312–321, Figure 1, © 1998, the American Physiological Society; Figures 6.5 and 6.6 from P. van Dijk, P. M. Narins, and J. Wang, Spontaneous otoacoustic emissions in seven frog species, Hearing Research 101:102–112, Figures 3 and 5, © 1996, Elsevier Science; Figure 7.2 from H. C. Gerhardt, Acoustic communication in treefrogs, Proceedings German Zoological Society, Gustav Fischer Verlag, 1983, 25–35, Figure 3, © Spektrum Akademischer Verlag, Heidelberg; Figure 8.3 from M. J. Ryan and A. S. Rand, Female responses to ancestral advertisement calls in the túngara frog, Science 269:390–392, Figure 1, 1995; Figure 12.4 from G. E. Wever, The Amphibian Ear, © 1985 by Princeton University Press, and A. Elepfandt, Underwater acoustics and hearing in the clawed frog, *Xenopus, in* R. Tinsley and H. Kobel (eds.), The Biology of *Xenopus,* © 1996 by Oxford University Press, Oxford; Figure 14.3 from J. J. Schwartz, Male calling behavior, female discrimination and acoustic interference in the Neotropical treefrog *Hyla microcephala* under realistic acoustic conditions, Behavioral Ecology and Sociobiology 32:401–414, Figure 5, © 1993 by Springer-Verlag GmbH and Co. KG; Figure 16.3 from R. Márquez and J. Bosch, Female preference in complex acoustical environments in the midwife toads *Alytes obstetricans* and *Alytes cisternasii,* Behavioral Ecology 8:588–594, © 1997; Figure 16.4 from R. Márquez and J. Bosch, Male advertisement call and female preference in sympatric and allopatric midwife toads (*Alytes obstetricans* and *Alytes cisternasii*), Animal Behaviour 54:1333–1345, © 1997; Figure 16.5 from J. Bosch and R. Márquez, Acoustic competition in male midwife toads *Alytes obstetricans* and *A. cisternasii* (Amphibia, Anura, Discoglossidae): Response to neighbour size and calling rate, Implications for female choice, Ethology 102:841–855, © 1996; Figures 17.1, 17.2, and 17.3 from B. Waldman, J. E. Rice, and R. L. Honeycutt, Kin recognition and incest avoidance in toads, American Zoologist 32:18–30, © 1992; and Figure 17.4 from B. Waldman and M. Tocher, Behavioral ecology, genetic diversity, and declining amphibian populations, *in* T. Caro (ed.), Behavioral Ecology and Conservation Biology, © 1998, Oxford University Press, New York.

Copy Editor: Fran Aitkens
Production Editor: Ruth G. Thomson
Designer: Janice Wheeler
Library of Congress Cataloging-in-Publication Data
Anuran communication / edited by Michael J. Ryan.
p. cm.
Includes bibliographical references.
ISBN 1-56098-973-4 (alk. paper)
1. Anura—Behavior. 2. Animal communication. I. Ryan, Michael J. (Michael Joseph), 1953–
QL668.E2 A58 2001
597.8′159—dc21 00-047006

British Library Cataloguing-in-Publication Data available

∞ The paper used in this publication meets the minimum requirements of the American National Standard for Information Sciences—Permanence of Paper for Printed Library Materials ANSI Z39.48-1984.

Contents

Preface

When I was a graduate student intent on studying frog communication, I was asked: If you are interested in behavior why not study an animal that behaves? Times have changed. Not only have studies of anuran communication made critical contributions to our understanding of behavior, evolution, and neurobiology, but much of the world has become concerned with the demise of many of these placid gnomes of the night. One of the purposes of this book is to review recent studies of how these animals communicate with each other. This is a field that has been rapidly expanding but has not been reviewed in book form since the publication of *The Evolution of the Amphibian Auditory System* by Fritzsch et al. (1988, John Wiley and Sons, New York).

The authors contributing to this volume come from diverse backgrounds: animal behavior, developmental biology, endocrinology, evolution, ecology, and neurobiology. This diversity reflects the strength of frogs as study organisms. The biology of most frogs lends itself to scrutiny by a variety of disciplines and thus encourages integrative studies of how and why these animals communicate as they do. As the following chapters show, we now have some understanding of how the frog's brain recognizes sound, how its larynx makes sound, and how both of these processes are influenced by the animal's internal physiological state. We have information on the strategies that males use to call to one another; how female preferences for call variation contribute to sexual selection, speciation, and hybridization; and how the inherent structure of the auditory system might generate sensory biases that direct signal evolution. The wealth and integration of these studies have led to more fruitful insights than would be possible from a more narrow hypothesis-testing approach, which sometimes leads researchers to concentrate only on the biology that is superficially relevant to the question at hand. If nothing else, these chapters should make it clear that every aspect of an animal's biology is relevant to some biological questions.

The second purpose of this book is to celebrate the distinguished career of Austin Stanley (Stan) Rand. The chapters are from a symposium in his honor at the joint meeting of the American Society of Ichthyologists and Herpetologists, the Herpetologists' League, and the Society for the Study of Amphibians and Reptiles held at the University of Guelph in Ontario in 1998. Stan has contributed substantially to tropical biology, and especially to herpetology. As importantly, he has made an indelible mark on the careers of many tropical biologists, including most of the contributors to this book. For many of us he has opened the door to the wonders of the New World tropics and the behavior and ecology of so many of the animals that reside there. His contributions and his friendship are appreciated by all.

Acknowledgments

I am grateful to the Smithsonian Tropical Research Institute in Balboa, Panama, and especially its director, Ira Rubinoff, for financial support of this endeavor, as well as to the Herpetologists' League for underwriting some of the expenses of the symposium from which this volume was birthed. Every researcher invited to participate in this project agreed to do so. I thank all of them for their enthusiasm throughout—a tribute, no doubt, to the esteem in which they hold Stan Rand. I especially thank Nicole Kime whose assiduous attention to detail has greatly improved the quality of this final product.

Michael J. Ryan

PART ONE

Introduction to Anuran Communication

1

MARY JANE WEST-EBERHARD

The Importance of Taxon-Centered Research in Biology

Readers who choose this as their first book on frogs and who have never visited the Smithsonian Tropical Research Institute in Panama may well ask why there is a whole book dedicated to A. S. Rand. That question has both a particular and a general answer. The particular answer is that this person, Stan Rand, is one of the finest tropical field biologists in the world, and probably one of the best tropical herpetologists of all time, respected for his uncanny ability to find and understand living organisms in the field, as well as for his technical know-how and understanding of evolutionary biology. Armed with a vast knowledge of natural history and a formidable sense of humor, he has inspired people of all ages and backgrounds to study tropical reptiles and amphibians in the field. Some of those people—the authors of this book—gathered for a two-day symposium at the 1998 meeting of the American Society of Ichthyologists and Herpetologists. This was an appropriate tribute to Rand, for he has published prominently on so many of the groups covered under "Herps" at the "Ichs and Herps" meeting: crocodiles, snakes, geckos, iguanas, *Anolis* lizards, and of course frogs.

The more general answer to the question, "Why a whole book?" has to do with the value such a biologist has for a field of science. This book brings together an outstanding group of scientists drawn from a variety of disciplines between which there is often little interaction. All of them are interested in some aspect of animal communication, and all do some work on frogs. But they are united by a third factor, and that factor is Stan Rand. Because Rand knows the organisms and the issues well enough to connect with all of these fields, he ends up being the hub of a broad interconnected research enterprise. He is the one who, by his unparalleled understanding of the organisms, can bring collaborators together and make a cohesive group out of a diverse collection of individual scientists, and can make a model organism out of a frog—or, in some of his earlier work, a lizard. Ernest Williams, with considerable input from Stan, was similarly the hub for a group of scientists focused on the genus *Anolis*, making it into a model group for studies of ecology, evolution, and biogeography. The "Anolis Newsletter" was a famous sign of the enthusiasm and unifying synthetic power of what I am calling taxon-centered research.

When a taxon becomes the focal organism for research, there is a tremendous multiplier effect on the increase in knowledge that can result. Taxon-centered research is a self-accelerating process. The more you know about a particular species or a particular group of organisms the more you can find out about it, and the more valuable it is as a resource for research of general interest.

A taxon-centered biologist is a person who sets out to learn everything he or she can about a particular species or group, a person who is interested in every aspect of its biology and who lets the organism suggest the important questions.

I worry that biology today has too few taxon-centered biologists. The scarcity of such people is a kind of crisis in modern biology. Why does it matter? Consider what can happen in fields where taxon-centered biologists are scarce, fields like genetics, psychology, and developmental biology. These fields can easily get fixated on a narrow set of species subjected to a narrow set of questions—*Drosophila* and *E. coli* in genetics, the white rat in psychology, nematodes, *Xenopus, Drosophila,* and "the mouse" in developmental biology, for example. When this happens biologists in these fields are in a trap. A student could not dare work on another species and hope to compete, not only because of the backlog of information on the traditional species, but because there are unlikely to be taxon-centered biologists around to consult regarding other potentially useful species. No one is likely to know enough about any other organism to start in a new direction or to help a student do so.

Someone could argue that these taxon-poor, narrowly focused fields are the most successful of all. But eventually such a field runs into a wall where limited vision is a handicap. Developmental biologists are running into such a wall in their limited ability to answer questions about evolution. This has driven them, for example, to consult entomologists about the comparative biology of segmentation in insects in an effort to understand the evolution of the Hox genes that regulate segmental diversification. How typical is *Xenopus* of amphibians, and the mouse of mammals? What is the natural history of learning in wild rats? It takes a broadly informed taxon-centered biologist to answer these questions, which are crucial for a deeper understanding of both mechanisms and ideas.

There is a long list of specialized frog users who would have run into a wall if they had not first run into Stan Rand. In the last 10 years he has collaborated with molecular phylogeneticists, spectral analysts of sound, alkaloid chemists studying acquisition of defensive chemicals, physiologists studying the physical properties of air and water for flight and sound production, students of nutrition and growth, researchers of diurnal activity patterns, biodiversity experts doing surveys of neotropical vertebrates, climate experts interested in the effects of El Niño and seasonality, and a morphologist working on how to tell the age of a vertebrate from its bones.

What has caused the shortage of taxon-centered biologists, and how can we do something about it? One cause is the emphasis on hypothesis testing in graduate education. Hypothesis testing has improved the level of research in biology and raised the quality of doctoral theses, but it has driven students away from organisms into a belief that meaningful questions can be, or even must be, formulated without knowing some organism first. So people lose something—the feeling, or the conviction, that you can be a first-rate biologist and even get a job if you start with an organism and aim to find out all you can about it—everything it does and is, testing hypotheses, yes, but letting the organism lead you to the important questions.

Of course you would not choose a research organism without reference to concepts or questions. First you might decide in general what concept or field or idea you are interested in. Then you would pick a group or a species that has at least certain key qualities of interest, and is reasonably common and accessible. Then you would set out to become the world's expert on the organism, always, of course, asking questions but not too narrowly at first.

Isn't this risky? Is it bad advice for a beginner? Absolutely not. There is a secret, though, to doing this successfully. The secret is to select your own organism, one not too much studied, especially in your field and especially not by your major professor. Sounds risky again? It is true that some of the old guys—professors—would have to approve, but perhaps they could learn that if you start with a new organism, originality is assured. Quick status as a world authority is assured. Important new ideas are guaranteed, because the organism will lead you to new ideas if you just give it a chance.

Many people think that reading theory is a quicker route to a conceptual splash. You can criticize old theories and make a new model to replace them, and then find a suitable organism and devise a test. That is what I would call risky. If, instead, you test and modify theory by reflecting on an organism—your favorite organism, the one you know well—you are assured of a biologically meaningful answer. If you just think abstractly and then pick what you optimistically think will be a convenient or easy organism, you are far more likely to go off the track or along the same track others would take.

Beyond these practical scientific reasons for focusing on an organism, taxon-centered research is more fun. Organisms are beautiful, intricate, full of secrets that are surprisingly easy to discover if you just look. Taxon-centered research is endlessly interesting because you can make it whatever you want it to be. If you like lab work you can work on your organism in the lab. If you like library research you can do that. If you like to travel you can do a comparative study of a Brazilian species with one in India. And if you like to teach you can teach, with the assurance that everyone will be fascinated by what you say, whether they are university students, members of the local Rotary club, or the kids next door. You can make a drastic change in the topic of your research without losing momentum, because a new direction will capitalize on what you already know and will carry you to a new level of expertise.

How easy is it to find "your own" species or group not already being studied by someone else? Not as hard as you might think. Greene and Losos (1988) list several species of reptiles and amphibians recently described for the first time in California, an area explored in great detail by herpetologists for more than a century. You would think that with Stan Rand and his mafia hanging around Barro Colorado Island and Gamboa in Panama, everything would be known about the local frogs. But Rand and Myers (1990, p. 387) say that "most biologists would be genuinely shocked at how little we really know about the biota" of this long-studied region. More important is what they say next: "Many species are amenable to study in ways limited only by the imagination and energy of the investigator," and they hope that their review will encourage further work. They then list 52 species of amphibians and 82 species of reptiles. For each species they list what is known about their abundance, habitat, activity cycles, food, seasonality, reproductive mode, breeding places, and more, with references. They then list what is completely or almost completely unknown. Such a review is a ready-made setup for dozens of research careers on reptiles and amphibians.

An organism does not have to seem perfect for some preconceived question to be a good choice. The so-called ideal organism of grant proposals and research reports may give the impression that the project started with a question and proceeded to find the ideal test organism. As likely as not the process was the reverse, and the phrase should read "the organism for which this is an ideal question." In an article in praise of taxon-centered research Wilson (1989) invented what he called the "inverse Krogh's Rule." Krogh's Rule, attributed to the physiologist August Krogh, states that for every biological problem there exists an organism ideal for its solution. E. O. Wilson's version states that for every organism there exists a problem for the solution of which it is ideally suited. Even then the statement is an exaggeration. Consider the frog. This book shows how useful frogs are for many kinds of research on animal sounds. Is the frog the "ideal organism" for studies of animal communication? Certainly some other organism could do as well. Hundreds of species of beetles make sounds (Alexander and Moore 1963). Many beetles can be bred in captivity, and are common and easy to find in a pleasant stroll through any sunlit field. Frogs, on the other hand, have fairly long and delicate life cycles, and few species are bred in the lab. They live in snake- and mosquito-infested swamps. You have to go skulking after them at night, and they hop and slip out of your hand. Why frogs? Why do so many scientists with diverse interests, like the authors of this book, find frogs to be suitable research organisms? Because frogs are a solid subject of taxon-centered research by a number of biologists broadly interested in their biology. Frogs did not have to start as "ideal organisms." They became extremely useful organisms for the topics of this book because taxon-centered work made it possible to follow a diversity of lines of research without running into walls of ignorance and impracticability.

The history of biology is replete with monuments to the value of taxon-based research. An overly casual account of Darwin's life might suggest that his successful career in biology began with the voyage of the *Beagle* and culminated in a visit to the Galápagos, where the nature of species was suddenly revealed. In fact Darwin was ignorant of the significance of many of the Galápagos specimens until they were examined by taxon-centered biologists back in England, and he failed to label properly the localities of the Galápagos finches, believing them, in his innocence regarding birds, to be all of a single species. Darwin's friend Hooker once described Darwin's scientific life as divided into three stages: the first, as mere collector during his student days at Cambridge; the second, as collector and observer during the voyage of the *Beagle*; and the third as "trained naturalist" after eight years of concentrated work on barnacles. During the barnacle period he mastered not only the taxonomy of this enormous group of marine invertebrates, but engaged in a broad study of variation within and between species that involved extensive dissections and studies of embryology, sexual dimorphisms, and general natural history—work that qualified him as the world expert of his day on barnacles (see Darwin 1892). The barnacle work served as a solid base for all of Darwin's later thinking on the causes of variation and selection in evolution, even though his thoughts ranged over all of nature. Of the long concentration on barnacles Huxley said that he "never did a wiser thing" (Darwin 1892, p. 166).

A taxon-centered biologist is interested in the organism per se. Among scientists this is a special breed. You can recognize a genuine taxon-centered biologist easily, by several key attitudes:

(1) Recognition of the organism as an individual. This kind of biologist does not call an organism "it" as if the organism were an inanimate thing. Such a biologist refers to an organism as he or she—unless of course it is a genuine hermaphrodite, castrated, or truly asexual.
(2) Proprietary recognition of the species or group. The taxon-centered biologist refers to his or her taxon somewhat possessively—as "my snakes" or "my

wasps"—and is likely to do the same with respect to organisms studied by others ("Stan's frogs," "Bill's spiders").

(3) Respect for the organism as a living thing. The organism is proudly referred to as an actual organism—a frog, a wasp, or a spider—no matter how quaint or trivial this may sound to others. The organism is not called "my system" as if it were a dead object, a diagram, or a machine.

(4) Recognition of the organism as an aesthetically pleasing, beautiful object. No matter how ugly the organism may seem to others, the taxon-centered biologist considers it beautiful. A prominent comparative morphologist, for example, revealed herself to be a true taxon-centered biologist by introducing a lecture with a slide showing the most nondescript, colorless organism I have ever seen and then speaking of it as "this beautiful animal."

(5) A feeling for the organism. The organism-centered biologist has a certain empathy for the organism that is not the same as regard for a person or a pet, but that allows sufficient identification for deep understanding. The most striking example of this I have witnessed was that of a woman student of social wasps who described the swollen, bilateral ovaries of a queen wasp while unconsciously pressing her hands over her own ovaries! This type of biologist often embellishes lectures with lifelike imitations of animal behavior, and may show startling accuracy in mimicking animal postures and sounds.

This kind of empathy is not restricted to zoologists or to people working on whole organisms. A perfect example of a taxon-centered biologist was the maize geneticist Barbara McClintock, a molecular biologist of plants who was awarded the Nobel prize for her discovery of transposable elements in corn. McClintock identified so strongly with her corn plants that she would not leave them for a day while they were seedlings, and she was alerted to the effects of transposable elements by variegated abnormalities in leaves, which she considered responses of her plants to some sort of trauma suffered during growth. She did not see corn plants as genetically programmed objects, but as living things capable of responses to what she called "insults" from their environments. A biographer emphasized the reason for her special talent for intuitive insights as "a feeling for the organism," a phrase taken from interviews with McClintock herself, and quotes a colleague engaged in cancer research as saying, "If you want to really understand about a tumor, you've got to *be* a tumor" (Keller 1983, p. 207). Not surprisingly, some unusual insights on the behavior of yeasts, overlooked by laboratory scientists for decades, have come from the laboratory of Gerald Fink, a geneticist who has dedicated more than 30 years to studying brewer's yeasts (*Saccharomyces cerevisiae*). Fink reveals himself to be a true taxon-centered biologist by referring to those particular yeasts as "our guys" (Radetsky 1994).

There are probably many such organism- or taxon-centered researchers who hide their true identities when they write grant requests and research reports. They too often succumb to pressure to use terms like "systems" for species, and to give the impression that ideas have come from hard quantitative theory rather than a feeling for the organism. Many large-scale research programs, such as ecosystem or biodiversity initiatives, depend upon taxon-centered biologists for facts and understanding, yet they do not budget for taxon-centered fieldwork and museum work to back them up. Can we as biologists and citizens afford to openly promote this taxon-centered kind of research? We can ill afford not to support it.

It does not take much reflection to realize how crucial taxon-centered expertise is for biology and for the public good. Funding administrators need to be reminded of the successes and needs of this kind of research. When U.S. troops were suddenly transplanted into the Middle East during the 1991 Gulf War, there was a military emergency because of the threat of insect-borne disease. Army entomologists in Panama were pressed into service to provide expert information on vectors of disease, and they scrambled to see what could be found in their experience and knowledge of a pre-existing literature. Whenever there is a taxon-centered emergency, whether due to a virus like HIV, a disease of cattle in Britain, worries about biting insects in the Persian Gulf, or disappearing amphibians around the world, the public turns to taxon-centered biologists for answers and solutions. If the answers are not immediately forthcoming, as in the widespread and alarming die-offs of amphibian populations, someone should point out that answers might be available if basic research on these organisms and other taxonomic groups had been given higher priority in the research budgets of the past.

An old book called *Historian Naturalis Ranarum* by August Roesel von Rosenhof shows that taxon-centered research and "a feeling for the organism" has a long tradition in herpetology. It was printed in 1758, exactly 200 years before the publication of Stan Rand's ninth scientific paper at the end of his first year of graduate school. The frontispiece from von Rosenhof's frog book, reproduced as the frontispiece of this book, is a classic of biological illustration, one of the earliest to depart from fanciful depictions of semimythical beasts to show the true beauty of real frogs (Ford 1992). I like to imagine that this hand-colored engraving shows Barro Colorado Island, Panama, in the eighteenth century, and I note that there is an observation bench with space for sound-recording equipment already in place among the frogs, in anticipation of the glorious era of frog-centered research ahead. On the side of the bench is an inscription:

ADMIRANDA LEVIUM SPECTACULA RERUM

or "Wonders of small things for your admiration." The inscription captures the spirit of Stan Rand's career and is a good introduction for this book.

References

Alexander, R. D., and T. E. Moore. 1963. The evolutionary differentiation of stridulatory signals in beetles (Insecta: Coleoptera). Animal Behaviour 11(1):111–115.

Darwin, F. (ed.) 1892 [1958]. The Autobiography of Charles Darwin and Selected Letters. Dover, New York.

Ford, B. J. 1992. Images of Science: A History of Scientific Illustration. Oxford University Press, New York.

Greene, H. W., and J. B. Losos. 1988. Systematics, natural history, and conservation. BioScience 38:458–462.

Keller, E. F. 1983. A Feeling for the Organism: The Life and Work of Barbara McClintock. W. H. Freeman, New York.

Radetsky, P. 1994. The yeast within. Discover March 1994:45–49.

Rand, A. S., and C. W. Myers. 1990. The herpetofauna of the Barro Colorado area. Pp. 386–409. *In* A. Gentry (ed.), Four Neotropical Forests. Yale University Press, New Haven, CT.

Wilson, E. O. 1989. The coming pluralization of biology and the stewardship of systematics. BioScience 39(4):242–245.

2

A. STANLEY RAND

A History of Frog Call Studies 405 B.C. to 1980

Introduction

In the days before computer literature searches, researching a topic meant long hours over dusty volumes, after first wandering down the hall to ask your neighbors what they knew. Currently electronic searches yield not only titles but abstracts as well; e-mail exchanges replace conversations with colleagues down the hall. Far faster than the old system, the new one does not extend far into the past.

Does it matter? Why should anyone be interested in the old literature? In part, because it contains information that has been overlooked but may have relevance today. For example, Donaldson in 1911 reported that the relative size of the leopard frog central nervous system changes seasonally. This is currently relevant because of recent work on birds relating changes in brain size to song production. Why was it ignored? I don't know, but I can report that I found a reprint of Donaldson's paper in the herpetological library of the Smithsonian National Museum of Natural History with the pages still uncut.

It is also useful to be aware that paradigms, viewpoints, and even fashions in research change. In the field of anuran communication, an example is the switch in emphasis from reproductive isolating mechanisms to sexual selection. Others are evident in the following pages.

In preparing this paper I have browsed in my own eclectic reprint collection and in those of the herpetological libraries at the Museum of Comparative Zoology and the Smithsonian National Museum of Natural History. I have drawn on historical accounts such as that in the Duellman and Trueb textbook (1985). Adler in his book for the First World Congress of Herpetology (1989) focused on people and their institutions. Altig's (1989) list of academic lineages of herpetologists includes many who have worked with anuran communication. Also, I consulted two papers by professional historians (Burkhardt 1988; Mittman and Burkhardt 1988) that provide a vivid sense of the intellectual climate from 1890 to 1945. Several review papers, particularly Bogert (1960) and Wells (1988) discuss the history of their field. Finally, I have drawn on conversations with colleagues, particularly Mike Ryan and colleagues at the University of Texas, and colleagues at the Smithsonian in Washington and in Panama.

The first published reference to frog calls is by Aristophanes in 405 B.C. in the croaking chorus from *The Frogs:* "brekekekex ko-ax ko-ax." Perhaps the significance of a frog chorus in this satire on politicians is that frogs say the same thing over and over and over again. The model is probably *Rana ridibunda,* whose call is described as "croax croax" by Arnold and Burton (1978).

"These foul and loathsome animals . . . are abhorrent because of their cold body, pale color, cartilaginous skeleton, filthy skin, fierce aspect, calculating eye, offensive smell, harsh voice, squalid habitation, and terrible venom, and so their Creator has not exerted his powers to make many of them," Carolus Linnaeus (1758) *Systema Naturae.* Although Linnaeus did not seem to think much of frogs, and the above description includes several rather egregious errors, his catalog included them and encouraged their description.

Darwin (1859) did not draw on frog communication for examples of natural selection in *The Origin of Species.* But in a later volume (1871), in a section on selection in relation to sex, he says:

> Frogs and toads offer one interesting sexual difference, namely, in the musical powers possessed by the males; but to speak of music, when applied to the discordant and overwhelming sounds emitted by male bullfrogs and some other species seems, according to our taste, a singularly inappropriate expression. Nevertheless, certain frogs sing in a decidedly pleasing manner. Near Rio Janeiro I used often to sit in the evening to listen to a number of little Hylae, perched on blades of grass close to the water, which sent forth sweet chirping notes in harmony.

Early travelers to the United States reported on the voices of frogs. A 1671 history of New England says: "There be also store of frogs, which in the spring time will chirp, whistle like birds; there be also toads, that will creep to the tops of trees, and sit there croaking, to the wonderment of strangers" (quoted by Wright and Wright 1932). Their surprise at these North American frog songs is because the calls of English anura are so unimpressive. This in turn may help explain why the songs of British frogs were so little studied.

Main Threads 1900 to 1980

Science is done by people and the history of science can be written in terms of people, but I have tried also to include the importance of changes in ideas and technology.

Modern study of anuran communication starts about 1900 and I have continued my survey to 1980. In this period I see four main threads: (1) concepts or questions that stimulated and oriented research programs; (2) technological advances that made new kinds of research possible; (3) people that conducted the research, and the institutions that supported them; and (4) academic disciplines of which researchers were a part.

Concepts and Questions

Perhaps the first of the main questions that guided researches was, "What songs do frogs sing?" It was undoubtedly noticed long before Aristophanes that species had distinctive calls and repeated them over and over again at certain seasons and that different species gave different calls. Efforts to describe frog calls and calling behavior still continue.

Ideas that have shaped the study of anuran communication have come from several sources. Important among these have been evolutionary theory, ethology and linguistics, and engineering.

Ideas from the Evolutionists

The concept of evolution, developed in the first half of the 1800s, not only gave impetus to cataloging nature, but encouraged the search for underlying patterns of relationships.

The suggestion that behavior, including displays such as calls, might reflect relationships, or the lack thereof, was made by Heinroth (1911) for ducks and geese. Noble and Noble (1923) used differences in calls as evidence that the North American *Hyla andersonii* was not closely related to the superficially similar European *Hyla arborea.*

Schiøtz (1967) pointed out that calls are more useful as indicators of relationships at some levels than at others. He said that there are sometimes differences between subspecies within a species, that there are clear differences between most species within a genus, and in most cases there are similarities between the species in one genus that distinguish them from species of other genera; whereas it is impossible to find characters common to the genera in one family that are not shared by genera in other families.

Straughan (1972) provided another view when he stated that "given the usual approach of comparing detailed descriptions of vocalizations, their usefulness dilutes rapidly at higher taxonomic levels. However, as with any other attribute of taxonomic importance, relationships on a larger scale may be more readily derived from fundamental form, rather than from detailed structure."

Martin (1971) used call characteristics and the structures that produce them to group species of *Bufo* into related assemblages. Calls are still used in phylogenetic analyses, but currently molecular evidence is given more weight in constructing phylogenies (Cannatella et al. 1998).

As the idea of descent with modification leads one to search for relationships, the idea of natural selection leads one to search for adaptations. An adaptationist believes that anything an animal does must be good for something. Darwin himself wrote that birds sing "to charm the female" (1859). Bogert (1960) quotes him disapprovingly, perhaps because at that time Darwin's concept of sexual selection was not generally accepted.

Scant attention was given to call function in the years after Darwin. Wright (1914) describes in detail the calls of the eight anurans in Ithaca, but does not even mention their function. Abbot in 1884 wrote that spadefoot toad songs are "expressions of delight at meeting" after spending the winter underground. This expresses a temperate bias against which we in the tropics are still fighting. It is clear to me that, in Panama, frogs sing to express delight that the rains have come. Dickerson (1906) asserts that frogs "feel physical joy and express it in song." As late as 1926 Barbour in his popular book entitled *Reptiles and Amphibians: Their Habits and Adaptations* devotes a page or so to describing frog calls, but makes no suggestion of their function.

As early as 1907 Courtis reported that toads were attracted to the calls of other toads. In 1923 Noble and Noble say: "Voice is stated [by Boulenger and Cummins] NOT to control direction of migration towards the breeding grounds, or the movements of individuals on the grounds." The Nobles assert that in *Hyla andersonii* "the voice plays a considerable role in bringing the two sexes in contact." However they had no experimental evidence that other cues were not involved. As late as 1956 Cagle asserted that the function of the anuran voice was not clear and further that "The specific function of the call differs in various anurans." As Bogert (1960) pointed out, part of the disagreement was because females that are attracted to calls at close range may ignore a distant chorus.

One of Bogert's major contributions in 1958 and 1960 was to provide a functional classification of the sounds of frogs. It is still used, though *mating calls* are now often referred to as *advertisement calls.* His classification provided a framework for thinking about frog calls and focused attention on the different functions of diverse calls within the repertoire.

In 1937 Dobzhansky introduced the term *species isolating mechanism.* Frog calls were quickly recognized as possible isolating mechanisms. A. P. Blair in 1941 published a paper entitled "Isolating Mechanisms in Tree Frogs," among which he included their calls. W. F. Blair (1955) used the concept to define species in *Microhyla* (= *Gastrophryne*) in the southern United States and championed this approach to anuran communication. The idea that calls were important as isolating mechanisms explained the conspicuous call differences reported for mixed species choruses around the world (e.g., Carr [1934] in Florida; Breder [1946] in Panama; Duellman [1967] in Costa Rica; Littlejohn and Martin [1969] in Southwest Australia; Heyer [1971] in Thailand). Littlejohn (1969) showed that the temporal patterns of calls (such as pulse rate) of sympatric relatives usually differ by a factor of two or more.

With the development of techniques for playing back calls to receptive females, Martof and Thompson (1958) and Bogert (1960) could show that receptive females were attracted to calls of their own species independent of any other cues.

Littlejohn and Michaud (1959) reported the first discrimination experiments between two species of *Pseudacris,* demonstrating that calls could function as species isolating mechanisms. Calls as reproductive isolating mechanisms were brought to the attention of the biological community in Blair's critically important reviews in 1963, 1964, and 1968. For a long time the best examples of premating reproductive isolating mechanisms came from frog studies.

Spatial or temporal differences, as well as differences in calls, separate sympatric species. Call differences alone do not always prevent mixed matings, and species with quite different calls may hybridize (e.g., *Geocrinia laevis* complex Littlejohn and Roberts 1975). Indeed, Martof (1961) found that *Pseudacris crucifer* females did not discriminate between their own calls and those of *P. ornata.* But Gerhardt (1973), in experiments in the field instead of in a reflective discrimination tank that would distort the main difference in the calls—duration—showed excellent discrimination by *P. crucifer.*

How call differences evolve has not been completely resolved, at least to my satisfaction. One process, character displacement, proposes that closely related species evolve more different calls where they breed sympatrically because of selection against hybridization. The logic is compelling and the data in at least some, but perhaps surprisingly few, cases seem convincing. Littlejohn (1965) has shown displacement in calls in the *Littoria ewingii* complex, Gerhardt (1994) in female preferences in the North American gray treefrogs.

Paterson (1978) challenged the existence of species isolating mechanisms and of character displacement with his Specific Mate Recognition concept. Restating Paterson, Passmore (1981) said that species with different calls coexist because the calls differ, rather than that the calls differ because the species coexist. This is the clearest statement of a position that is not currently widely accepted but remains provocative.

A very different sort of idea about how communication evolves is that of operational sex ratio (Emlen and Oring 1977). Noble and Noble (1923) recognized two kinds of frog breeding seasons: explosive and prolonged. Wells (1977) pointed out that prolonged breeding systems allowed males to mate more than once and this could produce a higher variance in male mating success and more opportunity for female choice and thus stronger sexual selection on calls than would explosive breeding systems.

An idea introduced in the 1970s to evolutionists from game theory by Maynard Smith and Price (1973) was the concept of Evolutionary Stable Strategies. This idea that two strategies could be maintained in a population if each was superior to the other when it was uncommon was applied by Perrill et al. (1978) to explain how both calling and satellite male strategies could occur in one population of *Hyla cinerea*.

The ideas behind Evolutionary Stable Strategies have been important in evaluating evolutionary tradeoffs such as those between predatory risk or energetic costs and mating success. The predatory bats that locate túngara frogs by their calls are familiar (Ryan et al. 1982). Bullfrogs in northern Kansas are reported to approach the alarm calls of smaller frogs and eat them (Smith 1977).

Several authors have measured rates of oxygen consumption to show that calling is highly energetically expensive (see Wells, this volume). Energy limitations may be involved in patterns of calling, both on a nightly and on a seasonal scale.

A dominant theme in current studies of anuran communication is that of sexual selection. This is not a new idea but one that has resurfaced. Darwin (1871) wrote that in anurans "the vocal organs differ considerably in structure, and their development in all cases may be attributed to sexual selection." Observations that females may choose among calling males were early reported in the literature. Noble and Noble (1923) described a female *Hyla andersonii* bypassing calling males to reach a more distant caller, and Axtell (1958) reports a female *Bufo speciosus* passing several small calling males to reach a bigger one. The implications were not explored.

A number of papers contributed to growing appreciation of the importance of sexual selection and mating systems in the evolution of animal communication. They include Hamilton (1964), Williams (1966), Trivers (1972), Wilson (1975), and West-Eberhard (1979). Frog mating systems were reviewed in this context by Wells (1977).

Release calls and sex recognition were studied extensively in the first half of our period. The subsequent emphasis on calls first as isolating mechanisms and later as objects of sexual selection, however, resulted in greater focus on mating calls. Consequently the other aspects of frog communication received relatively little attention.

Ideas from Ethology and Linguistics

Although behavior journals regularly publish papers on frog communication today, neither comparative psychologists in the United States nor ethologists in Europe worked much with frog communication in the first half of the twentieth century. Nor were the ideas developed by them much used in studies of frog communication. An exception is the concept of Fixed Action Patterns, which was used by those who wanted genetically based behavioral units such as displays to use as characters in a systematic study.

Linguistic concepts, such as semantics, semiotics, and John Smith's (1965) very useful distinction between message and meaning, have seldom been applied to anuran communication, although they are sometimes mentioned (Pough et al. 1998).

Ideas from Engineering

Although animal behaviorists played a relatively small role in early studies of anuran communication, people with an engineering background, particularly in acoustics, were much more influential. Many frog papers have been published in the *Journal of the Acoustical Society of America*. The first and most influential of the engineers was Bob Capranica.

To complement developing neurophysiological studies of frog hearing, Capranica (1965) wanted a behavioral assay for what a frog could hear. Heart rate, he found, was too irregular, and female phonotaxis too seasonal. He selected evoked calling by captive bullfrogs. He continued to study biologically relevant sounds throughout his career, though neurophysiology replaced evoked calling as an assay of what frogs could hear. His demonstration that the dominant frequencies in the call match the peak sensitivities in the ear argued that call and hearing coevolved. Perhaps the classic study of this is Nevo and Capranica (1985) on the correlated geographic variation in calls and hearing in *Acris*. The tight coevolution of male call and female preference is being questioned (Ryan and Rand 1993), but the idea of a match between voice and ear is still important (see Gerhardt and Schwartz, this volume).

Also ultimately from the acoustic engineers was the idea of communication as information transfer. Blair (1968) defined the function of communication as "conveyance of information." The standard communication picture of sender-signal-receiver was widely adopted. The idea of communication as information transfer has been challenged by the concept of communication as manipulation (e.g., Dawkins and Krebs 1978; Owings and Morton 1998), a return to Wallace Craig's view in 1921 that "emotional expression . . . serves as a means of communication whereby one individual can influence or control another."

Ideas about sound transmission and the problems of signaling in noise also come from engineers. The behavior of sound in different environments has not received the attention in frogs that it has in birds (e.g., Morton 1975), but it has not been ignored. Noble (1931) asserts that species breeding in temporary ponds have louder voices. Schiøtz (1967) writing about West African rachophorids described the different characteristics of forest and open country calls. Calls adapted to special environments have been described for

Himalayan torrents by Dubois (1976), and for calling from underground burrows in Papua-New Guinea by Menzies and Tyler (1977). Straughan and Heyer (1976) analyzed the calls of the frogs of the genus *Leptodactylus* from a functional point of view relating call structure to environment and mating system.

In addition to their function as species isolating mechanisms, differences of calls in mixed-species choruses have been viewed as ways of avoiding acoustic interference (Straughan 1972). Interspecific interference can be avoided by using different frequency channels, much as radio stations do, or by calling at different times or from different places.

In summary the most important ideas seem to have been: calls as reproductive isolating mechanisms, sexual selection, and the coevolution of call and ear.

Technology—Critical Equipment

Several inventions have advanced studies of frog behavior greatly since 1900. Particularly important are flashlights, portable tape recorders, and sound spectrographs.

Battery-powered flashlights were invented in the 1890s and probably were generally available by 1900 (Barney Finn, personal communication). They largely replaced carbide lamps, though Noble was using both in Santo Domingo in 1922, as well as magnesium flares (Noble 1923). Flashlights allowed observation of nocturnal frog behavior in the field in ways that had never been possible before. Frogs are quite tolerant of flashlights, though not, in my experience, as tolerant as early reports assert. If nothing else the illuminated frogs can see their own surroundings better in the light.

A comparison of two reports on the reptiles and amphibians of Puerto Rico illustrates the effect of flashlights. Stejneger in 1904 reporting on collections made before flashlights listed five frog species from the island. In 1924 K. P. Schmidt described six new species of frogs from Puerto Rico, saying "All but one of these were captured by the aid of an electric hand-lamp while they were singing at night."

Most of the remaining technological advances important in the context of anuran communication studies depend on the small size and light weight of transistors and printed circuits.

Sounds of captive monkeys were recorded by Garner (1892) on wax cylinders. Frogs lagged. Goin in 1949 used a musical friend to record the calls of different males on a musical score in his "Peep Order in Peepers: Swamp Water Serenade." Verbal descriptions transcribing vocal imitations of calls are unsatisfactory. Different people may describe the same call very differently. Fitch (1956) quotes descriptions of *Gastrophryne olivacea* calls that include "high shrill buzz," "bleating baa," and a "whistled whee followed by a bleat." These may give an idea of the call but they make comparison between different calls difficult. A. P. Blair (1947) recorded toad release vibrations by attaching the toad to a lever arm that scratched a trace on the rotating smoked drum of a kymograph; it was elegant but it did not work for recording sound. A. A. Allen and P. P. Kellogg recorded leopard frog calls on 34 feet of movie film as an optical sound track in 1935 (K. Adler, personal communication).

Bogert began to record frogs in 1953 with a portable tape recorder lent him by Folkways records. By the late 1950s Don Griffin was willing to lend me, then a graduate student, a tape recorder that was portable in the sense that it had handles. It was a windup machine with a monster flywheel. I did use it to record peepers in the laboratory but did not take it into the field, in part because of its great weight and in part from fear of what Griffin would do to me if I dropped it in a swamp. In 1961, in Brazil, when I appealed to Paulo Vanzolini for a portable tape recorder to record frogs at the Boraceia field station, he had a power line run into the forest with sockets on the poles where we could plug in a tape recorder. It was not really satisfactory since the frogs were always calling just beyond the reach of our extension cord.

The electronics that allowed the development of tape recorders also allowed portable speakers and amplifiers so that calls could be played back. Bogert (1960) demonstrated that when toads were released in the center of the Archibold Station parking lot, reproductively active males and females would go toward a speaker playing a toad chorus.

A long tradition of female choice experiments based on phonotaxis started with Bogert and with Martof and Thompson (1958) who put female *Pseudacris* in a large tank—but one considerably smaller than a parking lot—and observed that they approached a loudspeaker playing a male call much as they approached a cloth bag with calling males.

Electronics made possible the easy and accurate analysis of sounds. Where details of temporal patterning of a constant frequency sound are important, as they are in insects, an oscilloscope is appropriate. However, when changes in frequency are important, as they are in birds and frogs, the sound spectrograph is the instrument of choice. Littlejohn in 1958 in pre-sonagraph days reports using stopwatch, oscilloscope, and "ferrogram" to measure call variables (Littlejohn 1998). A ferrogram (Frings and Frings 1956) was produced by dusting fine iron powder onto the magnetic tape surface where it adhered to the places where a call was recorded. The excess powder was shaken off and the evidence of the recorded signal was removed with transparent tape and transferred to a card for examination and storage.

The Kay sonagraph, developed almost as early as the portable tape recorder, was quickly adopted by students of bird and frog vocalization. Blair and Pettus (1954) described its use. The sonagrams could be measured and differences could be described quantitatively. Measurements could be made with a precision that often exceeded accuracy.

Early sonagraphs were expensive and it took considerable time to produce each sonagram so that "typical" calls were often selected by ear, sonagraphed, and assumed to represent a species or population. Whatever the drawbacks it was exciting to watch as, in a cloud of ozone, the spark leapt from needle to revolving drum to scorch the intervening paper and visualize your sound.

The artificial sounds that Capranica (1965) used to evoke bullfrog calls, and Gerhardt (1974a, 1974b) used with treefrogs were synthesized electronically with complex Bell Labs type arrays of variable capacitors, resistors, and coils.

Electronic equipment also made possible the nerve recordings that underlie modern neurophysiology. Much of our understanding of hearing in frogs is based on eighth nerve recordings and on multicellular recordings in the brain.

In 1980 computers for general use in call analysis and synthesis and neurophysiology were still in the future.

Setting aside four-wheel drive vehicles, outboard motors, and jet aircraft that facilitate access to distant habitats, the technological advances that most affected anuran communication studies were electric portable flashlights, portable tape recorders and amplifiers, and sound spectrographs.

It was these technological advances that allowed investigators to take advantage of the many virtues of frogs for studies of communication. Wilczynski and Ryan (1988) argue cogently that "anuran communication signals are species-specific and highly stereotyped, and each species repertoire is small. . . . The social behavior that acoustic signals elicits in frogs also consists of a small stereotyped repertoire. . . . The consequences of acoustic behavior—mating success—can also be measured." Further, many frogs gather in choruses, pairs form amplexus, and eggs are fertilized externally so that paternity can be determined and eggs counted.

People and Institutions

Ideas and technology are important, but only when they are used by people. Our era opens with Boulenger's papers on "Tailless Batrachians of Europe" in 1897 and 1898. Boulenger was primarily a systematist at the British Museum but with an interest in natural history. The following is a list of the people that I think had the greatest influence, between 1900 and 1980, particularly in the early years. I have grouped them by institution.

Few people work in isolation, and anyone who teaches at a small college and tries to do research understands the importance of institutional support. Three institutions have contributed disproportionately to anuran communication research in North America: the American Museum of Natural History, Cornell University, and the University of Texas at Austin.

The American Museum curator of herpetology, Mary Dickerson, published *The Frog Book* in 1906. This popular description of the appearance and natural history of frogs in the United States is still in print. Dickerson hired G. K. Noble.

Noble was curator of herpetology and founder of a new department of animal behavior at the American Museum of Natural History (AMNH). His *Biology of the Amphibia* (1931) was a standard reference for years. He both conducted and sponsored field observations and experimental studies of anuran communication, particularly of sex recognition.

Charles Bogert succeeded Noble as curator of herpetology at the AMNH. Bogert seized the opportunity presented by the development of portable tape recorders to travel around the United States and Mexico recording frog voices. The resultant recording from Folkways Records in 1958 has just been reissued on CD by the Smithsonian. It is called "Sounds of North American Frogs," but more important is the subtitle, "The Biological Significance of Voice in Frogs." The extensive liner notes were elaborated in 1960 into "The Influence of Sound on the Behavior of Amphibians and Reptiles." Bogert established the archives of amphibian sounds at the AMNH. Also at the AMNH, Richard Zweifel conducted several important studies of mating calls of *Bufo* and published particularly important papers relating size and temperature to call characteristics (Zweifel 1959, 1968).

A. H. Wright and A. A. Wright at Cornell wrote the first *Handbook of North American Frogs* (1932) that brought together published natural history information and personal observations. A. A. Allen at the Cornell Laboratory of Ornithology's Library of Natural Sounds with P. P. Kellogg produced the first phonograph record of frog calls, "Voices of the Night," in 1948.

Cornell has continued institutional support for people doing studies in anuran communication, but Bob Capranica was not an academic descendent of the Wrights, instead he had done his Ph.D. at Massachusetts Institute of Technology on a fellowship from Bell Labs.

Capranica brought together frog behavior and acoustic engineering to the benefit of both. Most of the participants in this symposium knew Capranica as graduate or postdoctoral students or paid homage, as I did, at Cornell. Capranica's course on animal communication had wide and continuing influence and much of his influence is apparent in a recent book on animal communication (Bradbury and Vehrencamp 1998). Peter Narins, a student of Capranica now at the University of California at Los Angeles, continues the tradition of bringing an engineering background to bear on communication among anurans in the wild.

In contrast to Capranica, Kent Wells, in his Ph.D. thesis at Cornell on green frogs (*Rana clamitans*), emphasized observations in the field. His greatest influence came from his reviews published in 1977–1978. Toward the end of our period a number of the researchers represented in this volume (Brenowitz, Gerhardt, Rose, Ryan, Waldman, and Wilczynski) were graduate or postdoctoral students at Cornell.

At the University of Texas at Austin, W. Frank Blair worked with acoustic isolating mechanisms and evolution and wrote several important reviews (1963, 1968). The volume *Evolution in the Genus* Bufo that he edited brought together a variety of aspects of toad biology, including communication. Blair was particularly notable for the large number of students and postdocs who worked in his laboratory. Prominent among them were J. Bogart, Gerhardt, Littlejohn, and W. Martin. The tradition has been reactivated by Ryan and his colleagues, so that Texas is now again a leading center of study and training in anuran communication, rivaled only by the University of Missouri under Carl Gerhardt.

Gerhardt represents a synergism between the University of Texas where he did his Ph.D. with Blair and Cornell where he did a postdoc with Capranica. Particularly important during the 1970s were his studies using synthetic stimuli to examine in detail the characteristics of calls that were important in discrimination between the calls of closely related species of *Hyla* (1974a, 1974b). Perhaps even more important was the scientific rigor that he brought to experimentation with frog phonotaxis.

Altig's (1989) analysis of the academic family trees of herpetologists traces a disproportionate number of those working on anuran communication, indeed in much of zoology, back to Agassiz at Harvard, but none ever worked there. A number of institutions have supported or trained individuals or clusters of practitioners in our field, including the Smithsonian Tropical Research Institute, and particularly the University of Michigan with Blair, Martof, Rabb, Duellman, and Howard, but none has supported a dynasty comparable to that at the American Museum of Natural History.

This list has largely focused on researchers in the United States; even there it is incomplete. I should mention several important non-US frog people: Avelino Barrio in Argentina, Madeleine Paillette and Jean Lescure in France, Murray Littlejohn in Australia, Nevile Passmore in South Africa, Hans Schneider in Germany, and Arne Schiøtz in Denmark. This is still not a complete list.

Academic Disciplines—Approaches to Anuran Communication

There has never been a recognized discipline of amphibian communication. There is no journal, no society, no textbook. It is in symposia such as this that students of anuran communication come together to exchange ideas.

In the period 1900 to 1980 people who have worked with frog communication worked in three overlapping fields: systematics, natural history and behavior, and physiology.

Systematists are basically interested in frogs; in their taxonomy, systematics, phylogenetics; and in the diversity of frog biology. Systematists find calls important because they facilitate the discrimination of otherwise very similar species, and because they may indicate relationships between species.

Systematic herpetologists often collect and observe anurans in the field. They often record behavior, including communication. For example, Dixon in a 1957 paper entitled "Geographic Variation and Distribution of the Genus *Tomodactylus* in Mexico" reports that males and females counter call. He says, "Axtell and I took advantage of this behavior by enclosing several calling females in a sack and placing it on the ground. The captive females continued to call and as unsuspecting males answered they were easily located by flashlight and captured." This strongly suggests that these frogs could be manipulated in field experiments but, to my knowledge, it was never followed up.

Naturalists and animal behaviorists are interested in what animals do and why. They view behavior as an adaptation rather than as evidence of relationship or the manifestation of a physiological process.

Field studies of individual species often contain important information about communication. Toward the end of our period there was a cluster of synergistic papers on the behavior of free-living bullfrogs by Emlen (1968, 1976), Wiewandt (1969), Howard (1978), and Ryan (1980).

There have been studies of single species conducted on captive animals. These often report more detail on nonacoustic signaling, particularly during oviposition, than do field studies, and less on acoustic behavior. These include the Rabbs' work on Pipids (1961, 1964), and R. M. Savage's (1932) study of *Bombina variegata*.

Frogs were not much studied in the psychology departments where most animal behavior in the United States was being done in the second quarter of the twentieth century. Not even toads learn to run mazes very well.

In general, physiologists are not primarily interested in frogs as such, but instead use frogs to investigate mechanisms experimentally. Physiology as used here includes studies of the neurophysiology of hearing, energetics and biomechanics of calling, and the role of hormones in development and expression of calling behavior.

Frogs are good vertebrates to work with in physiological experiments because they do not need artificial respiration even under deep anesthesia because of their effective cutaneous respiration. Anurans are "small, hearty, and tolerate anesthesia and surgery well" (Wilczynski and Ryan 1988).

The earliest neurophysiology of frogs is reported by Ewert (1980, p. 4): "Galvani (1786) hung some frog legs on his balcony railing one day and noticed that the legs twitched when they touched the metal fence." This does not add much to our understanding of frog communication, but the image of Galvani hanging some frog legs on his balcony railing demands to be included.

Much physiology is done in the laboratory, but many of the people I am classifying as physiologists studied frogs extensively in the field as well. Peter Narins' work on *Eleutherodactylus coqui* (Narins and Capranica 1976 et seq.) is an outstanding example of a synergistic combination of field and laboratory studies, as is the work of Gerhardt on *Hyla* (1978).

The earliest relevant experimental study that I know of is that of Yerkes (1903, 1905). From observations at local ponds, Yerkes was convinced that frogs could hear. To demonstrate this he suspended a blindfolded green frog on a stand with its hind legs hanging free (Figure 2.1). He then hit the frog on the head with a little rubber hammer. A bell was rung with the blow, or before, or after it. The frog jerked its leg more strongly if the sound occurred at the same time as the blow or within 0.35 seconds before than it did if the sound came earlier or later. This showed, for the first time, that frogs could hear.

Noble and his collaborators at the American Museum did experimental manipulations that showed that girth, firmness, and warning vibrations were used by frogs and toads to determine the sex of their mating partners (e.g., Noble and Farris 1929).

In 1938 Adrian, Craik, and Sturdy recorded electric potentials from the frog eighth nerve using a large electrode that picked up responses from many nerve fibers at once. In 1960 Axelrod, and Glekin and Erdman independently reported recordings from individual nerve fibers in the eighth nerve. Frishkopf and Goldstein (1963) recorded single units in the eighth nerve of the bullfrog that responded to acoustic stimuli.

Zakon and Wilczynski (1988) state that "the early studies of Capranica and colleagues . . . on coding of communication calls in the auditory periphery of anurans had an impact on auditory physiology similar to the work of Lettvin and his laboratory on visual physiology."

The basic outlines of how frogs produce sound were described by Boulenger (1898). Air is squeezed from the lungs through the larynx into the vocal sac. As the air passes the vocal folds, they vibrate and produce sound. The basics are clear, but the details are still being elucidated.

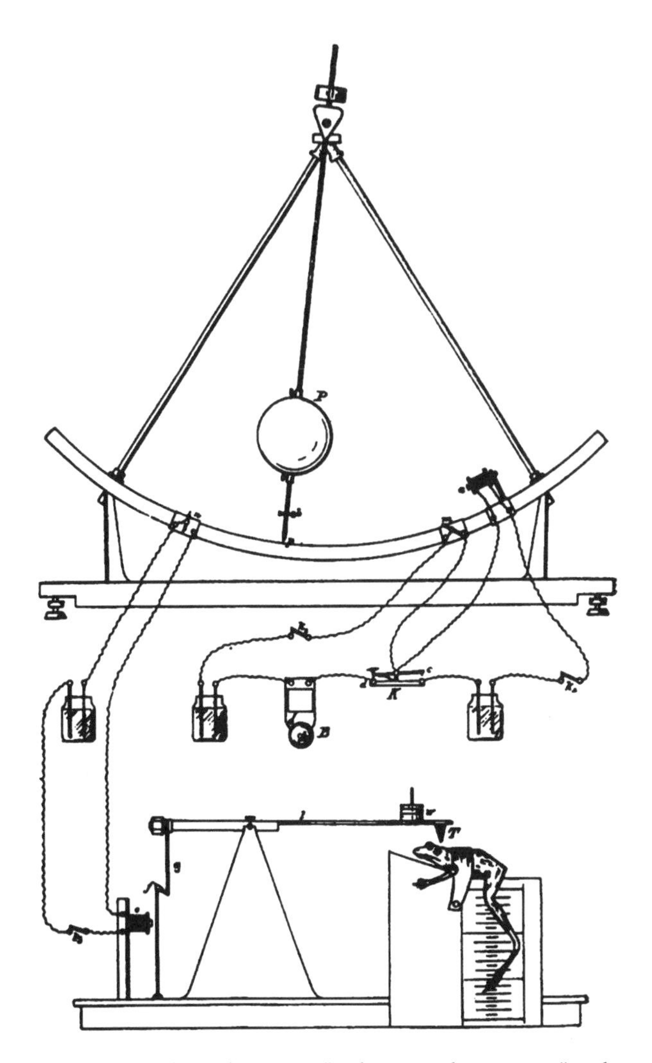

Figure 2.1. The Yerkes (1905) "auditory-tactile apparatus" with which he demonstrated experimentally that frogs could hear.

The anatomy of the sound-producing apparatus was described relatively early. Trewavas in 1933 published a detailed description of the hyoid and larynx of a wide variety of anurans. She did not relate the great variation that she found in morphology to function, and that connection is only recently being explored (Wilczynski et al. 1993). Robert Schmidt (1966) is one of the few who has directly observed laryngeal movements. He replaced one of a frog's eyeballs with a transparent plug that allowed him to watch what happened when the frog, hopped up on hormones, called. Martin (1971, 1972) described calling in *Bufo*, relating differences in morphology to differences in call characteristics. With Carl Gans, he further detailed the role of muscle contractions in release calling in *Bufo valliceps* (Martin and Gans 1972).

Liu (1935) described the diversity of vocal sacs. Again there was little attempt to relate form to function. Tyler (1971a) used vocal sacs as a character to investigate the phylogeny of hylid frogs. McAlister (1961) made plaster of Paris models of *Bufo* vocal sacs to study size and shape differences between species, but not the function of these differences. The function of the vocal sac is still not fully settled. The old and often repeated idea that the vocal sac acts as a cavity resonator amplifying the call is almost certainly not true (Rand and Dudley 1993), though other kinds of resonance may well be important in call production (Purgue 1995).

A report from the past that tempts reexamination was published by Tyler in 1971 (Tyler 1971b). He says that in *Limnodynastes tasmaniensis* a call may be given with the vocal sac fully inflated, uniformly partially inflated, or deflated anteriorly and inflated posteriorly. Tyler goes on: "There was no audibly detectable difference between the calls emitted with the vocal sac maximally inflated or incompletely deflated."

Hormones were used early to induce frogs to call. Greenberg in 1942 injected testosterone into cricket frogs, Schmidt (1966) used both *Rana* pituitaries or chorionic gonadotropin to bring *Hyla* into calling condition, but the endocrinology of frog communication has developed greatly since then. For example, by the end of our chosen period, Kelley (1980) was studying neuroendocrinology of calling centers in frog brains.

Status 1980

In 1980 when this review ends, the scene was changing. The dominant figures of earlier years were largely gone. Dickerson, Wright, and Noble were dead; Bogert, Capranica, and Blair were not actively doing research. Littlejohn, Gerhardt, Kelley, Narins, and Wells were active, and still are. A number of people who would be important in the years after 1980 were still in graduate school or on postdoctoral fellowships, among them several represented in this volume: Brenowitz, Rose, Ryan, Schwartz, Waldman, and Wilczynski. Nineteen hundred and eighty is a good place to stop.

Conclusions

I have reviewed the development of frog communication studies from 1900 to 1980. I followed four major threads—concepts, technology, people, and academic disciplines—and tried to identify the major components in each. This review suggests that the study of anuran communication, during the period under consideration, made its major contributions in the following areas: (1) species-isolating mechanisms/recognition, (2) neuroethological paradigm match of signaler and receiver, and (3) sexual selection by female choice.

Much of the intellectual vitality of anuran communication studies has come from the diversity of inputs to it: conceptual, technological, personality, and academic, and the interactions among them. This symposium demonstrates that this diversity and the strength it brings continues today. People from different backgrounds and diverse interests can interact to learn to appreciate the strengths of the work of others, as well as to point out the weaknesses. These interactions should improve the work of the individuals involved and thus improve the field as a whole.

References

Abbot, C. C. 1884. Recent studies of the spade-foot toad. American Naturalist 18:1076–1080.

Adler, K. 1989. Herpetologists of the past. Pp. 5–141. *In* K. Adler (ed.), Contributions to the History of Herpetology, Contributions to Herpetology 5. Society for the Study of Amphibians and Reptiles, Milwaukee.

Adrian, E. D., K. J. W. Craik, and R. S. Sturdy. 1938. The electrical response of the auditory mechanism in cold-blooded vertebrates. Proceedings of the Royal Society, London, Series B 125:435–455.

Allen, A. A. 1950. Voices of the night. National Geographic 97(4):507–522.

Allen, A. A., and P. P. Kellogg. 1948. Voices of the night: The calls of 34 frogs and toads in the United States and Canada. (phonograph record). Cornell Laboratory of Ornithology, Library of Natural Sounds, Ithaca, New York. Houghton Mifflin, Boston.

Altig, R. 1989. Academic lineages of doctoral degrees in herpetology. Pp.179–202. *In* K. Adler (ed.), Contributions to the History of Herpetology, Contributions to Herpetology 5. Society for the Study of Amphibians and Reptiles.

Aristophanes. 405 B.C. The Frogs. Translated by R. Lattimore. 1962. University of Michigan Press, Ann Arbor, MI.

Arnold, E. N., and J. A. Burton. 1978. A Field Guide to the Reptiles and Amphibians of Britain and Europe. Collins, London.

Axelrod, F. S. 1960. (as reported by Lettvin and Maturana) Hearing senses in the frog. MIT Research Laboratory Electronics Quarterly Progress Report 57:167–168.

Axtell, R. W. 1958. Female reaction to the male call in two anurans (Amphibia). Southwestern Naturalist 3:70–76.

Barbour, T. 1926. Reptiles and Amphibians: Their Habits and Adaptations. Houghton Mifflin, Boston.

Blair, A. P. 1941. Isolating mechanisms in tree frogs. Proceedings National Academy of Science 27:14–17.

Blair, A. P. 1947. The male warning vibration in *Bufo*. American Museum Novitates 1344:1–7.

Blair, W. F. 1955. Mating call and stage of speciation in the *Microhyla olivacea—M. carolinensis* complex. Evolution 9:469–480.

Blair, W. F. 1963. Acoustic behaviour of amphibia. Pp. 694–708. *In* R.-G. Busnell (ed.). Acoustic Behaviour of Animals. Elsevier, Amsterdam.

Blair, W. F. 1964. Isolating mechanisms and interspecific interactions in anuran amphibians. Quarterly Review of Biology 39:334–344.

Blair, W. F. 1968. Amphibians and reptiles. Pp. 289–310. *In* T. A. Sebeok (ed.), Animal Communication. Indiana University Press, Bloomington.

Blair, W. F. 1972. (ed.) Evolution in the genus *Bufo*. University of Texas Press, Austin.

Blair, W. F., and D. Pettus. 1954. The mating call and its significance in the Colorado river toad (*Bufo alvarius girard*). Texas Journal of Science 6:72–77.

Bogert, C. M. 1958. Sounds of North American frogs: The biological significance of voice in frogs. (Phonograph recording and commentary). Folkways Records and Service Corp. Science Series FX6166, New York.

Bogert, C. M. 1960. The influence of sound on amphibians and reptiles. Pp. 137–320. *In* W. E. Lanyon, W. E. and W. N. Tavolga (eds.), Animals Sounds and Communication. American Institute of Biological Sciences, Washington, DC.

Boulenger, G. A. 1897. The Tailless Batrachians of Europe. Part I. Pp. 1–219. The Ray Society, London.

Boulenger, G. A. 1898. The Tailless Batrachians of Europe. Part II. Pp. 211–376. The Ray Society, London.

Bradbury, J. W., and S. L. Vehrencamp. 1998. Principles of Animal Communication. Sinauer, Sunderland, MA.

Breder, C. M. 1946. Amphibians and reptiles of the Rio Chucunaque drainage, Darien, Panama, with notes on their life histories and habits. Bulletin of the American Museum of Natural History 86:375–436.

Burkhardt, R. W. Jr. 1988. Charles Otis Whitmann, Wallace Craig, and the biological study of animal behavior in the United States, 1898–1925. Pp. 185–218. *In* R. Rainger et al. (eds.), The American Development of Biology. University of Pennsylvania Press, Philadelphia.

Cagle, F. D. 1956. An outline for the study of an amphibian life history. Tulane Studies in Zoology 4:79–110.

Cannatella, D.C., D. M. Hillis, P. Chippindale, L. Weigt, A. S. Rand, and M. J. Ryan. 1998. Phylogeny of frogs of the *Physalaemus pustulosus* species group, with an examination of data incongruence. Systematic Biology 47:311–335.

Capranica, R. R. 1965. The Evoked Vocal Response of the Bullfrog: a Study of Communication by Sound. MIT Research Monograph 33.

Carr, A. F., Jr. 1934. A key to the breeding songs of Florida frogs. Florida Naturalist, New Series 7:19–23.

Courtis, S. A. 1907. Response of toads to sound stimuli. American Naturalist 41:677–682.

Craig, W. 1921–22. A note on Darwin's work on the expression of emotions in man and animals. Journal of Abnormal Psychology and Social Psychology 16:356–66.

Darwin, C. 1859. The Origin of Species by Means of Natural Selection. John Murray, London.

Darwin, C. 1871. The Descent of Man and Selection in Relation to Sex. John Murray, London.

Dawkins, R., and J. R. Krebs. 1978. Animal signals: information or manipulation? Pp. 282–309. *In* J. R. Krebs and N. B. Davies (eds.), Behavioural Ecology: An Evolutionary Approach. Sinauer, Sunderland, Massachusetts.

Dickerson, M. 1906. The Frog Book. Doubleday, Page and Company, New York.

Dixon, J. R. 1957. Geographic variation and distribution of the genus *Tomodactylus* in Mexico. Texas Journal of Science 9:379–409.

Dobzhansky, T. 1937. Genetic nature of species differences. American Naturalist 71:404–420.

Donaldson, H. H. 1911. On the regular seasonal changes in the relative weight of the central nervous system of the leopard frog. Journal of Morphology 22:663–694.

Dubois, A. 1976. Chants et ecologie chez les amphibiens du Nepal. Ecologie et Geologie de l'Himalaya. Colloques Internationaux du Centre National de la Recherche Scientifique, Paris. 268:109–118.

Duellman, W. E. 1963. Importance of breeding call in amphibian systematics. Proceedings of the XVI International Congress of Zoology 4:106–110.

Duellman, W. E. 1967. Courtship isolating mechanisms in Costa Rican hylid frogs. Herpetologica 23:169–183.

Duellman, W. E., and L. Trueb. 1985. Biology of Amphibians. McGraw-Hill, New York.

Emlen, S. T. 1968. Territoriality in the bullfrog, *Rana catesbeiana*. Copeia 1968:240–243.

Emlen, S. T. 1976. Lek organization and mating strategies in the bullfrog. Behavioral Ecology and Sociobiology 1:283–313.

Emlen, S. T., and L. W. Oring. 1977. Ecology, sexual selection, and the evolution of mating systems. Science 197:215–223.

Ewert, J.-P. 1980. Neuroethology: An Introduction to the Neurophysiological Fundamentals of Behavior. Springer-Verlag, Berlin, Heidelberg, New York.

Fitch, H. S. 1956. A field study of the Kansas ant-eating frog, *Gastrophryne olivacea*. University of Kansas Publication of the Museum of Natural History 8:275–306.

Frings, H., and M. Frings. 1956. A simple method of producing visible patterns of tape-recorded sounds. Nature 178:328–329.

Frishkopf, L. S., and M. M. Goldstein. 1963. Responses to acoustic stimuli from single units of the eighth nerve of the bull frog. Journal of the Acoustical Society of America 35:1219–1228.

Garner, R. I. 1892. The Speech of Monkeys. Heinemann, London.

Gerhardt, H. C. 1973. Reproductive interactions between *Hyla crucifer* and *Pseudacris ornata*. American Midland Naturalist 89:81–88.

Gerhardt, H. C. 1974a. Behavioral isolation of the tree frogs, *Hyla cinerea* and *Hyla andersonii*. American Midland Naturalist 91(2):424–433.

Gerhardt, H. C. 1974b. The significance of some spectral features in mating call recognition in the green treefrog (*Hyla cinerea*). Journal Experimental Biology 61:229–241.

Gerhardt, H. C. 1978. Temperature coupling in the vocal communication system of the tree frog *Hyla versicolor*. Science 199:992–994.

Gerhardt, H. C. 1994. Reproductive character displacement of female mate choice in the grey treefrog *Hyla chrysoscelis*. Animal Behaviour 47:959–969.

Glekin, G. V., and G. M. Erdman. 1960. Discrimination of a useful signal by the auditory analyzer; I. Potential is from the elements of the frog auditory nerve. Biophysics 5:412–419.

Goin, C. J. 1949. The peep order of peepers: A swamp water serenade. Quarterly Journal of the Florida Academy of Science 11:59–61.

Greenberg, B. 1942. Some effects of testosterone on the sexual pigmentation and other sex characters of the cricket frog (*Acris gryllus*). Journal of Experimental Zoology 91:435–446.

Hamilton, W. D. 1964. The genetical theory of social behaviour, I & II. Journal of Theoretical Biology 7:1–52.

Heinroth, O. 1911/1985. Contributions to the biology, especially the ethology and psychology of the Anatidae (English translation). Pp. 246–301. *In* G. M. Burghardt (ed.), Foundations of Comparative Ethology. Van Nostrand Reinhold, New York.

Heyer, W. R. 1971. Mating calls of some frogs from Thailand. Fieldiana, Zoology 58:61–82.

Howard, R. D. 1978. The evolution of mating strategies in bullfrogs, *Rana catesbeiana*. Evolution 32:850–871.

Kelley, D. 1980. Auditory vocal nuclei in the frog brain concentrate sex hormones. Science 207:553–555.

Linnaeus, C. 1758. Systema Naturae, Edit. 10. Tom 1. Pt. 1. Upsala.

Littlejohn, M. J. 1958. A new species of frog of the genus *Crinia tschudi* from South-eastern Australia. Proceedings of the Linnean Society of New South Wales 83:222–226.

Littlejohn, M. J. 1965. Premating isolating in the *Hyla ewingi* complex (Anura: Hylidae). Evolution 19:234–243.

Littlejohn, M. J. 1969. The systematic significance of isolating mechanisms. Pp. 459–482. *In* Systematic Biology: Proceedings of an International Conference, National Academy of Sciences, Washington, DC.

Littlejohn, M. J. 1998. Historical aspects of recording and analysis in anuran bioacoustics: 1954–1997. Bioacoustics 9:69–80.

Littlejohn, M. J., and A. A. Martin. 1969. Acoustic interaction between two species of leptodactylid frogs. Animal Behaviour 17:785–791.

Littlejohn, M. J., and T. C. Michaud. 1959. Mating call discrimination by females of Strecker's chorus frog (*Pseudacris streckeri*). Texas Journal of Science 11:86–92.

Littlejohn, M. J., and J. D. Roberts. 1975. Acoustic analysis of an intergrade zone between two call races of the *Limnodynastes tasmaniensis* complex (Anura: Leptodactylidae) in South-eastern Australia. Australian Journal of Zoology 23:113–122.

Liu, C. C. 1935. Types of vocal sac in the Salientia. Proceedings of the Boston Society of Natural History 41:19–40.

Martin, W. F. 1971. Mechanics of sound production in toads of the genus *Bufo:* Passive elements. Journal of Experimental Zoology 176:273–294.

Martin, W. F. 1972. Evolution of vocalization in the genus *Bufo*. Pp. 279–309. *In* W. F. Blair, Evolution in the Genus *Bufo*. University Texas Press, Austin.

Martin, W. F., and C. Gans. 1972. Muscular control of the vocal tract during release signaling in the toad *Bufo valliceps*. Journal of Morphology 137:1–28.

Martof, B. S. 1961. Vocalization as an isolating mechanism in frogs. American Midland Naturalist 65:118–126.

Martof, B. S., and E. F. Thompson, Jr. 1958. Reproductive behavior of the chorus frog, *Pseudacris nigrita*. Behaviour 13:243–258.

Maynard Smith, J., and G. R. Price. 1973. The logic of animal conflict. Nature 246:15–18.

McAlister. W. H. 1961. The mechanics of sound production in North American *Bufo*. Copeia 1961:86–95.

Menzies, J. I., and M. J. Tyler. 1977. The systematics and adaptations of some Papuan microhylid frogs which live underground. Journal of Zoology, London 183:431–464.

Mittman G., and R. W. Burkhardt, Jr. 1988. Struggling for identity: The study of animal behavior in America, 1930–1945. Pp.164–194. *In* R. Rainger et al. (eds.), The American Development of Biology. University of Pennsylvania Press, Philadelphia, Pennsylvania.

Morton, E. S. 1975. Ecological sources of selection on avian sounds. American Naturalist 109:17–34.

Narins, P. M., and R. R. Capranica. 1976. Sexual differences in the auditory system of the treefrog, *Eleutherodactylus coqui*. Science 192:378–380.

Nevo, E., and R. R. Capranica. 1985. Evolutionary origin of ethological reproductive isolation in cricket frogs, *Acris*. Pp. 147–214. *In* M. K. Hecht et al. (eds.), Evolutionary Biology. Plenum, New York.

Noble, G. K. 1923. In pursuit of the giant tree frog: Night hunting in Santo Domingo by the Angelo Heilprin expedition. Natural History 23:105–116.

Noble, G. K. 1931. The Biology of the Amphibia. McGraw-Hill, New York.

Noble, G. K., and E. J. Farris. 1929. The method of sex recognition in the wood frog *Rana sylvatica* Le Conte. American Museum Novitates 363:1–17.

Noble, G. K., and R. C. Noble. 1923. The Anderson tree frog (*Hyla andersonii* Baird): Observations on its habits and life history. Zoologica 2:416–455.

Owings, D. H., and E. S. Morton. 1998. Animal Communication: A New Approach. Cambridge University Press, Cambridge, MA.

Passmore, N. I. 1981. The relevance of the specific mate recognition concept to anuran reproductive biology. Monitore Zoologica Italiano 6:93–108.

Paterson, H. E. H. 1978. More evidence against speciation by reinforcement. South African Journal of Science 74:369–371.

Perrill, S. A., H. C. Gerhardt, and R. Daniel. 1978. Sexual parasitism in the green treefrog (*Hyla cinerea*). Science 209:523–525.

Pough, F. H., R. M. Andrews, J. E. Cadle, M. L. Crump, A. H. Savitzky, and K. D. Wells. 1998. Herpetology. Prentice-Hall, Upper Saddle River, NJ.

Purgue, A. P. 1995. The sound broadcasting system of the bullfrog (*Rana catesbeiana*). Ph.D. dissertation. University of Utah, Salt Lake City.

Rabb, G. B., and M. S. Rabb. 1961. On the mating and egg-laying behavior of the Surinam toad, *Pipa pipa*. Copeia 1960:271–276.

Rabb, G. B., and M. S. Rabb. 1963. On the behavior and breeding biology of the African pipid frog *Hymenochirus boettgeri*. Zeitschrift fur Tierpsychologie 20:215–241.

Rand, A. S., and R. Dudley. 1993. Frogs in helium: The anuran vocal sac is not a cavity resonator. Physiological Zoology 66:793–806.

Ryan, M. J. 1980. The reproductive behavior of the bullfrog, *Rana catesbeiana*. Copeia 1980:108–114.

Ryan, M. J., and A. S. Rand. 1993. Sexual selection and signal evolution: The ghost of biases past. Philosophical Transactions of the Royal Society of London B 338:187–195.

Ryan, M. J., M. D. Tuttle, and A. S. Rand. 1982. Bat predation and sexual advertisement in a neotropical anuran. American Naturalist 119:136–139.

Savage, R. M. 1932. The spawning, voice, and sexual behaviour of *Bombina variegata variegata*. Proceedings of the Zoological Society, London 4:889–898.

Schiøtz, A. 1967. The treefrogs (Racophoridae) of West Africa. Spolia zoologica Musei hauniensis 25:1–346.

Schmidt, K. P. 1924. Amphibians and land reptiles from Porto Rico, with a list of those reported from the Virgin Islands. Scientific Survey of Porto Rico and the Virgin Islands 10:1–161.

Schmidt, R. S. 1966. Central mechanisms of frog calling. Behaviour 26:251–285.

Smith, A. 1977. Attraction of bullfrogs (Amphibia, Anura, Ranidae) to distress calls of immature frogs. Journal of Herpetology 11:232–234.

Smith, W. J. 1965. Message, meaning, and context in ethology. American Naturalist 97:117–126.

Stejneger, L. 1904. The herpetology of Porto Rico. Report of the United States National Museum 1902.

Straughan, I. R. 1972. Evolution of anuran mating calls: bioacoustical aspects. Pp. 321–327. In J. L. Vial (ed.), Evolutionary Biology of the

Anurans: Contemporary Research on Major Problems. University of Missouri Press, Columbia.

Straughan, I. R., and W. R. Heyer. 1976. A functional analysis of the mating calls of the neotropical frog genera of the *Leptodactylus* complex (Amphibia, Leptodactylidae). Papeis Avulsos Zoologie, Sao Paulo 29:221–245.

Trewavas, E. 1933. The hyoid and larynx of the Anura. Philosophical Transactions of the Royal Society of London B 222:401–527.

Trivers, R. 1972. Parental investment and sexual selection. Pp. 136–179. In B. Campbell (ed.), Sexual selection and the descent of man. Aldine, Chicago, IL.

Tyler, M. J. 1971a. The phylogenetic significance of vocal sac structure in hylid frogs. University of Kansas Publication, Natural History 19:318–360.

Tyler, M. J. 1971b. Voluntary control of the shape of the inflated vocal sac by the Australian leptodactylid frog *Limnodynastes tasmaniensis*. Transactions of the Royal Society of South Australia 95:49–52.

Wells, K. D. 1977. The social behaviour of anuran amphibians. Animal Behaviour 25:666–693.

Wells, K. D. 1988. The effect of social interactions on anuran vocal behavior. Pp. 433–454. *In* B. Fritzsch et al. (eds.), The Evolution of the Amphibian Auditory System. John Wiley and Sons, New York.

West-Eberhard, M. J. 1979. Sexual selection, social competition, and evolution. Proceedings of the American Philosophical Society 123:222–234.

Wiewandt, T. A. 1969. Vocalization, aggressive behavior and territoriality in the bull frog, *Rana catesbeiana*. Copeia 1969:276–285.

Wilczynski, W., B. E. McClelland, and A. S. Rand. 1993. Acoustic, auditory, and morphological divergence in three species of neotropical frog. Journal of Comparative Physiology series A 172:425–438.

Wilczynski, W., and M. J. Ryan. 1988. The amphibian auditory system as a model for neurobiology, behavior, and evolution. Pp. 3–13. *In* B. Fritzsch et al. (eds.), The Evolution of the Amphibian Auditory System. John Wiley and Sons, New York.

Williams, G. C. 1966. Adaptation and Natural Selection. Princeton University Press, Princeton, NJ.

Wilson, E. O. 1975. Sociobiology: The New Synthesis. Harvard University Press, Cambridge, MA.

Wright, A. H. 1914. Life-Histories of the Anura of Ithaca, New York. Carnegie Institution of Washington. Washington, DC.

Wright, A. H., and A. A. Wright. 1932. Handbook of frogs and toads: The frogs and toads of the United States and Canada. Comstock, Ithaca, NY.

Yerkes, R. M. 1903. The instincts, habits and reactions of the frog. Psychology Reviews 4:579–638.

Yerkes, R. M. 1905. The sense of hearing in frogs. Comparative Neurological Psychology 15:279–304.

Zakon, H. H., and W. Wilczynski. 1988. The physiology of the anuran eighth nerve. Pp. 125–155. *In* B. Fritzsch et al. (eds.), The Evolution of the Amphibian Auditory System. John Wiley and Sons, New York.

Zweifel, R. G. 1959. Effect of temperature on call of the frog, *Bombina variegata*. Copeia 1959:322–327.

Zweifel, R. G. 1968. Effects of temperature, body size and hybridization on mating calls of toads, *Bufo a. americanus* and *Bufo woodhousii fowleri*. Copeia 1968:269–285.

PART TWO

Physiology and Energetics

3

WALTER WILCZYNSKI AND JOANNE CHU

Acoustic Communication, Endocrine Control, and the Neurochemical Systems of the Brain

Introduction

Much of the communication behavior of anuran amphibians, as well as of many other vertebrates, takes place in the context of reproduction. The most obvious displays of animals are often mate attraction signals that serve to identify the signaler's species, sex, and reproductive state to conspecifics. Communication signals are also important components of territoriality and intrasexual competition, but even in this domain these signals are produced in the broader context of reproduction. The territories defended often represent display sites, nesting sites, or resources necessary for mate attraction and defense. The associated intrasexual competition often represents competition for mating.

The relationship between communication and reproduction manifests itself in predictable ways. Signal production is routinely sexually dimorphic. Similarly responses elicited by a signal are often completely different in males and females. Both production and response can be seasonal and linked to the circannual waxing and waning of physiological reproductive readiness. In animals that manifest an estrous cycle, communication behavior may track that cycle. Neither signal production nor signal reception is a constant feature throughout an organism's life; neither is static during its adult phase. Rather the propensity or ability to engage in communication behavior varies over a variety of time scales from annual to daily to hourly in such a way as to track the reproductive readiness of the signaler and the recipient of those signals.

Sex differences in communication and the links between behavior and physiological reproductive state allow for some obvious predictions about the development and control of communication behavior. One would expect the morphological structures and neural control systems underlying the sexually dimorphic production of communication signals to differentiate under the control of gonadal steroid hormones (androgens and estrogens). Similarly gonadal steroids are obvious candidates for gating or modulating the expression of signal production or reception seasonally or under shorter time scales such as intraseasonal estrous cycles.

The obvious role of gonadal steroids in influencing communication behavior should not obscure the fact that other factors also influence reproductive behaviors and reproductive physiology. Endocrine products and neuromodulators, such as adrenal steroids and peptide hormones, may influence communication behavior along with gonadal steroids. Conspecific communication signals can influence reproductive social behavior, as can other environmental cues. The levels of gonadal steroids themselves are regulated by some of these factors, and so in principle reproductive behaviors may be influenced by these other external and internal factors either directly or indirectly via effects on circulating gonadal steroids. Viewed in this light the control of reproductive behavior and physiology can be seen as a complex multidimensional problem, with behavior, physiology, and endocrine state mutually influencing one another.

Acoustic Communication in Frogs and Toads

In anurans the tie between communication and reproduction is clear and unambiguous. The most obvious and typical vocal signal produced by these vertebrates is an advertisement call (Wells 1977; Rand 1988; Ryan 1991). The advertisement call is produced during the breeding season and serves as the species-typical mate attraction signal (Wells 1977; Gerhardt 1988; Rand 1988). Call production is sexually dimorphic. In most species described, males produce the advertisement call (Rand 1988). Researchers have recently described a number of species in which females also produce a call of some kind, but even where both species produce some vocal signal, the form, amplitude, and/or function of the call remains sexually dimorphic (reviewed in Emerson and Boyd 1999; see also Emerson, this volume, and Kelley et al., this volume). The advertisement call, some modification of it, or occasionally a distinctly different call can serve an intrasexual aggressive function among males as well (Wells 1988; Wagner 1989; Burmeister et al. 1999; see also Brenowitz et al., this volume). But even in this function, aggressive signaling is sexually dimorphic and the aggression seems to serve to defend a calling site from which males advertise their availability to females.

In keeping with the link to reproductive state, the production of male advertisement calls depends on the presence of circulating androgens (Emerson and Boyd 1999). The anuran vocal apparatus, the larynx, is sexually dimorphic both in size and form (Schneider 1988; McClelland and Wilczynski 1989; McClelland et al. 1997), as are the physiological characteristics of the laryngeal muscles (Kelley and Tobias 1989; Catz et al. 1992; see also Kelley et al., this volume). This structural dimorphism is due to the presence of circulating androgens during postmetamorphic maturation, which masculinize the larynx and indirectly the motor neurons innervating it (Kelley 1986; Sassoon and Kelley 1986; Catz et al. 1992). Androgens also appear to have an activating effect after sexual maturity (Wada et al. 1976; Mendonca et al. 1985; Emerson and Hess 1996; Marler and Ryan 1996; Solis and Penna 1997). Gonadectomy causes males to cease calling, whereas treatment with androgens can reverse this effect (Wetzel and Kelley 1983). Gonadal steroids vary seasonally in the temperate zone frogs in which they have been measured (Licht et al. 1983; Mendonca et al. 1985; Herman 1992), and the production of advertisement calls is correlated with this variation.

The activational effects of gonadal steroids on communication and other aspects of reproduction must somehow be related to their effects on the central nervous system. Indications of sexual dimorphism should also be present in these areas, reflecting the developmental organizing effects of sex steroids that lead to the sex differences in behavior and physiology. Furthermore, communication behavior, reproductive physiology, and endocrine state are sensitive to a range of external influences (Herman 1992; Jorgensen 1992), including the communication signals produced by conspecifics (Brozska and Obert 1980; Wells 1988; Propper and Moore 1991). We have turned our attention to three neurochemical systems in frogs that have been implicated in reproductive function in frogs and other vertebrates, with the long-term goal of understanding how gonadal hormones and external cues interact. One thrust of this investigation is to describe these systems and their sensitivity to circulating gonadal steroid hormones. A second thrust is to investigate how external sensory cues, specifically conspecific communication signals, may influence the levels of hormones that these systems control.

Forebrain Neurochemical Systems and Reproductive Behavior and Physiology

Two chemical systems are firmly established as influencing acoustic communication and other reproductive behaviors in anurans. One system involves a peptide, arginine vasotocin (AVT), produced by neurons in the telencephalon and diencephalon of the vertebrate brain. The other system involves the gonadal steroids, particularly androgens, whose production by the testes is ultimately regulated by another forebrain neuropeptide, gonadotropin releasing hormone (GnRH). We have been investigating both the AVT and the GnRH systems in the limbic/basal forebrain regions of the frog brain, as well as a third system composed of cells containing the neurotransmitter dopamine, marked by immunoreactivity for the synthetic enzyme tyrosine hydroxylase (TH). There remain significant gaps in understanding these systems. Even at this point, however, a review of what is known about these systems indicates interesting features of basal forebrain neuromodulator systems and potential avenues for further research.

Arginine Vasotocin

Arginine vasotocin (AVT) is the amphibian equivalent of the mammalian neuropeptide arginine vasopressin (AVP). Numerous studies have shown that AVP/AVT, presumably acting as a neuromodulator within the central nervous system, has a significant influence on intraspecific communication, social behavior, and aggression in mammals (e.g., Roche and Leshner 1979; Winslow et al. 1993; Ferris et al. 1997), birds (e.g., Voorhuis et al. 1991; Maney et al. 1997; Riters and Panksepp 1997; Goodson 1998), and fish (Wilhelmi et al. 1955). The effects are particularly obvious in males. Moore (1983) was the first to demonstrate that AVT influences mating behavior in amphibians by demonstrating that it stimulates clasping in male newts (*Taricha granulosa*). More recent work in several labs has demonstrated that AVT treatment increases calling in male frogs (Penna et al. 1992; Boyd 1994a; Marler et al. 1995; Propper and Dixon 1997; Chu et al. 1998; Emerson and Boyd 1999). Although fewer studies examine female behavior, AVT seems to stimulate female receptivity (Raimondi and Diakow 1981; Boyd 1992, 1994a).

Immunohistochemical studies have revealed several populations of AVT-containing cells in the anuran forebrain, as well as a substantial network of AVT fibers throughout much of the brain (Boyd et al. 1992; Gonzalez and Smeets 1992a, 1992b; Boyd 1997; Marin et al. 1998; Marler et al. 1999). One major population exists in the magnocellular preoptic area. Neuroanatomical tracing studies by Burmeister and Wilczynski (1996, 1997) suggest that these are neurosecretory neurons. We presume these are the neurons involved in the peripheral kidney and vasoregulation functions of AVT. A second and third set of AVT-containing neurons are more or less continuous with each other in the telencephalon (Figure 3.1). The more rostral of these is a line of cells extending along the nucleus accumbens close to the bed nucleus of the stria terminalis. This population grades into another clumped in the area of the amygdala. Smaller populations exist in the suprachiasmatic nucleus and in the periventricular hypothalamus. There is no evidence that any of these are neurosecretory neurons. They may rather serve a central nervous system neuromodulatory role, and are therefore candidates for the behavioral influence that AVT exerts on calling and other aspects of reproductive behavior.

The idea that the more rostral AVT cells are involved in reproductive behavior is supported by the fact that these extrahypothalamic, telencephalic AVT populations are sexually dimorphic in frogs (Boyd et al. 1992; Boyd 1994b; Emerson and Boyd 1999; Marler et al. 1999), with less AVT staining in females than in males. This pattern is seen in extrahypothalamic AVP/AVT populations in numerous taxa (e.g., Stoll and Voorn 1985; Wang and De Vries 1995; Jurkevich et al. 1996; Wang et al. 1997). More direct evidence comes from recent work by Marler et al. (1999). They assessed AVT staining in females, calling males, and satellite males in cricket frogs (*Acris crepitans*). Satellites are noncalling males that adopt a characteristic low posture around calling males and attempt to intercept females attracted to the caller's vocalization. It is a facultative strategy in that satellites can become calling males very quickly if a call site becomes available. Females, which do not call, have less AVT staining in the nucleus accumbens than either calling or satellite males. This is consistent with findings in other anurans (e.g., Boyd et al. 1992). Marler et al. (1999) also found a difference in AVT staining density in the nucleus accumbens between calling and satellite males. Satellite males

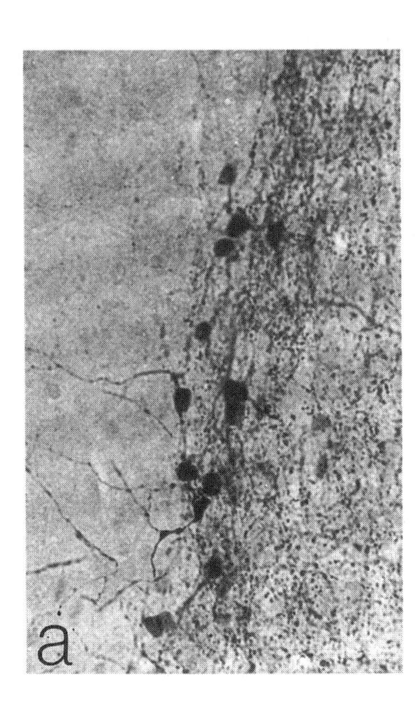

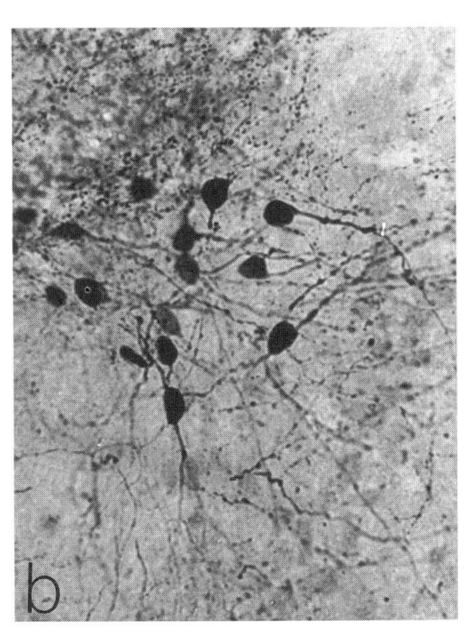

Figure 3.1. Arginine vasotocin (AVT) immunoreactive cells in (a) the nucleus accumbens and (b) the amygdala in the cricket frog (*Acris crepitans*). Both are seen in horizontal section, with rostral to the top and lateral to the left of each photograph. Sections are counterstained with cresyl violet.

had a higher staining density than calling males. Although counterintuitive, this staining difference would be consistent with the behavioral differences in calling and satellite males if the lower accumbens AVT level reflects the release of the peptide from storage there to promote calling. A similar result has been obtained in prairie voles, in which the onset of AVP-dependent social behavior is correlated with decreased AVP staining in telencephalic AVP populations (Winslow et al. 1993; Bamshad et al. 1994). Presumably, the low AVT level in females indicates that they have no natural AVT reserves in this area with which to stimulate calling. Significantly, treatment with exogenous AVT, if coupled with testosterone implants, will stimulate calling in female treefrogs, *Hyla cinerea* (although such calling is abnormal, likely due to the undeveloped female larynx; Penna et al. 1992).

In keeping with the observation of a sex dimorphism in this system, some populations of AVT neurons in amphibians are sensitive to sex steroids. In bullfrogs, staining in the cells of the amygdala population is influenced by treatment with androgens and estrogens, and fiber staining in more caudal regions, including those associated with vocal production, are also influenced by androgen treatment (Boyd 1994b). Whereas gonadal steroids are very likely responsible for the sex dimorphisms in this system, and surely influence AVP populations in adult amphibians, it is unlikely that the more rapid changes in this system over a single evening are mediated by the steroid hormones, because the changes are more rapid than the generally understood time course of the genomic actions of sex steroid hormones. It is possible that there are more rapid, nongenomic steroid effects, although membrane receptors for sex steroids have not been unequivocally demonstrated. A reasonable model is that gonadal steroids and AVT work together to adjust the calling behavior of male frogs on different time scales, with testosterone responsible for changes on the scale of days to weeks and AVT modifying calling within minutes (Marler et al. 1999). What then are the stimuli that affect AVT system function and thus the enhanced calling behavior that its release triggers? One hypothesis is that this system is influenced by social signals (i.e., calls) produced by conspecifics. At present there is no direct evidence for this idea, but it is one that is important to address if the neural control of amphibian reproductive behavior is to be understood.

Gonadotropin Releasing Hormone

The control of gonadal steroids in amphibians and other vertebrates operates according to a common pattern (Ball 1981; Crews and Silver 1985; Jorgensen 1992). Neurosecretory cells in the basal forebrain release gonadotropin releasing hormone (GnRH) into the hypophyseal portal system at the median eminence. This in turn stimulates the release of gonadotropin from the pituitary, which ultimately regulates gonadal steroid secretion. Because gonadal steroids, and hence the reproductive state of the gonads, are ultimately under the control of these GnRH cells, any central nervous system-mediated influence of exogenous cues such as social signals, or endogenous cues such as brain neuromodulators, must at some point be directed toward these cells.

Immunohistochemical staining for different forms of GnRH reveals two distinct populations in the anuran brain (Alpert et al. 1976; Jokura and Urano 1986; Wilczynski and Northcutt 1994). One population stains for the mammalian luteinizing hormone releasing hormone (LHRH) form of GnRH (Figure 3.2). It begins as a thin line of neurons along the very median wall of the septal area of the telencephalon and continues caudally to become continuous with a large cluster of LHRH-ir neurons forming a cap over the rostral pole of the anterior preoptic area. Many LHRH-ir fibers can be seen investing wide areas of the brain from the caudal telencephalon through the diencephalon to the rostral midbrain. The neurons also give rise to a large bundle of axons that travel along the ventrolateral margin of the preoptic area and hypothalamus and into the median eminence as far as the intermediate lobe of the pituitary. Based on this distribution and our investigations of neurosecretory cells (Burmeister and Wilczynski 1997), we suggest that these LHRH neurons are the neurosecretory cells that release GnRH into the pituitary portal system to stimulate gonadotropin and therefore gonadal steroids. The location of these neurons also fits well with the work of McCreery (1984) that showed that electrical stimulation of the septal area and rostral preoptic area induced pituitary gonadotropin release.

A second more caudal GnRH population characterized by the Chicken II (CII) form of GnRH also exists in frogs (Wilczynski and Northcutt 1994) and other amphibians (Muske 1993). It lies in the rostral midbrain between the levels of the third and fourth cranial nerve nuclei. Although some CII-ir fibers are apparent in the hypothalamus and median eminence, most are contained in the midbrain, isthmal region, and medulla. It is not known whether CII fibers contribute to the neurosecretory system, but the descending brainstem CII fibers are unlikely to be involved with that process. The brainstem CII system, and the widespread forebrain LHRH fiber system, must perform some sort of neuromodulatory function with the central nervous system. A substantial literature implicates GnRH as a central nervous system modulator in mammals. Propper and Dixon (1997) showed that systemic injections of LHRH stimulated amplectic clasping in frogs, but the mechanism

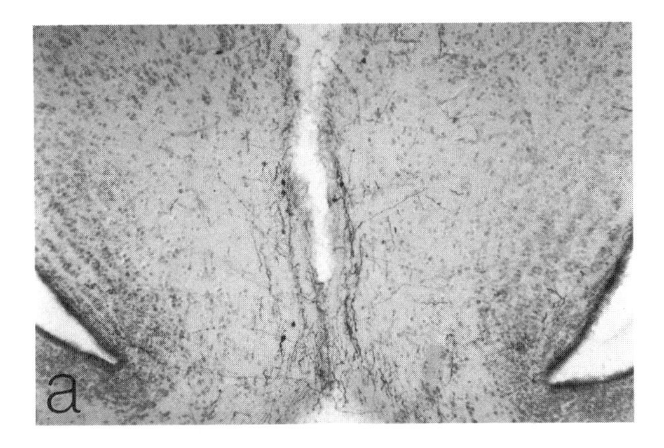

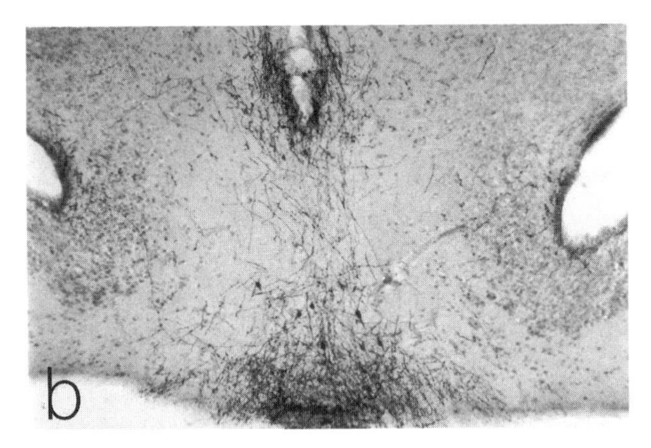

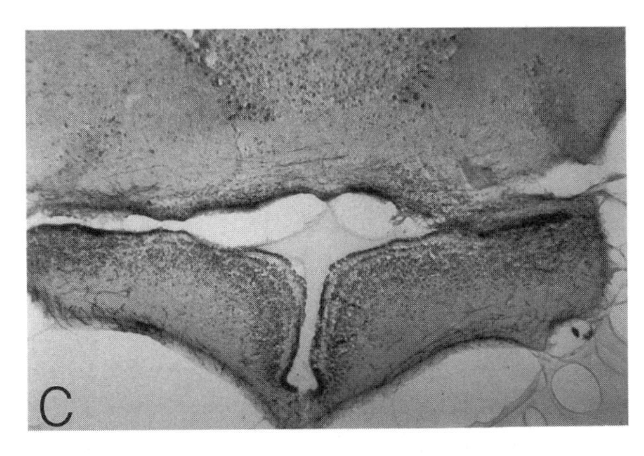

Figure 3.2. LHRH immunoreactive cells and fibers seen in frontal sections of the brain of the green treefrog (*Hyla cinerea*). (a) Immunoreactive septal neurons and their fibers are seen along the median edge of each telencephalic hemisphere. (b) In a more caudal section, immunoreactive septal neurons and fibers appear at the top of the photograph and also at the rostral pole of the preoptic area at the bottom of the photograph. LHRH fibers can be seen between the septal and preoptic populations. (c) In a more caudal section, LHRH fibers appear along the ventral edge of the median eminence.

of this effect remains unclear. Their study is the only one suggesting any behavioral function, independent of sex steroid release, of GnRH in amphibians.

Changes in the forebrain GnRH system as a function of sex steroid levels are well documented in mammals (Simerly 1995) and fish (Grober et al. 1991). These changes include increases in the number of GnRH immunoreactive neurons similar to those Boyd (1994b) reported for anuran AVT cells. As yet, such effects on GnRH cells are unexplored in frogs and toads. In fact it is not even known if GnRH cells are sexually dimorphic in amphibians.

Earlier lesion and stimulation studies (Ball 1981; Crews and Silver 1985) identified an area in the caudal hypothalamus that participates in the control of gonadal steroid secretion, along with the more rostral sites where we have identified GnRH cells. Our immunohistochemical studies show no GnRH cells located in that region; therefore this center does not contain a second, auxiliary GnRH neurosecretory population of cells. Instead, our immunohistochemical studies and others indicate that the caudal hypothalamus contains a significant population of TH immunoreactive neurons (Gonzalez and Smeets 1991; Gonzalez et al. 1993; Chu et al. 1995; Marin et al. 1997). These appear to be dopamine-containing cells (Gonzalez and Smeets 1991; Gonzalez et al. 1993). This is significant, as dopamine systems have been implicated in the control of both GnRH secretion (and the secretion of several other hypothalamic peptide hormones) and reproductive behavior in mammals (Bitran and Hull 1987; Melis and Argiolas 1995) and birds (Balthazart et al. 1997).

Tyrosine Hydroxylase

Tyrosine hydroxylase (TH) is one of the enzymes in the synthetic pathway for the production of the catecholamine neurotransmitters dopamine and norepinephrine. Amphibian catecholamine systems have been described via several methods, and in general these systems are similar to those in other vertebrates (Gonzalez and Smeets 1991; Gonzalez et al. 1993; Chu et al. 1995; Marin et al. 1997). Within-species comparisons of TH and dopamine localization using immunohistochemical staining show that TH is a good marker for forebrain dopamine populations in frogs (Gonzalez and Smeets 1991; Gonzalez et al. 1993).

The majority of TH-containing neurons are found in three populations in the frog basal forebrain. Most rostrally, these cells occupy a periventricular position in the anterior preoptic area. Many of these cells have short processes extending through the ependymal lining of the ventricle and appear to be cerebrospinal fluid-contacting cells. Caudal to this a population of large TH cells appears in the very rostral and ventral portion of the suprachiasmatic nucleus (SCN). These cells continue in the ventrolateral part of the SCN throughout much of its length. Finally a large population of medium-sized TH-containing neurons is present in the caudal hypothalamus spanning the posterior tuberculum and dorsal hypothalamus. Comparisons of TH staining in different ranid species and *Xenopus* indicate some minor differences among them, but these three populations are the largest and most consistent.

As yet it remains difficult to assign homologies between the amphibian dopaminergic cell populations and the those of the mammalian midbrain and forebrain. The caudal TH population in the posterior tuberculum and dorsal hypothalamus is responsible for ascending dopaminergic projections to the striatum and nucleus accumbens (Marin et al. 1997), so is at least in part equivalent to the substantia nigra and ventral tegmental area of mammals. Gonzalez and Smeets (1991) suggested that the anuran SCN population is homologous to the mammalian dopaminergic neurons of the arcuate nucleus of the hypothalamus, which gives rise to the tuberoinfundibular dopaminergic pathway terminating in the median eminence of the hypothalamus. That pathway regulates neuroendocrine release in mammals. Our TH immunohistochemical results do show a very dense dopaminergic innervation of the median eminence in frogs, but we cannot determine with certainty from which dopaminergic population it originates. It is significant, however, that the retrograde neuroanatomical tracer horseradish peroxidase, placed in the median eminence or into the blood, labels cells in the TH-containing region of the SCN, suggesting that cells there are neurosecretory (Burmeister and Wilczynski 1997). Other dopaminergic populations in this region in mammals include the anteroventral periventricular nucleus of the hypothalamus and neurons in the zona incerta that give rise to the incertohypothalamic dopamine system. These latter two dopamine populations regulate GnRH secretion (Weber et al. 1983; MacKenzie et al. 1989). The caudal dopaminergic population contributing to the dopaminergic innervation of the striatum and nucleus accumbens is closest in position to the incertohypothalamic system. What we can say at present is that, as a whole, the dopaminergic cell populations we and others have observed do provide a dopaminergic innervation to all the targets seen in mammals: the striatum, the nucleus accumbens and associated limbic structures, and the preoptic and hypothalamic regions, including the median eminence.

In mammals different dopaminergic systems are considered to have different functions. The striatal system is involved with motor control, the nucleus accumbens with motivated behaviors of all kinds (including reproductive behaviors), and the preoptic/hypothalamic connections with

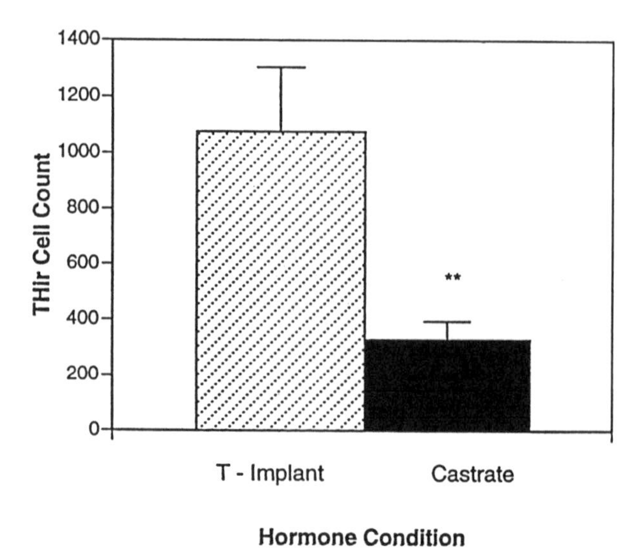

Figure 3.3. Tyrosine hydroxylase cell number in the forebrain of male *Rana pipiens* following castration and castration with testosterone propionate implant (*T-implant*). ** $p < 0.01$.

the control of various neuroendocrine functions, including the control of gonadal steroid secretion. The only functional investigations of amphibian dopamine systems to date have assessed dopamine's role in motor control (Ablordeppy et al. 1992; Glagow and Ewert 1996; Wilczynski et al. 1997). As with mammals, interfering with dopamine transmission impairs motor performance, including righting reflexes, locomotion, and orientation movements.

Although nothing is known about the role that the dopaminergic systems play in sexual behavior, communication, or reproductive physiology in frogs, our recent work has shown that gonadal steroids, specifically androgens, have a substantial influence on the amphibian dopaminergic system, just as they do in mammals (Chu and Wilczynski 1996; Chu et al. 1998). Gonadectomized male frogs (*Rana pipiens*) have substantially fewer TH-ir neurons in all three of the major forebrain TH populations than do gonadectomized frogs implanted with testosterone propionate or dihydrotestosterone (Figure 3.3). This decrease likely represents a decrease in TH activity, and hence dopamine production and activity, throughout the forebrain. Thus forebrain dopaminergic function is sensitive to circulating androgen levels, at least in male frogs. Brain areas manifesting sensitivity to circulating gonadal steroids are often sexually dimorphic. Some of the mammalian forebrain dopamine systems are sexually dimorphic, but whether such a dimorphism exists in amphibians is not known.

The behavioral and physiological significance of the TH cells' androgen sensitivity remains to be investigated. It is significant, however, that we observed these effects in sev-

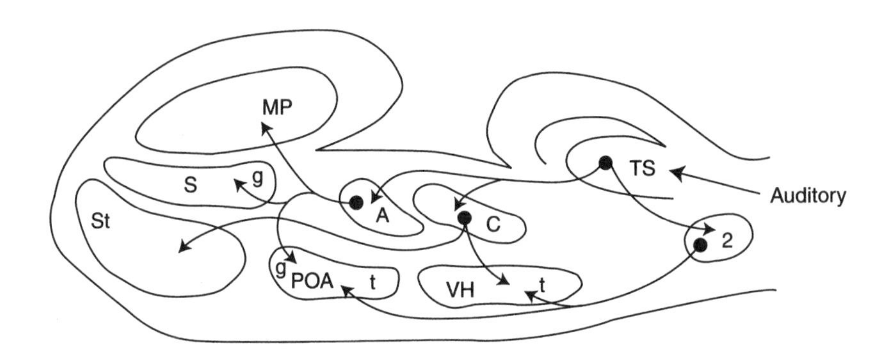

Figure 3.4. Schematic diagram of auditory pathways to endocrine control regions and other areas of the forebrain. *g* = location of gonadotropic releasing hormone (LHRH immunoreactive) neurons; *t* = location of tyrosine hydroxylase immunoreactive neurons; *A* = anterior thalamic nucleus; *C* = central thalamic nucleus; *MP* = medial pallium; *POA* = preoptic area; *S* = septal area; *St* = striatum; *TS* = torus semicircularis; *VH* = ventral hypothalamus; *2* = secondary isthmal nucleus.

eral major dopaminergic populations. This suggests that circulating androgen levels could potentially affect a host of behaviorally important systems, from locomotion to the expression of sexual behavior and from call production to the adjustment of gonadal function.

Social Stimuli and Endocrine Control Systems

Behavioral interactions and the receipt of communication signals influence reproductive hormone state in many different vertebrates (reviewed in Wilczynski et al. 1993; Runfeldt and Wingfield 1985; Francis et al. 1993; Rissman et al. 1997). Although sparse, evidence does exist indicating that social stimulation influences gonadal steroid levels and reproductive physiology in amphibians. Propper and Moore (1991) found that engaging in courtship raised gonadal steroid levels in newts, and Orchinik et al. (1988) found that androgen levels increased in males with increasing duration of amplexus. Using male *Rana temporaria,* Brozska and Obert (1980) found that lengthy exposure to conspecific calls resulted in larger testes compared with control animals, caused by a decrease in the normal gonadal recrudescence seen in the fall. That is, simply hearing conspecific calls maintained the reproductive state of the gonads.

Both neuroanatomical and neurophysiological data indicate substantial auditory input to the endocrine control centers for GnRH secretion described above (Neary 1988; Wilczynski and Allison 1989; Allison and Wilczynski 1991; Allison 1992; Wilczynski 1992; Wilczynski et al. 1993). The septal area, preoptic area, and caudal hypothalamus are all targets of ascending auditory pathways, with each receiving a characteristic mix of thalamic and midbrain auditory input (Figure 3.4). In frogs, ascending auditory pathways from the hindbrain converge on a large midbrain center, the torus semicircularis (inferior colliculus) (Wilczynski 1981, 1988). From this midbrain auditory center, ascending connections reach two thalamic nuclei, the central and anterior, as well as a large nucleus in the isthmal tegmentum, the secondary isthmal nucleus (Neary 1988). The central thalamic nucleus relays auditory information primarily to two areas. One is the striatum, the presumed basal ganglia homologue and therefore likely a motor center (Wilczynski and Northcutt 1983). The other target is the ventral hypothalamus, one of the key endocrine control zones of the frog brain, where its auditory inputs converge with a projection from the secondary isthmal nucleus (Neary and Wilczynski 1986; Neary 1988; Allison and Wilczynski 1991). The anterior thalamic nucleus relays auditory information, and probably somatosensory and visual information as well, to the medial pallium, which is thought to be homologous to the hippocampus and related limbic cortices (Neary 1988; Northcutt and Ronan 1992). Like the central thalamic nucleus, the anterior thalamic nucleus has additional fiber pathways to endocrine control regions, in this case the preoptic area (Allison and Wilczynski 1991) and septal area (Wilczynski, unpublished observations). In the preoptic area, secondary isthmal connections converge with the thalamic connections, just as they do in the hypothalamus. Overall the connections to basal forebrain regions implicated in GnRH control and other neuroendocrine functions are so numerous and robust that it is not inaccurate to say that these basal forebrain centers are a major target of ascending auditory pathways in anuran amphibians. (See Neary 1988 and Wilczynski et al. 1993 for a more detailed description of these nuclei.)

Neurophysiological studies have confirmed that these pathways do carry auditory information that influences the activity of preoptic and hypothalamic neurons (Urano and Gorbman 1981; Wilczynski and Allison 1989; Allison 1992). In the preoptic area, more than one-third, and in the hypothalamus, nearly half of the neurons change their firing rate in response to stimulation of the auditory system by conspecific advertisement calls (Allison 1992). Neural responsiveness in these areas takes the form of relatively slow changes in baseline firing after repeated stimulation over several minutes (Figure 3.5). Often the increased firing outlasts the stimulus presentation. Such responses indicate that acoustic stimulation acts as a modulator of ongoing activity.

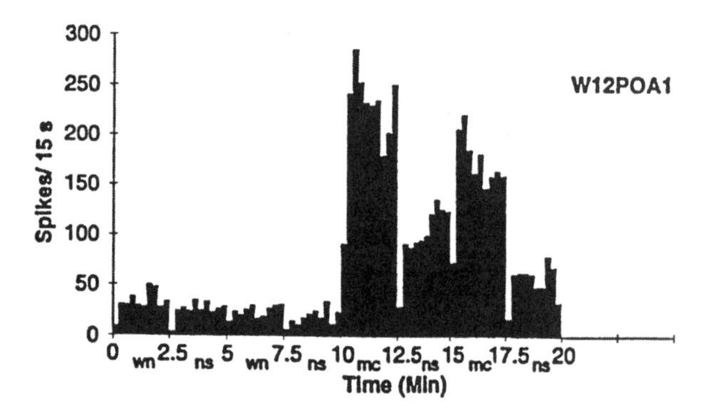

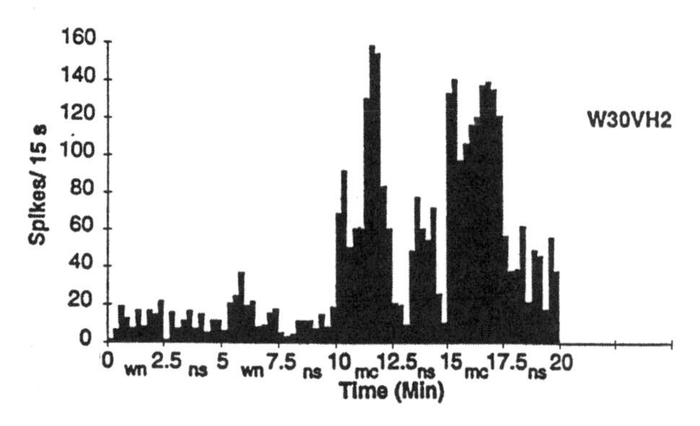

Figure 3.5. Poststimulus time histograms of firing rate in a preoptic (top) and ventral hypothalamic (bottom) neuron sensitive to acoustic stimulation in the green treefrog, *Hyla cinerea*. Both cells showed far greater response to a mating call (*mc*) than a white noise (*wn*) stimulus. Firing during the interstimulus interval is seen in the period marked *ns* (no stimulus).

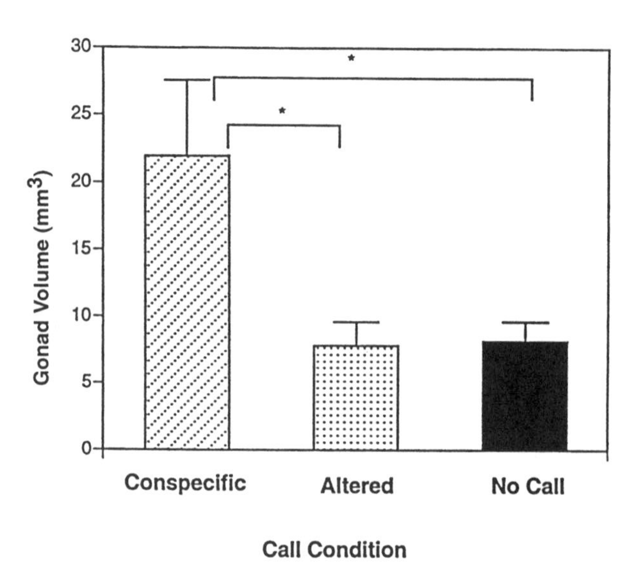

Figure 3.6. Testes volume in male *Rana sphenocephala* exposed for 12 days to conspecific calls, spectrally altered calls, and no acoustic stimulus. * $p < 0.05$.

The ultimate question involving these auditory inputs is whether acoustic stimuli of the type modulating neural activity there can influence circulating gonadal steroid levels. The report of Brozska and Obert (1980) provides indirect evidence that it does. After testing male *Rana temporaria* at the end of their breeding period, they found an increase in overall gonadal size as well as interstitial cell size in males exposed to conspecific calls compared with males exposed to a control noise stimulus. Brozska and Obert did not measure plasma steroid levels. However a change in sex steroid levels after exposure to conspecific communication signals would not be unprecedented, given the direct and indirect evidence for such effects in several taxa (reviewed in Wilczynski et al. 1993).

We repeated the experiment of Brozska and Obert under more controlled conditions and measured directly plasma androgen levels and gonadal size and histology in *Rana sphenocephala*. We performed the experiment twice, once with frogs collected during the spring mating season and once with frogs collected during the more variable fall mating aggregation. The animals were collected near Nashville, Tennessee, with the help of the Charles Sullivan Co. Males were exposed nightly for 10 nights to either chorus sounds composed of conspecific calls or a control stimulus created by spectral altering of the component calls in the chorus. Males exposed to the natural chorus stimulus during the spring had significantly higher levels of plasma androgen. During the fall experiment, mean plasma androgen levels were also higher in the natural chorus group, although the variance in all groups was so high that no statistically significant difference could be demonstrated. The gonadal profiles were quite different in the spring and fall experiment (Figures 3.6 and 3.7). Testes size was significantly larger in the males exposed to the natural chorus during the fall. Those animals had larger seminiferous tubule diameter and more shallow germinal cell layers, both of which denote increased spermatogenesis (Lofts 1984). In the spring, however, the frogs' testes size was not significantly different between groups, nor was the histology noticeably different, despite the significant difference in plasma androgen. These results are not inconsistent, because testes size can be determined largely by seminiferous tubule size, which by itself has no relationship to gonadal steroid production.

The effect of calls on testes size in the fall animals is consistent with the results of Brozska and Obert. Our results may, however, be due to acceleration of testes maturation rather than prevention of recrudescence. The natural annual pattern of testes and hormonal state has not been documented in this species. Where we collected animals, *Rana sphenocephala* breed very early in the spring (often late February), then become refractory during the summer. The fall may represent the beginning of testes growth in preparation for the spring mating, with the opportunistic fall mating,

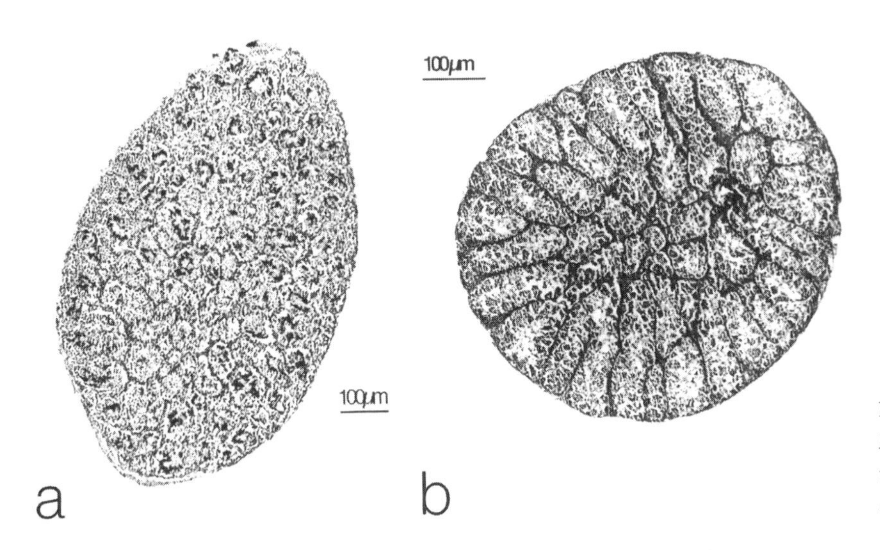

Figure 3.7. Cross sections of testes of *Rana sphenocephala* presented with (a) altered calls or (b) conspecific calls. Note the expanded seminiferous tubules in b.

triggered apparently by cool temperatures and heavy rain, reflecting the onset of gonadal maturation. We interpret the spring results on testes size as indicating that the testes reached a seasonal peak before we collected animals (all of which were collected while already in breeding aggregations) so that the calls could not add to testes growth. However the calls were revealed as a powerful modulator of steroidogenesis in these fully functional testes. In the fall frogs we could not statistically demonstrate an increase in plasma androgen level as a function of hearing calls, despite the fact that group mean differences were at least as large as in the spring. We interpret the high variability in the fall population to reflect our capture of frogs with different degrees of seasonal gonadal maturation.

Work in other vertebrates indicates that the changes in sex steroid secretion and gonadal histology triggered by exposure to the calls may be related to changes in the structure and function of GnRH cells. Cheng et al. (1998) recently provided direct evidence that acoustic communication signals stimulate LHRH secretion in ring doves, which her laboratory earlier established have auditory-hypothalamic connections very similar to those in frogs. Francis et al. (1993) showed that in cichlid fish social interactions associated with intermale aggression influence the size of GnRH neurons in the preoptic area along with the size of the gonads. Rissman et al. (1997) found that GnRH cell number decreased in female musk shrews after behavioral interactions with males or exposure to their odors. It is possible that such changes in GnRH-ir cell size or number, or some other metric that indicates increased production of GnRH, might also occur in frogs in response to appropriate social stimulation, but this has not been demonstrated.

It is also possible that acoustic stimulation may change AVT and catecholamine (TH) systems either directly through activation of central nervous system pathways or indirectly through the increased plasma androgen levels. Certainly the indirect effects are plausible, given that we and others have shown that these systems are sensitive to androgens in frogs (Boyd 1994b; Chu and Wilczynski 1996; Chu et al. 1998) and other vertebrates (e.g., Simerly 1995; Wang and de Vries 1995). Under conditions resulting in androgen changes within physiological limits, such effects may or may not include such gross changes as increases in immunoreactive cell number.

Conclusions

The three neurochemical systems reviewed above have been experimentally linked to reproductive behavior and communication in frogs either directly or by inference from their sensitivity to gonadal steroids and the precedents set by work in other vertebrates. The first challenge is to understand how sensory inputs associated with intraspecific communication and the influence of circulating gonadal steroid hormones affect each system, and how this interaction of external and internal cues changes internal physiological and hormonal state, as well as behavioral expression, in a coordinated way. As discussed, for each system the challenge has only been partially met. The subsequent challenge is to understand how these three systems interact with each other, mutually influencing the function of each to arrive at the end state we observe as reproductive social behavior and communication. In frogs and toads, work on this level of understanding has not even begun.

There must undoubtedly be other neurochemical systems that participate in reproductive social behavior and communication, from neurotransmitters like serotonin, linked to aggressive social behavior in a wide range of vertebrate and invertebrate organisms, to neuropeptides like enkephalin, which is found in the pathways that connect the auditory system to the hypothalamus in ring doves (Cheng and Zuo 1994), to hormones like corticosterone, which influences calling in frogs (Marler and Ryan 1996). In amphibians we barely know the basic information about these systems, such as whether there are sexual or seasonal differences in them, or whether they are sensitive to circulating gonadal steroids.

A vast territory of anatomical, physiological, and behavioral questions remains to be explored. But the acoustic communication system of frogs and toads is a good vehicle with which to undertake that exploration. With a relatively simple well-described communication system that readily lends itself to rigorous experimental manipulation and quantitative analysis, that also includes a significant range of interspecific and intraspecific variation, anurans have historically provided important model systems for a variety of fields, from evolutionary biology to neurophysiology (Wilczynski and Ryan 1988). These same strengths make frogs and toads candidates for studies aimed at understanding the mutual interactions among social behavior, environmental influences, and endocrine state. Understanding these mutual interactions and their underlying neurochemical systems in frogs and toads would, of course, lead to a deeper understanding of anuran acoustic communication and reproduction. But such processes, in the general sense, are fundamental to natural behavior of all kinds in all vertebrates. The mechanisms by which gonadal steroid secretions are controlled by forebrain GnRH neuroendocrine populations, the modulation of reproductive social behavior by vasotocin/vasopressin and catecholamines like dopamine, the sensitivity of these neurochemical systems to circulating sex steroid hormones, and the phenomenon that social interactions and communication signals have profound influences on endocrine state as well as behavior are all conserved features of vertebrate brain-behavior interactions manifest in various ways in numerous taxa. By exploring these interactive systems and processes in frogs and toads, we are exploring a model for the most basic processes that underlie vertebrate reproduction and social behavior.

Acknowledgments

Much of the recent research described in this chapter was supported by NIMH grant R01 MH57066 to WW and by NIMH Training Grant T32 MH18837, which supported work by JC.

References

Ablordeppy, S. Y., R. F. Borne, and W. M. Davis. 1992. Freezing action of a metabolite of haloperidol in frogs. Biochemical Pharmacology 43:218–217.

Allison, J. D. 1992. Acoustic modulation of neural activity in the preoptic area and ventral hypothalamus in the green treefrog (*Hyla cinerea*). Journal of Comparative Physiology 171:387–395.

Allison, J. D., and W. Wilczynski. 1991. Thalamic and midbrain auditory projections to the preoptic area and ventral hypothalamus in the green treefrog (*Hyla cinerea*). Brain, Behavior and Evolution 37:322–331.

Alpert, L. C., J. M. Brouwer, I. M. D. Jackson, and S. Reichlin. 1976. Localisation of LHRH in neurons in the frog brain (*Rana pipiens* and *Rana catesbeiana*). Endocrinology 98:910–921.

Ball, J. N. 1981. Hypothalamic control of the pars distalis in fishes, amphibians, and reptiles. General and Comparative Endocrinology 44:135–170.

Balthazart, J., C. Castagna, and G. F. Ball. 1997. Differential effects of D1 and D2 dopamine-receptor agonists and antagonists on appetitive and consummatory aspects of male sexual behavior in Japanese quail. Physiology and Behavior 62:571–580.

Bamshad, M., M. A. Novak, and G. J. De Vries. 1994. Cohabitation alters vasopressin innervation and paternal behavior in prairie voles. Physiology and Behavior 56:751–758.

Bitran, D., and E. M. Hull. 1987. Pharmacological analysis of male rat sexual behavior. Neuroscience and Biobehavioral Review 11:365–389.

Boyd, S. K. 1992. Sexual differences in hormonal control of release calls in bullfrogs. Hormones and Behavior 26:522–535.

Boyd, S. K. 1994a. Arginine vasotocin facilitation of advertisement calling and call phonotaxis in bullfrogs. Hormones and Behavior 28:232–240.

Boyd, S. K. 1994b. Gonadal steroid modulation of vasotocin concentrations in the bullfrog brain. Neuroendocrinology 60:150–156.

Boyd, S. K. 1997. Brain vasotocin pathways and the control of sexual behavior in the bullfrog. Brain Research Bulletin 44:345–350.

Boyd, S. K., C. J. Tyler, and G. J. De Vries. 1992. Sexual dimorphism in the vasotocin system of the bullfrog (*Rana catesbeiana*). Journal of Comparative Neurology 325:313–325.

Brozska, J., and H. J. Obert. 1980. Acoustic signals influencing the hormone production of the testes in the grass frog. Journal of Comparative Physiology 140:25–29.

Burmeister, S., and W. Wilczynski. 1996. Projections from forebrain and midbrain areas to the median eminence and neurohypophysis in the frog, *Rana pipiens*. Society for Neuroscience Abstracts 22:1414.

Burmeister, S., and W. Wilczynski. 1997. Identification of neurosecretory cell populations by injection of HRP into the blood of the frog, *Rana pipiens*. Society for Neuroscience Abstracts 23:1088.

Burmeister, S., W. Wilczynski, and M. J. Ryan. 1999. Temporal call changes and social context affect graded signaling in the cricket frog. Animal Behaviour 57:611–618.

Catz, D. S., L. M. Fisher, M. C. Moschella, M. L. Tobias, and D. B. Kelley. 1992. Sexually dimorphic expression of a laryngeal-specific, androgen-regulated myosin heavy chain gene during *Xenopus laevis* development. Developmental Biology 154:366–376.

Cheng. M. F., J. P. Peng, and P. Johnson. 1998. Hypothalamic neurons preferentially respond to female nest coo stimulation: Demonstration of direct acoustic stimulation of luteinizing hormone release. Journal of Neuroscience 18:5477–5489.

Cheng, M. F., and M. Zuo. 1994. Proposed pathways for vocal self-stimulation: Met-enkephalinergic projections linking the midbrain vocal nucleus, auditory-responsive thalamic regions, and neurosecretory hypothalamus. Journal of Neurobiology 25:361–179.

Chu, J., C. A. Marler, and W. Wilczynski. 1998. The effects of arginine vasotocin on the calling behavior of male cricket frogs in changing social contexts. Hormones and Behavior 34:248–261.

Chu, J., M. S. Rand, and W. Wilczynski. 1995. Distribution of immunoreactive tyrosine hydroxylase in the fore- and midbrain of *Rana pipiens* using whole mount immunohistochemistry. Society for Neuroscience Abstracts 21:437.

Chu, J., J. Whitman, B. Brudermanns, A. Clark, S. Burmeister, and W. Wilczynski. 1998. Effects of androgens on tyrosine hydroxylase immunoreactive cells in the CNS of male leopard frogs, *Rana pipiens*. Society for Neuroscience Abstracts 24:1705.

Chu, J., and W. Wilczynski. 1996. The effect of testosterone on tyrosine hydroxylase immunoreactive cells in the CNS of male leopard frogs, *Rana pipiens*. Society for Neuroscience Abstracts 22:1414.

Crews, D., and R. Silver. 1985. Reproductive physiology and behavior interactions in non-mammalian vertebrates. Pp. 101–182. *In* N. Adler, D. Pfaff, and R. W. Goy (eds.), Handbook of Behavioral Neurobiology, vol.7. Plenum Press, New York.

Emerson, S., and S. K. Boyd. 1999. Mating vocalizations of female frogs: Control and evolutionary mechanisms. Brain, Behavior and Evolution 53:187–197.

Emerson, S., and D. Hess. 1996. The role of androgens in opportunistic breeding, tropical frogs. General and Comparative Endocrinology103:220–230.

Ferris, C. F., R. H. Melloni, G. Koppel, K. W. Perry, R. W. Fuller, and Y. Delville. 1997. Vasopressin/serotonin interactions in the anterior hypothalamus control aggressive behavior in golden hamsters. Journal of Neuroscience 17:4331–4340.

Francis, R. C., K. Soma, and R. D. Fernald. 1993. Social regulation of the brain-pituitary-gonadal axis. Proceedings of the National Academy of Science USA 90:7794–7798.

Gerhardt, H. C. 1988. Acoustic properties used in call recognition by frogs and toads. Pp. 455–483. *In* B. Fritzsch, M. J. Ryan, W. Wilczynski, T. E. Hetherington, and W. Walkowiak (eds.), The Evolution of the Amphibian Auditory System. John Wiley, New York.

Glagow, M., and J. P. Ewert. 1996. Apomorphine-induced suppression of prey oriented turning in toads is correlated with activity changes in pretectum and tectum [^{14}C] 2DG studies and single cell recordings. Neuroscience Letters 220:215–218.

Gonzalez, A., and W. J. A. J. Smeets. 1991. Comparative analysis of dopamine and tyrosine hydroxylase immunoreactivities in the brain of two amphibians, the anuran *Rana ridibunda* and urodele *Pleurodeles waltii*. Journal of Comparative Neurology 303:457–477.

Gonzalez, A., and W. J. A. J. Smeets. 1992a. Comparative analysis of the vasotocinergic and mesotocinergic cells and fibers in the brain of two amphibians, the anuran *Rana ridibunda* and the urodele *Pleurodeles waltii*. Journal of Comparative Neurology 315:53–73.

Gonzalez, A., and W. J. A. J. Smeets. 1992b. Distribution of vasotocin- and mesotocin-like immunoreactivities in the brain of the South African clawed frog, *Xenopus laevis*. Journal of Chemical Anatomy 5:465–479.

Gonzalez, A., R. Tuinhof, and W. J. A. J Smeets. 1993. Distribution of tyrosine hydroxylase and dopamine immunoreactivities in the brain of the South African clawed frog *Xenopus laevis*. Anatomy and Embryology 187:193–201.

Goodson, J. L. 1998. Territorial aggression and dawn song are modulated by septal vasotocin and vasoactive intestinal polypeptide in male field sparrows (*Spizella pusilla*). Hormones and Behavior 34:67–77.

Grober, M. S., I. M. D. Jackson, and A. H. Bass. 1991. Gonadal steroids affect LHRH preoptic number in a sex/role changing fish. Journal of Neurobiology 22:734–741.

Herman, C. A. 1992. Endocrinology. Pp. 40–54. *In* M. E. Feder and W. Bruggren (eds.), Environmental Physiology of the Amphibians. University of Chicago Press, Chicago.

Jokura, Y., and A. Urano. 1986. Extrahypothalamic projections of luteinizing hormone-releasing hormone fibers in the brain of the toad, *Bufo japonicus*. General and Comparative Endocrinology 62:80–88.

Jorgensen, C. B. 1992. Growth and reproduction. Pp. 439–466. *In* M. E. Feder and W. Bruggren (eds.), Environmental Physiology of the Amphibians. University of Chicago Press, Chicago.

Jurkevich, A., S. W. Barth, N. Aste, G. Panzica, and R. Grossmann. 1996. Intracerbral sex differences in the vasotocin system in birds: Possible implication in behavioral and autonomic functions. Hormones and Behavior 30:673–681.

Kelley, D. B. 1986. Neuroeffectors for vocalization in *Xenopus laevis:* Hormonal regulation of sexual dimorphism. Journal of Neurobiology 17:231–248.

Kelley, D. B., and M. L. Tobias. 1989. The genesis of courtship song: cellular and molecular control of a sexually differentiated behavior. Pp. 175–194. *In* T. J. Carew and D. B. Kelley (eds.), Perspectives in Neural Systems and Behavior. Liss, New York.

Licht, P., B. R. McCreery, R. Barnes, and R. Pang. 1983. Seasonal and stress related changes in plasma gonadotropin, sex steroids, and corticosterone in the bullfrog, *Rana catesbeiana*. General and Comparative Endocrinology 50:124–145.

Lofts, B. 1984. Amphibians. Pp. 172–191. *In* G. E. Lamming (ed.), Marshall's Physiology of Reproduction. Churchill Livingstone, New York.

MacKenzie, F. J., M. D. James, and C. A. Wilson. 1989. Evidence that the dopaminergic incerto-hypothalamic tract has a stimulatory effect on ovulation and gonadotropin release. Neuroendocrinology 39:289–295.

Maney, D. L., C. T. Goode, and J. C. Wingfield. 1997. Intraventricular infusion of arginine vasotocin induces singing in a female songbird. Journal of Neuroendocrinology 9:487–491.

Marin, O., W. J. A. J. Smeets, and A. Gonzalez. 1997. Basal ganglia organization in amphibians: Catecholamine innervation of the striatum and nucleus accumbens. Journal of Comparative Neurology 378:50–69.

Marin, O., W. J. A. J. Smeets, and A. Gonzalez. 1998. Basal ganglia organization in amphibians: Chemoarchitecture. Journal of Comparative Neurology 392:285–312.

Marler, C. A., S. K. Boyd, and W. Wilczynski. 1999. Forebrain neuropeptide correlates of alternative male mating strategies under field conditions. Hormones and Behavior 36:53–61.

Marler, C. A., J. Chu, and W. Wilczynski. 1995. Arginine vasotocin injection increases probability of calling in cricket frogs, but causes call changes characteristic of less aggressive males. Hormones and Behavior 29:554–570.

Marler, C. A., and M. J. Ryan. 1996. Energetic constraints and steroid hormone correlates of male calling behaviour in the túngara frog. Journal of Zoology, London 240:397–409.

McClelland, B. E., and W. Wilczynski. 1989. Sexually dimorphic laryngeal morphology in *Rana pipiens*. Journal of Morphology 201:293–299.

McClelland, B. E., W. Wilczynski, and A. S. Rand. 1997. Sexual dimorphism and species differences in the neurophysiology and morphology of the acoustic communication system of two neotropical hylids. Journal of Comparative Physiology 180:415–462.

McCreery, B. R. 1984. Pituitary gonadotropic release by graded elec-

trical stimulation of the preoptic area in the male bullfrog, *Rana catesbeiana*. General and Comparative Endocrinology 55:367–372.
Melis, M., and A. Argiolas. 1995. Dopamine and sexual behavior. Neuroscience and Biobehavioral Reviews 19:19–38.
Mendonca, M. T., P. Licht, M. J. Ryan, and R. Barnes. 1985. Changes in hormone level in relation to breeding behavior in male bullfrogs (*Rana catesbeiana*) at the individual and population levels. General and Comparative Endocrinology 58:270–279.
Moore, F. L. 1983. Behavioral endocrinology of amphibian reproduction. Bioscience 33:557–561.
Muske, L. E. 1993. Evolution of gonadotropin releasing hormone (GnRH) neuronal systems. Brain, Behavior and Evolution 42:215–230.
Neary, T. J. 1988. Forebrain auditory pathways in ranid frogs. Pp. 233–252. *In* B. Fritzsch, M. J. Ryan, W. Wilczynski, T. E. Hetherington, and W. Walkowiak (eds.), The Evolution of the Amphibian Auditory System. John Wiley, New York.
Neary, T. J., and W. Wilczynski. 1986. Auditory pathways to the hypothalamus in ranid frogs. Neuroscience Letters 71:142–146.
Northcutt, R. G., and M. Ronan. 1992. Afferent and efferent connections of the bullfrog medial pallium. Brain, Behavior and Evolution 40:1–16.
Orchinik, M., P. Licht, and D. Crews. 1988. Plasma steroid concentrations change in response to sexual behavior in *Bufo marinus*. Hormones and Behavior 22:338–350.
Penna, M., R. R. Capranica, and J. Somers. 1992. Hormone-induced vocal behavior and midbrain auditory sensitivity in the green treefrog, *Hyla cinerea*. Journal of Comparative Physiology 170:73–82.
Propper, C. A., and T. B. Dixon. 1997. Differential effects of arginine vasotocin and gonadotropin-releasing hormone on sexual behavior in an anuran amphibian. Hormones and Behavior 32:99–104.
Propper, C. R., and F. L. Moore. 1991. Effects of courtship on brain gonadotropin hormone-releasing hormone and plasma steroid concentrations in a female amphibian (*Taricha granulosa*). General and Comparative Endocrinology 81:304–312.
Raimondi, D., and C. Diakow. 1981. Sex dimorphism in responsiveness to hormonal induction of female behavior in frogs. Physiology and Behavior 27:167–170.
Rand, A. S. 1988. An overview of acoustic communication. Pp. 415–431. *In* B. Fritzsch, M. J. Ryan, W. Wilczynski, T. E. Hetherington, and W. Walkowiak (eds.), The Evolution of the Amphibian Auditory System. John Wiley, New York.
Rissman, E. F., X. Li, J. A. King, and R. P. Millar. 1997. Behavioral regulation of gonadotropin-releasing hormone production. Brain Research Bulletin 44:459–464.
Riters, L. V., and J. Panksepp. 1997. Effects of vasotocin on aggressive behavior in male Japanese quail. Annals of the New York Academy of Science 807:478–80.
Roche, K. E., and A. I. Leshner. 1979. ACTH and vasopressin treatments immediately after a defeat increase future submissiveness in male mice. Science 204:1343–1344.
Runfeldt, S., and J. C. Wingfield. 1985. Experimentally prolonged sexual activity in female sparrows delays termination of reproductive activity in their untreated mates. Animal Behaviour 33:403–410.
Ryan, M. J. 1991. Sexual selection and communication in frogs: some recent advances. Trends in Ecology and Evolution 6:351–354.
Sassoon, D., and D. B. Kelley. 1986. The sexually dimorphic larynx of *Xenopus laevis:* Development and androgen regulation. American Journal of Anatomy 177:457–472.
Schneider, H. 1988. Peripheral and central mechanisms of vocalization. Pp. 537–558. *In* B. Fritzsch, M. J. Ryan, W. Wilczynski, T. E. Hetherington, and W. Walkowiak (eds.), The Evolution of the Amphibian Auditory System. John Wiley, New York.
Simerly, R. B. 1995. Hormonal regulation of limbic and hypothalamic pathways. Pp. 85–114. *In* P. E. Micevych and R. P. Hammer (eds.), Neurobiological Effects of Sex Steroid Hormones. Cambridge University Press, Cambridge.
Solis, R., and M. Penna. 1997. Testosterone levels and evoked vocal responses in a natural population of the frog *Batrachyla taeniata*. Hormones and Behavior 31:101–109.
Stoll, C. J., and P. Voorn. 1985. The distribution of hypothalamic and extrahypothalamic vasotocinergic cells and fibers in the brain of a lizard, *Gecko gecko:* Presence of a sex difference. Journal of Comparative Neurology 239:193–204.
Urano, A., and A. Gorbman. 1981. Effects of pituitary hormonal treatment on responsiveness of anterior preoptic neurons in male leopard frog *Rana pipiens*. Journal of Comparative Physiology 141:163–171.
Voorhuis, T. A. M., E. R. De Kloet, and D. De Weid. 1991. Effect of a vasotocin analog on singing behavior in the canary. Hormones and Behavior 25:549–559.
Wada, M., J. C. Wingfield, and A. Gorbman. 1976. Correlation between blood levels of androgens and sexual behavior in male leopard frogs, *Rana pipiens*. General and Comparative Endocrinology 29:72–77.
Wagner, W. E. 1989. Social correlates of variation in male calling behavior in Blanchard's cricket frog, *Acris crepitans blanchardi*. Ethology 82:27–45.
Wang, Z., and G. J. De Vries. 1995. Androgen and estrogen effects on vasopressin messenger RNA expression in the medial amygdaloid nucleus in male and female rats. Journal of Neuroendocrinology 7:827–831.
Wang, Z., K. Moody, J. D. Newman, and T. R. Insel. 1997. Vasopressin and oxytocin immunoreactive neurons and fibers in the forebrain of male and female common marmosets (*Callithrix jacchus*). Synapse 27:14–25.
Weber, R. F. A., W. J. De Greef, J. de Koning, and J. T. M. Vreeburg. 1983. LHRH and dopamine levels in the hypophysial stalk plasma and their relation to plasma gonadotropins and prolactin levels in male rats bearing a prolactin- and adrenocorticotropin-secreting tumor. Neuroendocrinology 36:205–210.
Wells, K. D. 1977. The social behavior of anuran amphibians. Animal Behaviour 25:666–693.
Wells, K. D. 1988. The effect of social interactions on anuran vocal behavior. Pp. 433–454. *In* B. Fritzsch, M. J. Ryan, W. Wilczynski, T. E. Hetherington, and W. Walkowiak (eds.), The Evolution of the Amphibian Auditory System. John Wiley, New York.
Wetzel, D. M., and D. B. Kelley. 1983. Androgen and gonadotropin effects on male mate calls in South African clawed frogs, *Xenopus laevis*. Hormones and Behavior 17:388–404.
Wilczynski, W. 1981. Afferents to the midbrain auditory center in the bullfrog, *Rana catesbeiana*. Journal of Comparative Neurology 198:421–434.
Wilczynski, W. 1988. Brainstem auditory pathways in anuran amphibians. Pp.209–231. *In* B. Fritzsch, M. J. Ryan, W. Wilczynski, T. E. Hetherington, and W. Walkowiak (eds.), The Evolution of the Amphibian Auditory System. John Wiley, New York.

Wilczynski, W. 1992. Auditory and endocrine inputs to forebrain centers in anuran amphibians. Ethology, Ecology and Evolution 4:75–87.

Wilczynski, W., and J. D. Allison. 1989. Acoustic modulation of neural activity in the hypothalamus of the leopard frog. Brain, Behavior and Evolution 33:317–324.

Wilczynski, W., J. D. Allison, and C. A. Marler. 1993. Sensory pathways linking social and environmental cues to endocrine control regions of amphibian forebrains. Brain, Behavior and Evolution 42:252–264.

Wilczynski, W., J. Chu, and M. J. Hannaman. 1997. Apomorphine impairs complex locomotor performance and decreases the number of tyrosine hydroxylase immunoreactive cells in the brain of the frog, *Rana pipiens*. Society for Neuroscience Abstracts 23:2135.

Wilczynski, W., and R. G. Northcutt. 1983. Connections of the bullfrog striatum: Afferent organization. Journal of Comparative Neurology 214:321–332.

Wilczynski, W., and R. G. Northcutt. 1994. LHRH and FMRF-amine immunoreactivity in the green treefrog. Society for Neuroscience Abstracts 20:1420.

Wilczynski, W., and M. J. Ryan. 1988. The amphibian auditory system as a model for neurobiology, behavior, and evolution. Pp.3–12. *In* B. Fritzsch, M. J. Ryan, W. Wilczynski, T. E. Hetherington, and W. Walkowiak (eds.), The Evolution of the Amphibian Auditory System. John Wiley, New York.

Wilhelmi, A. E., G. E. Pickford, and W. H. Sawyer. 1955. Initiation of the spawning reflex response in *Fundulus* by the administration of fish and mammalian neurohypothysial preparations and synthetic oxytocin. Endocrinology 57:243–252.

Winslow, J. T., N. Hastings, C. S. Carter, C. R. Harbaugh, and T. R. Insel. 1993. A role for central vasopressin in pair bonding in monogamous prairie voles. Nature 365:545–548.

4

SHARON B. EMERSON

Male Advertisement Calls

Behavioral Variation and Physiological Processes

Introduction

An increasing sophistication in both theory and technique now offers the opportunity for truly integrative studies of some major questions in evolutionary biology, including the origin and elaboration of morphological and behavioral novelties. For at least some systems the morphology-performance paradigm (Arnold 1983) can be intercalated between the proximate mechanistic basis for phenotypic expression and its evolutionary origin and maintenance (e.g., Ketterson et al. 1992). Such a broad interdisciplinary approach holds the promise of uncovering new and important connections among hierarchical levels of biological organization, as well as providing a more complete understanding of the proximate and ultimate processes affecting morphology and behavior (e.g., Ryan and Rand 1995).

Systems that focus on sexually dimorphic features are especially promising for this type of broad research. Predictions from sexual selection theory can be coupled to investigations of the endocrinological basis for secondary sexual characteristics at the organismal, cellular, and molecular levels (Crews 1987; Catz et al. 1995; Emerson 1996). Here I examine the relationship between steroid hormones and male advertisement vocalization in anurans from such a dual perspective.

Androgen levels vary in vertebrates (Crews 1987, and references therein). For example, in frogs, dihydrotestosterone and testosterone can span an order of magnitude intraspecifically and two orders of magnitude interspecifically (e.g., Licht et al. 1983; Orchinik et al. 1988; Emerson et al. 1993; Emerson and Hess 1996; Harvey et al. 1997). Wingfield et al. (1990) proposed the Challenge Hypothesis to account for this variation. The model posits that androgen levels, above that required for breeding, are related to social instability, male–male aggression, and male parental care. Although the Challenge Hypothesis is consistent with variation in androgen levels and mating systems among birds (Wingfield et al. 1990), it has not been well corroborated in the few studies on territorial aggression and male parental care in frogs (Mendonca et al. 1985; Townsend and Moger 1987). If, however, another social behavior of frogs—the male advertisement call—is considered as a form of male–male physical aggression, variation in androgen levels in frogs may be more consistent with the Challenge Hypothesis of Wingfield and co-workers.

Consideration of the energetically expensive male advertisement call may also provide insights into the highly variable levels of metabolic hormones called glucocorticosteroids that also occur in frogs (e.g., Emerson and Hess 1996). In adult males, corticosteroid levels vary within and among species even when daily variation is controlled by sampling individuals at the same time of day or night (Emerson and Hess 1996). Also elevated glucocorticosteroid levels often occur in breeding males as they are actively calling (Emerson and Hess 1996). Yet many other calling males of the same species in the same vicinity, at the same time, lack such elevated corticosteroid levels.

An extension of the Challenge Hypothesis that I will subsequently refer to as the Energetics–Hormone Vocalization Model provides a cohesive explanation for the wide variation in androgen and corticosteroid hormone levels in breeding male frogs. Specifically this model posits that variation in the net power output of the advertisement call as represented in the call characteristics, such as note repetition rate, call intensity, call length, and call effort (rate × duration of calling) is related to the intra- and interspecific variation in frog androgen and corticosterone levels above (the variable) breeding baseline. The basic idea is that variation in hormone levels reflects variation in calling performance among individuals and species and in the energetic costs of vocalization.

This model is similar to the Challenge Hypothesis in that it postulates endocrine responsiveness to behavioral cues and a positive feedback loop between testosterone levels and behavioral interactions (Wingfield et al. 1990). It extends the Challenge Hypothesis by linking variation in glucocorticosteroid and androgen levels to each other and by explicitly suggesting that in frogs it is the high energetic cost of male advertisement calling that is driving hormone levels.

The Male Advertisement Call

Male frog advertisement vocalization is one of the most energetically expensive activities that has ever been recorded in ectothermic vertebrates (MacNally 1981; Bucher et al. 1982; Taigen and Wells 1985; Taigen et al. 1985; Prestwich et al. 1989; Wells and Taigen 1989). Males, when calling, experience energetic costs that are 10 to 25 times those at rest, and a number of studies have documented a direct, positive relationship between metabolic rate and calling rate (reviewed by Pough et al. 1992). Energy conservation appears to be a major factor shaping the call strategies of individual frogs (Schwartz et al. 1995; Grafe 1997).

The muscles responsible for generating the male advertisement call include the external and internal obliques (Gans 1974) as well as the muscles of the larynx (Martins and Gans 1972). In some species, the lungs are inflated with air before the beginning of the call. The oblique muscles then go through cycles of contraction and relaxation, resulting in the passing of a volume of air back and forth between the vocal sacs and lungs. When the external and internal obliques contract, they push air out of the thoracic cavity and through the larynx. This air stream makes the vocal cords and associated cartilages vibrate and a pulse of sound is produced. The larynx may open passively when a critical air pressure is built up in the lungs or, alternatively, laryngeal muscles may actively control laryngeal opening, contracting with a constant phase relationship to the activation of the obliques (Girgenrath and Marsh 1997).

Call-producing muscles operate at high frequencies and produce high power output (Girgenrath and Marsh 1997). The body wall and laryngeal muscles have a number of unusual properties that correlate with their high-energy capabilities: 100% fast oxidative fibers, high citrate synthase (CS) and β-hydroxyacyl-CoA-dehydrogenase (HOAD) enzyme activities, high adenosine triphosphatase (ATPase) activity, and high mitochondrial and capillary densities (Taigen et al. 1985; Marsh and Taigen 1987; Bevier 1995; Ressel 1996). The muscles are sexually dimorphic in these physiological parameters and in mass, with males having muscles that are an order of magnitude larger in some species (reviewed in Pough et al. 1992). There is also considerable intraspecific variation in the degree of development of the calling muscles among adult breeding males (Given and McKay 1990).

In addition to variation in the calling musculature, male frogs show substantial intra- and interspecific variation in the characteristics of advertisement calls (e.g., Sullivan and Hinshaw 1990; Runkle et al. 1994; Bevier 1995; Gerhardt and Watson 1995). Pulse repetition rate, call intensity, and call length can vary among individual males even when variables such as temperature and body size are held constant (e.g., Wells et al. 1996). A study on 91 populations including 41 species of four different families revealed that coefficients of variation for pulse rate, call length, and call rate typically exceeded 20% (Gerhardt 1991).

The temporal distribution of calling also shows intra- and interspecific variation. Some species are explosive breeders, spending only one to a few nights calling during an entire breeding season (e.g., Bevier 1997). Other species may call over protracted time periods of days, weeks, or even months (e.g., Bevier 1997). Within species with prolonged breeding and calling, individuals vary considerably in the number of days spent calling. On a nightly time scale, call effort also varies within and between species (e.g., Wells et al. 1995). In some species, males call in distinct bouts punctuated by periods of silence (Wells 1977; Schwartz 1991). Often males adjust the timing and number of their call notes in response to the calls of other males (e.g., Wells and Schwartz 1984; Ryan 1985).

These call parameters, as well as the amount of time a male spends in a chorus, are primary factors determining male mating success (e.g., Murphy 1994). Additionally female choice has been demonstrated in a number of frog species (for summaries see Andersson 1994; Sullivan et al. 1995). Females of a wide diversity of species prefer high call rates, high call effort (rate × duration of calling), and/or high intensity (Gerhardt 1991; Sullivan et al. 1995)—variables that are controlled by the oblique and laryngeal muscles (Pough et al. 1992; Das Munshi and Marsh 1996).

The Role of Hormones

Steroid hormones regulate the expression of many sexually dimorphic characters, some reproductive behaviors, immunocompetence, muscle contraction characteristics, and metabolic function. Androgens, including testosterone and dihydrotestosterone, can promote energy consumption, stimulate respiratory metabolism (Nelson 1995), and have known protein anabolic effects that increase muscle mass (Gibbs et al. 1989), including heart size (John-Alder et al. 1996). Androgens can also alter the expression of myosin heavy chain isoforms in muscles (e.g., Morano et al. 1990; Catz et al. 1992). Muscle size and myosin heavy chain composition are related to the performance capabilities of muscles because they determine the force of contraction (McMahon 1984) and contraction velocities (Lannergren 1987; Edman et al. 1988; Reiser et al. 1988; Hermanson et al. 1993).

Steroids also interact with the immune system. There is some evidence that testosterone is immunosuppressive in vertebrates (Grossman 1985, 1990; Slater and Shreck 1997), and corticosteroids such as corticosterone play an important role in the regulation of the immune system. Corticosteroids may also decrease androgen sensitivity in certain cells and tissues (Burnstein et al. 1995). Elevated levels of glucocorticosteroids are often associated with a stress response (Sapolsky 1985, 1994). Generally such "flight or fight" responses have been viewed as occurring over a very short time period, on the order of seconds or minutes. More recently, however, work on birds indicates that elevated levels of glucocorticosteroids may also be associated with an emergency response that operates over a longer time scale of hours to days (Wingfield et al. 1997). Chronically elevated corticosteroids are thought to suppress reproductive behavior, regulate the immune system, increase gluconeogenesis, and increase foraging behavior (Wingfield 1994; Wingfield et al. 1997).

Corticosteroids most commonly show a negative correlation with testosterone levels (Sapolsky 1994). However some work has highlighted cases where animals with high testosterone levels also have high corticosteroid levels (Morell 1996). In mongoose and Red-winged Blackbirds, significantly higher testosterone and corticosteroid levels were found in dominant individuals involved in aggressive interactions (Beletsky et al. 1989; Creel et al. 1996). In *Iguana iguana* dominant males had significantly higher testosterone levels and showed a trend toward higher corticosteroid levels as well (Pratt et al. 1994).

Interspecifically, corticosteroid and androgen levels measured in calling male frogs have also been found to be positively correlated with each other (Emerson and Hess 1996). In parallel, intraspecific corticosteroid and testosterone levels also show a significant positive correlation in wild caught, calling frogs (e.g., Orchinik et al. 1988; Harvey et al. 1997). Furthermore there appears to be a relationship between male advertisement vocalization and hormone level. In some Southeast Asian frogs, species that have the highest average androgen and corticosteroid levels are also characterized, qualitatively, by the greatest call effort (Emerson and Hess 1996). Townsend and Moger (1987) found that calling male *Eleutherodactylus coqui* had higher testosterone levels than noncalling males. Moreover, in one species of South American leptodactylid, *Pleurodema thaul*, testosterone levels were found to be positively correlated with the number of pulses in the call of the male (Solis and Penna 1997).

The Energetics–Hormone Vocalization Model

Figure 4.1 outlines the Energetics–Hormone Vocalization Model showing the relationship between intra- and interspecific variation in male advertisement call characteristics in frogs and intra- and interspecific variation in corticosteroid and androgen levels. This model builds on the well-established physiological link between steroid hormone levels and muscle size, performance, and energetics, and operates over three time scales: a breeding season, a breeding bout, and a single night of breeding activity.

Experimental data indicate the external and internal oblique muscles are, at least in part, under the activational effects of androgens (Blair 1946; Emerson et al. 1999). Therefore, as androgen levels increase to a breeding baseline at the beginning of a breeding season, the call-producing muscles of the males increase in size. Muscle size is related to the performance capability of the muscle and hence call characteristics, because muscle size determines force of contraction and thus call intensity.

Elevated androgen levels may also alter the myosin heavy chain isoform in the call-producing muscles, depending on the species and the particular characteristics of the call. Myosin heavy chain composition is related to contraction velocity and thus pulse repetition rates of a call. A seasonal change in the contractile properties of the trunk muscles is

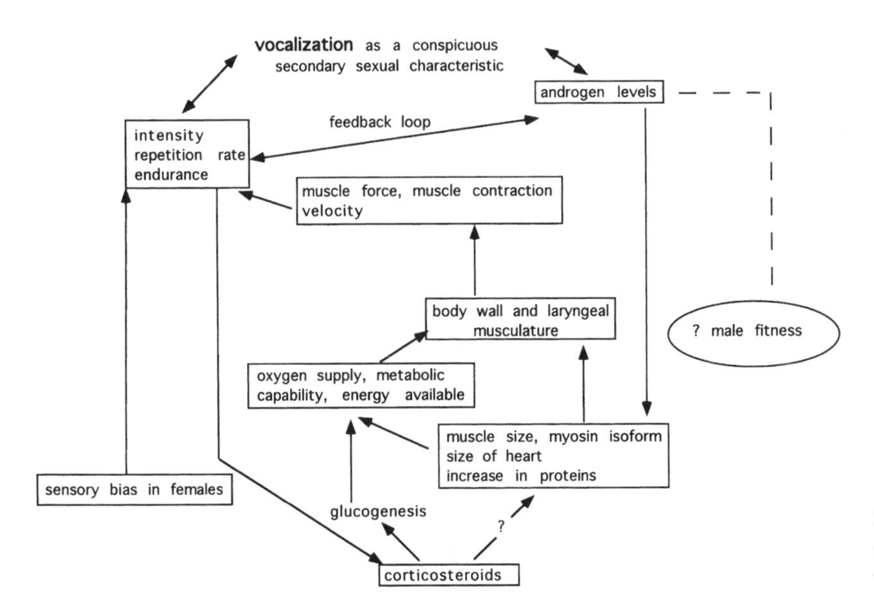

Figure 4.1. Flow chart showing the major aspects of the Energetics–Hormone Vocalization Model.

known to occur in at least one frog species (Das Munshi and Marsh 1996). No direct data are yet available on whether androgen levels actually alter the myosin heavy chain isoform in male internal and external obliques, but previous work has shown that to be the case for sexually dimorphic, androgen-dependent arm flexor muscles of male frogs. The muscles that males use to clasp females, similar to the oblique muscles, are under the activational effects of hormones and show seasonal shifts in contraction properties in relation to changing androgen levels. The muscles also stain differently for ATPase activity and show different muscle performance characteristics than females (Muller et al. 1969; Melichna et al.1972; Oka et al. 1984; Regnier and Herrera 1993; Dorlochter et al. 1994). These data are consistent with other work on vertebrates that shows a relationship between androgens, the expression of specific myosin heavy chain isoforms, and contractile properties of muscle fibers (Sheer and Morkin 1984; Lyons et al. 1986; Mahdavi et al. 1987; Morano et al. 1990).

According to the Energetics–Hormone Vocalization Model there should be a direct relationship between the properties of the call-producing muscles and the characteristics of the advertisement call. For example, contractions of the oblique muscles should correlate with call pulses, and species with faster pulse repetition rates should have muscles with faster contraction times than species with slower calls. In a recent study on hylid frogs, the call characteristics and call musculature were compared in males of two sister taxa whose calls differed primarily in pulse repetition rate. In both species the sound pulse frequency directly correlated with the contraction frequency of the external and internal oblique muscles, and, as expected, the oblique muscles of the species with the higher pulse repetition rate had a higher shortening velocity (Girgenrath and Marsh 1997).

Nightly, as males call, energy is expended and corticosteroid levels rise. Depending on the length and intensity of the vocalizations, high corticosteroid levels may induce gluconeogenesis as an additional energy source (Norris 1985; Wingfield 1994). In the model the nightly calling of the males has a feedback effect on the production of androgens so that vocalization itself increases androgen levels. Vocal self-stimulation is known to occur in birds (e.g., Cheng 1992) and such a feedback loop is consistent with experimental work on frogs indicating an increase in testis size and androgen level with species-specific vocal stimulation (Brzoska and Obert 1980; Chu and Wilcynski 1997). In the model, as males call each night, corticosteroid and androgen levels are elevated and positively correlated (Figure 4.2).

After nightly calling, corticosteroid and androgen levels drop during daytime inactivity (LeBoulenger et al. 1982), but then hormone levels rise again during the next night of calling. Importantly, although hormone levels decline with inactivity during the day, they do not return to the season breeding baseline (Figure 4.2, point *s*). As a result, over some number of nights of calling, depending on the level of power output, androgen and corticosteroid levels continue to rise. During this time corticosteroid levels are (probably) independent of the hypothalamus-pituitary-adrenal (HPA) axis (Astheimer et al. 1994). However if corticosterone becomes sufficiently elevated (level *a* in Figure 4.2) a short-term stress reaction (sensu Wingfield 1994) is evoked. At this point the elevated corticosterone suppresses reproductive behavior, the frogs stop calling, and begin to forage. During this part of the "breeding-bout cycle," corticosteroids are negatively

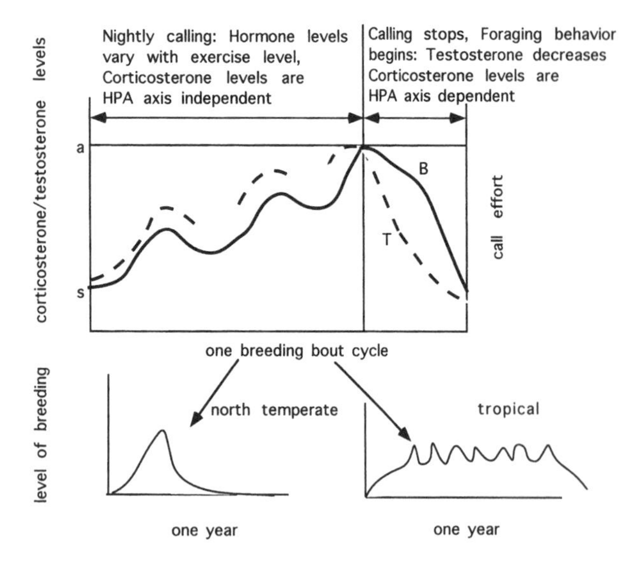

Figure 4.2. Shifts in hormone level of breeding male frogs as predicted from the Energetics–Hormone Vocalization Model. Note that the model operates over three time scales: the breeding season, a breeding bout, and a single night of breeding activity. *B* = corticosterone; *T* = testosterone; *s* = seasonal breeding baseline; *HPA* = hypothalamus-pituitary-adrenal axis.

Table 4.1 Interspecific differences as predicted by the Energetics–Hormone Vocalization Model

	Species 1	Species 2
Call characteristics	Low calling rate	High calling rate
	Low calling time	High calling time
	Low intensity	High intensity
Call energetic costs	Low	High
Call-producing muscles (obliques, laryngeal)		
Myosin heavy chain	Different in the two species	
Contraction velocity	Slower	Faster
Cross-sectional area	Smaller	Larger
Mitochondrial volume	Low	High
Capillary length	Lower	Higher
Intercellular lipid volume	Lower	Higher
ATPase activity	Lower	Higher
Relative B and T levels[a]	Lower	Higher
Breeding behavior		
Consecutive nights calling	More	Fewer
Hours calling/night	More	Fewer

[a] B = corticosterone; T = testosterone.

correlated with androgens (Figure 4.2). In frogs that show repeated breeding bouts throughout a protracted breeding season, this cycle may be repeated a number of times. Alternatively some frogs are known to feed while calling. These species may have advertisement calls that are energetically less expensive because of a lower net power output. As a consequence the males of these species may never experience corticosteroid levels of sufficient elevation to trigger a short-term stress response that would suppress reproductive activity.

This model suggests that frogs are limited in calling effort because of elevated corticosteroid levels rather than by absolute total energy reserves, and that the relationship between corticosteroids and androgens in frogs will vary depending on where individuals are in the breeding-bout cycle. Males will have a positive correlation between corticosteroids and androgens when calling and a negative correlation between corticosteroids and androgens when in the short-term stress response part of the breeding-bout cycle and not regularly calling. These predictions are consistent with literature data indicating that (1) when males stop calling their energy reserves are not completely exhausted (Grafe 1997 and references therein), (2) male *Physalaemus pustulosus* given corticosteroid supplements (and having very high corticosterone levels) are significantly less likely to call (Marler and Ryan 1996), (3) testosterone and corticosterone are negatively correlated in male *Physalaemus pustulosus* when corticosteroid levels are high (Marler and Ryan 1996), (4) males often do not feed when calling (e.g., MacNally 1981; Ryan 1985), and (5) in tropical species with prolonged breeding periods, males show repeated breeding bouts throughout a season (e.g., Bevier 1997).

The Energetics–Hormone Vocalization Model produces a multitude of testable predictions at both the intra- and interspecific levels regarding aspects of anuran advertisement vocalization. The basis for the predictions is an understanding of the complex interactions among call intensity, pulse repetition rate, and call length in determining the energetic costs of an advertisement call. All of these call characteristics are the result of the muscular work of oblique and laryngeal muscle contractions. Together they set the energetic costs of a single call. In turn, this cost shapes the temporal patterning of calling duration within and between species and, ultimately, the overall energetic costs of male calling performance.

Within a species, experimentally increasing the call effort of individual males should result in these males expending more energy calling and thus showing an increase in their androgen and corticosteroid levels. Food deprivation coupled with nightly calling in males with a high-energy call should result in individual males reaching corticosteroid levels that would trigger a short-term stress response, suppressing reproductive activity. The animals would then stop calling, leading to a decline in testosterone levels, and would begin foraging. These same shifts in behavior and testosterone levels should also occur if calling males are given corticosteroid supplements to artificially raise their corticosteroid levels.

Table 4.1 illustrates how interspecific differences in call

characteristics and thus energy costs of two hypothetical species might be expected to affect their morphology, physiology, and reproductive behavior. The importance of this example is that it draws attention to the extremely broad explanatory potential of the Energetics–Hormone Vocalization Model. Predictions span multiple levels of resolution from the myosin heavy chain in the muscle to the temporal patterning of breeding activity.

Sexual Selection Theory and the Energetics–Hormone Vocalization Model

The male frog advertisement call is a conspicuous secondary sexual characteristic and the propensity for calling appears to be directly related to androgen level (e.g., Solis and Penna 1997). Females often choose males on the basis of call characteristics (Andersson 1994; Sullivan et al. 1995), but there is only a limited understanding of the biological significance of such preferences (Sullivan et al. 1995). Some workers have suggested that male call characteristics and effort may have evolved to match a pre-existing female sensory bias for stronger stimuli (e.g., louder calls, faster repetition rates, and/or longer stimuli) (Ryan 1990). If this were the case, the higher androgen and corticosteroid levels in males that are predicted to occur with the most intense, fastest, and/or longest calls might be considered to be correlated effects.

An alternative idea that follows from the Energetics–Hormone Vocalization Model is that females may be using call characteristics and call effort as honest indicators of male fitness. Females might select males with the highest androgen and corticosteroid levels and, concomitantly, best overall condition. This alternative perspective is an application of the "good genes" model of parasite-mediated sexual selection (Hamilton and Zuk 1982; Folstad and Karter 1992) that posits a relationship between female choice and the expression of secondary sexual characteristics, androgen (and corticosteroid; Moller 1995) levels, and immunocompetence. This model assumes heritable variation in male condition and suggests that to the degree that testosterone is immunosuppressive, males with high hormone levels may have better genetic resistance to parasites and disease. That is, males with higher testosterone levels must be in better condition or they would not be able to tolerate the (presumed) immunosuppressive effects of the elevated androgen levels (Grossman 1985; Zuk 1996).

Although this good genes model has attracted considerable attention, few organisms outside of birds have been used to test this hypothetical relationship among secondary sexual characteristics, hormone level, and immunocompetence. If the hypothesized connection between call energetics and steroid hormone levels in frogs is supported, anurans may provide excellent test cases for this controversial model of sexual selection.

Conclusions

The Energetics–Hormone Vocalization Model links aspects of intra- and interspecific variation in corticosteroid and androgen levels, advertisement call characteristics, and reproductive behavior of male frogs. The basic idea is that variation in hormone levels reflects variation in the calling performance among individuals and species and in the energetic costs of vocalization. This connection is the result of hormone-related mechanistic relationships with muscle size, muscle contraction properties, and energy metabolism.

Although male advertisement calling has been well studied from ecological, behavioral, and neurophysiological perspectives, until now little work has been done to examine the endocrine aspects of advertisement vocalization in relation to muscle physiology. The Energetics–Hormone Vocalization Model builds on data from recent field endocrinology studies to outline a plausible mechanistic link between these variables. The linkage makes it possible for the model to make testable predictions regarding specific aspects of muscle structure and breeding behavior of male frogs based on the characteristics and energetic costs of the call.

Female frogs often choose males on the basis of call characteristics and performance, but the biological significance of such choices has remained an enigma. If, however, male call characteristics and performance reflect androgen and corticosteroid levels of individuals, then the Energetics–Hormone Vocalization Model also suggests a testable explanation for such female choices based on a good genes model of sexual selection.

References

Andersson, M. 1994. Sexual Selection. Princeton University Press, Princeton, NJ.

Arnold, S. 1983. Morphology, performance, and fitness. American Zoologist 23:347–361.

Astheimer, L. B., W. A. Buttemer, and J. C. Wingfield. 1994. Gender and seasonal differences in the adrenocortical response to ACTH challenge in an Arctic Passerine, *Zonotrichia leucophrys gambelii*. General and Comparative Endocrinology 94:33–43.

Beletsky, L., G. Orians, and J. Wingfield. 1989. Relationships of steroid hormones and polygyny to territorial status, breeding experience, and reproductive success in male red-winged blackbirds. Auk 106:107–117.

Bevier, C. 1995. Biochemical correlates of calling activity in neotropical frogs. Physiological Zoology 68:1118–1142.

Bevier, C. 1997. Breeding activity and chorus tenure of two neotropical hylid frogs. Herpetologica 53:297–311.

Blair, A. 1946. The effects of various hormones on primary and secondary sex characters of juvenile *Bufo fowleri*. Journal of Experimental Zoology 103:365–400.

Brzoska, J., and H. Obert. 1980. Acoustic signals influencing the hormone production of the testes in the grass frog. Journal of Comparative Physiology 140:25–29.

Bucher, T., M. Ryan, and G. Bartholomew. 1982. Oxygen consumption during resting, calling and nest building in the frog, *Physalaemus pustulosus*. Physiological Zoology 55:10–22.

Burnstein, K., C. Maiorino, J. Dai, and D. Cameron. 1995. Androgen and glucocorticoid regulation of androgen receptor cDNA expression. Molecular and Cellular Endocrinology 115:177–186.

Catz, D., L. Fischer, and D. Kelley. 1995. Androgen regulation of a laryngeal-specific myosin heavy chain mRNA isoform whose expression is sexually differentiated. Developmental Biology 171:448–457.

Catz, D., L. Fischer, M. Moschella, M. Tobias, and D. Kelley. 1992. Sexually dimorphic expression of a laryngeal-specific, androgen-regulated myosin heavy chain gene during *Xenopus laevis* development. Developmental Biology 154:366–376.

Cheng, M. 1992. For whom does the female dove coo? A case for the role of vocal self-stimulation. Animal Behaviour 43:1035–1044.

Chu, J., and W. Wilcynski. 1997. Conspecific social signals influence testes size and androgen levels in the southern leopard frog. Neuroscience Abstracts (pt. II) 23:10088.

Creel, S., Creel, N., and S. Monfort. 1996. Social stress and dominance. Nature 379:212.

Crews, D. 1987. Psychobiology of Reproductive Behavior: An evolutionary perspective. Prentice-Hall, New Jersey.

Das Munshi, M., and R. Marsh. 1996. Seasonal changes in contractile properties of the trunk muscles in *Hyla chrysoscelis*. American Zoologist 36:16A.

Dorlochter, M., S. Astrow, and A. Herrera. 1994. Effects of testosterone on a sexually dimorphic frog muscle: Repeated in vivo observations and androgen receptor distribution. Journal of Neurobiology 25:897–916.

Edman, K., C. Reggiani, S. Schiaffino, and G. teKronnie. 1988. Maximum velocity of shortening related to myosin isoform composition in frog skeletal muscle fibres. Journal of Physiology 395:679–694.

Emerson, S. 1996. Phylogenies and physiological processes—the evolution of sexual dimorphism in Southeast Asian frogs. Systematic Biology 45:278–289.

Emerson, S., A. Greig, L. Carroll, and G. Prins. 1999. Androgen receptors in two androgen-mediated, sexually dimorphic characters of frogs. General and Comparative Endocrinology 114:173–180.

Emerson, S., and D. Hess. 1996. The role of androgens in opportunistic breeding, tropical frogs. General and Comparative Endocrinology 103:220–230.

Emerson, S., C. Roswemitt, and D. Hess. 1993. Androgen levels in a Bornean voiceless frog, *Rana blythi*. Canadian Journal of Zoology 71:196–203.

Folstad, I., and A. Karter. 1992. Parasites, bright males, and the immunocompetence handicap. American Naturalist 139:603–622.

Gans, C. 1974. Biomechanics: An Approach to Vertebrate Biology. Lippincott, Philadelphia.

Gerhardt, H. 1991. Female mate choice in treefrogs: Static and dynamic acoustic criteria. Animal Behaviour 42:615–635.

Gerhardt, H., and G. Watson. 1995. Within-male variability in call properties and female preference in the grey treefrog. Animal Behaviour 50:1187–1191.

Gibbs, R., W. Kingston, R. Jozefowick, B. Herr, G. Forbes, and D. Halliday. 1989. Effect of testosterone on muscle mass and muscle protein synthesis. Journal of Applied Physiology 66:498–503.

Girgenrath, M., and R. Marsh. 1997. In vivo performance of trunk muscles in tree frogs during calling. Journal of Experimental Biology 200:3101–3108.

Given, M., and D. McKay. 1990. Variation in citrate synthase activity in calling muscles of carpenter frogs, *Rana virgatipes*. Copeia 1990:411–421.

Grafe, U. 1997. Use of metabolic substrates in the gray treefrog *Hyla versicolor:* Implications for calling behavior. Copeia 1997:356–362.

Grossman, C. 1985. Interactions between the gonadal steroids and the immune system. Science 227:257–261.

Grossman, C. 1990. Are there underlying immune-neuroendocrine interactions responsible for immunological sexual dimorphism? Progress in NeuroEndocrinImmunology 3:75–82.

Hamilton, W., and M. Zuk. 1982. Heritable true fitness and bright birds: A role for parasites? Science 218:384–387.

Harvey, L., C. Propper, S. Woodley, and M. Moore. 1997. Reproductive endocrinology of the explosively breeding desert spadefoot toad, *Scaphiopus couchii*. General and Comparative Endocrinology 105:102–113.

Hermanson, J., M. Cobb, W. Schutt, F. Muradali, and J. Ryan. 1993. Histochemical and myosin composition of vampire bat (*Desmodus rotundus*) pectoralis muscle targets a unique locomotory niche. Journal of Morphology 217:347–356.

John-Alder, H., S. McMann, L. Katz, A. Gross, and D. Barton. 1996. Social modulation of exercise endurance in a lizard (*Sceloporus undulatus*). Physiological Zoology 69:547–567.

Ketterson, E., V. Nolan, L. Wolf, and C. Ziegenfus. 1992. Testosterone and avian life histories: Effects of experimentally elevated testosterone on behavior and correlates of fitness in the dark-eyed junco (*Junco hyemalis*). American Naturalist 140:980–999.

Lannergren, J. 1987. Contractile properties and myosin isoenzymes of various kinds of *Xenopus* twitch muscle fibres. Journal of Muscle Research and Cell Motility 8:260–273.

LeBoulenger, F., C. Delarue, A. Belanger, L. Perroteau, P. Netchitailo, S. JeGou, M. Tonon, and H. Vaudry. 1982. Direct radioimmunoassays for plasma corticosterone and aldosterone in frog. I. Validation of the methods and evidence for daily rhythms in a natural environment. General and Comparative Endocrinology 46:521–532.

Licht, P., B. McCreery, R. Barnes, and R. Pang. 1983. Seasonal and stress-related changes in plasma gonadotropins, sex steroids, and corticosterone in the bullfrog *Rana catesbeiana*. General and Comparative Endocrinology 50:124–145.

Lyons, G., A. Kelly, and N. Rubenstein. 1986. Testosterone-induced changes in contractile protein isoforms in the sexually dimorphic temporalis muscle of the guinea pig. Journal of Biological Chemistry 261:13278–13284.

MacNally, R. 1981. On the reproductive energetics of chorusing males: Energy depletion profiles, restoration, and growth in two sympatric species of *Ranidella* (Anura). Oecologica 51:181–188.

Mahdavi, V., S. Izumo, and B. Nadal-Ginard. 1987. Developmental and hormonal regulation of sarcomeric myosin heavy chain gene family. Circulation Research 60:804–814.

Marler, C., and M. Ryan. 1996. Energetic constraints and steroid hormone correlates of male calling behavior in the túngara frog. Journal of Zoology London 240:397–409.

Marsh, R., and T. Taigen. 1987. Properties enhancing aerobic capacity of calling muscles in gray tree frogs *Hyla versicolor*. American Journal of Physiology 21:R786–R793.

Martin, W. and C. Gans. 1972. Muscular control of the vocal tract during release signaling in the toad *Bufo valliceps*. Journal of Morphology 137:1–28.

McMahon, T. 1984. Muscles, reflexes and locomotion. Princeton University Press, Princeton, NJ.

Melichna, J., E. Gutmann, A. Hebrychova, and J. Stichova. 1972. Sexual dimorphism in contraction properties and the fibre pattern of the flexor carpi radialis muscle of the frog. Experientia Basel 28:89–91.

Mendonca, M., P. Licht, M. Ryan, and R. Barnes. 1985. Changes in hormone levels in relation to breeding behavior in male bullfrogs (*Rana catesbeiana*) at the individual and population levels. General and Comparative Endocrinology 58:270–279.

Møller, A. 1995. Hormones, handicaps and bright birds. Trends in Ecology and Evolution 10:121.

Morano, I., J. Gerstner, C. Ruegg, U. Ganten, D. Ganten, and H. Vosberg. 1990. Regulation of myosin heavy chain expression in the hearts of hypertensive rats by testosterone. Circulation Research 66:1585–1590.

Morell, V. 1996. Life at the top: Animals pay the high price of dominance. Science 271:292.

Muller, E., G. Galavasi, and J. Szirmai. 1969. Effects of castration and testosterone treatment on fiber width of the flexor carpi radialis muscle in the male frog. General and Comparative Endocrinology 133:275–284.

Murphy, C. 1994. Chorus tenure of male barking treefrogs, *Hyla gratiosa*. Animal Behaviour 48:763–777.

Nelson, R. 1995. An Introduction to Behavioral Endocrinology. Sinauer Press, Sunderland, MA.

Norris, D. 1985. Vertebrate Endocrinology. Lea and Febiger, Philadelphia.

Oka, Y., R. Ohtani, M. Satou, and K. Ueda. 1984. Sexually dimorphic muscles in the forelimb of the Japanese toad, *Bufo japonicus*. Journal of Morphology 180:297–308.

Orchinik, M., P. Licht, and D. Crews. 1988. Plasma steroid concentrations change in response to sexual behavior in *Bufo marinus*. Hormones and Behavior 22:338–350.

Pough, F., W. Magnusson, M. Ryan, K. Wells, and T. Taigen. 1992. Behavioral energetics. Pp. 395–436. *In* Feder, M., and W. Burggren (eds.), Environmental Physiology of the Amphibians. University of Chicago Press, Chicago.

Pratt, N. C., J. A. Phillips, A. C. Alberts, and K. S. Bolda. 1994. Sexual bimaturation in green iguanas: Functional versus physiological puberty. Animal Behaviour 47:1101–1114.

Prestwich, K., K. Brugger, and M. Topping. 1989. Energy and communication in three species of hylid frogs: Power input, power output and efficiency. Journal of Experimental Biology 144:53–80.

Regnier, M., and A. Herrera. 1993. Changes in contractile properties by androgen hormones in sexually dimorphic muscles of male frogs (*Xenopus laevis*). Journal of Physiology 461:565–581.

Reiser, P., M. Greaser, and R. Moss. 1988. Myosin heavy chain composition of single cells from avian skeletal muscle is strongly correlated with velocity of shortening during development. Developmental Biology 129:509–516.

Ressel, S. 1996. Ultrastructural properties of muscles used for call production in Neotropical frogs. Physiological Zoology 69:952–973.

Runkle, L., K. Wells, C. Robb, and S. Lance. 1994. Individual, nightly, and seasonal variation in calling behavior of the gray tree frog, *Hyla versicolor:* Implications for energy expenditure. Behavioral Ecology 5:318–325.

Ryan, M. 1985. The Túngara Frog: A Study in Sexual Selection and Communication. University of Chicago Press, Chicago.

Ryan, M. 1990. Sexual selection, sensory systems, and sensory exploitation. Oxford Surveys of Evolutionary Biology 7:157–195.

Ryan, M., and A. Rand. 1995. Female responses to ancestral advertisement calls in túngara frogs. Science 269:390–392.

Sapolsky, R. 1985. Stress-induced suppression of testicular function in the wild baboon: Role of glucocorticoids. Endocrinology 116:2273–2278.

Sapolsky, R. 1994. Why Zebras Don't Get Ulcers. W. H. Freeman and Company, New York.

Schwartz, J. 1991. Why stop calling? A study of unison bout singing in a neotropical treefrog. Animal Behaviour 42:565–577.

Schwartz, J., S. Ressel, and C. Bevier. 1995. Carbohydrate and calling: Depletion of muscle glycogen and the chorusing dynamics of the neotropical treefrog *Hyla microcephala*. Behavioral Ecology and Sociobiology 37:125–135.

Sheer, D., and E. Morkin. 1984. Myosin isoenzyme expression in rat ventricle: Effects of thyroid hormone analogs, catecholamines, glucocorticoids, and high carbohydrate diet. Journal of Pharmacology and Experimental Therapeutics 229:872–879.

Slater, C., and C. Shreck. 1997. Physiological levels of testosterone kill salmonid leukocytes in vitro. General and Comparative Endocrinology 106:113–119.

Solis, R., and M. Penna. 1997. Testosterone levels and evoked vocal responses in a natural population of the frog, *Batrachyla taeniata*. Hormones and Behavior 31:101–109.

Sullivan, B., and C. Hinshaw. 1990. Variation in advertisement calls and male calling behavior in the spring peeper (*Pseudacris crucifer*). Copeia 1990:1146–1150.

Sullivan, B., M. Ryan, and P. Verrell. 1995. Female choice and mating system structure. Pp. 470–517. *In* J. Heatwole, and B. Sullivan (eds.), Amphibian Biology Vol. 2. Surrey Beatty and Sons, Chipping Norton, New South Wales.

Taigen, T., and K. Wells. 1985. Energetics of vocalization by an anuran amphibian (*Hyla versicolor*). Journal of Comparative Physiology B 155:163–170.

Taigen, T., K. Wells, and R. Marsh. 1985. The enzymatic basis of high metabolic rates in calling frogs. Physiological Zoology 58:719–726.

Townsend, D., and W. Moger. 1987. Plasma androgen levels during male parental care in a tropical frog (*Eleutherodactylus*). Hormones and Behavior 21:93–99.

Wells, K. 1977. The social behaviour of anuran amphibians. Animal Behaviour 25:666–693.

Wells, K., and J. Schwartz. 1984. Vocal communication in a neotropical treefrog, *Hyla ebraccata:* Advertisement calls. Animal Behaviour 32:405–420.

Wells, K., and T. Taigen. 1989. Calling energetics of a neotropical treefrog, *Hyla microcephala*. Behavioral Ecology and Sociobiology 25:13–22.

Wells, K., T. Taigen, and J. O'Brien. 1996. The effect of temperature on calling energetics of the spring peeper (*Pseudacris crucifer*). Amphibia-Reptilia 17:149–158.

Wells, K., T. Taigen, S. Rusch, and C. Robb. 1995. Seasonal and nightly variation in glycogen reserves of calling gray treefrogs (*Hyla versicolor*). Herpetologica 51:359–376.

Wingfield, J. 1994. Modulation of the adrenocortical response to stress in birds. Pp. 520–528. *In* K. G. Davey, R. E. Peter, and S. S. Tob (eds.), Perspectives in Comparative Endocrinology, National Research Council, Ottawa, Canada.

Wingfield, J., R. Hegner, A. Dufty, and G. Ball. 1990. The "Challenge Hypothesis": Theoretical implications for patterns of testosterone secretion, mating systems and breeding strategies. American Naturalist 136:829–846.

Wingfield, J., D. Maney, C. Breuner, J. Jacobs, S. Lynn, M. Ramenofsky, and R. Richardson. 1997. Ecological bases of hormone-behavior interaction: The emergency life history stage. American Zoologist 338:191–206.

Zuk, M. 1996. Disease, endocrine-immune interactions, and sexual selection. Ecology 77:1037–1042.

5

KENTWOOD D. WELLS

The Energetics of Calling in Frogs

For biologists interested in the vocal communication of frogs and toads, a nocturnal trip to a temporary pond in the tropics can be an overwhelming experience. Often more than a dozen and a half species can be found calling together on a single night, producing a confusing and almost deafening cacophony (Hödl 1977; Cardoso and Haddad 1992; Gerhardt and Schwartz 1995). Closer study often reveals that different species, although sharing the same general habitat, have very different seasonal and nightly patterns of calling activity. Some species are continuous breeders that call whenever water is available for oviposition. Others are opportunistic breeders that form choruses only after heavy rains. Each chorus may last only a single night, but new choruses form after each major storm. Still other species are explosive breeders that appear only once each year, usually at the beginning of the rainy season. Males call vigorously for a few nights and then disappear (Crump 1974; Aichinger 1987; Donnelly and Guyer 1994; Bevier 1997a; Moreira and Barreto 1997). Similar variation in patterns of calling activity can be found in temperate-zone frogs. Some species are explosive breeders that appear for a few nights in early spring or after heavy rains, whereas others have prolonged breeding seasons that can last several months (Wells 1977a). Males of different species vary not only in their overall calling patterns, but also in their rate of calling, the structure of their calls, and call intensity.

This variability in patterns of calling activity among different species of frogs and toads provides an opportunity to investigate variation in the energetic costs of calling and the relationship of calling behavior to overall patterns of energy use. Previous work has shown that the cost of calling in frogs can be very high, with some species having metabolic rates while calling that are more than 20 times higher than resting metabolism (Pough et al. 1992; Prestwich 1994). This chapter reviews the energetics of calling in frogs, with particular attention to four general questions: (1) What are the costs of calling and how are these costs related to inter- and intraspecific variation in rates of call production? (2) Is variation in calling behavior reflected in variation in the morphology and biochemical characteristics of the muscles involved in call production? (3) What sources of energy are used to support call production, and how does variation in patterns of energy use relate to hourly, nightly, and seasonal differences in calling behavior? (4) To what extent is the performance of individual males constrained by depletion of energy reserves?

Measuring the Cost of Calling in Frogs

Many frogs call at relatively high rates for several hours at a time. Such sustained activity normally must be supported entirely by aerobic metabolism. Indeed several studies have shown that anaerobic metabolism does not contribute significantly to call production (Ryan et al. 1983; Taigen and Wells 1985; Prestwich et al. 1989; Grafe et al. 1992). Most measurements of the aerobic costs of calling have been made using closed respirometer systems. A male frog is placed in a small container and air samples are taken before and after a period of calling. The oxygen content of the samples is then measured with an oxygen analyzer and the amount of oxygen used for call production is calculated. Microphones placed in or near the chambers enable investigators to record the calling rate of each male and to correlate the rate of oxygen consumption with calling rate (Bucher et al. 1982; Taigen et al. 1985; Taigen and Wells 1985; Wells and Taigen 1986, 1989; Prestwich et al. 1989; Grafe et al. 1992; Grafe 1996, 1997). Flow-through respirometer systems, which have been used to measure the cost of signaling in insects (Lighton 1987; Lee and Loher 1993; Rheinhold et al. 1998), would provide a more detailed record of instantaneous changes in metabolic rate in relation to changes in calling rate, but this technique has not yet been used with frogs.

In all anurans studied so far, maximum oxygen consumption while calling is more than 10 times resting metabolism. Estimates of the cost of calling derived from field measurements of calling rates range from 6 to more than 20 times resting metabolic rates (Pough et al. 1992; Prestwich 1994; Grafe 1996, 1997). These values may not be representative of frogs as a whole, however, because all of the species studied so far are small hylids, leptodactylids, or hyperoliids that have relatively high calling rates compared with some other frogs. Many ranids, for example, have calling rates about 10 times lower than many of these species and probably devote a smaller proportion of their energy budget to call production (Pough et al. 1992).

Within species, oxygen consumption generally is a linear function of calling effort (Figure 5.1). For species such as *Pseudacris crucifer, Hyla microcephala, Hyperolius viridiflavus,* and *Hyperolius marmoratus,* there is little variation in call note duration, and note rate alone explains most of the variation in oxygen consumption (Taigen et al. 1985; Wells and Taigen 1989; Grafe et al. 1992; Grafe 1996; Wells et al. 1996). In contrast *Hyla versicolor* males give relatively long trills composed of a series of distinct pulses. Calls vary considerably in duration, and the best predictor of oxygen consumption is the product of note rate and duration (calling effort) (Taigen and Wells 1985; Wells and Taigen 1986; Grafe 1997), or the total number of pulses produced per hour (Figure 5.1C).

Social Interactions and the Cost of Calling

Social interactions among males can have a major effect on the cost of calling, but not necessarily in the same way in all species. Male frogs often respond to the calls of other individuals by increasing calling rate, call duration, call complexity, or the number of notes in calls (Wells 1988). In *Hyla microcephala,* males add notes to their calls as vocal competition increases, leading to a linear increase in metabolic rate (Wells and Taigen 1989). Males appear to conserve energy by maintaining a low calling effort when only a few males are present, but increase calling effort and energetic expenditures when competition among males is intense. This species calls in distinct bouts in which several males respond to each other's calls, followed by periods of silence (Schwartz 1991). These periodic pauses appear to be essential for conserving energy, because the calling rate of this species is so high. Males regularly achieve sustained calling rates of 6,000 notes per hour. Some individuals produce up to 10,000 notes per hour, but usually do so for only a few minutes at a time. Schwartz et al. (1995) estimated that males would deplete most of the glycogen reserves in their trunk muscles in less than 2 hours if they called continuously at such high rates (see "Energy Substrates for Call Production" for further discussion).

Male gray tree frogs (*Hyla versicolor*) increase the duration of their calls in response to the calls of other males, but they simultaneously decrease calling rate. The result is a relatively constant level of energy expenditure, regardless of chorus density (Wells and Taigen 1986; Grafe 1997). Playback experiments have shown that females prefer long calls to short calls, and high rates of calling to low rates (Klump and Gerhardt 1987). More importantly, as predicted by Wells and Taigen (1986), females prefer long calls delivered at slow rates to short calls delivered at fast rates, even when total calling effort is equal (Klump and Gerhardt 1987; Gerhardt et al. 1996). Therefore males in dense choruses alter their calls in ways that enhance their attractiveness to females. Most individuals appear to be calling near maximum sustainable levels, even when calling in isolation, and males may be unable to increase calling rate and call duration simultaneously because of energetic constraints. Individual males tend to maintain roughly the same relative hourly calling effort throughout the season, but the duration of choruses decreases from about 4 hours early in the season to only about 1.5 hours toward the end. Consequently total energy expenditure for calling decreases significantly as the season progresses (Runkle et al. 1994). This decreased calling effort probably is related to decreased chances of encountering females late in the season and reduced stimulation from other males in the chorus, rather than depletion of energy reserves (Wells et al. 1995).

Despite the plasticity of calling rate and call duration seen in individual males, there are also repeatable differences among males in a chorus in calling rate, call duration, and total calling effort (Gerhardt 1991; Sullivan and Hinshaw 1992; Runkle et al. 1994; Gerhardt et al. 1996). Some males

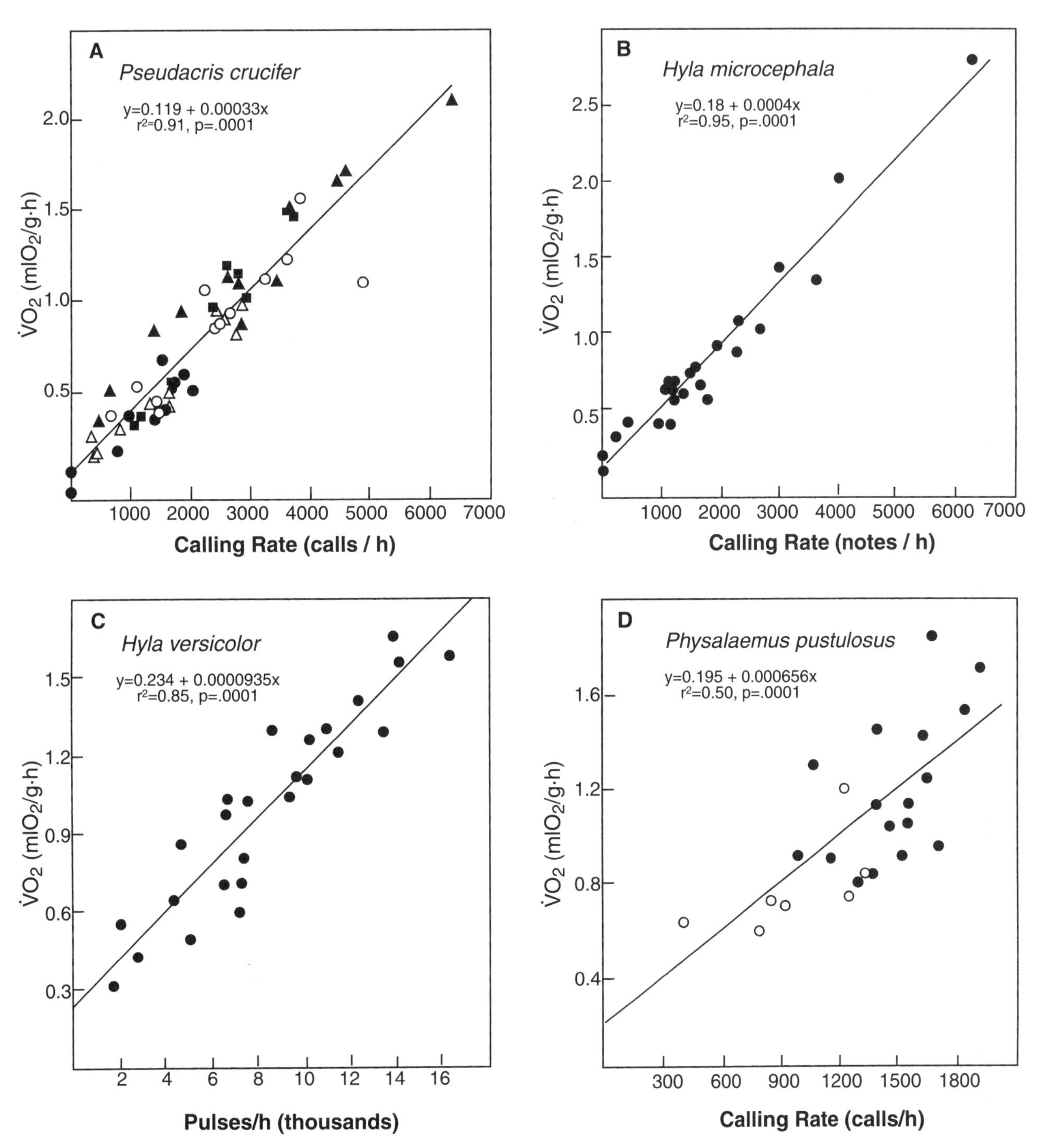

Figure 5.1. Rates of oxygen consumption as a function of calling rate in four species of frogs. (A) *Pseudacris crucifer*, measured at five different temperatures: 7°C (●), 10°C (△), 15°C (■), 19°C (○), and 23°C (▲). Modified from Wells et al. (1996). (B) *Hyla microcephala*, measured at 26–29°C. Modified from Wells and Taigen (1989). (C) *Hyla versicolor*, measured at 16–21°C. Data from Taigen and Wells (1985) and Wells and Taigen (1986). (D) *Physalaemus pustulosus*, measured at 24–26°C. ○, data from Bucher et al. (1982), omitting the two low outlier values; ●, data of Taigen and Wells from Pough et al. (1992). Each point represents a single individual, except that some *Hyla versicolor* males were measured twice on the same night.

consistently give slightly longer calls than do other males exposed to the same acoustic conditions. A recent study by Welch et al. (1998) suggested that females might use differences in calling performance as indicators of the genetic quality of individual males. These investigators selected males that consistently gave relatively long or short calls and used them to sire the eggs of the same females. They raised tadpoles from these crosses at both high and low densities of food and quantified several measures of larval performance, including larval growth rate, length of the larval period, mass at metamorphosis, larval survival, and growth rate of metamorphosed frogs. In all cases in which there was a significant difference in offspring performance, the offspring of males that gave long calls exhibited high larval growth rates, better larval survival, or both. The precise connection between the physiological ability to produce long calls for sustained periods of time and determinants of offspring performance is unknown, but the results of this experiment suggest that at least some of the variation in calling performance among males is heritable. An interesting follow-up experiment, but one that would be much more difficult, would be to raise male offspring to adulthood and compare the calling performance of frogs sired by males with long and short calls.

The effect of social interactions on the cost of calling has also been examined in *Physalaemus pustulosus*. This species has an unusual call consisting of a frequency-modulated "whine" note that may be combined with one or more secondary "chuck" notes. The latter are given simultaneously with the whine note and are not added onto the end of the call as in other species with complex multinote calls (Ryan 1985). Males calling in low-density choruses generally give calls with relatively few chucks. In dense choruses males call at faster rates and add more chuck notes to their calls. Females prefer males producing calls at high rates to those calling at low rates, and they prefer calls with chuck notes to whines alone (Rand and Ryan 1981). Although energetic expenditures increase with increasing calling rate, there is no evidence from the data of Bucher et al. (1982) or the data collected by Taigen and Wells (unpublished data) that number of chucks is related to the cost of calling. Probably this is because chucks are produced by passive vibration of a fibrous mass in the air stream passing through the larynx, not by additional contractions of the trunk muscles (Drewry et al. 1982).

Morphological Correlates of Calling Performance

The muscles involved in call production by frogs are quite distinct in morphology, histochemistry, and contractile properties from anuran leg muscles (Marsh 1999), but until recently these muscles were largely ignored by muscle physiologists. For example, there is no mention of call-producing muscles in a recent review of amphibian muscle physiology and functional morphology (Gans and De Gueldre 1992). Muscles in the trunk region, including the rectus abdominus, external oblique, and internal oblique (= transverse) muscles, provide the power for call production. When these muscles contract, air is forced out of the lungs, through the larynx, and into the buccal cavity and vocal sac. In a species such as *Pseudacris crucifer*, which has a simple single-note call, each call represents a contraction of the trunk muscles. In *Hyla microcephala*, each note in a multinote call represents a single contraction of these muscles. This is why the cost of calling is so tightly correlated with calling rate in these species (Figure 5.1A, B). The action of the trunk muscles is similar in *Physalaemus pustulosus*. Each long whine call is produced by a single contraction of these muscles (Dudley and Rand 1991; Jaramillo et al. 1997).

In species with longer, pulsed calls that have relatively low pulse repetition rates, the trunk muscles can also be involved in production of individual pulses (Martin 1972). For example, in the two species of North American gray tree frogs (*Hyla versicolor* and *H. chrysoscelis*), each pulse in the call corresponds to a separate cycle of contraction and relaxation of the trunk muscles, as shown by electromyographic recordings of muscle activity (Girgenrath and Marsh 1997). The two species differ in that *H. chrysoscelis*, a diploid species, has a pulse repetition rate about twice that of *H. versicolor*, a tetraploid species, at the same temperature. This difference corresponds to differences in the frequency of trunk muscle contraction (Marsh 1999). The cycle frequencies that maximize the power output of the muscles of these two species at 20–25°C (44 Hz in *H. chrysoscelis* and 21 Hz in *H. versicolor*) are similar to the frequencies used by the two species during calling (Girgenrath and Marsh 1999). The role of the trunk muscles in producing individual pulses is consistent with the observation that the cost of calling in *Hyla versicolor* is tightly correlated with the total number of pulses produced (Figure 5.1C).

The trunk muscles of male frogs generally are much larger (3–15% of body mass) than those of females (1–3%) (Pough et al. 1992; Bevier 1995b). In male frogs, the rectus abdominus muscle is separated from the other trunk muscles by a band of fibrous tissue, the linea masculina. In at least one species, *Physalaemus pustulosus*, this band of tissue is highly elastic and may provide elastic recoil during deflation of the lungs that increases the efficiency of sound production (Jaramillo et al. 1997). Elastic fibers are also found in the vocal sac of *Physalaemus*, and these may increase the efficiency with which air is pumped back into the lungs after a call is produced. Females lack vocal sacs, and the band of elastic tissue in the trunk muscles is absent as well. Relative trunk muscle mass tends to be greater in species with high calling rates than in species with low calling rates (Bevier 1995a). Within species, larger individuals have larger trunk muscles and may have higher calling rates or louder calls than small individuals. For example Zimmitti (1999) reported that male spring peepers (*Pseudacris crucifer*) that called at higher than average rates were larger and had larger trunk muscles than those that called at lower rates. However there was no difference in relative trunk-muscle mass among males once differences in body size were accounted for.

The other muscles involved in call production are the laryngeal muscles that open and close the larynx during call production. In some species these muscles also change the tension on the vocal cords to produce frequency-modulated calls (Martin 1972; Schneider 1988). Like the trunk muscles, the laryngeal muscles are considerably larger in males than in females (Marsh and Taigen 1987; Schneider 1988; McClelland and Wilczynski 1989; Kelley 1996; McClelland et al. 1997, 1998). Nevertheless their contribution to the total cost of call production probably is small, because these muscles are much less massive than the trunk muscles (Marsh and Taigen 1987). The role of the laryngeal muscles in producing calls varies among families of anurans. In most frogs the laryngeal dilator muscles actively contract to open the larynx and allow air to flow into the lungs (Schmidt 1965; Schneider 1977, 1988). Removal of these muscles makes vocalization impossible, because the frogs cannot generate the necessary pulmonary pressure (Weber 1975, 1976). In hylids and ranids the laryngeal dilator muscles also open the larynx during calling, allowing a burst of air to pass through the larynx and over the vocal cords. In bufonids the larynx apparently is opened passively by pulmonary air pressure, but a call is terminated when the laryngeal dilators move the arytenoid cartilages out of the air stream (Martin 1972).

In hylids that have calls consisting of a series of repeated pulses, the laryngeal muscles actively open and close the larynx in synchrony with the trunk muscles to produce the highly stereotyped, regularly spaced pulses that often are important for species recognition (Gerhardt 1991). When these muscles are removed, the spacing and duration of pulses is disrupted (Weber 1976). McLister et al. (1995) compared the contractile properties of one of the laryngeal muscles, the tensor chordarum, involved in call production in three hylids. Two of these species, *Hyla versicolor* and *H. chrysoscelis,* have calls consisting of long trains of pulses. The third, *H. cinerea,* has relatively short calls that lack pulses. The laryngeal muscles of the two species with pulsed calls can contract much more rapidly than the muscles of *H. cinerea.* The muscles of *H. chrysoscelis,* which has a fast pulse repetition rate, are faster than those of *H. versicolor,* with a slower pulse repetition rate. Marsh (1999) examined the contractile properties of a different laryngeal muscle, the laryngeal dilator, in *H. versicolor* and *H. chrysoscelis,* and he again found that *H. chrysoscelis* has a faster contraction rate. Maximum shortening velocities of the laryngeal dilator muscles reported by Marsh were considerably higher than values for the tensor chordarum reported by McLister et al., but it is not clear whether this represents a real difference between the two muscles or a difference in experimental techniques.

Both the trunk and laryngeal muscles of frogs are composed mainly of fast oxidative fibers (> 90%) along with a few tonic fibers and have rapid shortening velocities (Eichelberg and Schneider 1973, 1974; Schneider 1977, 1988; Marsh and Taigen 1987; McLister et al. 1995). Trunk and laryngeal muscles are dramatically different from leg muscles in ultrastructural characteristics. The muscles involved in call production are well supplied with sarcoplasmic reticulum and have several times as many mitochondria per unit volume as do the leg muscles. Capillary densities are also much higher in trunk and laryngeal muscles than in leg muscles, reflecting the much higher oxygen demand of the trunk muscles (Eichelberg and Schneider 1973, 1974; Schneider 1977, 1988; Marsh and Taigen 1987; Ressel 1993, 1996; McLister et al. 1995).

In a detailed analysis of muscle ultrastructure in relation to calling behavior, Ressel (1996) found that both mitochondrial volume density and capillary density were positively correlated with calling rate among tropical frogs from two families that call at similar temperatures. Indeed comparison of mitochondrial volume density for a variety of tropical and temperate-zone species reveals a positive correlation with calling rate that does not appear to be related to phylogeny. Ranid frogs with very low calling rates have mitochondrial volume densities similar to those of hylids with low calling rates, such as *Agalychnis callidryas* (Figure 5.2A). Two leptodactylid frogs, *Physalaemus pustulosus* and *Eleutherodactylus coqui,* have muscle characteristics similar to those of hylids with comparable calling rates.

Knowledge of the phylogenetic relationships of frogs is not sufficiently detailed (Ford and Cannatella 1993) to allow a complete comparative analysis of muscle structure. We do not, for example, have a good phylogeny for all of the genera of hylid frogs. Nevertheless paired comparisons of closely related species consistently show that those with high calling rate have higher mitochondrial densities than do those with low calling rates. For example *Hyla microcephala* and *H. ebraccata* are thought to be closely related (Duellman 1970). The former has a higher calling rate and

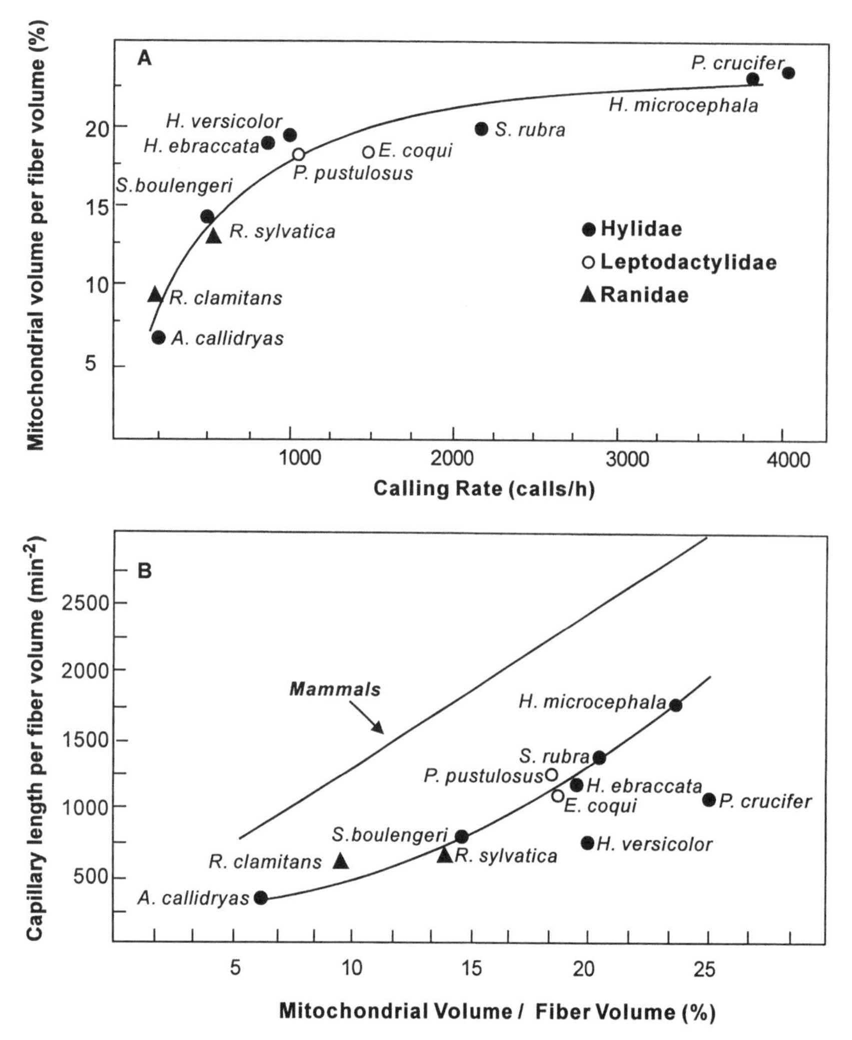

Figure 5.2. (A) Relationship of mitochondrial volume density of trunk muscle tissue to calling rate for anurans in three families. The curve is from Ressel (1996) for tropical hylids and leptodactylids only (*A* = *Agalychnis*, *E* = *Eleutherodactylus*, *H* = *Hyla*, *L* = *Leptodactylus*, *P. pustulosus* = *Physalaemus pustulosus*, *S* = *Scinax*). Additional data for temperate-zone hylids (*Pseudacris crucifer* and *Hyla versicolor*) and ranids (*Rana clamitans* and *R. sylvatica*) are from Ressel (1993). (B) Relationship of trunk-muscle capillary length to mitochondrial volume density for anuran trunk muscles and mammalian locomotor muscles. The curve is for tropical hylids and leptodactylids only (from Ressel 1996). Additional data for temperate-zone frogs (*P. crucifer*, *H. versicolor*, *R. clamitans*, and *R. sylvatica*) are from Ressel (1993).

higher mitochondrial density than does the latter. Two other tropical hylids, *Scinax rubra* and *S. boulengeri*, are more closely related to each other than to the other hylids in Ressel's study (Duellman and Wiens 1992), and they show the same relationship between calling rate and mitochondrial density. The two temperate-zone hylids, *Pseudacris crucifer* and *Hyla versicolor*, are thought to be more closely related to each other than either is to the tropical hylids (the genus *Hyla* is paraphyletic as presently constituted; Hedges 1986; Cocroft 1994). Again, calling rate is positively related to mitochondrial density. Finally the same relationship is seen in the two species of ranids, *Rana sylvatica* and *R. clamitans* (Figure 5.2A).

In mammals, the number of capillaries supplying the highly aerobic skeletal muscles is closely correlated with mitochondrial volume density, because oxygen supply must match metabolic demand (Conley et al. 1987). The same relationship appears to be true for frogs, but the trunk muscles of frogs have fewer capillaries for a given mitochondrial volume than do the locomotor muscles of mammals (Figure 5.2B). Presumably this reflects the lower activity temperatures of frogs, which in turn results in lower metabolic demands (Ressel 1996). Most frogs are active at temperatures below 25°C, whereas mammals are endothermic and operate at about 38°C. Points for temperate-zone frogs that call at lower temperatures than do most tropical frogs fall below the curve for tropical frogs alone (Figure 5.2B). This is particularly true for species with high calling rates, such as *Pseudacris crucifer*. Again this suggests that oxygen supply is matched to metabolic demand at the typical activity temperature of each species.

Other parts of the circulatory system involved in delivering oxygen to the trunk muscles are likely to vary with calling rate as well, but there have not been any comparative studies of the circulatory system in relation to calling performance. In spring peepers (*Pseudacris crucifer*), males have much larger heart ventricles and higher hemoglobin content in the blood than do females (Zimmitti 1999). Presumably this is related to the greater oxygen demands of the much larger trunk muscles of males. It seems likely that interspecific comparisons among species with different calling rates will also reveal a relationship of ventricle mass and hemoglobin concentration to calling performance.

Biochemical Correlates of Calling Performance

Both citrate synthase (CS) and β-hydroxyacyl-CoA-dehydrogenase (HOAD) activities have been used as indicators of aerobic metabolism in the trunk muscles of frogs (Taigen et al. 1985; Marsh and Taigen 1987; Given and McKay 1990; Lance and Wells 1993; Bevier 1995b). Activities of both enzymes are consistently higher in trunk muscles than in leg muscles (Figure 5.3), and variation in the activities of these enzymes is closely correlated with calling rate (Figure 5.4). The activities of these enzymes are closely correlated with each other in both trunk and leg muscles. This suggests that species with high calling rates not only have high aerobic capacities, but also depend more heavily on lipids to fuel call production (Bevier 1995b; see "Energy Substrates for Call Production" for further discussion).

A comparison of two neotropical hylids in the same genus, *Scinax rubra* and *S. boulengeri,* nicely illustrates the relationship between calling behavior and trunk muscle characteristics (Bevier 1997a). These species are about the same size (40-mm snout–vent length) and mass (3.5 g), and they call at the same temperatures. They differ dramatically in calling behavior, however. *S. rubra* is an explosive breeder that calls only after heavy rains, and individual males are seldom present in a chorus for more than one or two nights. Males call at relatively high rates (more than 2,000 calls per hour) for up to 7 hours per night, and they move actively about the chorus area in search of females. *S. boulengeri* is a continuous breeder, with some males present in choruses for weeks at a time. Males have relatively low calling rates. They usually call for only about 3 hours per night, and they reduce calling rates after the first hour. The trunk muscles of

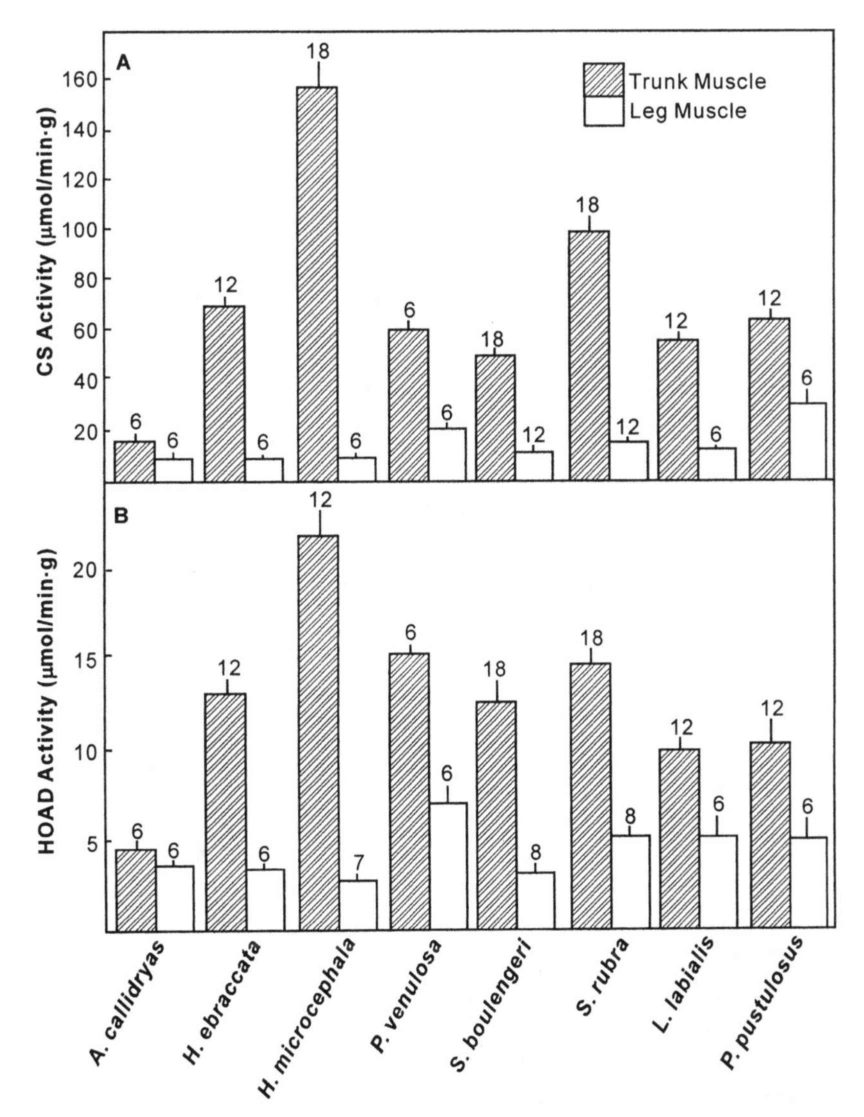

Figure 5.3. Comparison of the activities of (A) citrate synthase (*CS*) and (B) β-hydroxyacyl-coA-dehydrogenase (*HOAD*) enzymes in trunk and leg muscle tissues of tropical hylids and leptodactylids. Numbers above the bars are sample sizes. Species names as in Figure 5.2, with the addition of *Phrynohyas venulosa*. Modified from Bevier (1995b).

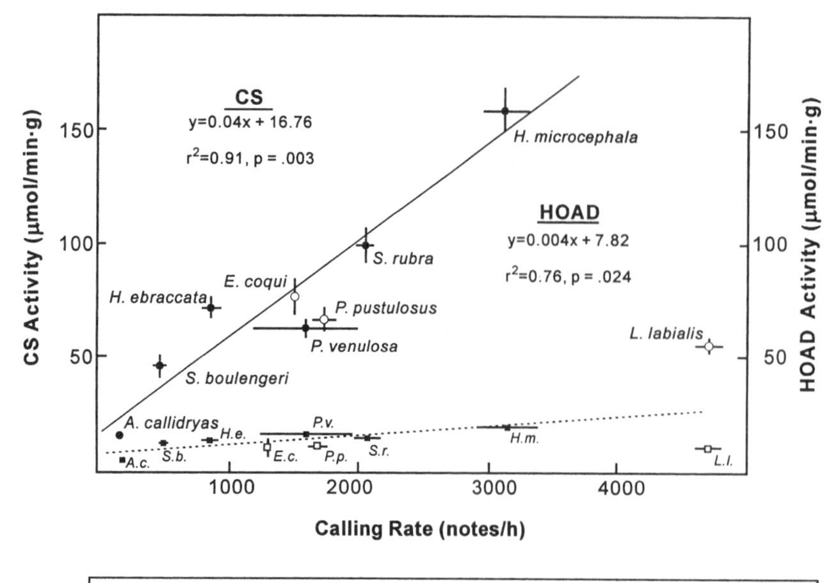

Figure 5.4. Activities of CS and HOAD enzymes in trunk muscles of tropical frogs in relation to calling rate. Data are given as means ± 1 SE. The regression lines are for the hylids (●) only, but two leptodactylids (○) are included for comparison. Species names as in Figure 5.2 and 5.3. Modified from Bevier (1995b).

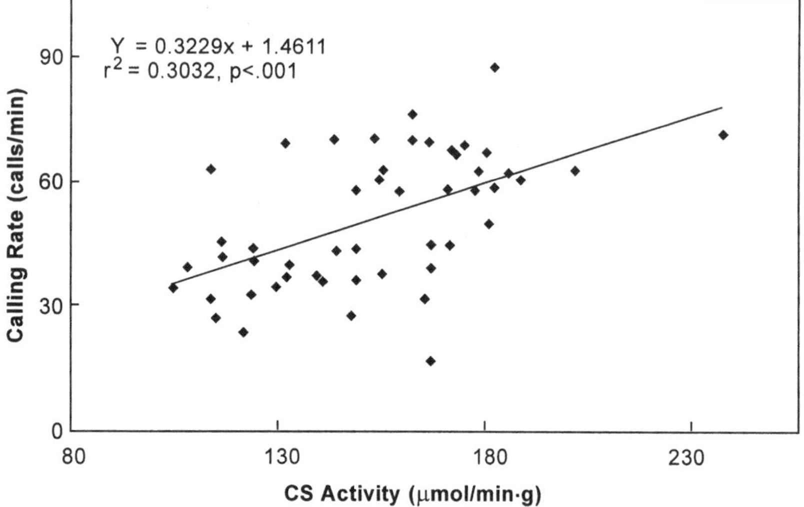

Figure 5.5. Relationship of calling rate, corrected for temperature and body mass, to CS activity in the trunk muscles of male spring peepers (*Pseudacris crucifer*). Data from Zimmitti (1998).

S. rubra are designed for high levels of calling activity during the short periods when females are available, whereas those of *S. boulengeri* are designed for slow but sustained activity over much longer periods of time. *S. rubra* males have larger trunk muscles (10% of body mass) than do *S. boulengeri* males (6.6%), and their muscles are more aerobic, with a much higher density of mitochondria and capillaries and higher enzyme activities, especially CS activity. These behavioral and physiological differences are reflected in their use of energy substrates as well. *S. rubra* has much larger stores of lipids and glycogen in the trunk muscles, but also depletes them more rapidly than does *S. boulengeri*.

A study of spring peepers (*Pseudacris crucifer*) by Zimmitti (1998, 1999) was the first to examine variation in trunk muscle enzyme activities among males in the same population. Studies of this type of individual variation have been largely neglected by physiologists (Bennett 1987), but they are essential for understanding how physiological traits evolve through natural or sexual selection. Zimmitti collected pairs of males on the same night, choosing individuals with unusually high or low calling rates. He then determined the activities of CS and HOAD in their trunk muscles. He found that males with high calling rates consistently had higher activities of both enzymes in their trunk muscles than did males with low calling rates, and the activities of the two enzymes were correlated with each other. In a multiple regression analysis, trunk muscle CS activity was positively correlated with calling rate after accounting for the effects of temperature and body mass (Figure 5.5). Previous studies have shown that there are consistent differences in calling rate among males of this species (Sullivan and Hinshaw 1990), and females are most likely to be attracted to males with high calling rates (Forester et al. 1989). It seems likely that some of the variation in physiological traits that underlies variation in calling performance is heritable, so a female preference for high calling rate is likely to result in selection for morphological and physiological traits that enable males to call at high rates.

Energy Substrates for Call Production

The enzymatic profiles of muscles involved in call production indicate that both lipids and carbohydrates are important sources of energy for calling frogs. In other vertebrates both lipids and carbohydrates supply energy for sustained exercise (Roberts et al. 1996). Most of the energy comes from lipid and glycogen stores in the muscles themselves, with a relatively small proportion (< 25%) coming from glucose and free fatty acids transported in the blood (Weber, Brichon et al. 1996; Weber, Roberts et al. 1996). Only at relatively low levels of activity, or low calling rates in frogs, are extramuscular stores of lipids and carbohydrates likely to be important, because of limitations on the rate of substrate delivery by the circulatory system. Glycogen provides energy at a faster rate than lipids and therefore provides greater power output. Glycogen stores are more limited, however, and provide less energy per gram of storage product, so glycogen reserves are depleted much more rapidly. Lipids provide more energy per gram, so stores of lipids are depleted more slowly. Depletion of muscle glycogen reduces muscle performance, even when abundant lipid stores remain, so glycogen represents a short-term constraint on muscular activity (Guppy 1988; Weber 1988).

The relative importance of glycogen and lipids as substrates to support activity varies among species, with lipids being most important in animals that sustain high levels of activity over long periods of time. For example highly aerobic mammals that are capable of long-distance running, such as dogs, rely much more on intramuscular lipid stores to support vigorous exercise than do less aerobic species that do not move rapidly over long distances, such as opossums (Weber 1992). The same is true for insects that produce sound. Species with high calling rates typically derive most of their energy for sound production from lipids, whereas those with low calling rates sometimes rely entirely on carbohydrates (Lee and Loher 1993; Prestwich 1994).

Both the laryngeal and trunk muscles of frogs are well supplied with lipid stores in the form of droplets closely associated with clusters of mitochondria (Eichelberg and Schneider 1973; Eichelberg and Obert 1976; Schneider 1977, 1988; Marsh and Taigen 1987; Ressel 1993, 1996; McLister et al. 1995). A comparative study of several neotropical hylids and leptodactylids (Ressel 1996) showed that lipid content of trunk muscles is positively correlated with calling rate, and this appears to be true for frogs in general (Figure 5.6). One neotropical hylid with a very low calling rate, *Agalychnis callidryas,* has virtually no lipid in its trunk muscles, and the same is true for two North American ranids that also have low calling rates, *Rana sylvatica* and *R. clamitans.* In contrast male spring peepers (*Pseudacris crucifer*) have more lipid in their trunk muscles than is found in most other highly aerobic vertebrate muscles (Ressel 1993).

In species with abundant lipid reserves, changes in lipid content of trunk muscles generally are not detectable on a nightly basis (such changes must be estimated from samples of males collected early and late in the evening, because animals are sacrificed to measure lipid content of muscles). Of the seven neotropical hylids and leptodactylids studied by Bevier (1997b), only *Phrynohyas venulosa* exhibited a significant reduction in trunk-muscle lipid stores after 2 to 3 hours of calling, and this species started with relatively small lipid stores. In contrast *Hyla microcephala,* which had initial lipid stores about three times as large, showed no change after 2 hours of calling. Spring peepers (*Pseudacris crucifer*) undergo a dramatic seasonal reduction in trunk-muscle lipid reserves. Males collected early in the season have larger reserves than any other species of frog that has been studied, but most of these reserves are gone by the end of the season (McKay 1989; Ressel 1993). This is not true for *Hyla versicolor,* however, perhaps because this species can replenish depleted energy reserves by feeding (Walker 1989). Less is known about depletion of energy substrates in laryngeal muscles. Eichelberg and Obert (1976) reported that lipid reserves in the laryngeal muscles of *Bombina bombina* were not significantly reduced by several hours of electrical stimulation, but were reduced after nearly two weeks of calling activity.

Both the trunk and laryngeal muscles of male frogs contain substantial quantities of glycogen (Schneider 1977; Schwartz et al. 1995; Wells et al. 1995; Bevier 1997b; Wells and Bevier 1997). Among neotropical hylids and leptodactylids, initial trunk-muscle glycogen stores were positively correlated with both hourly calling rates and the total number of calls produced in an evening (Bevier 1997b). Rates of glycogen depletion are also correlated with calling rate. Species with relatively high calling rates, such as *Hyla microcephala, H. versicolor, Scinax rubra,* and *Physalaemus pustulosus,* can deplete 50–60% of their trunk-muscle glycogen reserves after only 2 hours of calling. Species with lower calling rates, such as *Scinax boulengeri,* use up their trunk-muscle reserves more slowly (Schwartz et al. 1995; Wells et al. 1995; Bevier 1997b). In most of these species, liver glycogen reserves show little or no change after 2 hours of calling, suggesting that these are not used to support call production.

The rapid depletion of glycogen reserves relative to lipid reserves might lead one to conclude that glycogen is a more important source of energy for call production. For species with high calling rates, however, this is not the case. Because lipids yield much more energy per gram of stored substrate,

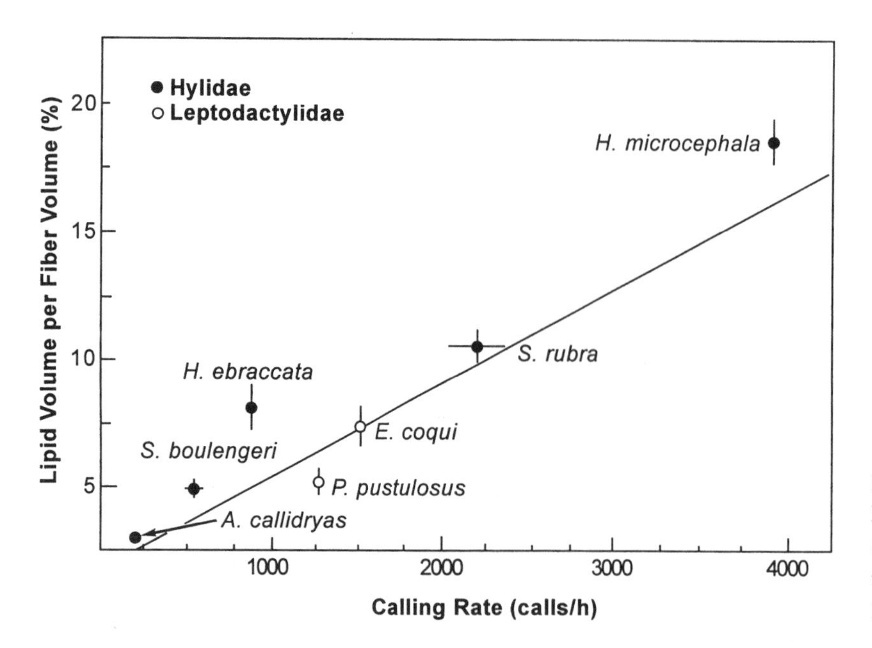

Figure 5.6. Relationship of trunk-muscle lipid volumes to calling rate in tropical hylids and leptodactylids. Species names as in Figure 5.2. Data are given as means ± 1 SE. Modified from Ressel (1996).

even a small amount of lipid can account for most of the energy used for call production. The relative importance of glycogen and lipids has been calculated for several species by measuring rates of glycogen depletion and calculating the amount of oxygen used to oxidize glycogen. It is then assumed that the remaining oxygen consumed by a calling frog is used to oxidize lipid. Such calculations have shown that frogs with relatively high calling rates derive 75–95% of their energy from lipids (Table 5.1), even though glycogen reserves can be depleted on a single night. These values are consistent with estimates derived from measurements of respiratory quotients (RQ = the ratio of carbon dioxide produced to oxygen consumed). Grafe (1997) measured RQs for calling *Hyla versicolor* males and estimated that they derive at least 60% of their energy for calling from lipids. In *Hyperolius marmoratus,* RQs varied dramatically depending on whether the frogs had recently fed. Calling males derived 37–89% of their energy from lipids, and they relied much more on stored lipids if they had not fed recently (Grafe 1996).

Frogs that breed in cold weather face a special energetic challenge because temperatures often are too low for their prey to be active or for frogs to digest food. Wood frogs (*Rana sylvatica*) breed in early spring and do not feed at all during the breeding period. They lack significant lipid reserves, either in the trunk muscles (Ressel 1993) or in abdominal fat bodies (Wells and Bevier 1997). Consequently, they must rely entirely on glycogen reserves to support calling. At the beginning of the breeding season, wood frogs have very large trunk-muscle glycogen reserves, but these are quickly depleted. In some years, all trunk-muscle reserves probably are gone within a week (Wells and Bevier 1997). Nevertheless, the apparent dependence of wood frogs on carbohydrates for calling is consistent with what one would predict for a frog with a low calling rate and explosive breeding period.

Spring peepers (*Pseudacris crucifer*) also begin breeding at temperatures too low for feeding, but in contrast to wood frogs, males collected early in the season have very little glycogen stored in either their trunk muscles or livers. Early-season males probably depend heavily on their enormous trunk-muscle lipid reserves (Wells and Bevier 1997). As the season progresses, and temperatures increase, calling rates and energy expenditure by male peepers increase (Wells et al. 1996), probably resulting in an accelerating rate of lipid depletion. Insects are also more readily available, and males can partially compensate for lost energy reserves by feeding. Wells and Bevier (1997) found that the amount of food in the stomachs of males increased late in the season, and this was paralleled by an increase in trunk-muscle and liver glycogen reserves (Figure 5.7). It seems likely that late-season males shift to a greater dependence on carbohydrates for calling as their lipid reserves are depleted, much as *Hyperolius* males shift to using carbohydrates after feeding (Grafe 1996).

The way in which these cold-weather breeders use energy contrasts with the energy use strategies of species in the same families that breed at warmer temperatures. Green frogs (*Rana clamitans*) are summer breeders, and some males defend territories for nearly two months (Wells 1977b). Like wood frogs, they lack lipid reserves in the trunk muscles (Ressel 1993) and they have much smaller glycogen reserves in their muscles than do wood frogs (Wells and Bevier, unpublished data). Green frogs have very low calling rates and deplete a relatively small proportion of their glycogen reserves each night. The slight seasonal decline in glyco-

Table 5.1 Estimated contributions of carbohydrates and lipids to call production in frogs

Species	Carbohydrates (%)	Lipids (%)
Hyla microcephala[a]	14	86
Hyla versicolor[b]	25	75
Hyla versicolor[c]*	40	60
Pseudacris crucifer[d]	10	90
Physalaemus pustulosus[e]	7	93
Hyperolius marmoratus (before feeding)[a]*	13	87
Hyperolius marmoratus (after feeding)[a]*	63	37

Sources: [a] Grafe 1996; [b] Wells et al. 1995; [c] Grafe 1997; [d] Wells and Bevier 1997; [e] Bevier 1997b.

* Derived from measurements of respiratory quotients; all others from measurements of metabolic rates and depletion of trunk-muscle glycogen reserves.

gen reserves contrasts with the very rapid depletion of reserves in wood frogs (Figure 5.8). In contrast to wood frogs, green frog males feed continuously during the breeding season and probably rely on food intake for much of the energy required for calling. They can be considered income breeders that depend on current food intake to support reproductive activities (Bonnet et al. 1998). Wood frogs, on the other hand, are capital breeders, relying entirely on stored reserves to support reproduction. Many species of frogs probably pursue a mixed strategy, relying on both stored reserves and food intake to support call production. This seems to be true of *Hyla versicolor,* for example, which has both lipid and glycogen reserves, but also feeds during the breeding season (Table 5.2). Capital breeders are likely to be constrained in their reproductive activities by the size of energy reserves laid down in the previous feeding season. Income breeders, on the other hand, are more likely to be constrained by food availability during the current breeding season.

The relative position of a given species of frog on the continuum of energy use strategies from pure capital breeders to pure income breeders is likely to affect its response to experimental food supplementation. Presumably species that rely mainly on stored energy reserves are likely to show weaker responses to increased food availability than are species that feed continuously during the breeding season. Several investigators have experimentally supplemented the food of individual male frogs to determine whether food availability limits call production. Sinsch (1988) studied a high-altitude population of *Gastrotheca marsupiata* in Peru and found that males given supplemental feedings in the laboratory called at higher rates than did those that were not fed. In contrast Green (1990) did not find any effect of supplemental feeding on calling activity or the time required for *Physalaemus pustulosus* males to return a chorus. Another study of the same

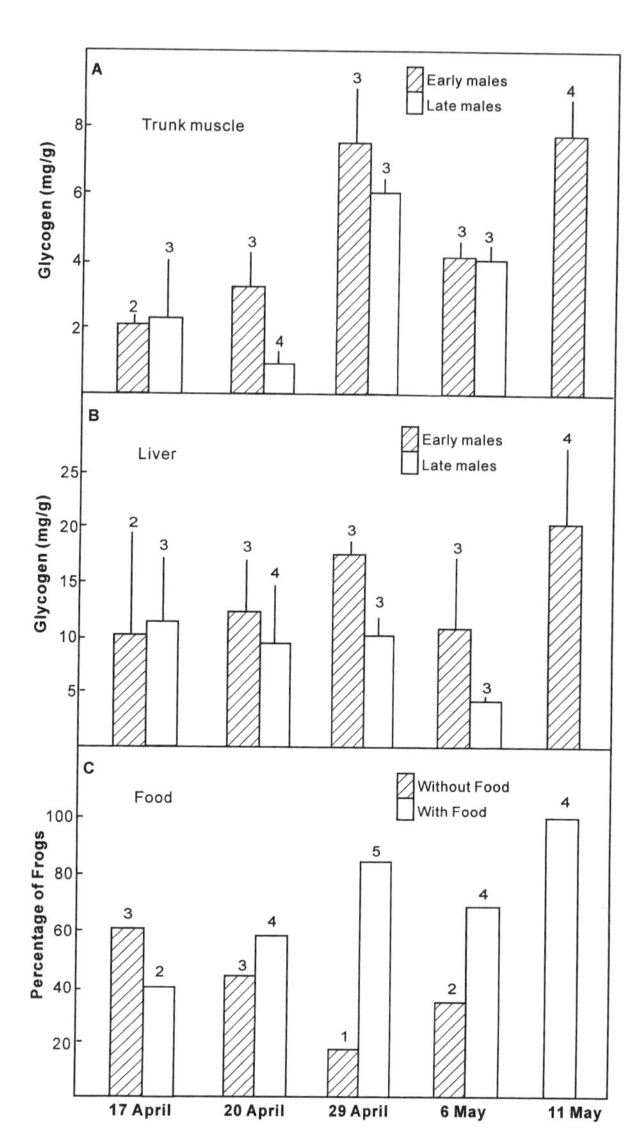

Figure 5.7. Seasonal changes in (A) trunk-muscle glycogen concentration, (B) liver glycogen concentration, and (C) presence of food in stomachs of male spring peepers (*Pseudacris crucifer*) that were collected early and late in the evening. Numbers above bars are sample sizes; length of line represents 1 SE. Modified from Wells and Bevier (1997).

species found that males given supplemental food were more likely to be calling during subsequent census periods than were males that were not fed (Marler and Ryan 1996). Marler and Ryan also found that males given supplemental food had increased levels of circulating testosterone, as did calling males in a chorus. This suggests that the effects of changes in available energy on calling behavior may be mediated by changes in hormone levels.

Unfortunately neither of these studies quantified rates of calling or total call production in males that were fed and not fed. Although *Physalaemus pustulosus* breeds more or less continuously throughout the rainy season in Panama, the

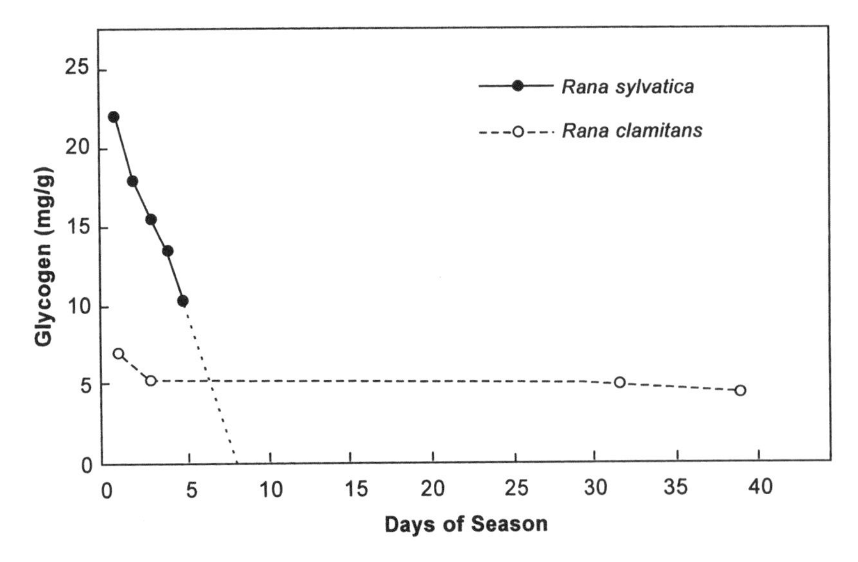

Figure 5.8. Seasonal changes in trunk-muscle glycogen concentrations for male wood frogs (*Rana sylvatica*) and green frogs (*Rana clamitans*) collected early in the evening. The dashed line for wood frogs projects complete depletion of glycogen by Day 8. Wood frog data from Wells and Bevier (1997). Green frog data from Wells and Bevier (unpublished data).

Table 5.2 Hypothesized strategies of energy use in four species of temperate-zone frogs

	Source of Energy[a]			
Species	Lipids	Glycogen	Food	Energy-use strategy
Rana sylvatica		++++		Rely on stored glycogen
Pseudacris crucifer	++++	++	+	Mostly use stored lipids
Hyla versicolor	+++	++	++	Mixed strategy: stored lipids, glycogen, and food
Rana clamitans		+	++	Mostly rely on current food intake

[a]The number of plus symbols gives a qualitative estimate of the relative importance of each source of energy for call production. A blank indicates that a particular source does not contribute significantly to call production.

most active choruses occur on nights after heavy rains (Ryan 1985). Males probably function as capital breeders on peak nights of chorusing, depending mainly on stored energy reserves to support calling. This species depletes a large proportion of its carbohydrate reserves on a single night of calling, but it may derive most of its energy for calling from stored lipids (Bevier 1997b). Feeding might enable males to conserve their lipid reserves by decreasing reliance on stored lipids, as shown for *Hyperolius* (Grafe 1996), while having only a small effect on call output.

Another frog with a prolonged breeding season is *Hyla gratiosa,* a summer-breeding tree frog from the southeastern United States. Choruses are formed from March through August, and some males remain in choruses for a month or more. Median chorus tenure, however, is much shorter, only 2 or 3 days in most years; about 75% of males are present for less than a week. This pattern of chorus attendance suggests that males are energy-limited. Indeed Murphy (1994a, 1994b) found that males suffered reductions in body mass throughout their time in a chorus. Males that remained longer in the chorus were in better condition and lost mass more slowly than those that were present for only a few nights. Males in good condition were also more successful in acquiring mates, both because they were present on more nights, and because they were more likely to attract mates when they were present. Feeding during the breeding season was important for maintaining condition, and males that received extra food returned to the chorus more quickly and stayed longer than those that were not fed (Murphy 1994a). These frogs appear to pursue a mixed strategy of supplementing stored energy reserves with food intake during the breeding season.

Conclusions

There is a general assumption in most discussions of sexual selection that signals given by males to attract females tend to be energetically expensive to produce. Such signals are likely to be honest indicators of the overall condition of individual males and may reflect differences in genetic quality as well (Kodric-Brown and Brown 1984; Ryan 1988; Andersson 1994). There is evidence for anuran amphibians (Sullivan et al. 1995) and other animals (Vehrencamp et al. 1989; Reinhold et al. 1998) that females often are attracted to males that give signals that are long, loud, or delivered at high rates, traits that are likely to be positively correlated with levels of energy expenditure. Yet not all signals are expensive to produce. Acoustic signals that are given infrequently, for example, may incur relatively little energetic cost (Horn et al. 1995). All of the species of anurans for which the cost of calling has been measured are ones that call at relatively high rates, usually in dense choruses in which competition among males is intense. In all of these species, metabolic rates of calling males are among the highest recorded for any ectothermic vertebrates, and metabolic scopes are 10 to 20 times resting metabolism. Caution is required, however, in extrapolating from these studies to all anurans. Many species, such as North American ranid frogs, have calling rates that are 10 to 100 times lower than those of hylids, leptodactylids, and hyperoliids. It seems likely that the cost of calling is far lower in these species. We need data from a wider variety of species to fully encompass the range of variation in strategies of energy use in anurans.

Morphometric and biochemical studies of the trunk muscles involved in call production have revealed significant variation in muscle aerobic capacity, as reflected in differences in mitochondrial density, capillary density, and activities of key catabolic enzymes. This variation in muscle characteristics is strongly correlated with differences in calling rate, even within a single family, such as the Hylidae. These studies have also shown that fuel supplies for calling muscles are correlated with calling activity. Species that call at high rates for several hours per night invariably have large stores of lipids in their trunk muscles, whereas species with lower calling rates or shorter calling periods do not. Frogs with low calling rates may rely mainly on carbohydrate reserves or current food intake to support call production. Similar variation in muscle ultrastructure, biochemistry, and fuel supplies have been found in studies of muscles of mammals that differ in locomotor capacity (Conley et al. 1987; Vock, Hoppeler et al. 1996; Vock, Weibel et al. 1996; Weber 1988, 1992; Weber, Brichon et al. 1996; Weber, Roberts et al. 1996). These patterns reflect a general tendency for structural features of animals to match physiological performance requirements (Weibel et al. 1991), although the precision of matching between structure and performance has been the subject of some debate (Garland and Huey 1987; Weibel et al. 1998). So far most of the work on muscle ultrastructure and biochemistry has been based on interspecific comparisons, but it is essential to have more information on how individual variation in such traits is related to variation in behavioral performance that may in turn affect individual male mating success (Bennett 1987; Zimmitti 1999).

At present our estimates of energy substrate use for anurans are relatively crude, based on measurements of substrate concentrations from successive collections of different animals. More sophisticated measurements of actual flux rates of carbohydrates and fatty acids in intact animals, like those obtained for mammals (Weber, Brichon et al. 1996; Weber, Roberts et al. 1996), would provide more detailed information on the use of energy substrates, but may be technically difficult to achieve with very small frogs. There is also a need for additional measurements of respiratory quotients in calling frogs, like those of Grafe (1996, 1997), including studies of nightly and seasonal shifts in respiratory quotients in individual males. This would provide more detailed information on changes in the use of energy substrates over time.

Finally the information currently available on depletion of energy substrates suggests that the calling performance of frogs may be constrained both nightly and seasonally by diminishing energy supplies. Unfortunately determining whether males actually leave a chorus because of limited energy supplies, or perhaps switch to low-cost behavioral tactics such as satellite behavior, can be difficult. Currently only destructive sampling has been used to measure energy reserves, so correlations between individual condition and behavior are difficult to make. Supplemental feeding experiments, like those of Murphy (1994a), offer a promising method for addressing questions about energetic constraints on behavioral performance. Such experiments, however, should be designed with a clear understanding of differences among species of frogs in the way in which different sources of energy are likely to be used to support call production.

Acknowledgments

This chapter is dedicated to Stan Rand in appreciation for his guidance and encouragement in my studies of tropical frogs, including work in his laboratory on calling energetics. Special thanks to Mike Ryan for inviting me to participate in the symposium in honor of Stan's retirement. The chapter would not have been possible without the efforts of colleagues and students at the University of Connecticut, including Ted Taigen, Cathy Bevier, Steve Ressel, Laura Runkle, Sharyn Rusch, Dawn McKay, Stacey Lance, Joshua Schwartz, Mac Given, and Sal Zimmitti.

References

Aichinger, M. 1987. Annual activity patterns of anurans in a seasonal neotropical environment. Oecologia 71:583–592.

Andersson, M. 1994. Sexual Selection. Princeton University Press, Princeton, NJ.

Bennett, A. F. 1987. Interindividual variability: An underutilized resource. Pp. 147–169. *In* M. E. Feder, A. F. Bennett, W. W. Burggren, and R. B. Huey (eds). New Directions in Ecological Physiology. Cambridge University Press, New York.

Bevier, C. R. 1995a. Physiological constraints on calling activity in neotropical frogs. Ph.D. dissertation, University of Connecticut, Storrs, CT.

Bevier, C. R. 1995b. Biochemical correlates of calling activity in neotropical frogs. Physiological Zoology 68:1118–1142.

Bevier, C. R. 1997a. Breeding activity and chorus tenure of two neotropical hylid frogs. Herpetologica 53:297–311.

Bevier, C. R. 1997b. Utilization of energy substrates during calling activity in tropical frogs. Behavioral Ecology and Sociobiology 41:343–352.

Bonnet, X., D. Bradshaw, and R. Shine. 1998. Capital versus income breeding: An ectothermic perspective. Oikos 83:333–342.

Bucher, T. L., M. J. Ryan, and G. A. Bartholomew. 1982. Oxygen consumption during resting, calling, and nest building in the frog *Physalaemus pustulosus*. Physiological Zoology 55:10–22.

Cardoso, A. J., and C. F. B. Haddad. 1992. Diversidade e turno de vocalizações de anuros em communidade neotropical. Acta Zoologia Lillonana 41:93–105.

Cocroft, R. B. 1994. A cladistic analysis of chorus frog phylogeny (Hylidae: *Pseudacris*). Herpetologica 50:420–437.

Conley, K. E., S. R. Kayar, K. Rösler, H. Hoppeler, E. R. Weibel, and C. R. Taylor. 1987. Adaptive variation in the mammalian respiratory system in relation to energetic demand. IV. Capillaries and their relationship to oxidative capacity. Respiration Physiology 69:47–64.

Crump, M. L. 1974. Reproductive strategies in a tropical anuran community. University of Kansas Museum of Natural History Miscellaneous Publication 61:1–68.

Donnelly, M. A. and C. Guyer. 1994. Patterns of reproduction and habitat use in an assemblage of neotropical hylid frogs. Oecologia 98:291–302.

Drewry, G. E., W. R. Heyer, and A. S. Rand. 1982. A functional analysis of the complex call of the frog *Physalaemus pustulosus*. Copeia 1982:636–645.

Dudley, R., and A. S. Rand. 1991. Sound production and vocal sac inflation in the túngara frog, *Physalaemus pustulosus* (Leptodactylidae). Copeia 1991:460–470.

Duellman, W. E. 1970. Hylid Frogs of Middle America. 2 vols. Monograph. 1, University of Kansas Museum of Natural History.

Duellman, W. E., and J. J. Wiens. 1992. The status of the hylid frog genus *Ololygon* and the recognition of *Scinax* Wagler, 1830. Occasional Papers of University of Kansas Museum of Natural History 151:1–23.

Eichelberg, H., and H. J. Obert. 1976. Fat and glycogen utilization in the larynx muscles of fire-bellied toads (*Bombina bombina* L.) during calling activity. Cell and Tissue Research 167:1–10.

Eichelberg, H., and H. Schneider. 1973. Die Feinstruktur der Kehlkopfmuskeln des Laubfrosches, *Hyla arborea arborea* (L.) im Vergleich zu eninem Skelettmuskel. Zeitscrhift für Zellforschung 141:223–233.

Eichelberg, H., and H. Schneider. 1974. The fine structure of the larynx muscles in female tree frogs, *Hyla a. arborea* L. (Anura, Amphibia). Cell and Tissue Research 152:185–191

Ford, L. S., and D.C. Cannatella. 1993. The major clades of frogs. Herpetological Monographs 7:94–117.

Forester, D.C., D. V. Lykens, and W. K. Harrison. 1989. The significance of persistent vocalization by the spring peeper, *Pseudacris crucifer* (Anura: Hylidae). Behaviour 108:197–208.

Gans, C., and G. De Gueldre. 1992. Striated muscle: Physiology and functional morphology. Pp. 277–313. *In* M. E. Feder and W. W. Burggren (eds). Environmental Physiology of the Amphibians. University of Chicago Press, Chicago, IL.

Garland, T., Jr., and R. B. Huey. 1987. Testing symmorphosis: Does structure match functional requirements? Evolution 41:1404–1409.

Gerhardt, H. C. 1991. Female mate choice in treefrogs: Static and dynamic acoustic criteria. Animal Behaviour 42:615–635.

Gerhardt, H. C., M. L. Dyson, and S. D. Tanner. 1996. Dynamic properties of the advertisement calls of gray tree frogs: Patterns of variability and female choice. Behavioral Ecology 7:7–18.

Gerhardt, H. C., and J. J. Schwartz. 1995. Interspecific interactions in anuran courtship. Pp. 603–632. *In* H. Heatwole and B. K. Sullivan (eds). Amphibian Biology, vol. 2: Social Behaviour. Surrey Beatty and Sons, Chipping Norton, New South Wales, Australia.

Girgenrath, M., and R. L. Marsh. 1997. In vivo performance of trunk muscles in tree frogs during calling. Journal of Experimental Biology 200:3101–3108.

Girgenrath, M., and R. L. Marsh. 1999. Power output of sound-producing muscles in the tree frogs *Hyla versicolor* and *Hyla chrysoscelis*. Journal of Experimental Biology 202:3225–3237.

Given, M. F., and D. M. McKay. 1990. Variation in citrate synthase activity in calling muscles of carpenter frogs, *Rana virgatipes*. Copeia 1990:863–870.

Grafe, T. U. 1996. Energetics of vocalization in the African reed frog (*Hyperolius marmoratus*). Comparative Biochemistry and Physiology 114A:235–243.

Grafe, T. U. 1997. Use of metabolic substrates in the gray treefrog *Hyla versicolor:* Implications for calling behavior. Copeia 1997:356–362.

Grafe, T. U., R. Schmuck, and K. E. Linsenmair. 1992. Reproductive energetics of the African reed frogs, *Hyperolius viridiflavus* and *Hyperolius marmoratus*. Physiological Zoology 65:153–171.

Green, A. J. 1990. Determinants of chorus participation and the effects of size, weight and competition on advertisement calling in the túngara frog, *Physalaemus pustulosus* (Leptodactylidae). Animal Behaviour 39:620–638.

Guppy, M. 1988. Limiting carbohydrate stores: Alleviating the problem in the marathon runner, a hibernating lizard, and the neonate brain. Canadian Journal of Zoology 66:1090–1097.

Hedges, S. B. 1986. An electrophoretic analysis of Holarctic hylid frog evolution. Systematic Zoology 35:1–21.

Hödl, W. 1977. Call differences and calling site segregation in anuran species from central Amazonian floating meadows. Oecologia 28:351–363.

Horn, A. G., M. L. Leonard, and D. W. Weary. 1995. Oxygen consumption during crowing by roosters: Talk is cheap. Animal Behaviour 50:1171–1175.

Jaramillo, C., A. S. Rand, R. Ibáñez, and R. Dudley. 1997. Elastic structures in the vocalization apparatus of the túngara frog *Physalaemus pustulosus* (Leptodactylidae). Journal of Morphology 233:287–295.

Kelley, D. B. 1996. Sexual differentiation in *Xenopus laevis*. Pp. 143–176. *In* R. C. Tinsley and H. R. Kobel (eds). The Biology of *Xenopus*. Oxford University Press, New York.

Klump, G. M., and H. C. Gerhardt. 1987. Use of non-arbitrary acoustic criteria in mate choice by female gray tree frogs. Nature 326:286–288.

Kodric-Brown, A., and J. H. Brown. 1984. Truth in advertising: The kinds of traits favored by sexual selection. American Naturalist 124:309–323.

Lance, S. L., and K. D. Wells. 1993. Are spring peeper satellite males physiologically inferior to calling males? Copeia 1993:1162–1166.

Lee, H.-J., and W. Loher. 1993. The mating strategy of the male short-tailed cricket *Anurogryllus muticus* de Geer. Ethology 95:327–344.

Lighton, J. R. B. 1987. Cost of tokking: The energetics of substrate communication in the tok-tok beetle, *Psammodes striatus*. Journal of Comparative Physiology B 157:11–20.

Marler, C. A., and M. J. Ryan. 1996. Energetic constraints and steroid hormone correlates of male calling behaviour in the túngara frog. Journal of Zoology, London 240:397–409.

Marsh, R. L. 1999. Contractile properties of muscles used in sound production and locomotion in two species of gray tree frog. Journal of Experimental Biology 202:3215–3223.

Marsh, R. L., and T. L. Taigen. 1987. Properties enhancing aerobic capacity of calling muscles in gray tree frogs *Hyla versicolor*. American Journal of Physiology 252:R786-R793.

Martin, W. F. 1972. Evolution of vocalizations in the genus Bufo. Pp. 279–309. *In* W. F. Blair (ed). Evolution in the genus *Bufo*. University of Texas Press, Austin.

McClelland, B. E., and W. Wilczynski. 1989. Sexually dimorphic laryngeal morphology in *Rana pipiens*. Journal of Morphology 201:293–299.

McClelland, B. E., W. Wilczynski, and A. S. Rand. 1997. Sexually dimorphism and species differences in the neurophysiology and morphology of the acoustic communication system of two neotropical hylids. Journal of Comparative Physiology A 180:451–462.

McClelland, B. E., W. Wilczynski, and M. J. Ryan. 1998. Intraspecific variation in laryngeal and ear morphology in male cricket frogs (*Acris crepitans*). Biological Journal of the Linnean Society 63:51–67.

McKay, D. M. 1989. Seasonal variation in calling behavior and muscle biochemistry in spring peepers (*Hyla crucifer*). M. S. Thesis, The University of Connecticut, Storrs, CT.

McLister, J. D., E. D. Stevens, and J. P. Bogart. 1995. Comparative contractile dynamics of calling and locomotor muscles in three hylid frogs. Journal of Experimental Biology 198:1527–1538.

Moreira, G., and L. Barreto. 1997. Seasonal variation in nocturnal calling activity of a savanna anuran community in central Brazil. Amphibia-Reptilia 18:49–57.

Murphy, C. G. 1994a. Determinants of chorus tenure in barking treefrogs (*Hyla gratiosa*). Behavioral Ecology and Sociobiology 34:285–294.

Murphy, C. G. 1994b. Chorus tenure of male barking treefrogs, *Hyla gratiosa*. Animal Behaviour 48:763–777.

Pough, F. H., W. E. Magnusson, M. J. Ryan, K. D. Wells, and T. L. Taigen. 1992. Behavioral energetics. Pp. 395–436. *In* M. E. Feder and W. W. Burggren (eds). Environmental Physiology of the Amphibians. University of Chicago Press, Chicago, IL.

Prestwich, K. N. 1994. The energetics of acoustic signaling in anurans and insects. American Zoologist 34:625–643.

Prestwich, K. N., K. E. Brugger, and M. Topping. 1989. Energy and communication in three species of hylid frogs: Power input, power output, and efficiency. Journal of Experimental Biology 144:53–80.

Rand, A. S., and M. J. Ryan. 1981. The adaptive significance of a complex vocal repertoire in a neotropical frog. Zeitschrift für Tierpsychologie 57:209–214.

Reinhold, K., M. D. Greenfield, Y. Jang, and A. Broce. 1998. Energetic cost of sexual attractiveness: Ultrasonic advertisement in wax moths. Animal Behaviour 55:905–913.

Ressel, S. J. 1993. A morphometric analysis of anuran skeletal muscle ultrastructure: Implications for the functional design of muscle in ectotherms. Ph.D. dissertation, University of Connecticut, Storrs, CT.

Ressel, S. J. 1996. Ultrastructural properties of muscles used for call production in neotropical frogs. Physiological Zoology 69:952–973.

Roberts, T. J., J.-M. Weber, H. Hoppelar, E. R. Weibel, and C. R. Taylor. 1996. Design of the oxygen and substrate pathways. II. Defining the upper limits of carbohydrate and fat oxidation. Journal of Experimental Biology 199:1651–1658.

Runkle, L. S., K. D. Wells, C. C. Robb, and S. L. Lance. 1994. Individual, nightly, and seasonal variation in calling behavior of the gray treefrog, *Hyla versicolor:* Implications for energy expenditure. Behavioral Ecology 5:318–325.

Ryan, M. J. 1985. The Túngara Frog: A Study in Sexual Selection and Communication. University of Chicago Press, Chicago, IL.

Ryan, M. J. 1988. Energy, calling, and selection. American Zoologist 28:885–898.

Ryan, M. J., G. A. Bartholomew, and A. S. Rand. 1983. Energetics of reproduction in a neotropical frog, *Physalaemus pustulosus*. Ecology 64:1456–1462.

Schmidt, R. S. 1965. Larynx control and call production in frogs. Copeia 1965:143–147.

Schneider, H. 1977. Acoustic behavior and physiology of vocalization in the European tree frog, *Hyla arborea* (L.). Pp. 295–336. *In* D. H. Taylor and S. I. Guttman (eds). The Reproductive Biology of Amphibians. Plenum Press, New York.

Schneider, H. 1988. Peripheral and central mechanisms of vocalization. Pp. 537–558. *In* B. Fritzsch, M. J. Ryan, W. Wilczynski, T. E. Hetherington, and W. Walkowiak (eds). The Evolution of the Amphibian Auditory System. John Wiley and Sons, New York.

Schwartz, J. J. 1991. Why stop calling? A study of unison bout singing in a neotropical treefrog. Animal Behaviour 42:565–577.

Schwartz, J. J., S. J. Ressel, and C. R. Bevier. 1995. Carbohydrates and calling: Depletion of muscle glycogen and the chorusing dynamics of the neotropical treefrog *Hyla microcephala*. Behavioral Ecology and Sociobiology 37:125–135.

Sinsch, U. 1988. Einfluss von Temperatur und Ernährung auf die diurnale Rufaktivität des Beutelfrosches, *Gastrotheca marsupiata*. Verhandelungen der Deutschen Zoologischen Gesellschaft 81:197.

Sullivan, B. K., and S. H. Hinshaw. 1990. Variation in the advertisement calls and male calling behavior in the spring peeper (*Pseudacris crucifer*). Copeia 1990:1146–1150.

Sullivan, B. K., and S. H. Hinshaw. 1992. Female choice and selection on male calling behaviour in the grey treefrog *Hyla versicolor*. Animal Behaviour 44:733–744.

Sullivan, B. K., M. J. Ryan, and P. A. Verrell. 1995. Female choice and mating system structure. Pp. 469–517. *In* H. Heatwole and B. K. Sullivan (eds.). Amphibian Biology. Vol. 2. Social Behaviour. Surrey Beatty and Sons, Chipping Norton, New South Wales, Australia.

Taigen, T. L., and K. D. Wells. 1985. Energetics of vocalization by an anuran amphibian (*Hyla versicolor*). Journal of Comparative Physiology B 155:163–170.

Taigen, T. L., K. D. Wells, and R. L. Marsh. 1985. The enzymatic basis of high metabolic rates in calling frogs. Physiological Zoology 58:719–726.

Vehrencamp, S. L., J. W. Bradbury, and R. M. Gibson. 1989. The energetic cost of display in male sage grouse. Animal Behaviour 38:885–896.

Vock, R., H. Hoppeler, H. Claassen, D. X. Y. Wu, R. Billeter, J.-M. Weber, C. R. Taylor, and E. R. Weibel. 1996. Design of the oxygen and substrate pathways. VI. Structural basis of intracellular substrate supply to mitochondria in muscle cells. Journal of Experimental Biology 199:1689–1697.

Vock, R., E. R. Weibel, H. Hoppeler, G. Ordway, J.-M. Weber, and C. R. Taylor. 1996. Design of the oxygen and substrate pathways. V. Structural basis of vascular substrate supply to muscle cells. Journal of Experimental Biology 199:1675–1688.

Walker, S. L. 1989. The effect of substrate depletion on the calling behavior of the gray treefrog. M. S. Thesis, The University of Connecticut, Storrs, CT.

Weber, E. 1975. Die Veränderung der Befreiungsreife bei sechs europäischen Anuren nach Ausschaltung von Kehlkopfmuskeln (Amphibia). Zoologische Jahrbucher. Abteilung für Allgemeine Zoologie und Physiologie der Tiere 79:311–320.

Weber, E. 1976. Die Veränderung der Paarungs- und Revierrufe von *Hyla arborea savignyi* Audouin (Anura) nach Ausschaltung von Kehlkopfmuskeln. Bonner Zoologishe Beitrage 27:87–97.

Weber, J.-M. 1988. Design of exogenous fuel supply systems: Adaptive strategies for endurance locomotion. Canadian Journal of Zoology 66:1116–1121.

Weber, J.-M. 1992. Pathways for oxidative fuel provision to working muscle: Ecological consequences of maximal supply limitations. Experientia 48:557–564.

Weber, J.-M., G. Brichon, G. Zwingelstein, G. McClelland, C. Saucedo, E. R. Weibel, and C. R. Taylor. 1996. Design of the oxygen and substrate pathways. IV. Partitioning energy provision from fatty acids. Journal of Experimental Biology 199:1667–1674.

Weber, J.-M., T. J. Roberts, R. Vock, E. R. Weibel, and C. R. Taylor. 1996. Design of the oxygen and substrate pathways. III. Partitioning energy provision from carbohydrates. Journal of Experimental Biology 199:1659–1666.

Weibel, E. R., C. R. Taylor, and L. Bolis (eds.). 1998. Principles of Animal Design: The Optimization and Symmorphosis Debate. Cambridge University Press, New York.

Weibel, E. R., C. R. Taylor, and H. Hoppeler. 1991. The concept of symmorphosis: A testable hypothesis of structure-function relationship. Proceedings of the National Academy of Sciences of the USA 88:10357–10361.

Welch, A. M., R. D. Semlitsch, and H. C. Gerhardt. 1998. Call duration as an indicator of genetic quality in male gray tree frogs. Science 280:1928–1930.

Wells, K. D. 1977a. The social behaviour of anuran amphibians. Animal Behaviour 25:666–693.

Wells, K. D. 1977b. Territoriality and male mating success in the green frog (*Rana clamitans*). Ecology 58:750–762.

Wells, K. D. 1988. The effect of social interactions on anuran vocal behavior. Pp. 433–454. *In* B. Fritzsch, M. J. Ryan, W. Wilczynski, T. E. Hetherington, and W. Walkowiak (eds). The Evolution of the Amphibian Auditory System. John Wiley and Sons, New York.

Wells, K. D., and C. R. Bevier. 1997. Contrasting patterns of energy substrate use in two species of frogs that breed in cold weather. Herpetologica 53:70–80.

Wells, K. D., and T. L. Taigen. 1986. The effect of social interactions on calling energetics in the gray treefrog (*Hyla versicolor*). Behavioral Ecology and Sociobiology 19:9–18.

Wells, K. D., and T. L. Taigen. 1989. Calling energetics of a neotropical treefrog, *Hyla microcephala*. Behavioral Ecology and Sociobiology 25:13–22.

Wells, K. D., T. L. Taigen, and J. A. O'Brien. 1996. The effect of temperature on calling energetics of the spring peeper (*Pseudacris crucifer*). Amphibia-Reptilia 17:149–158.

Wells, K. D., T. L. Taigen, S. W. Rusch, and C. C. Robb. 1995. Seasonal and nightly variation in glycogen reserves of calling gray treefrogs (*Hyla versicolor*). Herpetologica 51:359–368.

Zimmitti, S. J. 1998. Individual variation and sexual dimorphism in traits related to call production in the spring peeper (*Pseudacris crucifer*). M. S. thesis, University of Connecticut, Storrs, CT.

Zimmitti, S. J. 1999. Individual variation in morphological, physiological, and biochemical features associated with calling in spring peepers (*Pseudacris crucifer*). Physiological and Biochemical Zoology 72:666–676.

6

PETER M. NARINS

Ectothermy's Last Stand

Hearing in the Heat and Cold

Introduction

Ectothermy, or the process of obtaining body heat exclusively from the environment rather than from oxidative metabolism, is the norm among the majority of animal species. An inevitable consequence of ectothermy is the covariation of the ambient and auditory-system temperatures. It has long been known that aspects of both acoustic signal production (Blair 1958; Zweifel 1968; Navas 1996) and reception (Moffat and Capranica 1976; Stiebler and Narins 1990; van Dijk et al. 1990) in ectotherms are temperature-dependent. It is of interest as well that female preference in some ectotherms is also temperature-dependent (e.g., Gerhardt 1978), and thus it becomes crucial to understand the genesis of this dependence in detail. In this paper, I shall concentrate on the temperature dependence of auditory reception in frogs.

The inner ear of anurans (frogs and toads) is unusual in that it contains three organs that are specialized for sound reception. The amphibian papilla (AP), the basilar papilla (BP), and the sacculus (S) are anatomically distinct, spatially separate organs, each with its own complement of sensory hair cells and overlying tectorial structure. The BP and portions of the AP respond exclusively to airborne sounds, whereas other portions of the AP and the S exhibit sensitivity both to airborne sounds as well as to substrate-borne vibrations. Each of these organs displays a particular pattern of temperature-sensitivity. A goal of this paper is understanding these patterns and how they may be reflected in species-specific acoustic behavior.

Three types of measurements were made in the auditory periphery of several species of amphibians as a function of temperature: single fiber recordings from the eighth nerve, whole-cell patch clamp recordings from isolated hair cells under thermally controlled conditions, and recordings of spontaneous otoacoustic emissions from five frog species with derived inner ears, sensu Lewis (1978). We find that the tuning and thresholds of the AP fibers are highly temperature-dependent, whereas for the BP fibers they are not. Moreover we find that hair cells from the rostral AP and S show a clear temperature dependence of the oscillation frequency of their membrane potential in response to an intracellular current step. In contrast, although the ionic currents from caudal AP and BP hair cells are highly temperature-sensitive, the membrane potentials of these cells do not oscillate in response to a current step. In addition spontaneous otoacoustic emissions, which presumably reflect an active process in the inner ear, are highly temperature-dependent although emission frequencies range from 600 to 1600 Hz, the upper portion of this range being the sole domain of the BP.

The temperature dependencies observed in these three sets of measurements will be compared and related to those seen to govern calling behavior in several frog species. In addition a model based on the mechanical properties of the AP tectorial membrane is presented that may explain the observed shift in characteristic frequency of a single AP nerve fiber in response to changing temperatures. This model is currently being tested.

Experiment 1: Eighth Nerve Recordings

Background

The effect of temperature on single fiber tuning properties has been used to infer the contributions of electrical tuning in the gecko (Eatock and Manley 1981), caiman (Smolders and Klinke 1984), turtle (Wu et al. 1995), and pigeon (Schermuly and Klinke 1985). In each of these animals the characteristic frequency (CF) for single auditory nerve fibers increases with a thermal Q_{10} of 2.0 over its entire acoustical range. The thermal Q_{10} is a measure of the change in the rate of a biological process due to a 10° change in temperature. In contrast, in the goldfish sacculus (Fay and Ream 1992) and bullfrog sacculus (Egert and Lewis 1995), single fiber tuning properties are remarkably temperature-insensitive, exhibiting Q_{10}s below 1.1. In the frog auditory system, low-frequency AP fibers (CFs < 1 kHz) exhibit Q_{10}s of about 1.7, whereas high-frequency BP fibers are temperature-insensitive (Moffat and Capranica 1976; Stiebler and Narins 1990; van Dijk et al. 1990). The goal of this first set of experiments was to compare the temperature dependence of the characteristic frequencies of AP and BP fibers in two species of frogs.

Animals

Species chosen for the eighth nerve study were the Puerto Rican coqui (*Eleutherodactylus coqui*) and the Pacific tree frog (*Hyla regilla*), primarily because of the different range of temperatures to which each species is normally exposed. Although as a species *E. coqui* is subjected to a wide range of temperatures along an altitudinal gradient of over 1000 m (Narins and Smith 1986), local populations at a given altitude experience little temperature variation during the course of the year. Males of this arboreal species produce a two-note advertisement call (Co-Qui). At the higher elevations the Co note is a 100-ms, constant-frequency tone of about 1.1 kHz (Narins and Capranica 1978). The Co note is followed by a pause of 150–200 ms and a second (Qui) note that sweeps upward in frequency from about 1.8 kHz to 2.1 kHz in about 170 ms. The Co note functions in male–male territorial encounters, whereas the Qui note serves to attract females (Narins and Capranica 1976, 1978). Males give the two-note advertisement call every 2 to 4 seconds from sunset to shortly after midnight throughout 11 months of the year. In addition, in this species there is evidence that call note significance may be intensity-dependent (Stewart and Rand 1991; Narins, personal observations).

The Pacific tree frog is found in western North America, from southern British Columbia to Baja California (Stebbins 1966). Populations of this species occur from sea level up to 7000 feet in areas that experience large temperature variations throughout the year. Males of *H. regilla* produce three distinct calls, as described by Snyder and Jameson (1965). Playback experiments with females have shown that the species-specificity of the call obtains from its temporal parameters (Brenowitz and Rose 1994; Brenowitz et al., this volume).

Animal Preparation

Adult animals were anesthetized with an intramuscular injection of 1.5–2.0 μl/g of sodium pentobarbital (Nembutal, 50 mg/ml) 24–48 hours before recording. The eighth nerve was exposed by either a ventral or dorsal surgical approach. Details of the surgical procedure have been described elsewhere (Narins and Wagner 1989). Animals were only used for experiments if they exhibited normal postoperative behavior; that is, if they could sit upright and hop in a straight line without circling or listing. About an hour before the experiment, an animal was lightly anesthetized (Nembutal) and immobilized with an intramuscular injection of curare (1.75–2.5 μl/g). For the recordings the frog was wrapped in wet gauze and placed in a sound-insulated chamber (IAC 1202-A) on a Peltier plate located on a vibration-attenuating table.

Temperature Control

The body temperature of the animal was regulated by a Peltier plate (Cambion, fluid-cooled microscope stage), the surface of which could be heated or cooled with an accuracy of ± 0.1°C. Both the core temperature and that of the upper jaw close to the nerve were monitored using a two-channel thermocouple thermometer (Sensortek BAT-10). Several calibrations were made with one thermistor placed at the location of the nerve and the second one either in the cloaca, beneath the tongue, or on the roof of the mouth covered by wet gauze. During the experiments we routinely estimated eighth

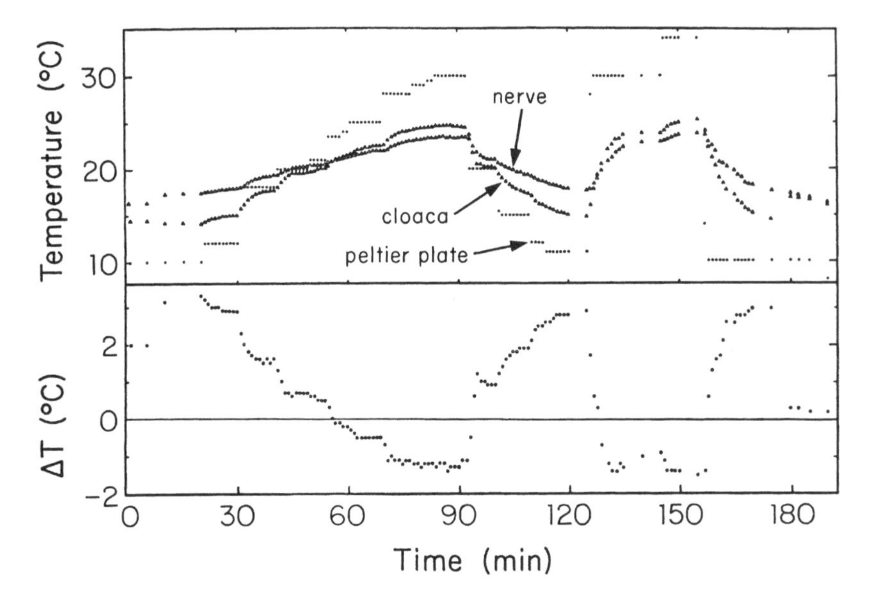

Figure 6.1. Temperature calibration experiment. The anesthetized frog was placed on a Peltier plate that was subject to a controlled temperature shift of ~17°C per hour. The temperature of a sensor placed in the animal's cloaca could differ from that of one placed at the location of the auditory nerve by up to 3°C during a trial run. Thus we estimated the eighth nerve temperature with a thermistor placed on the roof of the mouth and covered with wet gauze (see text for details).

nerve temperature with the thermistor placed on the roof of the mouth, because this location showed the minimum deviation (± 0.2°C) from auditory nerve temperature of all locations tested (Stiebler and Narins 1990). Our calibrations indicated that the cloacal temperature is a poor estimator of the nerve temperature; these two temperatures can differ by more than 3°C throughout an experiment (Figure 6.1).

Temperature was changed at a rate of approximately 0.5°C per minute, which did not appear to introduce significant error because similar results were obtained on repeated measures taken on single fibers at the same temperature during a heating cycle, a cooling cycle, or thermal stasis.

Results

In all frog species tested to date, increasing the temperature resulted in an upward shift in the CF and a concomitant reduction in CF-threshold of the tuning curves for fibers innervating the low- and mid-frequency fibers of the AP (Stiebler and Narins 1990; van Dijk et al. 1990). In contrast CFs and CF-thresholds of BP (high-frequency) fibers appear to be temperature-independent (Figure 6.2).

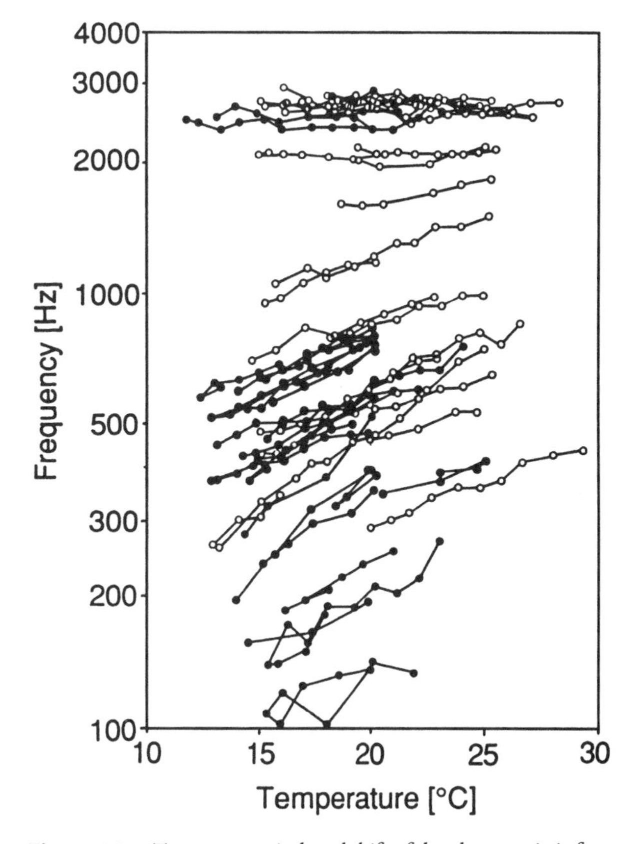

Figure 6.2. Temperature-induced shift of the characteristic frequency (CF) from all neurons that could be recorded over a range of 5°C or more; 24 neurons in *H. regilla* (●) and 27 neurons in *E. coqui* (○). Data points from the same neurons are connected by lines. The amount of frequency shift decreases with increasing temperature and with increasing CF (from Stiebler and Narins 1990).

Temperature-induced shifts of CFs varied from 0.08 octaves per °C for low-frequency fibers to no CF shift for high-frequency (BP) fibers. The frequency shift per °C may be expressed as a thermal Q_{10} value to simplify comparisons with other temperature-dependent processes (Eatock and Manley 1981). The thermal Q_{10} for a frequency shift of s octaves may be calculated by:

$$Q_{10} = e^{s(10/\Delta T)\ln 2}$$

where ΔT is the induced temperature shift in °C. From this relationship, the mean Q_{10} at 300 Hz for AP fibers of

H. regilla is 1.75 and for AP fibers of *E. coqui* is 1.73, whereas at 1000 Hz the mean Q_{10} for AP fibers of *H. regilla* is 1.43 and for AP fibers of *E. coqui* is 1.47 (Stiebler and Narins 1990). Clearly BP fibers have thermal Q_{10} values approximating 1.

Other response properties of auditory neurons also illustrate pronounced temperature dependence. For example the preferred phase of firing for auditory nerve fibers systematically and progressively advances with cooling, with phase locking diminishing rapidly below 13.0°C in *E. coqui*. In addition the response to short-duration clicks in this species exhibits increasing latency with cooling (Narins 1995).

Experiment 2: Hair Cell Recordings

Background

The frog sacculus is an inner ear organ that responds primarily to substrate-borne vibrations at frequencies extending up to about 160 Hz (Koyama et al. 1982; Lewis 1988). The organ is remarkably sensitive to seismic stimulation (Narins and Lewis 1984), but single nerve fibers are poorly tuned, exhibiting broad pass bands (Lewis 1988). Electrical resonances are exhibited by isolated saccular hair cells and are believed to be the primary mechanism of frequency selectivity for sounds or vibrations entering the sacculus (Hudspeth 1985; Hudspeth and Lewis 1988).

We recently quantified the effects of temperature on saccular hair cell electrical resonances and their underlying conductances (Smotherman and Narins 1998). The results bear directly upon whether temperature-dependent changes in the receptor potential can account for the temperature-dependent properties of the intact auditory system.

Animals

The Northern leopard frog (*Rana pipiens pipiens*) is the most widely distributed of the North American amphibians, being found as far north as the District of Mackenzie in Canada and as far south as northern Georgia, and from desert lowlands to high in the mountains (Stebbins 1966). The leopard frog endures active body temperature ranges of 8–35°C and is known to breed at temperatures below 15°C (Zug 1993). This species was chosen for the hair cell studies because it may experience a wide daily shift in its body temperature, and thus its auditory system, over much of its geographic range.

Hair Cell Preparation

Saccular maculae and amphibian papillae were dissected from 60- to 80-mm (snout–vent length; SVL) adult northern leopard frogs and were treated for 20 minutes at room temperature with papain (Calbiochem, USA; 500 μg/l) dissolved in a dissociation solution, pH 7.2. They were then transferred to a dissociation solution with BSA (500 μg/l) replacing papain for 15 to 30 minutes. Maculae were then transferred to a recording dish with dissociation solution alone, and hair cells were gently flicked free with a tungsten needle. The hair cells settled but did not adhere to the glass bottom of the recording chamber, after which the dissociation solution was replaced via perfusion with a standard external recording solution. The recording chamber was placed on the stage of an inverted microscope (Nikon Diaphot, Japan) with a 40× objective with phase contrast optics. The temperature of the recording solution was monitored with a thermocouple placed at the center of the bath, and the temperature was changed at a rate of 0.5–2°C per minute by a controller unit (Sensortek TS-4; Sensortek, Clifton NJ) attached to the stage.

Whole-Cell Recordings

Currents and voltages were recorded with the conventional whole-cell tight-seal patch-clamp technique (Hamill et al. 1981). Borosilicate glass pipettes were pulled with a Narishige two-stage vertical pipette puller (Narishige, Japan) to tip diameters of approximately 1 μm. Electrode resistances typically ranged from 2 to 10 MΩ. Series resistances during recordings ranged from 6 to 25 MΩ, and were compensated 60–95% using the compensation circuitry of the amplifier (Axon Instruments, Foster City, CA).

Electrical oscillations were induced in the hair cell's membrane potential by injecting a depolarizing current pulse (a 150-pA current pulse for 100 ms) at 15, 20, and 25°C. The effect of temperature on resonant frequency was investigated by recording the oscillations at the initiation of recording from hair cells that had equilibrated at a given temperature for more than 10 minutes, and averaging the oscillation frequencies of several hair cells recorded at the same temperature. Individual hair cell resonant frequencies were recorded over a range of temperatures, and thermal Q_{10}s were calculated for each hair cell. Whole-cell recordings were initiated at either 15, 20, or 25°C and the temperature was raised or lowered. Current-clamp recordings were made over as wide a temperature range as possible.

Results

The voltage waveform (Figure 6.3) induced by a current pulse typically exhibits a depolarized, damped onset resonance and may exhibit a damped offset resonance centered around the resting potential. The Q_{10} for averaged resonant frequencies between 15 and 35°C is 2.0 (Smotherman and Narins 1998).

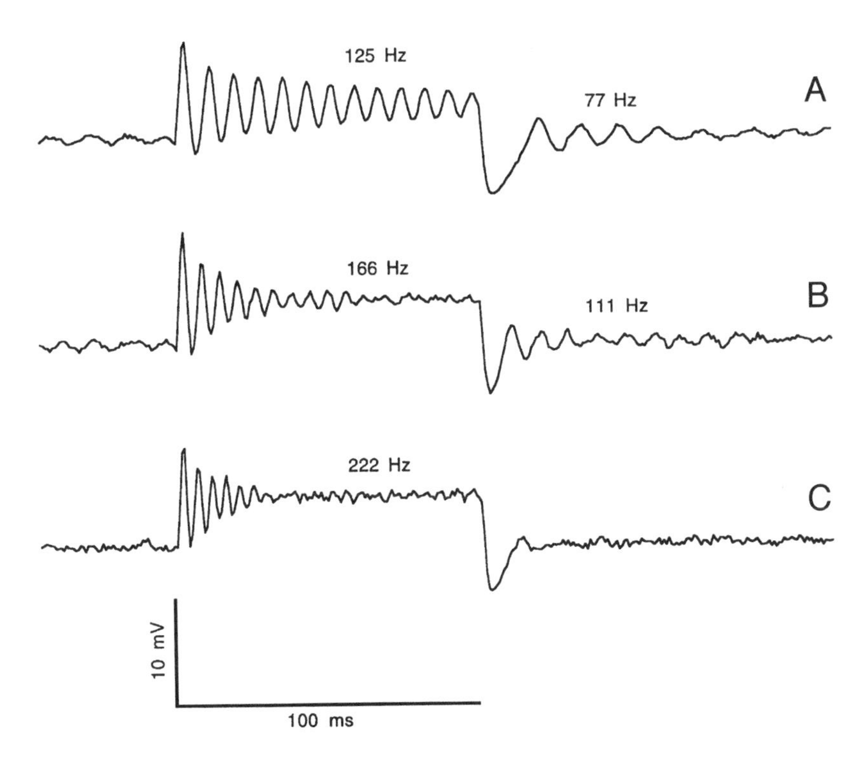

Figure 6.3. Electrical resonance at three temperatures. Membrane potential oscillations induced by a 150-pA current pulse for 100 ms at (A) 15°C, (B) 20°C, and (C) 25°C in a spherical saccular hair cell. Initial recordings were made at 15°C; subsequently the temperature was increased at a rate of 1°C per minute. The onset resonant frequency at 15°C was 125 Hz; at 20°C was 166 Hz; and at 25°C was 222 Hz. The Q_{10} of the onset oscillations is 1.8. Offset oscillations at 15 and 20°C were 77 and 111 Hz, respectively. Each record represents the average of 10 traces (from Smotherman and Narins 1998).

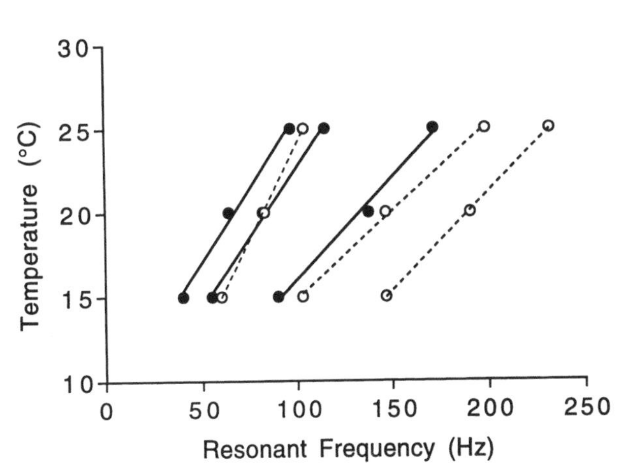

Figure 6.4. Resonant frequency of hair cells isolated from the leopard frog sacculus (●) and amphibian papilla (○), measured at three temperatures. Membrane potential oscillations were evoked using depolarizing current steps. For each hair cell, resonant frequencies displayed were measured starting from a common membrane potential of –53 mV, for which the highest quality resonances occur. These cells typically exhibited thermal Q_{10}s of about 2.0 between 15 and 25°C.

Moreover preliminary results indicate that hair cells from the rostral and medial region of the AP resemble saccular hair cells in their temperature dependence (Figure 6.4). This is not altogether surprising, given their similar biophysical properties (Smotherman and Narins 2000). Hair cells of the sacculus and rostral AP exhibit qualitatively similar resonant properties because the complement of their ionic currents is essentially identical. AP hair cells vary only in that their ionic currents have been adapted to a broader bandwidth than saccular hair cells. Neither high-frequency (caudally located) AP hair cells nor BP hair cells exhibit electrical resonance.

Experiment 3: Otoacoustic Emissions

Background

Kemp's (1978) discovery of "echoes" from the ear forever changed the way we think about the auditory system. Not only is the ear capable of detecting sound, but of producing it as well. Low-level sounds are detectable in the ear canal with no external source or stimulus, in which case they are referred to as spontaneous otoacoustic emissions (SOAEs). In mammals it has become well established that OAEs from the inner ear are considered evidence for an active (hair cell) mechanism in the ear (Probst et al. 1991). Köppl (1995) argued that in nonmammalian species hair cell motility is unlikely to be present, and thus other mechanisms must be responsible for the observed SOAEs. This has recently become clear with the discovery of evoked OAEs from a grasshopper, despite the complete lack of hair cells in the auditory system of this animal (Kössl and Boyan 1998a, 1998b).

Palmer and Wilson (1982) were the first to describe OAEs in a nonmammalian species—the European edible frog, *Rana esculenta.* In 1987 van Dijk and Wit demonstrated that the OAEs of *Rana temporaria* showed a strong temperature dependence. More recently van Dijk et al. (1989) and Long et al. (1996) demonstrated temperature sensitivity of the emission peaks of the European edible frog, *Rana esculenta*. The purpose of the present experiments was to compare the temperature sensitivity of spontaneous otoacoustic emission patterns of several frog species from five different families and to look for similarities and differences that might be correlated to the inner ear morphology. Clear temperature sensitivity of the SOAEs could implicate a hair cell contribution to SOAE generation, whereas temperature-independent SOAEs would mediate against such an interpretation.

Animals and SOAE Recording

We screened for the presence of emissions in the green tree frog, *Hyla cinerea*, the gray tree frogs, *Hyla chrysoscelis* and *Hyla versicolor*, the Northern leopard frog, *Rana pipiens pipiens*, the white-lipped frog, *Leptodactylus albilabris*, the African clawed frog, *Xenopus laevis*, and the oriental fire-bellied toad, *Bombina orientalis* (van Dijk et al. 1996). SOAEs in these frogs were measured with a Sennheiser 12–227 microphone. The microphone capsule was contained in a custom-built housing and was powered by a custom-built preamplifier. The microphone system was calibrated in a closed coupler system, containing a condenser microphone (Brüel and Kjaer 4134). A spectrum analyzer (Brüel and Kjaer 2033) was used to accumulate an average spectrum of the electrical signal from the microphone. Typically 512 spectra were averaged over a period of 3 to 4 minutes.

Temperature Control

The frog's temperature was controlled by placing a small bag filled with either crushed ice or lukewarm water on its back. In this way the frog's temperature could be gradually changed. The temperature was recorded every 2 to 3 minutes along with the counter setting of the tape deck. Thus an estimate of the temperature at each moment during the experiment could be obtained by interpolation.

Results

We demonstrated that emissions from five species of frogs all show strong temperature dependence, although emission frequencies range from 600 to 1600 Hz, the upper portion of this range being the sole domain of the BP (van Dijk et al. 1996). The pattern of changes caused by core temperature shifts was similar in all five of the "derived" frog species examined. In general we observed three SOAE peaks per ear. Spectral changes for *Hyla cinerea* as a function of temperature are illustrated in Figure 6.5. The pattern seen in this figure was typical for all species studied. At low temperatures

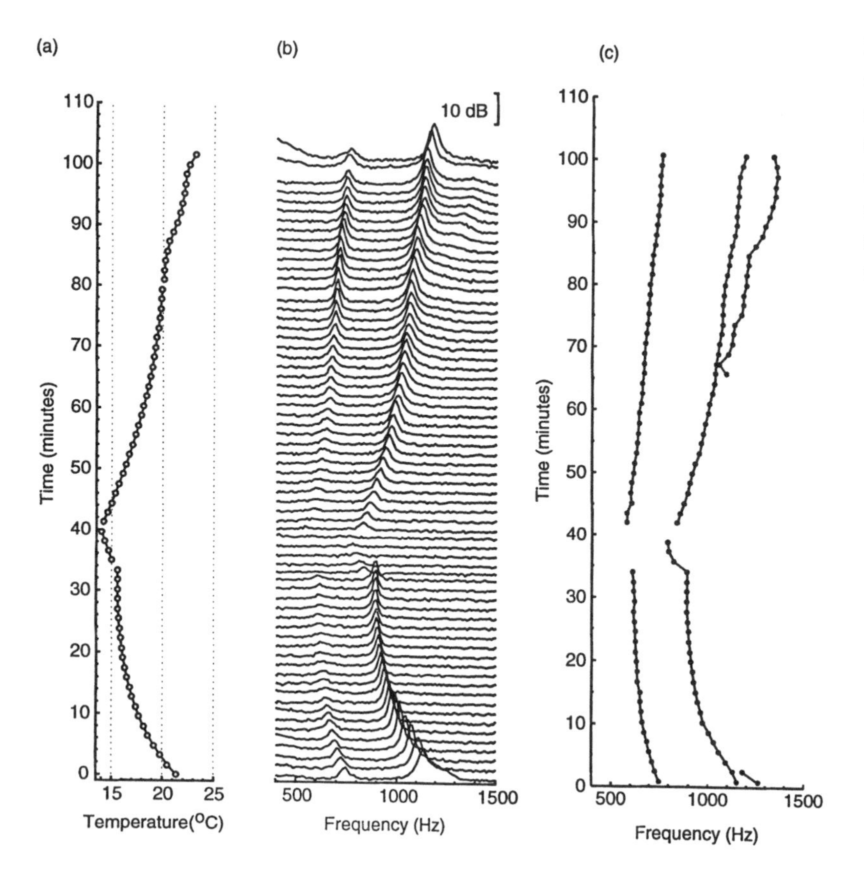

Figure 6.5. Spontaneous otoacoustic emissions (SOAEs) in a green treefrog, *Hyla cinerea*, during controlled core temperature shifts. (a) Temperature as a function of time during the experiment. (b) Waterfall plot of successive SOAE spectra. (c) Emission frequencies identified in the SOAE spectra in (b). From van Dijk et al. (1996).

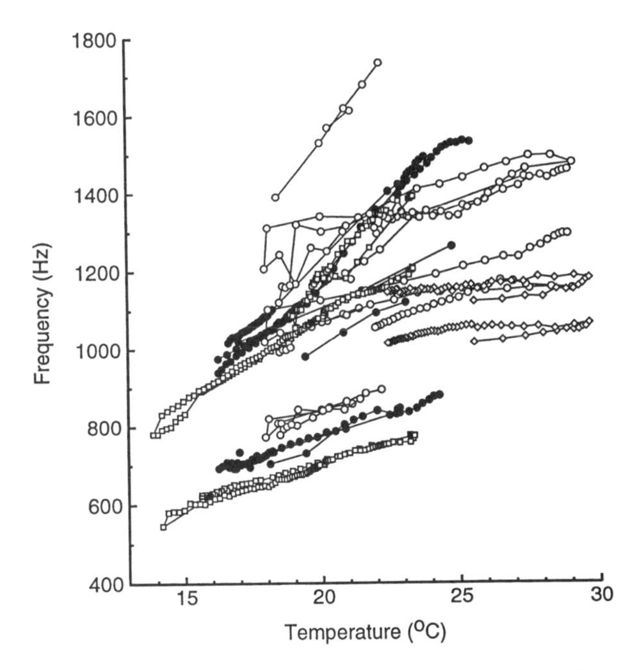

Figure 6.6. Emission frequency versus body temperature. Data collected from *Hyla cinerea* (3 ears, ○), *Hyla versicolor* (1 ear, □), *Hyla chrysoscelis* (3 ears, ●), and *Leptodactylus albilabris* (1 ear, ◇). From van Dijk et al. (1996).

(< 15°C) most SOAEs disappeared. As temperature was increased, the low- and mid-frequency emissions would first appear and would subsequently rise in frequency. With further increases in temperature, the high-frequency peak appeared, sometimes in parallel with a strong reduction of the mid-frequency peak. In some cases all SOAEs disappeared at higher temperatures (van Dijk et al. 1996).

The relationship between emission frequency and temperature was consistent: with increasing temperature, emission frequency increased at a rate between 0.009 and 0.09 octave per °C (Figure 6.6). Some hysteresis can be observed in Figure 6.6. When the changing of body temperature reversed from cooling to heating, the temperature increase measured in the frog's mouth lagged the increase of emission frequencies. The reverse effect was found when going from heating to cooling; in this case the decrease in emission frequencies lagged the change in body temperature.

Discussion

Model of CF Shift with Temperature

Although it is clear that auditory nerve tuning curves for AP fibers shift to lower CFs when the animal is cooled (Moffat and Capranica 1976; Stiebler and Narins 1990; van Dijk et al. 1990), the mechanisms underlying this CF-shift are unknown. A reduction in a fiber's CF on cooling could arise from changes in the mechanics of the tectorial membrane partition (a shift in the location of the vibration maximum [lvm] of the tectorial membrane (TM), stiffness changes of the hair cell bundle, etc.), and/or in the electrical properties (channel currents) of the hair cells that affect the cell's resonant frequency (Hudspeth 1985). Considering the mechanical changes first, both the TM stiffness and the viscosity of the fluids surrounding the TM would be expected to increase with decreasing temperature. These changes are incorporated into the schemes illustrated in Figure 6.7. If the basilar membrane (BM) is cooled (solid curves), all existing cochlear models predict increased stiffness, which results in a shift of the traveling wave (TW) envelope peak (dashed lines) toward the apex (lower frequencies). From the point of view of a recording electrode in a single fiber, the stimulus frequency would then have to be raised to return to the fiber's CF after cooling; thus we would say that the CF would increase with cooling. Our data and that of others show that exactly the opposite happens; thus, the stiffness-dominated model (e.g., Figure 6.2, top) alone is not consistent with the observed data. Now consider the effects of cooling the ear on the viscosity of the fluid surrounding the BM or TM. Viscosity is proportional to velocity—that is, it increases frictional forces and thus frictional losses on the BM or TM. Thus increased fluid viscosity due to cooling results in more of the energy in the TW being dissipated by friction into heat and thus the peak of the TW will occur closer to its source, namely the base. To return to the CF in this case, the stimulus frequency would have to be lowered; in other words, the CF drops upon cooling the ear. This is the effect we observe empirically at the level of the eighth nerve; thus, the viscosity-dominated model (Figure 6.2, bottom) is consistent with the observed data. It is likely that both membrane stiffness and fluid viscosity increase with cooling, but that the viscosity-dominated model more closely reflects the mechanism of CF-shift in the ear.

Considering next the temperature dependence of the hair cell currents that underlie the resonant behavior observed in the sacculus and in the rostromedial AP, we are faced with a dilemma. Saccular afferent fibers have been shown to be rather temperature-insensitive, with thermal Q_{10} values around 1.1 (Egert and Lewis 1995). This is in stark contrast to fibers innervating the rostral AP that exhibit clear temperature-dependencies with Q_{10}s of approximately 1.7. Temperature sensitivity of the rostral AP may be readily accounted for by the temperature sensitivity of the electrical tuning performed by its hair cells. Yet in the sacculus we are left with the question: How does this organ compensate for the temperature-dependent shift that must be occurring in its hair cells? A closer look at saccular tuning curves suggests

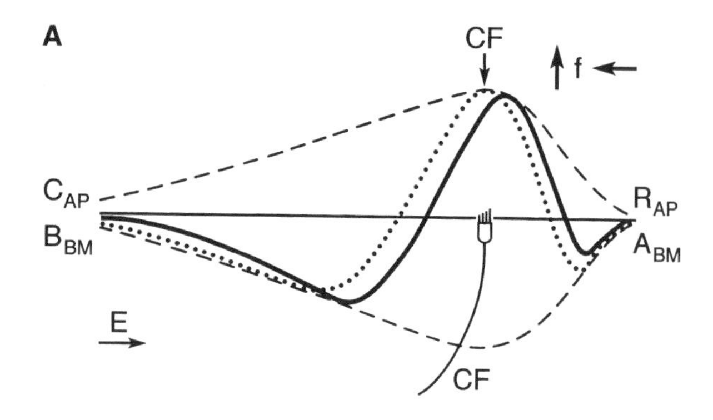

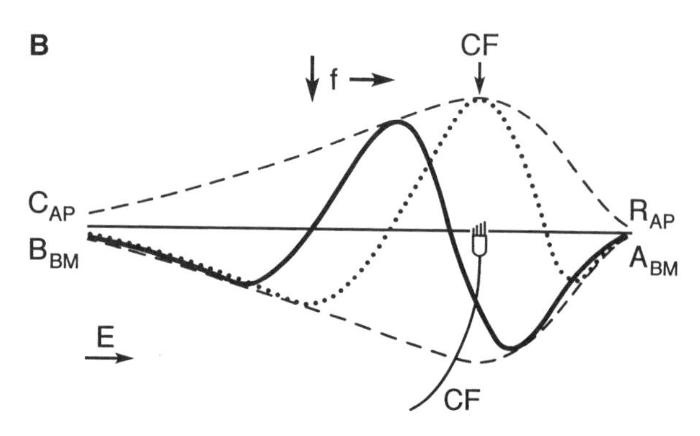

Figure 6.7. Alternative models of the mechanism underlying the temperature dependence of characteristic frequency (CF) in the vertebrate ear: (A) stiffness dominated and (B) viscosity dominated. (A) Sketch depicting how cooling the ear affects the presumed traveling wave (*TW*) on the frog tectorial membrane (*TM*) or on the mammalian basilar membrane (*BM*). C_{AP} and R_{AP} refer to the caudal and rostral poles of the amphibian papilla, respectively; B_{BM} and A_{BM} refer to the basal and apical portions of the mammalian basilar membrane. Energy (*E*) flows from left to right. Dotted curves = room temperature; heavy solid curves = cooled condition; dashed curve = TW displacement envelope; horizontal line = resting position for TM or BM; *CF* = characteristic frequency of the fiber and hair cell located at the position of the maximum TM or BM displacement at room temperature; *f* = stimulus frequency. See text for details.

that electrical tuning is likely to be only one of several filter components present in this organ (Egert and Lewis 1995; Lewis and Narins 1999). Although changing temperature may result in a shift in the electrical tuning characteristics of the hair cells, the response of the sacculus as a whole (including viscous and mechanical elements) remains insensitive to temperature.

Conclusions

The frog auditory system is unique among vertebrates in that it contains three spatially separate, anatomically distinct end organs that exhibit clear responses to airborne sound. Patches of sensory epithelial tissue in the amphibian papilla, basilar papilla, and sacculus each contain sensory hair cells that are innervated by the distal processes of the afferent eighth nerve fibers. In this chapter I reviewed the results of experiments in which the temperature sensitivity was reported for three physiological features of the anuran auditory system: (1) eighth nerve tuning properties, (2) hair cell resonant frequencies, and (3) spontaneous otoacoustic emissions.

Temperature affects nearly all biological processes. When individual components of a system exhibit unique temperature-dependent properties, it is a daunting challenge to reconstruct the behavior of the entire system from the individual components. But the task becomes insurmountable if some of the data are poorly known or missing. Clearly lacking in the case of anuran communication is a deeper understanding of the mechanics and temperature dependence of the mechanical structures of the frog inner ear. Whereas great progress in our knowledge of avian, reptilian, and mammalian inner ear mechanics has been made from direct measurements of the motion of the basilar membrane and other inner ear structures using laser Doppler vibrometry (Ruggero and Rich 1991; Lewis et al. 1997; Recio et al. 1998) and other modern techniques, similar work in the frog inner ear has not been forthcoming (but see Purgue and Narins 1997, 2000a, 2000b, for exceptions). When we learn how the AP tectorial membrane and contact membranes of the AP and BP move in response to sound and how this movement changes with temperature, we will begin to achieve a fundamental understanding of how the frog ear operates so efficiently over a wide range of ambient temperatures.

Another avenue for future research concerns the quantitative comparison of the temperature sensitivities of the temporal parameters of a frog's vocalizations with that of its auditory system. Temperature coupling (Gerhardt 1978) predicts that the temperature dependencies of the acoustic sensory and vocal motor systems should be matched, and yet this hypothesis has been tested only in insects (Pires and Hoy 1992), not as yet in vertebrates.

Acknowledgments

The author gratefully acknowledges the contributions of Michael Smotherman, Imme Stiebler, Pim van Dijk, and Jianxin Wang to this work. Useful discussions with Ted Lewis provided perspective on many of the issues discussed herein. Many thanks are also owed to Margaret Kowalczyk for assistance with the figure production. Finally I wish to express my deep gratitude to Stan Rand, whom I first met on Tuesday June 30 1981 in Gamboa, Panama. Stan inspired me to maintain a clear focus on potential auditory mechanisms underlying the acoustic behavior I was observing while wallowing in the swamp pursuing the elusive hylid or leptodactylid. This research was supported by NIH Grant no. DC00222 to PMN.

References

Blair, W. F. 1958. Mating call in the speciation of anuran amphibians. American Naturalist 92:27–51.

Brenowitz, E. A., and G. J. Rose. 1994. Behavioural plasticity mediates aggression in choruses of the Pacific treefrog. Animal Behaviour 47:633–641.

Eatock, R. A., and G. A. Manley. 1981. Auditory nerve fiber activity in the Tokay Gecko II. Temperature effect on tuning. Journal of Comparative Physiology 142:219–226.

Egert, D., and E. R. Lewis. 1995. Temperature-dependence of saccular nerve fiber response in the North American bullfrog. Hearing Research 84:72–80.

Fay, R. R., and T. J. Ream. 1992. The effects of temperature change and transient hypoxia on auditory nerve response in the goldfish (*Carassius auratus*). Hearing Research 58:9–18.

Gerhardt, H. C. 1978. Temperature coupling in the vocal communication system of the gray tree frog, *Hyla versicolor*. Science 199:992–994.

Hamill, O. P., A. Marty, E. Neher, B. Sakmann, and F. J. Sigworth. 1981. Improved patch-clamp techniques for high-resolution current recording from cells and cell-free membrane patches. Pflugers Archive 391:85–100.

Hudspeth, A. J. 1985. The cellular basis of hearing: The biophysics of hair cells. Science 230:745–752.

Hudspeth, A. J., and R. S. Lewis. 1988. Kinetic analysis of voltage- and ion-dependent conductances in saccular hair cells of the bullfrog, *Rana catesbeiana*. Journal of Physiology (London) 400:237–274.

Kemp, D. T. 1978. Stimulated acoustic emissions from the human auditory system. Journal of the Acoustical Society of America 64:1386–1391.

Köppl, C. 1995. Otoacoustic emissions as an indicator for active cochlear mechanics: A primitive property of vertebrate auditory organs. Pp. 207–216. *In* G. A. Manley, G. M. Klump, C. Köppl, H. Fastl, and H. Oeckinghaus (eds.), Advances in Hearing Research. World Scientific, Singapore.

Kössl, M., and G. S. Boyan. 1998a. Otoacoustic emissions from a nonvertebrate ear. Naturwissenschaften 85:124–127.

Kössl, M., and G. S. Boyan. 1998b. Acoustic distortion products from the ear of a grasshopper. Journal of the Acoustical Society of America 104:326–335.

Koyama, K., E. R. Lewis, E. L. Leverenz, and R. A. Baird. 1982. Acute seismic sensitivity in the bullfrog ear. Brain Research 250:168–172.

Lewis, E. R. 1978. Comparative studies of the anuran auditory papillae. Scanning Electron Microscopy 1978 (II):633–642.

Lewis, E. R. 1988. Tuning in the bullfrog ear. Biophysical Journal 53:441–447.

Lewis, E. R., R. Lyon, G. R. Long, P. M. Narins, and C. R. Steele. 1997. Diversity in Auditory Mechanics. World Scientific Publishers, Singapore.

Lewis, E. R., and P. M. Narins. 1999. The acoustic periphery of amphibians: Anatomy and physiology. Pp. 101–154. *In* R. R. Fay, and A. N. Popper (eds.), Comparative Hearing: Fish and Amphibians. Springer-Verlag, New York.

Long, G. R., P. van Dijk, and H. P. Wit. 1996. Temperature dependence of spontaneous otoacoustic emissions in the edible frog (*Rana esculenta*). Hearing Research 98:22–28.

Moffat, A. J. M., and R. R. Capranica. 1976. Effects of temperature on the response of the auditory nerve in the American toad (*Bufo americanus*). Journal of the Acoustical Society of America 60:Suppl. 1, S80.

Narins, P. M. 1995. Temperature dependence of auditory function in the frog. Pp. 198–206. *In* G. A. Manley, G. M. Klump, C. Köppl, H. Fastl, H. Oeckinghaus (eds.), Advances in Hearing Research. World Scientific Publishers, Singapore.

Narins, P. M., and R. R. Capranica. 1976. Sexual differences in the auditory system of the tree frog *Eleutherodactylus coqui*. Science 192:378–380.

Narins, P. M., and R. R. Capranica. 1978. Communicative significance of the two-note call of the treefrog *Eleutherodactylus coqui*. Journal of Comparative Physiology 127:1–9.

Narins, P. M., and E. R. Lewis. 1984. The vertebrate ear as an exquisite seismic sensor. Journal of the Acoustical Society of America 76:1384–1387.

Narins, P. M., and S. L. Smith. 1986. Clinal variation in anuran advertisement calls: Basis for acoustic isolation? Behavioral Ecology and Sociobiology 19:135–141.

Narins, P. M., and I. Wagner. 1989. Noise susceptibility and immunity of phase locking in amphibian auditory nerve fibers. Journal of the Acoustical Society of America 85:1255–1265.

Navas, C. A. 1996. The effect of temperature on the vocal activity of tropical anurans: A comparison of high and low-elevation species. Journal of Herpetology 30:488–495.

Palmer, A. R., and J. P. Wilson. 1982. Spontaneous and evoked otoacoustic emissions in the frog *Rana esculenta*. Journal of Physiology 324:66.

Pires, A., and R. R. Hoy. 1992. Temperature coupling in cricket acoustic communication II. Localization of temperature effects on song production and recognition networks in *Gryllus firmus*. Journal of Comparative Physiology 171:79–92.

Purgue, A. P., and P. M. Narins. 1997. The vibration patterns of the contact membrane: The key to the stimulation of the tectorial membrane in the bullfrog (*Rana catesbeiana*)? Abstract, 20th ARO Research Meeting 140.

Purgue, A. P., and P. M. Narins. 2000a. A model for energy flow in the inner ear of the bullfrog (*Rana catesbeiana*). Journal of Comparative Physiology 186:489–495.

Purgue, A. P., and P. M. Narins. 2000b. Mechanics of the inner ear of the bullfrog (*Rana catesbeiana*): The contact membranes and the periotic canal. Journal of Comparative Physiology 186:481–488.

Probst, R., B. L. Lonsbury-Martin, and G. K. Martin. 1991. A review of otoacoustic emissions. Journal of the Acoustical Society of America 89:2027–2067.

Recio, A., N. C. Rich, S. S. Narayan, and M. A. Ruggero. 1998. Basilar membrane responses to clicks at the base of the chinchilla cochlea. Journal of the Acoustical Society of America 103:1972–1989.

Ruggero, M. A., and N. C. Rich. 1991. Application of a commercially-manufactured Doppler shift laser velocimeter to the measurement of basilar-membrane vibrations. Hearing Research 51:215–230.

Schermuly, L., and R. Klinke. 1985. Change of characteristic frequency of pigeon primary auditory afferents with temperature. Journal of Comparative Physiology 156:209–211.

Smolders, J. W. T., and R. Klinke. 1984. Effects of temperature on the properties of primary auditory fibers of the spectacled caiman, *Caiman crocodilus* (L.). Journal of Comparative Physiology 155:19–30.

Smotherman, M. S., and P. M. Narins. 1998. The effect of temperature on electrical resonance in leopard frog saccular hair cells. Journal of Neurophysiology 79:312–321.

Smotherman, M. S., and P. M. Narins. 2000. Hair cells, hearing and hopping: Identifying the contributions of hair cells to auditory nerve response properties in the frog. Journal of Experimental Biology 203:2237–2246.

Snyder, W. F., and D. L. Jameson. 1965. Multivariate geographic variation of mating call in populations of the Pacific Tree Frog (*Hyla regilla*). Copeia 1965:129–142.

Stebbins, R. C. 1966. A Field Guide to Western Reptiles and Amphibians. Houghton Mifflin, Boston.

Stewart, M. M., and A. S. Rand. 1991. Vocalizations and the defense of retreat sites by male and female frogs, *Eleutherodactylus coqui*. Copeia 1991:1013–1024.

Stiebler, I. B., and P. M. Narins. 1990. Temperature-dependence of auditory nerve response properties in the frog. Hearing Research 46:63–82.

van Dijk, P., E. R. Lewis, and H. P. Wit. 1990. Temperature effects on auditory nerve fiber response in the American bullfrog. Hearing Research 44:231–240.

van Dijk, P., P. M. Narins, and J. Wang. 1996. Spontaneous otoacoustic emissions in seven frog species. Hearing Research 101:102–112.

van Dijk, P., and H. P. Wit. 1987. Temperature dependence of frog spontaneous otoacoustic emissions. Journal of the Acoustical Society of America 82:2147–2150.

van Dijk, P., H. P. Wit, and J. M. Segenhout. 1989. Spontaneous otoacoustic emissions in the European edible frog (*Rana esculenta*): Spectral details and temperature dependence. Hearing Research 42:273–282.

Wu, Y.-C., J. J. Art, M. B. Goodman, and R. Fettiplace. 1995. A kinetic description of the calcium-activated potassium channel and its application to electrical tuning of hair cells. Progress in Biophysics and Molecular Biology 63:131–158.

Zug, G. R. 1993. Herpetology. Academic Press, San Diego. pp 246–251.

Zweifel, R. G. 1968. Effects of temperature, body size, and hybridization on mating calls of toads, *Bufo a. americanus* and *Bufo woodhousii fowleri*. Copeia 1968:269–285.

PART THREE

Acoustic and Visual Signaling

7

H. CARL GERHARDT AND JOSHUA J. SCHWARTZ

Auditory Tuning and Frequency Preferences in Anurans

Introduction

Many studies have demonstrated a reasonably good match between auditory tuning and the band or bands of frequencies emphasized in the conspecific advertisement call, which is the main long-range signal in anurans (see Figure 7.1 and below). Such correlations are consistent with the idea, first introduced by Capranica and his colleagues, that the anuran auditory system acts as a matched filter, a concept borrowed from engineering (e.g., Karl 1989). One appeal of this hypothesis is that matched filtering can improve the detection of conspecific signals against a background of broad-band noise, and mixed-species choruses that produce broad-band noise are common in tropical and semitropical environments (e.g., Littlejohn 1977; Drewry and Rand 1983). At a finer level of analysis, the match between auditory tuning and carrier frequency is not so impressive. For example, the average frequency band to which the auditory system is most sensitive often differs by as much as 20% from the average dominant frequency in the advertisement calls of conspecific males in the same population (see Table 7.1 and below). Even small mismatches can potentially mediate directional sexual selection as emphasized by Ryan, Rand, and their colleagues (e.g., Ryan et al. 1990, 1992).

In this chapter we first document correlations between the spectral content of advertisement calls and maximum auditory sensitivity and discuss some factors that probably contribute to mismatches. We then address the following question: How well do neural estimates of maximum frequency sensitivity correlate with frequency preferences? We raise this issue because most estimates of neural sensitivity are based on audiograms derived from midbrain, multiunit activity or on tuning curves of single neurons, both of which are near-threshold measures. Natural communication and behavioral tests of frequency preferences, however, involve the processing of signals well above threshold, and the auditory system is notoriously nonlinear. Besides saturation, however, such factors as background noise, interactions between differently tuned neural channels, and temperature can affect frequency preferences in unpredictable ways, usually reducing the correlation between auditory tuning and preferences, but sometimes improving it. Finally we examine the assumption that size-dependent mating arises from frequency preferences.

Evidence for Matched Filtering: Comparative Data

In Figure 7.1 we plot estimates of average carrier frequency against estimates of auditory tuning, as estimated from audiograms (evoked potentials or multiunit recordings from the midbrain) or distributions of characteristic frequency (frequency at lowest threshold) of neurons recorded in the auditory nerve. As shown below, these two kinds of measurements yield similar estimates in the few species for which both kinds of data are available. We refer to these minimum thresholds as "best excitatory frequencies" (BEF) in the rest of this chapter. In those species whose calls have a bimodal spectrum, the two dominant frequencies and the two corresponding threshold-minima are plotted. We assume that sensitivity to relatively low frequencies is mediated by the amphibian papilla, one of the two inner ear organs of anurans, whereas sensitivity to relatively high frequencies is mediated by the basilar papilla, the other inner ear organ (review: Zakon and Wilczynski 1988). Neurons receiving inputs from both papillae contribute somewhat to the responses of some neurons recorded in the midbrain (Fuzessery 1988), but contamination is unlikely to distort midbrain audiograms (based on stimulation with a single frequency at a time) in the regions to which the two organs are maximally sensitive because these regions are usually separated by one to two octaves.

For many species, the data are based on threshold estimates from a sample of both males and females; in other species, which we identify in Figure 7.1, separate estimates are available for each sex. In three species (*Eleutherodactylus coqui, Pseudacris crucifer,* and *Acris crepitans*), good evidence exists for sexual differences in tuning, probably influenced by size differences (Narins and Capranica 1976; Wilczynski et al. 1984; Keddy-Hector et al. 1992, respectively; additional discussion below). Sample sizes, and the extent to which estimates for thresholds in individuals were based on data

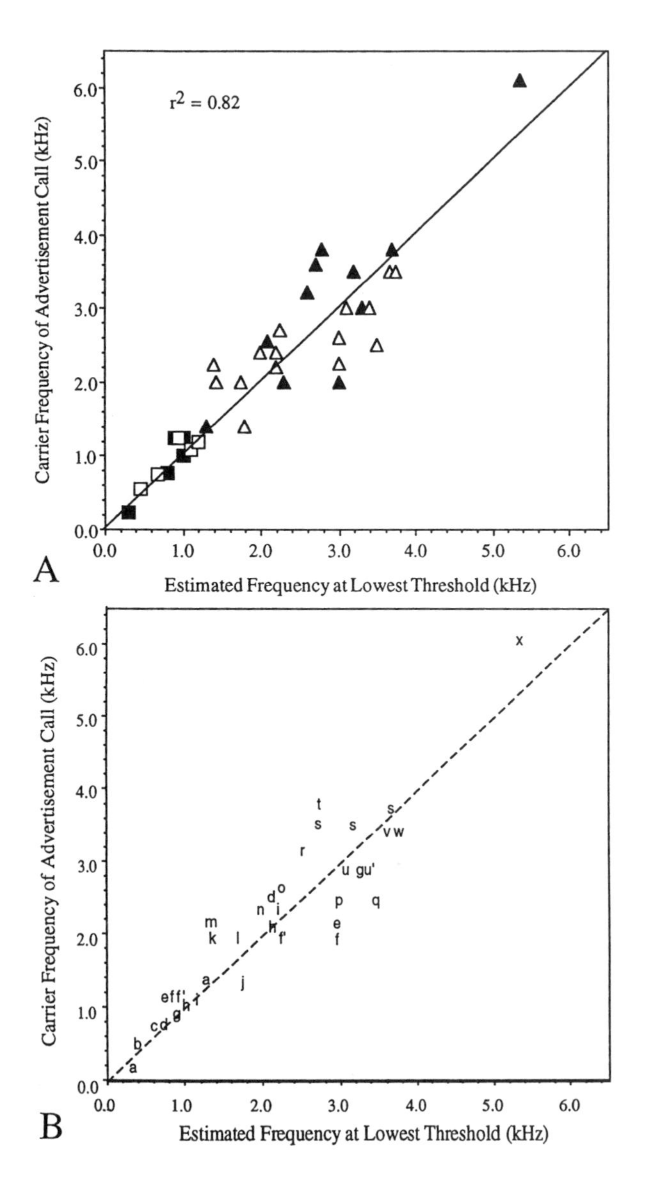

Figure 7.1. (A) Scatter diagram of spectra peaks in the advertisement calls of 24 species of anurans (plus three populations of one species) against estimates of minimum threshold (BEF). Solid symbols show data from studies in which closed-system stimulation was used, and open symbols show data from studies in which free-field stimulation was used. Squares indicate low-frequency sensitivity attributed to the amphibian papilla, and triangles indicate high-frequency sensitivity attributed to the basilar papilla. The line shows where all points would lie if the correlation were perfect. (B) Key to the data points plotted in A; *SU* = single-unit data (auditory nerve); *MU* = multi-unit spike data (torus semicircularis); *EP* = evoked potentials (torus semicircularis).
a - *Rana catesbeiana,* SU, Frishkopf et al. 1968; b - *Bombina variegata,* MU, Mohneke and Schneider 1979; c - *Limnodynastes dorsalis,* EP, Loftus-Hills and Johnstone 1970; d - *Physalaemus pustulosus,* MU, Ryan et al. 1990; e - *Hyla arborea,* MU, Hubl and Schneider 1979; f - Male *Eleutherodactylus coqui,* SU, Narins and Capranica 1980; f′ - Female *Eleutherodactylus coqui,* SU, Narins and Capranica 1980; g - *Hyla cinerea,* SU, Capranica and Moffat 1983; h - *Hyla versicolor,* MU, Schwartz and Gerhardt, unpublished; i - *Hyla chrysoscelis,* EP, Hillery 1984; j - *Alytes obstetricans,* MU, Mohneke and Schneider 1979; k - *Limnodynastes tasmaniensis,* EP, Loftus-Hills and Johnstone 1970; l - *Litoria verreauxii,* EP, Loftus-Hills and Johnstone 1970; m - *Rana ridibunda,* MU, Hubl and Schneider 1979; n - *Atelopus chiriquiensis,* MU, Jaslow and Lombard 1996; o - *Litoria ewingii,* EP, Loftus-Hills and Johnstone 1970; p - *Hyla regilla,* MU, Jaslow and Lombard 1996; q - *Hyla savignyi,* MU, Hubl and Schneider 1979; r - *Hyla ebraccata,* MU, Wilczynski et al. 1993; s - *Acris crepitans,* SU + MU, Ryan et al. 1992; t - *Hyla phlebodes,* MU, Wilczynski et al. 1993; u - Male *Pseudacris crucifer,* MU, Diekamp and Gerhardt 1992; u′ - Female *Pseudacris crucifer,* MU, Schwartz and Gerhardt 1998; v - *Ranidella parinsignifera,* EP, Loftus-Hills and Johnstone 1970; w - *Ranidella signifera,* EP, Loftus-Hills and Johnstone 1970; x - *Hyla microcephala,* MU, Wilczynski et al. 1993.

Table 7.1 Relatively large mismatches between auditory tuning and average carrier frequency (CF)

Species	Stimulation	Auditory Organ	BEF	
			< av. CF (%)	> av. CF (%)
Acris crepitans (Austin)[a]	CS	BP	25	—
Alytes obstetricans[b]	FF	BP	—	30
Eleutherodactylus coqui (male)[c]	CS	BP	—	50
Hyla arborea[d]	FF	AP	24	—
Hyla arborea[d]	FF	BP	—	33
Hyla phlebodes[e]	CS	BP	27	—
Hyla savignyi[d]	FF	BP	—	40
Limnodynastes tasmaniensis[f]	FF	BP	28	—
Rana ridibunda[d]	FF	BP	38	—

Sources: Data from [a]Ryan et al. 1992; [b]Mohneke and Schneider 1979; [c]Narins and Capranica 1980; [d]Hubl and Schneider 1979; [e]Wilczynski et al. 1993; [f]Loftus-Hills and Johnstone 1970.

BEF = best excitatory frequencies; CS = closed stimulation system; FF = free-field stimulation; BP = basilar papilla; AP = amphibian papilla.

from multiple recording sites, differ considerably among these studies, as does the way in which threshold estimates are presented. We usually used mean or median values of BEF presented in the text or tables, but sometimes we had to estimate these values from audiograms. As discussed below, tuning is often so shallow that deciding on a minimum is somewhat arbitrary; this procedure thus masks the fact that the range of near-maximum sensitivity usually covers a considerable range of frequency. Similarly by taking mean values for carrier frequencies or estimating the mean as the midpoint in the range of variation (in studies where mean values were not reported), the analysis ignores that the calls of many of the species considered have a broad-band spectrum, with substantial acoustic energy present over two or more octaves. The extent of some of the mismatches we report are likely to be reduced as better estimates of mean carrier frequency become available.

Some studies used free-field acoustic stimulation, whereas others used a closed-stimulation system sealed onto one of the tympanic membranes (see symbols in Figure 7.1). The latter arrangement might be expected to exaggerate low-frequency sensitivity in comparison with free-field stimulation (Pinder and Palmer 1983). Free-field stimulation adds another complication, however, because the anuran system acts as a pressure-difference system in which there are multiple inputs of sound, including internal pathways, to the tympanic membrane and probably directly to the inner ear organs (Hetherington 1992). Because the frequency-responses of the input surfaces and pathways are likely to be influenced by the sound direction, some variation among studies would be expected because of different placements of speakers relative to the orientation of the animal. Finally only two studies formally investigated the effect of temperature on auditory tuning as estimated by midbrain audiograms (Hubl and Schneider 1979; Mohneke and Schneider 1979). As discussed in a later section, if the temperature is significantly different from that experienced by the frogs during breeding periods, then relatively large mismatches between estimates of frequency sensitivity and emphasized frequencies in advertisement calls are possible (see also Narins, this volume). In the studies just cited, however, the main effect of increasing temperature was usually a decrease in threshold rather than a major shift in the BEF of the audiogram. In choosing an estimate of BEF from these studies for Figure 7.1, we used either values that were common to two or three of the temperatures used by the authors or the temperature at which most breeding occurs.

On the one hand, despite all of the experimental variation and sources of error, the correlation between BEF and call frequency in this sample of species (and populations) is remarkably good, with a coefficient of determination of 0.82. At this level of analysis, the matched filter hypothesis appears to be well supported, and we might expect auditory tuning to conspecific signals in many species to provide some measure of improved detection in mixed species choruses as well as mediating weak stabilizing selection on carrier frequency (see below). On the other hand, discrepancies between average carrier frequency and BEF of 15% or more were found for half of the 36 estimates in Figure 7.1; Table 7.1 provides details for mismatches of about 25% or more. Two-thirds (12 of 19) of the (15%) mismatches involved tuning to frequencies lower than the estimated average carrier frequency, but there was no trend for this kind of mismatch to be correlated with the type of stimulation: 6 of the estimates were obtained from closed-system stimulation and the other 6 from free-field stimulation. Substantial mismatches (in either direction) involved 4 of 10 estimates of low-frequency (amphibian papilla) tuning and 14 of 26 estimates of high-frequency (basilar papilla) tuning.

As previously suggested, however, many of these estimates could probably be improved by one or more of the following actions: increasing the sample size of animals, recording from multiple sites within each animal, and adjusting the temperature of preparations to the middle of the normal range of breeding temperatures. It will also be important to conduct systematic studies of the effects of the mode of stimulation and the direction of free-field stimulation using the same species, or better yet, the same animals. Separate analyses for the two sexes to check for sexual differences in tuning should also be conducted whenever significant sexual size dimorphism occurs. Size-dependent tuning and frequency preferences have been widely documented within females of the same species (Wilczynski et al. 1984; Keddy-Hector et al. 1992; Ryan et al. 1992; Jennions et al. 1995; Márquez and Bosch 1997). No sexual difference in tuning was reported in *Physalaemus pustulosus* despite the pronounced sexual size dimorphism in this species, but the sample of audiograms was only five, and the number of animals of each sex was unspecified (Ryan et al. 1990).

Estimates of Tuning and Frequency Preferences

Túngara Frogs

Several studies of anurans have correlated estimates of auditory tuning with the frequency preferences of females from the same populations (Gerhardt 1974, 1987; Ryan et al. 1990, 1993; Diekamp and Gerhardt 1992; Schwartz and Gerhardt 1998). In the túngara frog, *P. pustulosus*, audiograms based on averaging multiunit thresholds from five individuals suggested that these frogs would be more sensitive to chucks of lower-than-average frequency (about 2.13 kHz) than to chucks of average (2.55 kHz) and higher-than-average frequencies (Ryan et al. 1990). In behavioral tests, 63% of the females chose a synthetic call with a chuck-frequency that was lower than average (2.1 kHz) to an alternative with a chuck-frequency that was higher than average (3.0 kHz) (Wilczynski et al. 1995). This trend is consistent with the neural data even though it is not statistically significant.

Cricket Frogs

Estimates of auditory tuning in three populations of Blanchard's cricket frog (*Acris crepitans blanchardii*) were based on estimates of the characteristic frequency of primary auditory neurons and multiunit recordings (Ryan et al. 1992).

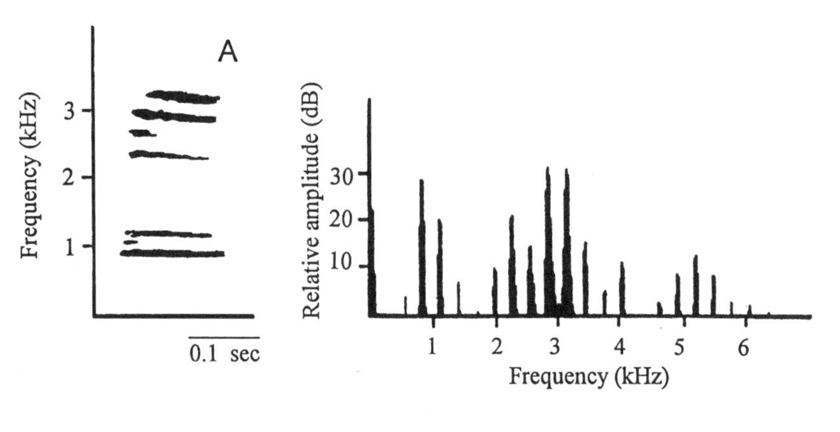

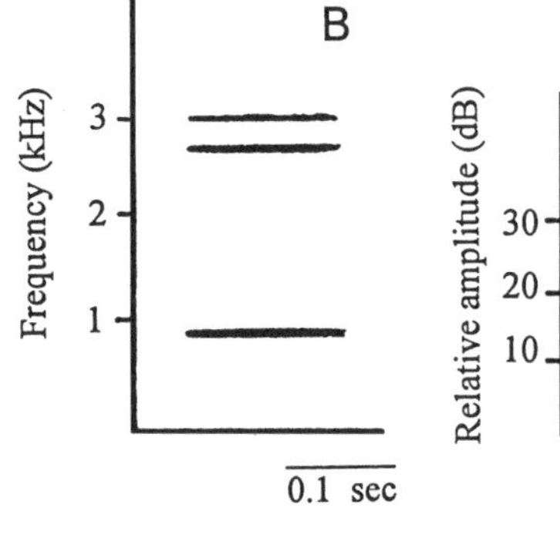

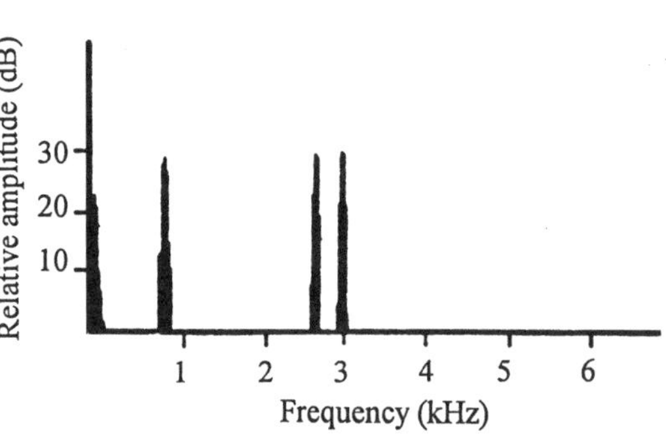

Figure 7.2. Sonograms (left) and power spectra (right) of (A) a typical advertisement call of a male green treefrog (*Hyla cinerea*), and (B) a synthetic advertisement call that was just as attractive as the natural call in a two-stimulus playback experiment (from Gerhardt 1983).

The mean BEF was always lower than the average carrier frequencies of advertisement calls in the same populations, although only barely so in a population from Bastrop, Texas. Females were tested with synthetic calls that differed only in carrier frequency and represented calls of about average frequency for the population and calls that differed from the average by more than one standard deviation. The results of three of six tests were qualitatively consistent with the neural data: statistically significant proportions of females preferred calls of average frequency to calls of higher frequency and preferred calls of lower-than-average frequency to calls of average frequency. Females from the Austin population surprisingly did not prefer the low-frequency call (3.2 kHz) to the call with a frequency of 3.5 kHz, which was close to the population average (3.6 kHz), although the mean BEF was 2.7 kHz. Females in the Bastrop population did not prefer calls of about average frequency (3.8 kHz) to calls of higher frequency (4.1 kHz), and they preferred the low-frequency call (3.5 kHz) to the call of average frequency (3.8 kHz) even though the mean BEF (3.72 kHz) better matched the latter frequency. By contrast, females from the Indiana population showed preferences in both tests that were predicted by the neural data; moreover, the preferences for the low-frequency call in one test was robust in the face of a 6-dB decrease in the sound pressure level (SPL) of this stimulus relative to the alternative.

Green Treefrogs

Males of the green treefrog (*Hyla cinerea*) produce a noisy advertisement call with the frequency components around 0.9 and 3 kHz having the greatest relative amplitude (Figure 7.2A). A synthetic call with just three components (0.9, 2.7, and 3.0 kHz) was as attractive to females as typical natural calls having many more components (Figure 7.2B; Gerhardt 1974). In Figure 7.3A, we show the distribution of characteristic frequency in a sample of auditory neurons recorded in the eighth nerve of the green treefrog by Capranica and Moffat (1983). The population of high-frequency neurons is made up of fibers innervating the basilar papilla, and the populations tuned to lower frequencies are representative of neurons innervating the amphibian papilla. In advanced anurans, such as treefrogs, the midfrequency population derives its sensitivity from hair cells found in a caudal extension of the amphibian papilla, whereas the low-frequency population, found in all anurans, derives its sensitivity from hair cells in the rostral part of the papilla (see Narins, this volume). Notice that the most sensitive neurons innervating the amphibian papilla have much lower thresholds than the most sensitive neurons innervating the basilar papilla (Figure 7.3B). The wide range of thresholds is not representative of any single individual, however, because the data were pooled from multiple recordings from a sample of frogs. What cannot be observed from such a figure is the nonlinear interaction occurring within the amphibian papilla: the response of a low-frequency neuron at its characteristic frequency can be suppressed by the addition of a

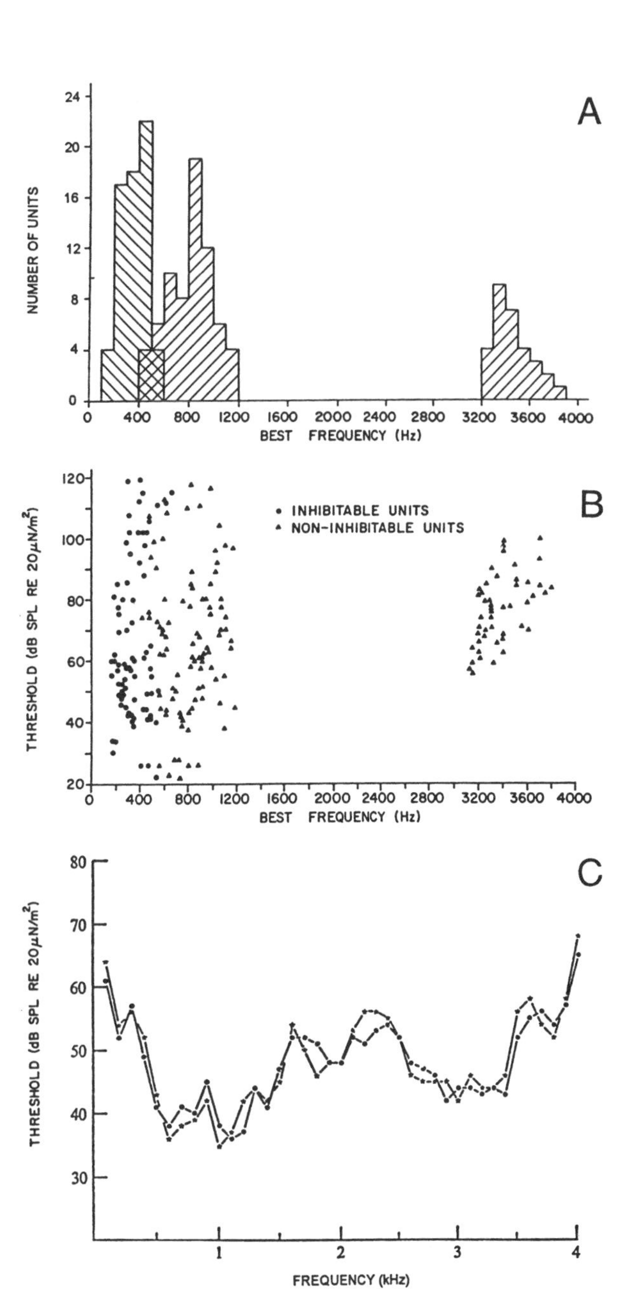

Figure 7.3. (A) Histogram showing the distribution of characteristic frequency (CF) in a sample of primary auditory neurons recorded from the green treefrog, *Hyla cinerea*. (B) Raw data from which the histogram of A was generated. (C) Audiogram based on evoked potentials recorded from the midbrain (torus semicircularis) of the green treefrog. A and B modified from Capranica and Moffat (1983); C modified from Lombard and Straughan (1974).

second tone of higher frequency (Capranica and Moffat 1983; review in Zakon and Wilczynski 1988).

As discussed above, most estimates of auditory sensitivity are not based on data from single auditory neurons but rather from multiunit recordings or evoked potentials. Such recordings are made using relatively large, low-impedance electrodes that register the spiking activity from many neurons. Typically the electrode is placed in the largest and most complex of the auditory nuclei in the ascending pathway, the torus semicircularis, which is considered a homolog of the inferior colliculus of mammals. As shown in Figure 7.3C, an audiogram derived from evoked potentials is consistent with the single-unit data (Figure 7.3A): lowest thresholds occur in the three frequency bands corresponding to the three populations of primary auditory neurons as defined by characteristic frequency. Moreover the relative sensitivity of the three populations is also correlated with the shape of the midbrain audiogram (Lombard and Straughan 1974).

How well do these estimates of frequency sensitivity predict the frequency preferences of females? At normal breeding temperatures of about 22–26°C, females from eastern Georgia tested at a playback level of 75 dB SPL (sound pressure level, re. 20 μPa) preferred a standard synthetic call with a low-frequency peak of 0.9 kHz, which was close to the mean in the population, to alternatives of 0.7 kHz and 1.1 kHz. Females also preferred a standard synthetic call with a high-frequency peak of 3.0 kHz (close to the mean in the population) to alternatives of 2.1 kHz and 3.6 kHz (Gerhardt 1987). Under these conditions then, female preferences are well correlated with estimates of the frequency sensitivity of the population of midfrequency neurons found in the amphibian papilla and that of the high-frequency neurons in the basilar papilla. These patterns of preference in green treefrogs are consistent with the matched-filter hypothesis, and thus should help females detect conspecific males in mixed-species choruses. These patterns also represent a possible basis for stabilizing selection that should not only result in the rejection of heterospecific calls with different emphasized frequency bands but also some conspecific calls at either end of the range of distribution (Gerhardt 1974, 1987, 1994). As we emphasize below, however, these two generalizations do not hold under other conditions.

In another respect, neural estimates of frequency selectivity fail to predict the preferences of females in the context of intraspecific communication if we assume that excitation of any primary auditory neuron contributes to the attractiveness of a signal as much as any other primary neuron. For example, a large population of sensitive amphibian-papilla neurons is tuned to frequencies below the low-frequency spectral peak in the advertisement call (Figures 7.2A and 7.3A, B). Yet at normal breeding temperatures females rejected synthetic calls with a low-frequency spectral peak of 0.5 or 0.6 kHz, which falls into a region to which many of these neurons are sensitive, in favor of the standard call with a low-frequency peak of 0.9 kHz. The estimated numbers of primary neurons tuned to about 0.7–1.2 kHz are about the same as those in the low-frequency population, and the lowest thresholds of the most sensitive neurons in both populations are also similar.

A stimulus containing components of 0.3 Hz and 0.9 kHz, which match the tuning of both populations of neurons, was neither more nor less attractive than a standard call lacking the 0.3 kHz-component (Gerhardt, unpublished data). The excitation of neurons of the low-frequency population was, however, likely to be negated somewhat by the tone-on-tone suppression (mediated by neurons tuned to the 0.9 kHz component) mentioned above. In general the biological function of auditory sensitivity to low-frequency regions of the spectrum that do not match acoustic energy in communication signals remains a mystery. The usual speculation is that such sensitivity serves to detect predators (e.g., Capranica and Moffat 1983).

Frequency Preferences and Stimulus Intensity: Examples from Studies of Green Treefrogs and Spring Peepers

In Figure 7.4B we show an example from *H. cinerea* of the saturation of an auditory nerve fiber whose tuning curve is shown in Figure 7.4A (Capranica 1992). At 65 dB SPL, about 10 dB above the absolute threshold, the maximum firing rate of this neuron is close to the characteristic frequency of 1020 Hz. At stimulus levels of 75 dB and 85 dB SPL, however, the firing rate of the neuron is uniform over a wide range of frequencies. The wide range of absolute thresholds (called range fractionation) of the population of auditory neurons serves to increase the linear (dynamic) operating range of the auditory system as a whole and hence to reduce its saturation. But clearly at some very high stimulus levels, most neurons would be expected to be saturated, and stimulus frequency could then only be encoded reliably by tonotopy or temporal coding schemes such as the volley principle. Tonotopy is the orderly spatial arrangement of hair cells or neurons according to their frequency sensitivity, and at a gross level, the different tuning of the two auditory organs can be considered an example. Tonotopy has also been documented in the amphibian papilla of advanced anurans (including the Hylidae), but its expression in higher centers such as the torus semicircularis is crude at best (review in Fuzessery 1988). No tonotopy has been described within basilar papilla of anurans. Temporal synchronization codes, in which the stimulus frequency is

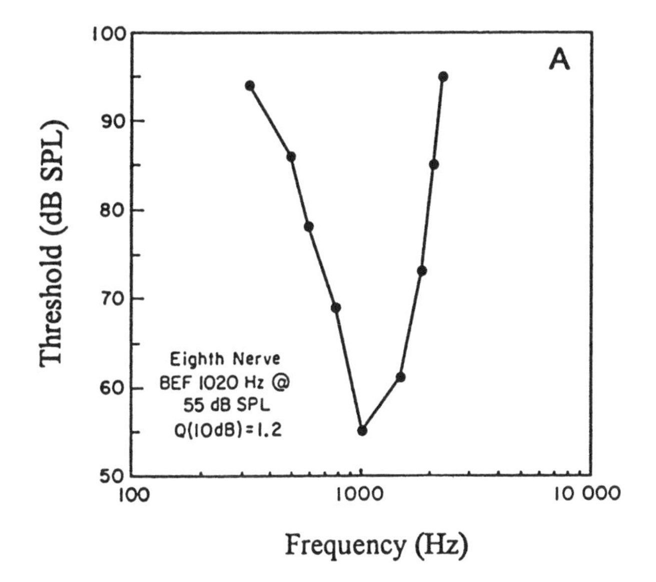

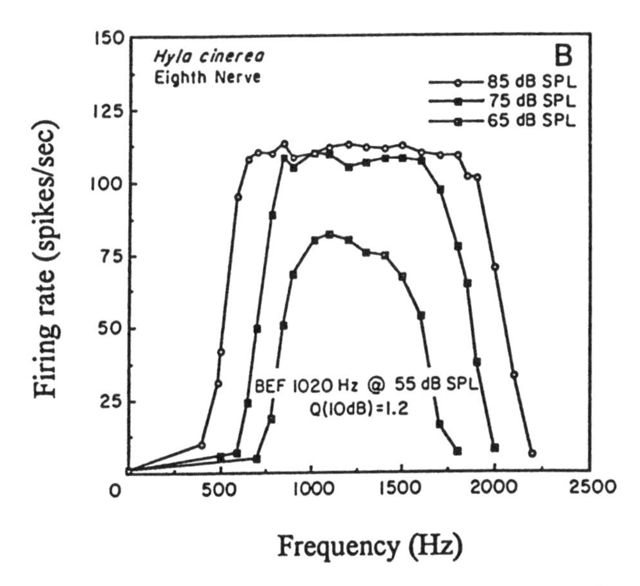

Figure 7.4. (A) Tuning curve of a primary auditory neuron of green treefrog, *Hyla cinerea*, with a characteristic frequency (CF) of about 1.0 kHz. (B) Iso-intensity functions for the same primary neuron. When stimulus intensity is held constant at a low sound-pressure level and frequency is varied, the highest firing rate is elicited at the neuron's CF. At higher sound-pressure levels the same (saturated) firing rate is observed over a frequency range of 1–2 octaves. Modified from Capranica (1992).

represented by the timing of action potentials, could potentially encode frequencies as high as 900 Hz in anurans (Narins and Hillery 1983).

The question of whether threshold estimates of tuning reliably predict frequency preferences at sound pressure levels well above threshold thus can only be tested by behavioral experiments at biologically realistic levels. Moreover this question requires systematic variation not only in frequency but also in absolute and relative intensity. Such studies have been conducted with green treefrogs and spring peepers, and in the latter species, correlations have been sought between auditory tuning and frequency preferences within the same individual females (Schwartz and Gerhardt 1998).

Behavioral studies of the green treefrog demonstrated changes in preferences with variation in the SPL to which alternative stimuli with different frequencies are equalized (Gerhardt 1987). As expected from the predicted effects of saturation, female green treefrogs became somewhat less selective for differences in frequency in the low-frequency (amphibian papilla) range as the SPL was increased from 65 to 85 dB, although females rejected higher-than-average frequencies at all playback levels (Figure 7.5). Inspection of Figure 7.3B indicates that at a playback level of 85 dB, SPL is 20 dB or more above the threshold of a large proportion of primary neurons innervating the amphibian papilla (Figure 7.3B). As shown in Figure 7.4B some neurons are already saturated at 20 dB above the threshold at the characteristic frequency. Although the distribution of thresholds of primary fibers tuned to the low-frequency spectral peak shows an enormous range of values, including neurons with estimated thresholds in excess of 110 dB SPL, we have already cautioned that this is a sample that was pooled from different animals and therefore exaggerates the dynamic range of any individual. Moreover we suspect that at least some of the thresholds higher than 100 dB reflect a deteriorating preparation rather than a truly insensitive neuron.

Contrary to the prediction that saturation would decrease frequency resolution, female green treefrogs became more selective for differences in frequency in the high-frequency (basilar papilla) range as SPL was increased from 65 to 85 dB (Figure 7.6). Estimates of the thresholds of primary auditory fibers (Figure 7.3B) innervating the basilar papilla are not only higher than those of many fibers innervating the amphibian papilla, but are also somewhat higher in comparison with the basilar papilla sensitivity of other species (compare with estimates for the spring peeper below and in Wilczynski et al. 1984). Thus one possible explanation for the increased selectivity at 85 dB SPL is that higher signal levels are needed to excite a sufficient number of these neurons to discriminate relatively small differences in stimulus frequency. In other words frequency discrimination is probably best at signal levels that fall within the linear operating ranges of the largest proportion of auditory neurons innervating a given inner ear organ.

In the spring peeper, *Pseudacris crucifer,* males produce nearly sinusoidal calls lasting, on average, about 165 ms; the mean frequency in populations in central Missouri is about 3 kHz (Doherty and Gerhardt 1984; Schwartz and Gerhardt

1998). In Figure 7.7A we show an audiogram based on data from 66 females. The frequency range of conspecific calls matches the maximum sensitivity of the basilar papilla, within which there is no evidence for tonotopy (e.g., Wilczynski et al. 1984). Notice that the lowest average thresholds are around 50 dB SPL over the range of about 3.2–3.6 kHz (median, 3.4 kHz). Moreover the bandwidth at which the average threshold was as low as 60 dB SPL was approximately 1.8 kHz and fully encompasses the range of variation in carrier frequency in the population. The Q_{10dB} value, a dimensionless measure of the sharpness of tuning, is about 1.91. (Q_{10dB} is calculated by dividing the BEF by the bandwidth at 10 dB above threshold.)

In Figure 7.7B we show a series of above-threshold estimates of frequency selectivity in the form of isointensity plots; these are the analogs of the iso-intensity spike-rate plots of Figure 7.4B. Here we measured the average size of the multiunit spike activity in the midbrain as stimulus frequency was varied and stimulus SPL was held constant. These plots roughly mirror the audiogram at levels below 85 dB SPL, and although the plot is much less peaked than at lower SPLs, the maximum activity at 75 dB SPL was elicited at about 3.3 kHz, which is very close to the median BEF. Notice however that at 85 dB SPL the iso-intensity function is very flat. If frequency preferences depend on differences in maximum activity elicited by many auditory neurons, and midbrain recordings accurately estimate such activity, these data predict that females should show few, if any, preferences between stimuli falling within the range of about 2.2 and 3.6 kHz at a playback level of 85 dB SPL.

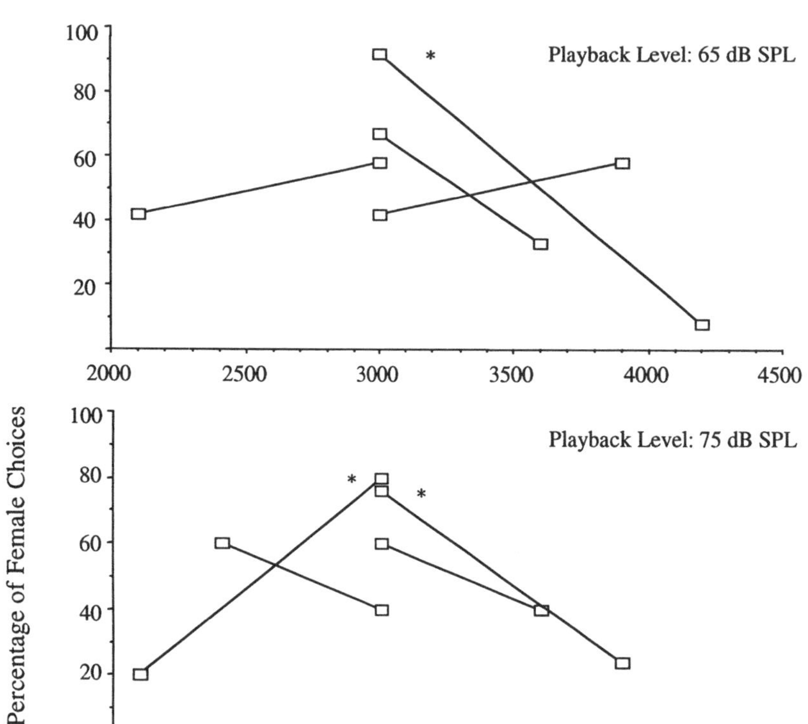

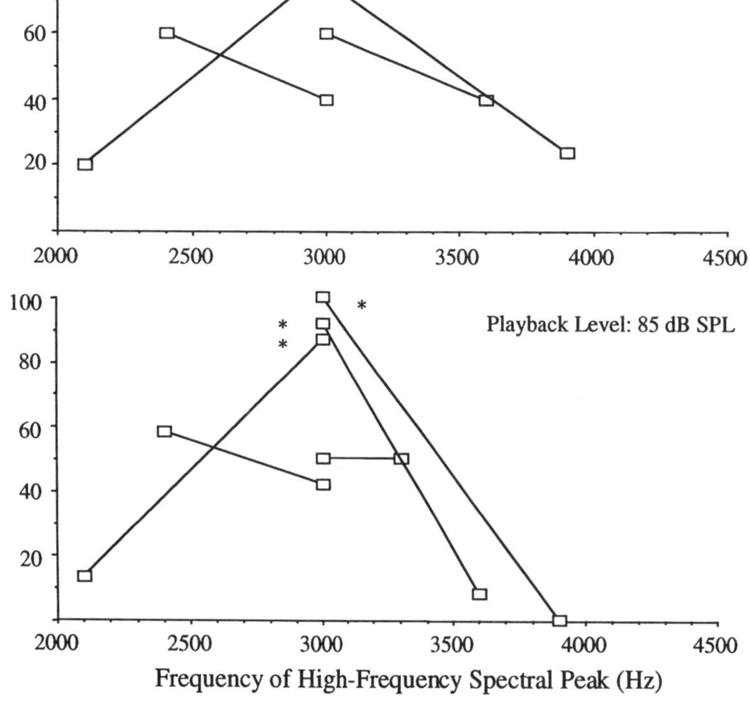

Figure 7.5. Effects of playback level on the behavioral selectivity of female green treefrogs for pairs of synthetic calls that differ in frequency in the high-frequency (basilar papilla) range. Each line connects pairs of points showing the percentages of females responding to two alternative calls with a different frequency. The symbol * indicates statistically significant ($p < 0.05$, two-tailed binomial tests) preferences for the standard high-frequency (3.0 kHz) peak.

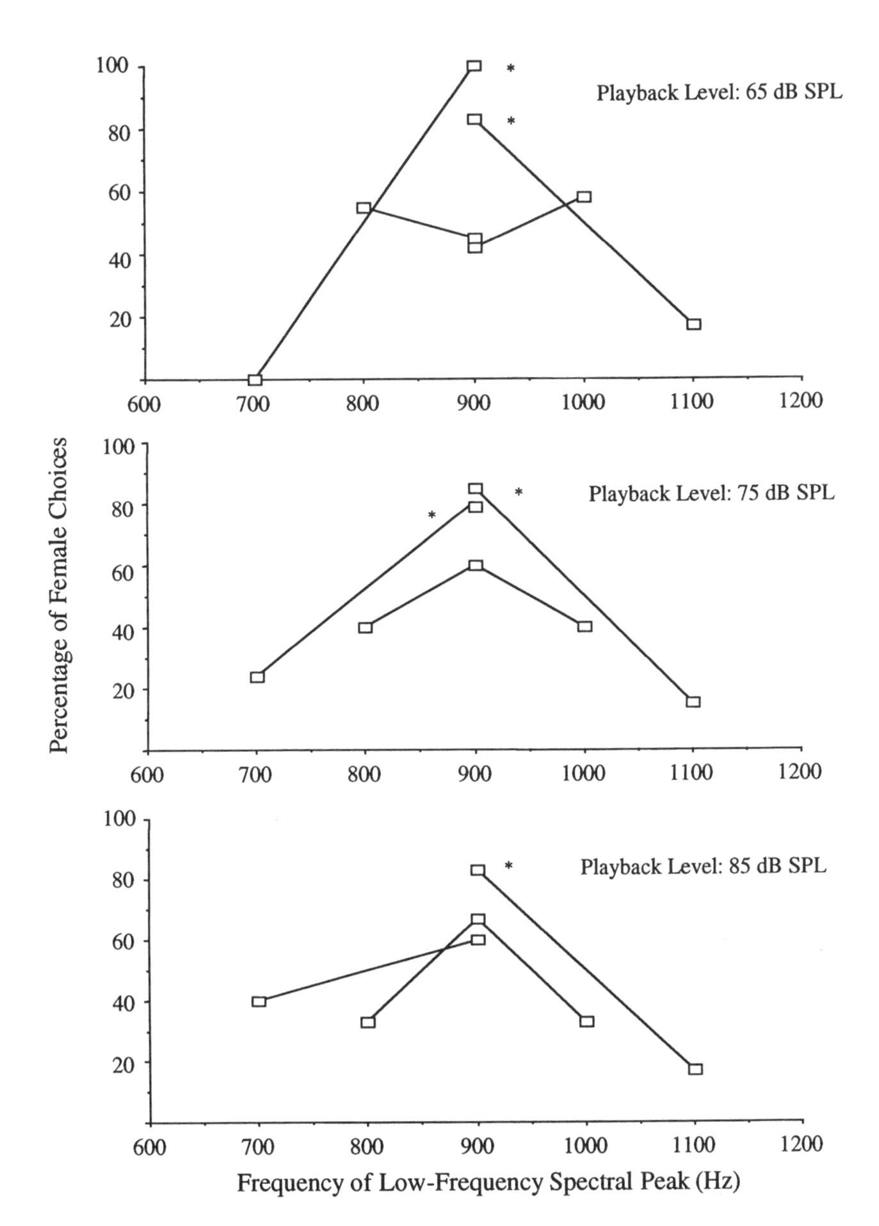

Figure 7.6. Effects of playback level on the behavioral selectivity of female green treefrogs for pairs of synthetic calls that differ in frequency in the low-frequency (amphibian papilla) range. The symbol * indicates statistically significant ($p < 0.05$, two-tailed binomial tests) preferences for the standard low-frequency (0.9 kHz) peak.

One possibility is that the flattening of the iso-intensity functions on the low-frequency side of the BEF is caused in part by stimulating some of the more sensitive neurons innervating the amphibian papilla. However the flattening of these curves is pronounced on both sides of the BEF. Moreover inspection of scatter plots of characteristic frequency versus threshold and typical tuning curves in *P. crucifer* from New York shows that the most sensitive midfrequency (amphibian papilla) fibers have characteristic frequencies below about 0.9 kHz and sharp cutoffs on the high-frequency side (Wilczynski et al. 1984). It is thus unlikely that these fibers would be excited by stimulation with frequencies of 1.8 kHz and higher, even at 85 dB SPL.

Female spring peepers from central Missouri were tested at 75 dB SPL with pairs of synthetic calls that differed only in frequency (Doherty and Gerhardt 1984). Females chose a standard call with a frequency just below the mean frequency in the calls of males in the population (2.88 kHz) over a lower-frequency alternative with a frequency falling just within the range of variation (2.60 kHz); they also chose the standard call over a high-frequency alternative (4.0 kHz) with a frequency above the population range of variation. Females did not, however, prefer the standard call to any higher-frequency alternative within the range of variation. Moreover when the SPL of the preferred ("average") stimulus was reduced by 6 dB, females failed to show preferences unless alternatives had frequencies well beyond the range of variation (1.8 and 4.6 kHz, respectively). Consistent with the trend for the iso-intensity plots to flatten with increasing SPL, more recent experiments conducted at a playback level of 85 dB SPL found no preferences between any alternatives with frequencies that fell within the range of variation

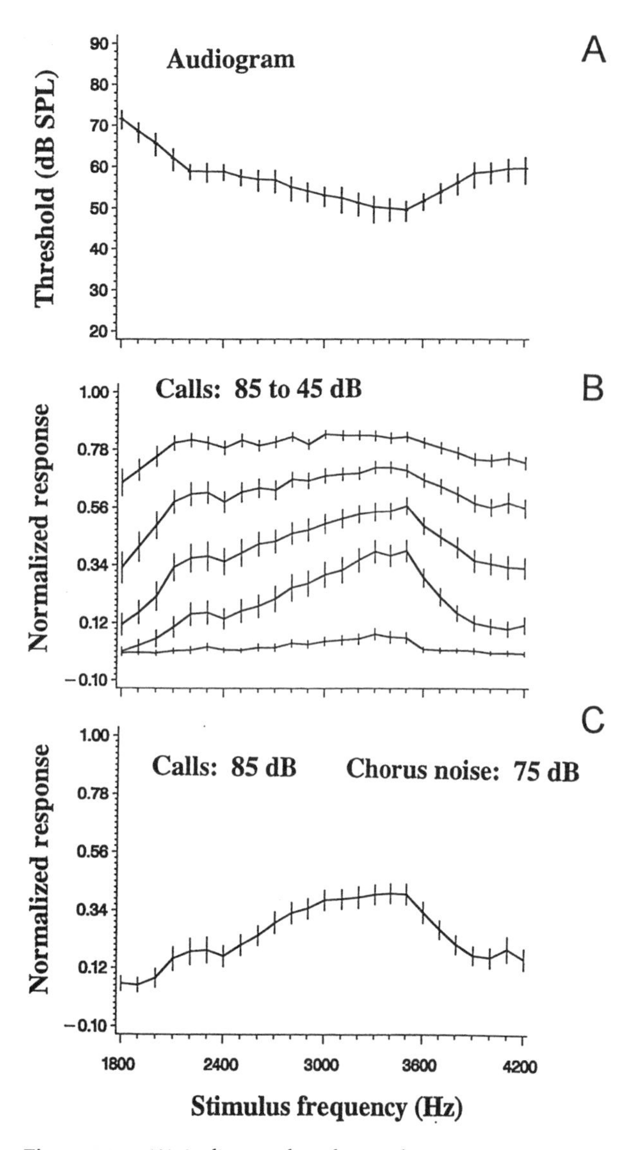

Figure 7.7. (A) Audiogram based on multiunit activity recorded from the torus semicircularis of 66 females of the spring peeper, *Pseudacris crucifer.* (B) Iso-intensity functions (see text) recorded from 65 females (top to bottom: 85, 75, 65, 55, and 45 dB SPL). (C) Iso-intensity function (85 dB SPL) recorded from 47 females that were simultaneously exposed to simulated chorus noise at 75 dB. All data are presented as means ± 2 SE.

(Schwartz and Gerhardt 1998). Forester and Czarnowsky (1985) reported that female spring peepers from Maryland, tested at 90 dB SPL, preferred calls with a frequency just below the mean value in the population to an alternative that was more than a standard deviation above the mean. Inspection of sonograms (Forester and Czarnowsky 1985) and oscillograms (Forester and Harrison 1987) of the natural calls used in these tests, however, indicates that the alternatives also differed significantly in their amplitude-time envelopes. These temporal differences could have mediated the choices of females either directly or in conjunction with the differences in frequency.

The results of Schwartz and Gerhardt (1998) just discussed were based on population-level analyses: median values of lowest thresholds in audiograms or maximum activity in iso-intensity plots were compared with the observed proportions of females that chose synthetic calls of one frequency over alternatives of another frequency. For this reason Schwartz and Gerhardt (1998) analyzed neural and behavioral data from the same individuals to determine whether the lack of selectivity at the population level could be masking significant between-female variation. For small differences in frequency within the conspecific range of variation, neither audiograms nor iso-intensity plots reliably predicted the preferences of the same individuals tested at 85 dB SPL. Although the neural estimates might have been improved by recording from multiple sites in the torus, recordings of single basilar-papilla neurons show very little variation in BEF. Similarly, Diekamp and Gerhardt (1992) found that electrode position had little effect on audiograms based on multiunit recordings from the midbrain. Repeated testing of females indicated that many individuals were inconsistent in their choices of synthetic calls that differed in frequency. Estimates of frequency preferences could perhaps be improved by testing females multiple times. Nevertheless the presently available neural and behavioral data both suggest that tuning in *P. crucifer* is broad and weak.

In summary audiograms and iso-intensity plots have been only partially successful in predicting frequency preferences and the effects of stimulus intensity in population-level studies. Future studies are likely to be more successful if better estimates of the selectivity of the entire central auditory system (e.g., by use of multiple electrodes) can be correlated with behavioral preferences in the same individuals. The chances for success will also be increased by picking species that show reliable preferences based on relatively small differences in frequency that fall within the range of variation in the population.

Chorus Noise and Frequency Preferences in Spring Peepers

Whereas midbrain iso-intensity plots did not reliably predict preferences when frequency differences were relatively small, the usefulness of such plots when differences in frequency were more substantial has been supported by a counter-intuitive result. Because spring peepers, like many other anuran species, communicate in large choruses of conspecifics, background noise levels as well as signal levels are well above threshold. Moreover the conspecific chorus is potentially a far more serious source of noise than that generated by other species, because its spectrum will more broadly overlap that of any particular conspecific signal (Gerhardt and Klump 1988). Thus we might predict that by adding simulated chorus background noise females would not only be less able to detect individual males but would also be less capable of discriminating between signals that differ in carrier frequency. In fact the single discrimination task that females accomplished at a playback level of 85 dB SPL—a preference for 3.5 kHz over an alternative of 2.6 kHz—occurred only in the presence of chorus background noise. In Figure 7.7C we show the mean iso-intensity function at 85 dB SPL recorded in the presence of simulated chorus noise played back at 75 dB SPL (range, 75–86 dB in natural choruses). Clearly this plot shows significantly higher activity at 3.5 kHz than at 2.6 kHz. Correlations between iso-intensity plots (85 dB SPL) in the presence of noise and the preferences of individuals tested at the same level with a chorus-noise background were also highly significant (Schwartz and Gerhardt 1998).

Speculation about the mechanisms underlying the improved frequency discrimination and more-peaked iso-intensity response plots center on temporary elevations of threshold in response to background noise (Schwartz and Gerhardt 1998). Threshold shifts in eighth nerve fibers of *Eleutherodactylus coqui* have also been reported by Narins (1987). This phenomenon changes the linear operating range of the auditory system to a higher range of signal intensity so that the intensity at which saturation occurs is also elevated.

Other Nonlinear Effects on Frequency Selectivity

Interaction between the Two Spectral Peaks in the Gray Treefrog

Gerhardt and Doherty (1988) reported that females of the gray treefrog, *Hyla versicolor,* tested at 85 dB SPL preferred synthetic calls with a frequency of 1.9 kHz to those of 2.2 kHz, which is close to the mean in the calls of males in the same population in central Missouri. In these experiments the synthetic calls had a single carrier frequency, whereas natural calls have a secondary band that is, on average, about 4–12 dB lower in relative amplitude. More recently females were tested with synthetic calls that had both spectral peaks, and at the same playback level females preferred a stimulus of 1.1 kHz (−6 dB) + 2.2 kHz to an alternative of 0.95 kHz (−6 dB) + 1.9 kHz (Tanner and Gerhardt, in preparation). Further research is required to define the neural bases for the difference in the results of these two experiments, but the assumption that the frequency selectivity of the two auditory organs is independent of one another is clearly unwarranted.

A Gross-Level Mismatch: Temperature Effects on Frequency Preferences in the Green Treefrog

Temporal properties such as pulse rate are highly temperature-dependent in the calls of anurans. Preferences based on differences in pulse rate change with temperature in a parallel fashion to that in the calls of males, a phenomenon that has been termed temperature coupling (Gerhardt 1978). In contrast typical changes in carrier frequency with temperature in anurans are relatively small (Gerhardt and Mudry 1980). Because the characteristic frequency of the primary auditory fibers innervating the amphibian papilla are temperature-dependent (for a review see Lewis and Lombard 1988; Narins, this volume), the possibility exists that the match between low-frequency auditory tuning and the carrier frequency of conspecific calls might be poorer at some temperatures than at others.

Behavioral correlates of these physiological properties were demonstrated in the green treefrog (Gerhardt and Mudry 1980). Although females prefer synthetic calls with a low-frequency peak of 0.9 kHz to an alternative of 0.6 kHz at normal breeding temperatures (22–26°C), females tested at about 18°C reversed their preference. Indeed, in this temperature range, females even preferred an alternative of 0.5 kHz, which falls within the range of variation of the sympatric congener, the barking treefrog, *H. gratiosa.* Preferences based on differences in the high-frequency peak showed relatively little temperature dependence, as did the characteristic frequencies of neurons innervating the basilar papilla (Narins, this volume).

The shift in frequency preferences is not matched by a parallel drop in the frequency of the low-frequency spectral peak in the advertisement call of *H. cinerea.* Males produced calls with nearly the same low-frequency peaks as they did at normal breeding temperatures after they were acclimated to lower temperatures over the same time course as females that changed their frequency preferences (Gerhardt and Mudry 1980). Thus the maximum sensitivity of the auditory system at low temperatures occurs at frequencies nearly an octave lower than the frequencies emphasized in the advertisement call. Field studies suggest that although males sometimes call at these low temperatures, few females arrive for mating (Richard E. Daniel, personal communication). Whereas females are thus unlikely to prefer the calls of *H. gratiosa* to conspecific males (there are also behaviorally relevant, fine-temporal differences in the calls; Gerhardt 1981), females breeding at somewhat lower-than-average temperatures might be expected to show a bias for conspecific calls with lower-than-average frequencies. Additional field studies are needed to test this prediction.

Conclusions

Enhanced regions of auditory sensitivity correspond to one or two frequency bands emphasized in the long-range acoustic signals of anurans, thus supporting the matched filter hypothesis at the interspecific level of analysis. Such tuning is broad and relatively weak but could also mediate frequency preferences that represent stabilizing or directional selection, depending on how well the tuning matches the average carrier frequency or frequencies in conspecific calls. This conclusion must be qualified by considering changes in frequency selectivity under different conditions. In some species, frequency selectivity (iso-intensity plots and behavior) is reduced as stimulus intensity increases; in other species, the opposite trend can occur. Additional nonlinear effects that confound predictions about frequency preferences from threshold estimates include: (1) the enhancement of selectivity based on relatively large differences in frequency by the addition of chorus noise; (2) interactions between preferences mediated by the two inner ear organs; and (3) temperature effects. These phenomena, as well as behavioral studies designed to explore the significance of enhanced sensitivity to frequencies that do not occur in communication signals, deserve much additional research.

References

Capranica, R. R.1992. Untuning of the tuning curve: Is it time? Seminars in the Neurosciences 4:401–408.

Capranica, R. R., and A. J. M. Moffat. 1983. Neurobehavioral correlates of sound communication in anurans. *In* J. P. Ewert, R. R. Capranica, and D. J. Ingle (eds.), Advances in Vertebrate Neuroethology. Plenum Press, New York.

Diekamp, B. M., and H. C. Gerhardt. 1992. Midbrain auditory sensitivity in the spring peeper (*Pseudacris crucifer*): Correlations with behavioral studies. Journal of Comparative Physiology A 171:245–250.

Doherty, J. A., and H. C. Gerhardt. 1984. Evolutionary and neurobiological implications of selective phonotaxis in the spring peeper (*Hyla crucifer*). Animal Behaviour 32:875–881.

Drewry, G. E., and A. S. Rand. 1983. Characteristics of an acoustic community: Puerto Rican frogs of the genus *Eleutherodactylus*. Copeia 1983:941–953.

Forester, D.C., and R. Czarnowsky. 1985. Sexual selection in the spring peeper, *Hyla crucifer* (Amphibia, Anura): Role of the advertisement call. Behaviour 92:113–128.

Forester, D.C., and W. K. Harrison. 1987. The significance of antiphonal vocalisation by the spring peeper, *Pseudacris crucifer* (Amphibia, Anura). Behaviour 103:1–15.

Frishkopf, L. S., R. R. Capranica, and M. H. Goldstein, Jr. 1968. Neural coding in the bullfrog's auditory system: A teleological approach. Proceedings IEEE 56:968–979.

Fuzessery, Z. M. 1988. Frequency tuning in the anuran central auditory system. Pp. 253–273. *In* B. Fritszch, W. Wilczynski, M. J. Ryan, T. Hetherington, and W. Walkowiak (eds.), The Evolution of the Amphibian Auditory System. John Wiley, New York.

Gerhardt, H. C. 1974. The significance of some spectral features in mating call recognition in the green treefrog *Hyla cinerea*. Journal of Experimental Biology 61:229–241.

Gerhardt, H. C. 1978. Temperature coupling in the vocal communication system of the gray treefrog *Hyla versicolor*. Science 199:992–994.

Gerhardt, H. C. 1981. Mating call recognition in the barking treefrog (*Hyla gratiosa*): Responses to synthetic calls and comparisons with the green treefrog (*Hyla cinerea*). Journal of Comparative Physiology A 144:17–25.

Gerhardt, H. C. 1983. Acoustic communication in treefrogs. Proceedings German Zoological Society 25–35.

Gerhardt, H. C. 1987. Evolutionary and neurobiological implications of selective phonotaxis in the green treefrog (*Hyla cinerea*). Animal Behaviour 35:1479–1489.

Gerhardt, H. C. 1994. The evolution of vocalizations in frogs and toads. Annual Review of Ecology and Systematics 25:293–324.

Gerhardt, H. C., and J. A. Doherty. 1988. Acoustic communication in the gray treefrog, *Hyla versicolor:* Evolutionary and neurobiological implications. Journal of Comparative Physiology A 162:261–278.

Gerhardt, H. C., and G. M. Klump. 1988. Masking of acoustic signals by the chorus background noise in the green treefrog: A limitation on mate choice. Animal Behaviour 36:1247–1249.

Gerhardt, H. C., and K. M. Mudry. 1980. Temperature effects on frequency preferences and mating call frequencies in the green treefrog, *Hyla cinerea* (Anura: Hylidae). Journal of Comparative Physiology 137:1–6.

Hetherington, T. E. 1992. The effects of body size on functional properties of middle ear systems of anuran amphibians. Brain, Behavior, and Evolution 39:133–142.

Hillery, C. M. 1984. Seasonality of two midbrain auditory responses in the treefrog, *Hyla chrysoscelis*. Copeia 1984:844–852.

Hubl, L., and H. Schneider. 1979. Temperature and auditory thresholds: Bioacoustic studies of the frogs *Rana ridibunda, Hyla arborea* and *Hyla savignyi* (Anura, Amphibia) Journal of Comparative Physiology 130:17–27.

Jaslow, A. P., and R. E. Lombard. 1996. Hearing in the neotropical frog, *Atelopus chiriquiensis*. Copeia 1996:428–432.

Jennions, M. D., P. R. Y. Backwell, and N. I. Passmore. 1995. Repeatability of mate choice: The effect of size in the African painted reed frog, *Hyperolius marmoratus*. Animal Behaviour 49:181–186.

Karl, J. H. 1989. An Introduction to Digital Signal Processing. Academic Press, San Diego.

Keddy-Hector, A. C., W. Wilczynski, and M. J. Ryan. 1992. Call patterns and basilar papilla tuning in cricket frogs. II. Intrapopulational variation and allometry. Brain, Behavior and Evolution 39:238–246.

Lewis, E. R., and R. E. Lombard.1988. The amphibian inner ear. Pp. 93–124. *In* B. Fritszch, W. Wilczynski, M. J. Ryan, T. Hetherington, and W. Walkowiak (eds.), The Evolution of the Amphibian Auditory System. John Wiley, New York.

Littlejohn, M. J. 1977. Long-range acoustic communication in anurans: An integrated and evolutionary approach. Pp. 263–294. *In*

D. H. Taylor and S. I. Guttman (eds.), The Reproductive Biology of Amphibians. Plenum Press, New York.

Loftus-Hills, J. J., and B. M. Johnstone. 1970. Auditory function, communication, and the brain-evoked response in anuran amphibians. Journal of the Acoustical Society of America 47:1131–1138.

Lombard, R. E., and Straughan, I. R. 1974. Functional aspects of anuran middle ear structures. Journal of Experimental Biology 61:57–71.

Márquez, R., and J. Bosch. 1997. Female preference in complex acoustical environments in the midwife toads *Alytes obstetricans* and *Alytes cisternasii*. Behavioral Ecology 8:588–594.

Mohneke, R., and H. Schneider. 1979. Effect of temperature upon auditory thresholds in two anuran species, *Bombina variegata* and *Alytes obstetricans* (Amphibia, Discoglossidae). Journal of Comparative Physiology 130:9–16.

Narins, P. M. 1987. Coding of signals in noise by amphibian auditory nerve fibers. Hearing Research 26:145–154.

Narins, P. M., and R. R. Capranica. 1980. Neural adaptations for processing the two-note call of the Puerto Rican treefrog, *Eleutherodactylus coqui*. Brain, Behavior and Evolution 17:48–66.

Narins, P. M., and C. M. Hillery. 1983. Frequency coding in the inner ear of anuran amphibians. Pp. 70–76. *In* R. Klinke and R. Hartmann (eds.), Hearing-Physiological Bases and Psychophysics. Springer-Verlag, Berlin.

Pinder, A. C., and A. R. Palmer. 1983. Mechanical properties of the frog ear: Vibration measurements under free- and closed-field acoustic conditions. Proceedings of the Royal Society of London B 219:371–396.

Ryan, M. J., J. H. Fox, W. Wilczynski, and A. S. Rand. 1990. Sexual selection for sensory exploitation in the frog *Physalaemus pustulosus*. Nature, London 343:66–67.

Ryan, M. J., S. A. Perrill, and W. Wilczynski. 1992. Auditory tuning and call frequency predict population-based mating preferences in the cricket frog, *Acris crepitans*. American Naturalist 139:1370–1383.

Schwartz, J. J., and H. C. Gerhardt. 1998. The neuroethology of frequency preferences in the spring peeper. Animal Behaviour 56:55–69.

Wilczynski, W., B. E. McClelland, and A. S. Rand. 1993. Acoustic, auditory and morphological divergence in three species of neotropical frog. Journal of Comparative Physiology A 172:425–438.

Wilczynski, W., A. S. Rand, and M. J. Ryan. 1995. The processing of spectral skills by the call analysis system of the túngara frog, *Physalaemus pustulosus*. Animal Behaviour 49:911–929.

Wilczynski, W., H. H. Zakon, and E. A. Brenowitz. 1984. Acoustic communication in spring peepers: Call characteristics and neurophysical aspects. Journal of Comparative Physiology A 155:577–587.

Zakon, H. H., and W. Wilczynski. 1988. The physiology of the anuran eighth nerve. Pp. 125–155. *In* B. Fritszch, W. Wilczynski, M. J. Ryan, T. Hetherington, and W. Walkowiak (eds.), The Evolution of the Amphibian Auditory System. John Wiley, New York.

8

MICHAEL J. RYAN AND A. STANLEY RAND

Feature Weighting in Signal Recognition and Discrimination by Túngara Frogs

Introduction

Communication between the sexes to identify appropriate mates is a critical component of the mating process. Since heterospecific matings often do not result in viable offspring, there should be strong selection on females to mate with conspecific males. Thus one would expect signalers to accurately convey species status and receivers to accurately decode this information (Dobzhansky 1940; Mayr 1942; Andersson 1994). Studies of communication systems involved in species recognition continue to make important contributions to our understanding of the process of speciation (Gerhardt 1988, 1994; Doherty and Howard 1996; Martens 1996; Veech et al. 1996). Furthermore receivers can generate selection among potential conspecific mates if there is a preference for some signals over others, and thus can generate variance in male mating success and influence signal evolution (Kirkpatrick and Ryan 1991; Andersson 1994). Studies of these interactions have enhanced our understanding of the process of sexual selection (Ryan 1991; Andersson 1994; Searcy and Yasakawa 1996). Species recognition and sexual selection, however, are related processes. Strong selection to avoid heterospecifics can incidentally cause females to avoid conspecifics that most closely resemble heterospecifics, thus generating sexual selection. Alternatively sexual selection for signal variants within the conspecific range can incidentally yield species recognition (Fisher 1930; West-Eberhard 1979; Ryan and Rand 1993b; Gerhardt 1994; Pfennig 1998).

Recognition and Discrimination

Studies of communication systems that result in conspecific matings are often couched in terms of species recognition and are concerned with what signals are recognized as indicating appropriate mates. Sexual selection studies, on the other hand, usually address the degree to which females discriminate among what are recognized as acceptable signal variants. Thus it is helpful to precisely define these terms and to conduct experiments that address each. We operationally define *recognize* as "to treat as valid" (*Compact Oxford English Dictionary,* 2nd edition, 1993). We determine if a stimulus is recognized by determining if it elicits an appropriate bioassay from the receiver. We define *discriminate* as "to make or constitute a difference in or between" (*Compact Oxford English Dictionary*). We test for discrimination by determining if the receiver responds to one rather than the other signal. This usage also has some parallels and some

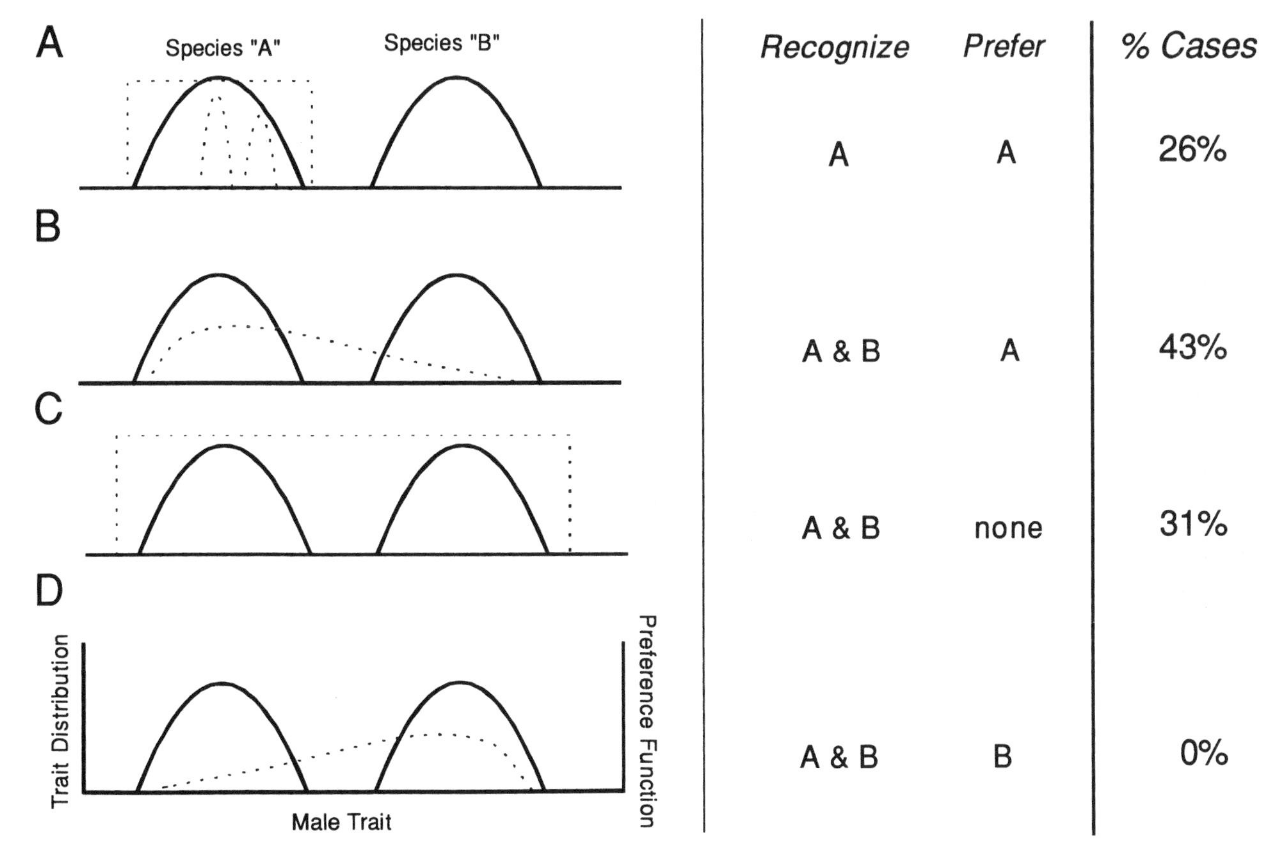

Figure 8.1. Possible patterns of signal recognition (test stimulus vs. white noise) and discrimination (test stimulus vs. conspecific call). We also show possible relationships between the trait distribution of male mating signals (solid lines) for two species, A and B, and the female preference function for species A (dashed line) that could result in each of the four patterns of recognition/discrimination. The right panels summarize the results of 36 phonotaxis experiments in which females were tested for recognition and discrimination with various heterospecific or ancestor calls.

differences with the terms *discrimination* and *identification* as used in the cognitive sciences (Harnad 1987).

As we use it, *recognition* is somewhat context-independent whereas *discrimination* is context-dependent. Recognition experiments do not allow us to determine if one signal would be preferred over another. Discrimination experiments tell us which signal of a pair is preferred, but do not tell us whether the unpreferred signal is recognized as appropriate but inferior or is not recognized at all. It is necessary to understand both of these factors to appreciate how receivers respond to signal variation.

We use these two classes of response to illustrate in Figure 8.1 how variation in preference functions and signal properties can result in different patterns of signal recognition and preference within and among species. The categories we show in Figure 8.1 are ranked (A–D) from a lesser to greater effect of the heterospecific signal on the female, and reflect what we think might be more or less likely outcomes when a female is confronted with the mate recognition signal of a heterospecific.

The situation we assume would occur most commonly is that only the conspecific signal is recognized (Figure 8.1A). Three possible preference functions might result in this pattern of response: (1) all conspecific signals might be equally attractive, (2) there might be stabilizing selection on the mean signal, or (3) there might be directional selection favoring signals that depart from the mean. What might be considered the next most likely alternative is when females recognize both conspecific and heterospecific signals as appropriate but prefer the conspecific signal (Figure 8.1B). A less likely alternative is when females (mistakenly) recognize both a conspecific and heterospecific signal as indicating an appropriate mate and do not discriminate between these two signals (Figure 8.1C). The least likely scenario is when females recognize both the conspecific and heterospecific signal and actually prefer the latter (Figure 8.1D).

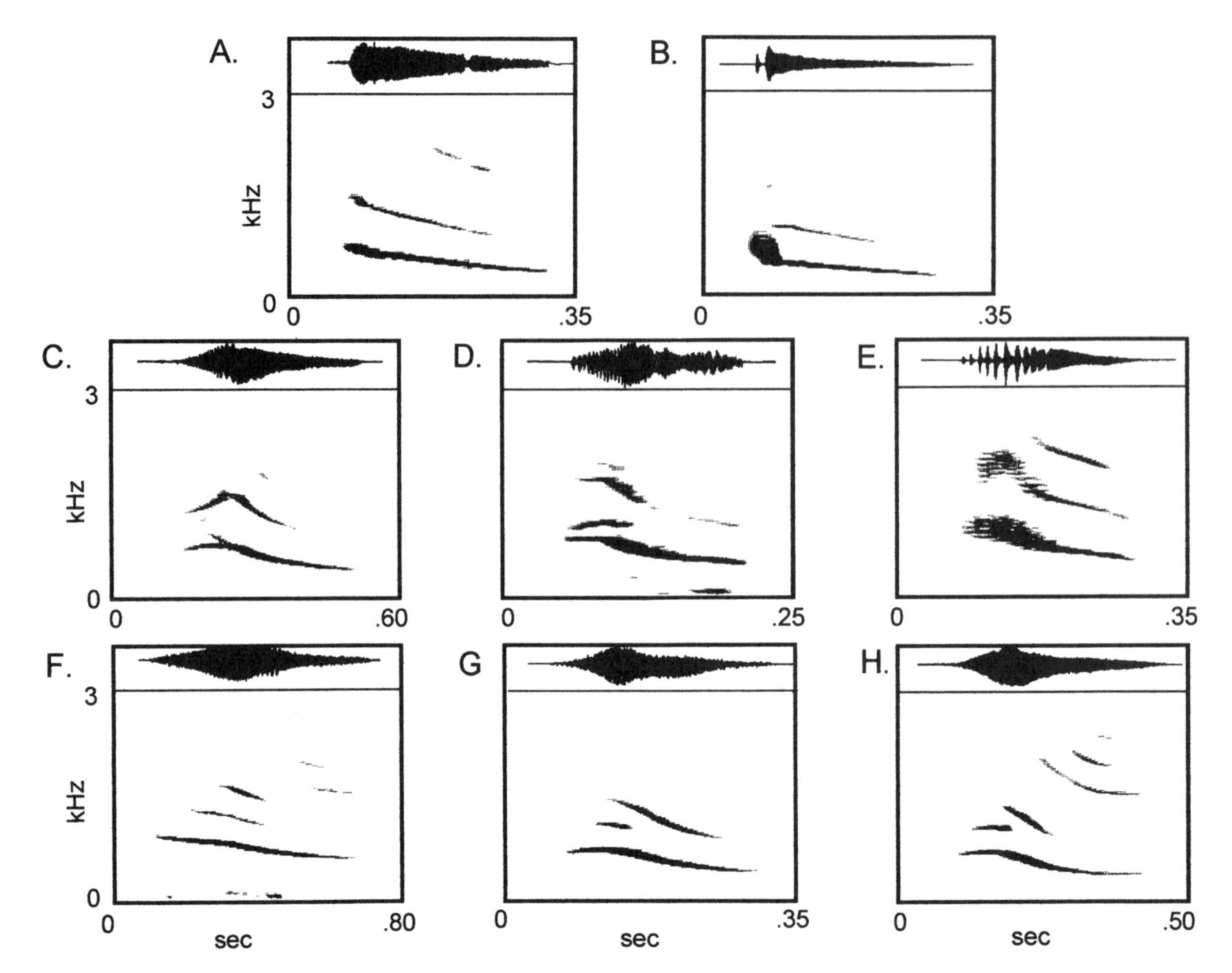

Figure 8.2. Calls of species of the *Physalaemus pustulosus* species group (A–E) and three closely related species (F–H) used as outgroups in the analysis (phylogenetic relationships are illustrated in Figure 8.3). Oscillograms (top) and sonogams (bottom): (A) *P. pustulosus,* (B) *P. petersi,* (C) *P. caicai,* (D) *P. coloradorum,* (E) *P. pustulatus,* (F) *P. enesefae,* (G) *P. ephippifer,* (H) *P. "roraima"* (an undescribed species). A species we tentatively identify as *P. freiberghi* (Cannatella et al. 1998) was not included in these studies.

This does not exhaust all possible interactions but we believe it presents the most likely ones.

Phylogenetic Comparisons

One of the goals of this study was to determine patterns of signal recognition and discrimination in female túngara frogs, *Physalaemus pustulosus.* We discerned these patterns over a wide range of calls of closely related heterospecifics and calls that we estimated at ancestral nodes of the phylogenetic tree of these species. For simplicity we refer to these calls as ancestral calls, but such shorthand should not suggest an arrogance of certainty about how ancestors of these frogs might have sounded. Testing female responses on this phylogenetic background is not a common approach but we feel it is a valid one because it might reveal how past history influences responses of females (e.g., Phelps and Ryan 2000; Phelps, this volume).

Túngara frogs are almost never found in choruses with congeners (the one exception is the small area of overlap between *P. pustulosus* and *P. enesefae* in the llanos of Venezuela), only occasionally with another species in a closely related genus (*Pleurodema*), and sometimes with other members of their subfamily (Leptodactylinae: Cannatella and Duellman 1984; Cannatella et al. 1998). The calls of the other syntopically breeding species are rarely similar to the túngara frog call. Although these current ecological interactions might influence the evolution of the túngara frog's communication system, we have chosen to emphasize historical factors. Signals and receivers are not deconstructed and reconstructed with each species event. Instead they are jury-rigged innovations of ancestral conditions. Thus we feel a phylogenetic approach to signal recognition can be a valuable addition to similar studies conducted in an ecological context, which might examine effects of habitat or acoustical competitors on signal evolution. An advantage of concentrating on either the phylogenetic or ecological approach is not to confound them.

In this study we determined how female túngara frogs recognize and discriminate calls of heterospecifics and ancestors, and how these patterns fit into the classification proposed in Figure 8.1.

Feature Weighting

Calls consist of a number of parameters or features. We cannot assume that variation in all of these features equally influences the females' responses (e.g., Tinbergen 1953). In a subsequent series of analyses, therefore, we asked how females weight these various features in making phonotactic decisions in the recognition and discrimination tests. We asked if their weighting schemes suggest they have been influenced by the need to discriminate closely related species, by phylogenetic divergence, and by the intraspecific variation in signal features.

The *Physalaemus pustulosus* Species Group

All members of the *Physalaemus pustulosus* species group and the three species we used as outgroups produce whine-like advertisement calls. These calls usually begin at around 1000 Hz and sweep to a frequency several hundred Hz lower in 200 to 800 ms (Figure 8.2, Table 8.1). Some calls are preceded by an amplitude-modulated segment of a few tens of milliseconds. The calls usually have several harmonics but the fundamental is the dominant. *P. pustulosus* and *P. freiberghi* are known to add suffixes facultatively to the whine (Ryan and Rand 1993a, 1993b, 1993c, 1999a).

The calls of frogs of the *Physalaemus pustulosus* species group and close relatives were recorded in the field between 1990 and 1993 using standard methods described in Ryan and Rand (1993a, 1993b, 1993c). The species group as defined by Cannatella and Duellman (1984) consists of two monophyletic groups: *P. pustulosus* and *P. petersi* are in one group, and *P. coloradorum* and *P. pustulatus* in the other. *P. pustulosus* occurs from north of Veracruz, Mexico, throughout much of Middle America and in northeastern South America in Venezuela and Guyana. *P. petersi* is distributed east of the Andes throughout much of the Amazon Basin. *P. coloradorum* and *P. pustulatus* are both found west of the Andes in Ecuador (Cannatella and Duellman 1984). Our more recent study (Cannatella et al. 1998) reaffirms the monophyly of the species group and each of the two clades within the species group, and also suggests that there are two additional species. One is a taxon thought to have been *P. petersi* in Amazonian southern Peru that we tentatively refer to as *P. freiberghi*. A taxon thought to be *P. pustulatus* in northeastern Peru also proves to be worthy of separate species status, and we tentatively refer to this species as *P. caicai* (Cannatella et al. 1998; Figure 8.3). Further analysis of these species is necessary. Our preliminary data suggest another species in western Ecuador, and the entire *P. petersi–P. freiberghi* clade requires a thorough study of geographic variation in molecular and advertisement call characters similar to the one we have conducted with *P. pustulosus* (Ryan et al. 1996).

We used three congeners for outgroup analysis: *P. enesefae*, which is sympatric with *P. pustulosus* in Venezuela; *P. ephippifer*, which was recorded in Belem, Brazil; and an unidentified species from northern Brazil, similar to *P. ephippifer*, which we tentatively refer to as a *Roraima* species.

General Methods

Call Analysis and Synthesis

We measured eight call variables for one call each of 10 individuals from each of the species in the species group and the three outgroup species (Figure 8.4, Table 8.1). Túngara frog females do not attend to upper harmonics of the whine, thus the fundamental frequency, which is also the dominant, is both necessary and sufficient for species recognition (Wilczynski et al. 1995). We measured the following spectral variables of the fundamental frequency: the initial frequency, final frequency, and a measure of frequency-sweep shape, which is the time from the initial frequency to midfrequency (Figure 8.4; see Ryan and Rand 1999b for details). The temporal variables were call duration, rise time, fall time, rise shape (time from call onset to midamplitude), and fall shape (time from midamplitude to call offset; Figure 8.4). Call duration was highly correlated with the sum of rise and fall times and thus was not used in the analysis.

We used these call variables of the species group and outgroup species, combined with our hypothesis of their phylogenetic relationships, branch lengths derived from DNA sequence changes, and a phylogenetic algorithm, either squared-change or local squared-change parsimony, to estimate each of the call variables at each ancestral node (Ryan and Rand 1995, 1999b). We then used these seven call variables to synthesize calls at the nodes; we refer to these as ancestral calls. We used the same variables to synthesize mean calls for all of the real species used.

We varied different parameters of the phylogenetic anal-

Table 8.1 Call parameters used in synthesis of phonotaxis stimuli under different assumptions of evolutionary history, and female responses to those stimuli in phonotaxis experiments

	Frequency (Hz)		Time (ms)			Shape			Female Response	
	Max	Final	Duration	Rise	Fall	Whine[a]	Fall[b]	Rise[c]	HvC[d]	HvN[e]
Species										
Physalaemus										
pustulosus	884	484	369.7	24.0	342.8	0.33	0.49	0.33		
petersi	1220	384	246.1	13.7	230.3	0.11	0.79	0.84	0	4
coloradorum	1180	628	209.3	53.4	161.7	0.39	0.71	0.44	1	11
pustulatus	964	676	206.0	99.5	104.3	0.43	0.49	0.95	0	2
caicai	888	444	394.5	105.1	293.7	0.29	0.68	0.66	3	10
ephippifer	944	576	266.4	83.5	177.4	0.53	0.66	0.47	1	7
roraima	876	460	339.1	94.6	251.6	0.47	0.72	0.60	1	16
enesefae	976	692	745.7	301.5	445.7	0.52	0.54	0.55	0	0
Assumptions[f]										
Node *a* [g]										
TLSG	910	518	302.0	89.0	213.0	0.51	0.69	0.54	1	3
Node *b*										
TLSG	949	622	568.0	216.0	353.0	0.51	0.58	0.55	1	1
LSG	937	589	483.0	176.0	308.0	0.51	0.60	0.54	0	0
Node *c*										
TLSG	974	466	333.0	32.0	300.0	0.29	0.57	0.53	14	15
LSG	1050	434	309.0	19.0	287.0	0.25	0.61	0.53	7	17
Node *d*										
TLSG	1120	439	274.0	44.0	230.0	0.21	0.75	0.73	4	13
LSG	1015	507	297.0	59.0	238.0	0.30	0.63	0.69	5	9
LSP	1008	479	314.0	55.0	259.0	0.28	0.64	0.67	3	11
LSG*	1015	507	297.0	59.0	238.0	0.30	0.63	0.69	6	13
LSP*	1008	479	314.0	55.0	259.0	0.28	0.64	0.67	8	19
SG	998	526	339.0	85.0	254.0	0.36	0.62	0.62	10	16
SP	995	512	372.0	92.0	280.0	0.36	0.62	0.60	11	19
Node *e*										
TLSG	961	572	320.0	94.0	230.0	0.32	0.67	0.69	3	13
LSG	988	564	287.0	90.0	200.0	0.34	0.65	0.72	0	10
LSP	962	527	320.0	94.0	229.0	0.32	0.67	0.69	0	14
LSG*	988	564	287.0	90.0	200.0	0.34	0.65	0.72	4	9
LSP*	962	527	320	94	229	0.32	0.67	0.69	6	16
SG	992	547	311	88	225	0.35	0.64	0.68	5	16
SP	974	521	339	93	249	0.34	0.65	0.67	3	15
Node *f*										
TLSG	1072	652	208	76	133	0.41	0.62	0.78	2	5
LSG	1062	654	208	78	130	0.41	0.61	0.79	1	4
Root										
TLSG	961	545	451	125	326	0.43	0.57	0.54	2	8
LSG	995	528	345	89	257	0.37	0.63	0.62	1	12
LSP	977	554	448	140	308	0.44	0.60	0.57	6	12
LSG*	990	510	298	66	233	0.34	0.64	0.65	0	12
LSP*	955	506	308	73	235	0.40	0.66	0.59	3	12
SG	995	528	345	89	257	0.37	0.63	0.62	3	19
SP	977	554	448	140	308	0.44	0.60	0.57	8	18

[a] Whine shape is the proportion of call duration when the frequency sweep reaches midfrequency.

[b] Fall shape is the proportion of the call duration when the call reaches half the amplitude from the peak amplitude to the end of the call.

[c] Rise shape is the proportion of the call duration when the call reaches half the amplitude from the initial to the peak amplitude.

[d] HvC = number of female responses to the heterospecific call when presented together with a conspecific call (max. $n = 20$).

[e] HvN = number of female responses to the heterospecific call when presented together with white noise (max. $n = 20$).

[f] Assumptions: T = pectinate tree, L = local, S = squared-change parsimony, G = gradual, P = punctuated evolution.

[g] Refers to nodes in Figure 8.3.

* Without *P. enesefae*.

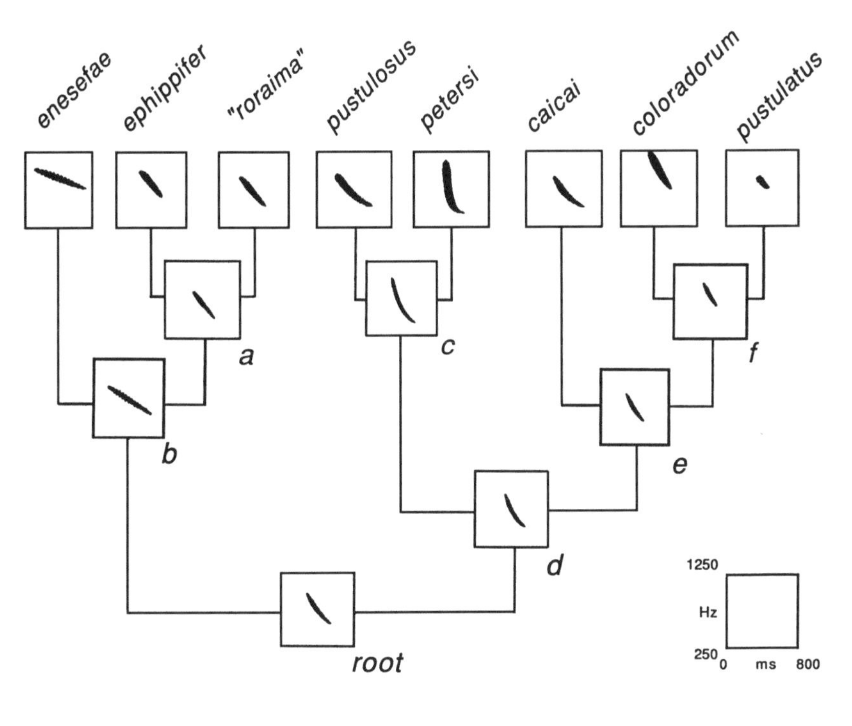

Figure 8.3. The phylogenetic tree illustrating the most parsimonious hypothesis for the relationships among members of the *Physalaemus pustulosus* species group and the three species we used as outgroups. *"roraima"* is an undescribed species. Sonograms illustrate the synthetic mating calls, which contain only the fundamental frequency of the whine (cf. Figure 8.2), for each taxon; the calls estimated for the ancestral nodes were derived from a local squared-change parsimony model assuming a gradual model of evolution (from Ryan and Rand 1995).

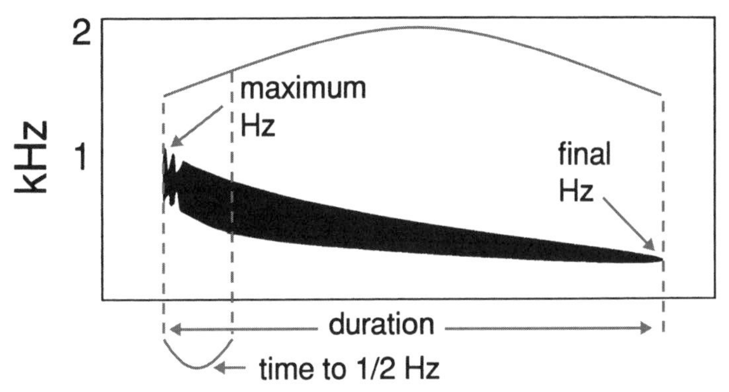

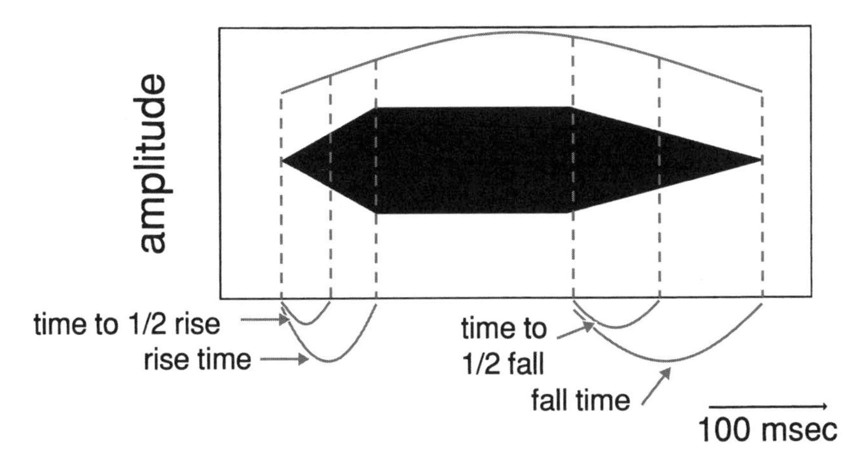

Figure 8.4. An illustration of the fundamental frequency of the túngara frog whine showing the various call parameters that were measured for analysis and synthesis; sonogram (top) and oscillogram (bottom). Measures of half frequency, half rise time, and half fall time were used to compute the shape of the frequency sweep, rise time, and fall time, respectively.

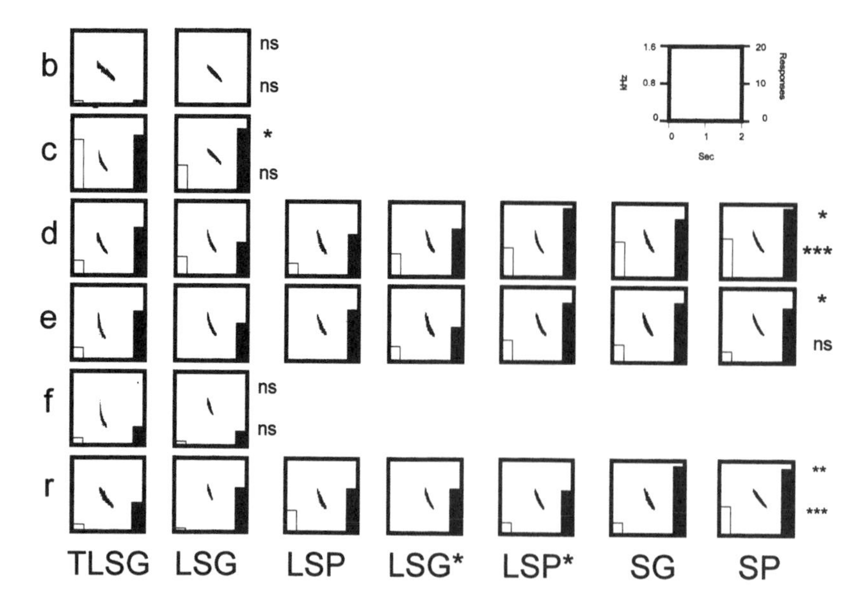

Figure 8.5. Sonograms of calls of the various nodes (rows; cf. Figure 8.3) that were estimated from different models of evolution (columns) that were considered different from one another and thus used in the phonotaxis experiments. (All estimates at node *a* were similar to one another given our criterion detailed in the text.) Open bars on the left of each box show the number of females responding to the call illustrated when it was presented with the túngara frog call. Closed bars on the right of each box show the number of females responding to that call when it was presented with a white noise stimulus. In each phonotaxis experiment, $n = 20$. Symbols at the end of each row indicate results of a *G* test, which tested the null hypothesis that the female responses to various calls at the same node did not differ for the discrimination tests (top symbol) and recognition tests (bottom symbol): ns = $p > 0.05$; * = $p < 0.05$; ** = $p < 0.01$; *** = $p < 0.001$. Abbreviations are defined in Table 8.1 (see also Ryan and Rand 1999).

ysis to determine if they would yield different estimates of calls at the same ancestral nodes. The parameters were tree topology (the most parsimonious tree and the next most parsimonious tree); rate of evolution (gradual, branch lengths coded to DNA sequence divergence; punctuated, branch lengths equal); and outgroup species (*P. enesefae,* which had the most divergent call, included or excluded). We combined these phylogenetic parameters to make seven sets of estimates of ancestral calls; thus each node of the phylogeny had seven separate call estimates. We considered call estimates at the same node to differ if any of the seven call variables were more than 10% different. Most call estimates at a single node were slightly different (Figure 8.5).

Phonotaxis Experiments

We tested female túngara frogs from Gamboa, Panama, near the laboratory facilities of the Smithsonian Tropical Research Institute. The phonotaxis experiments have been explained in detail elsewhere (Ryan and Rand 1995, 1999b). We placed a female in the center of an acoustic chamber measuring 3 by 3 m. We broadcast the stimuli antiphonally from speakers in the center of walls opposite one another. A positive phonotactic response was noted if a female approached within 10 cm of one of the speakers. A "no response" was noted if she remained motionless for 5 minutes at any time during the experiment, or if she did not approach a speaker within 15 minutes (see also Wilczynski et al. 1995; Ryan and Rand 1999a, 1999b).

In discrimination tests we presented a female with a túngara frog call synthesized from average values of the local Gamboa population versus one of the other synthetic heterospecific/ancestral calls. The null hypothesis of no preference was tested with an exact binomial probability. In recognition tests we presented a female with only one of the calls; the other speaker broadcast a white-noise control stimulus with the amplitude envelope of the túngara frog call. If a female did not respond in a recognition test, it could have been due to lack of attraction to the call being tested or a lack of motivation. To argue that a female's failure to respond was not due to lack of motivation, these tests were preceded and followed by discrimination tests in which the conspecific call was always one of the alternatives. Only if females responded in these discrimination tests did we consider the no-response in the recognition test as valid. No-responses included both lack of response by otherwise motivated females and approaches to the white noise stimulus, although the latter was rare. We used a Fisher's exact test to test the null hypothesis, which was the distribution of females approaching a silent speaker when it was paired with a white-noise stimulus: 18-no approach, 2-approach. This null distribution empirically estimated the number of times a females encountered a speaker without reference to the stimulus it was broadcasting. The sample size in all tests was 20. Females were tested in more than one experiment. This does not violate any statistical assumptions of independence, which pertain to independence of data within and not among experiments.

Results

How Are Heterospecific Calls Classified?

The call variables for each species in the species group and outgroup, all the ancestral calls, which includes multiple estimates at the same node, and the results of the recognition and discrimination tests are shown in Table 8.1 (see also Ryan and Rand 1999b). We summarized these results relative to the scheme shown in Figure 8.1 to describe the pattern by which females classify calls of heterospecifics and ancestors. Our criterion for significant recognition and discrimination is a probability level of 0.05. We realize, however, that this probability level, although widely accepted as "proving" a significant difference, is arbitrary, and that there can be a great risk of a type II error by assuming that higher p values indicate that the null hypothesis should be accepted. Nevertheless we feel this is a useful approach for this classification scheme. A more detailed analysis of strength of response, without categorizing responses by p values, is considered below.

Only 26% of the 36 heterospecific/ancestral calls did not elicit significant recognition from females; the conspecific calls were always preferred to these calls in the discrimination experiments. This is what we had assumed would be the most likely pattern of recognition/discrimination (Figure 8.1A). However, females recognized the other 74% of heterospecific/ancestral call as indicating an appropriate mate (Figure 8.1B–D). Although many heterospecific/ancestral calls were recognized as appropriate, in 69% of the discrimination experiments the females preferred the conspecific call to the heterospecific/ancestral call (Figure 8.1A–B). In 31% of all experiments, however, females recognized the heterospecific/ancestral call and did not significantly discriminate between a conspecific and heterospecific/ancestral call (Figure 8.1C). Considering only the seven heterospecific calls and not those of purported ancestors, females always preferred the conspecific call to the heterospecific call; in three of seven cases the heterospecific call elicited significant recognition and in four of seven cases it did not (Table 8.1).

Feature Weighting and Call Preferences

The previous section details the responses of female túngara frogs to a variety of heterospecific and ancestral calls. Here we ask how females weighted various call features in deciding their phonotactic responses in those experiments. This analysis could suggest if the features weighted by females are those expected if female preferences evolved to discriminate closely related species, were subject to phylogenetic influences, or were guided by the amount of intraspecific variation in signal features.

Species Discrimination and Feature Weighting

The first question we addressed was if females weight more heavily those call features that best discriminate among species (conspecific, heterospecifics, and ancestors). But first a caveat. Túngara frogs are currently allopatric with all of the heterospecifics we tested, except for a small zone of sympatry with *P. enesefae*. The geographic relationships with any purported ancestors can not be determined. Thus túngara frogs have not previously been under selection to make the conspecific–heterospecific contrasts we test here. It would seem, therefore, that the conspecific preferences versus these heterospecific calls are incidental consequences of selection for self recognition (Passmore 1981; Paterson 1985) or for discriminating between conspecifics and other heterospecifics (Dobzhansky 1940; Coyne and Orr 1989). Nevertheless given the above studies showing to what degree females discriminated against and recognized heterospecific/ancestral calls, it is of interest to understand the call features that informed these phonotactic decisions.

We used a principal component analysis (PCA) to determine the importance of each call feature in statistically discriminating among species along the first three principal components (i.e., without reference to how females respond to signals). The standardized mean value of each heterospecific/ancestral call was assigned the scores for the first three components, which explained 44%, 21%, and 14% of the variation among calls, respectively (Table 8.2). Final frequency, initial frequency, and rise time loaded most heavily onto principal component 1; fall time, frequency-sweep shape, and fall shape onto principal component 2; and frequency-sweep shape onto principal component 3. Ancestral calls were assigned principal component scores based on the z scores of their call variables relative to those of the real species (Figure 8.6).

We determined to what degree the three PCA scores explained the female's responses in discrimination and recognition tests. The PCA scores were used as independent variables in a multiple regression analysis, whereas the number of female responses was the dependent variable. Principal

Table 8.2 Principal component loadings from the analysis of call variation among species of the *Physalaemus pustulosus* group and three closely related species used in the outgroup analysis

Call Variable	Principal Component 1	2	3
Maximum frequency	0.858	0.068	−0.401
Final frequency	0.935	−0.046	−0.183
Rise time	0.833	0.220	0.056
Fall time	0.171	0.875	0.226
Frequency-sweep shape	0.376	−0.529	0.689
Fall shape	−0.682	0.476	0.163
Rise shape	−0.716	−0.245	−0.393

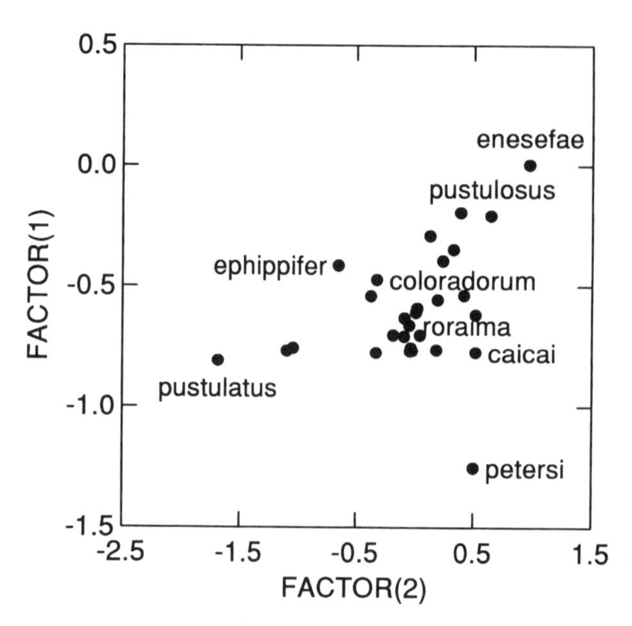

Figure 8.6. A plot of the first two axes of a principal components analysis of call variation among species of the *Physalaemus pustulosus* species group, the three closely related species used as outgroups, and the estimates of calls at ancestral nodes, including multiple estimates at the same node.

component 1 did not enter into the multiple regression equation for either recognition or discrimination responses. Principal components 2 and 3 explained only 26% and 32% of the variation in discrimination and recognition, respectively. These results suggest that females are not weighting features in a manner that would best discriminate among the mean calls of species.

We conducted a similar analysis based on call variables of individuals. We used a discriminant function analysis (DFA) to ask how well individual call variables can discriminate between conspecific (túngara frog) calls and those of the other seven real species. We used the 10 calls from each species that were used to calculate the species means for the analyses above. Data were standardized to z scores before discriminant function analysis.

A jack-knifed classification matrix based on the discriminant function analysis using all call variables correctly classified all túngara frog calls as conspecific and all of the 70 other calls as heterospecific (we were not concerned with whether the heterospecifics were correctly classified to species). A forward, stepwise discriminant function analysis revealed that rise shape, rise time, and fall time, in that order, were the variables that best discriminated species. Rise shape by itself had a jack-knifed classification accuracy of 70%, adding rise time increased the accuracy to 80%, and adding fall time increased accuracy to 96%. Thus there is sufficient information in the seven call variables to discriminate perfectly between the conspecific call and the other calls of heterospecific/ancestors, and fairly accurate discrimination can result from using only a few of these variables. The question is whether females use these features in their phonotactic decisions.

We used a multiple regression analysis of the female choice results to determine what variables best describe female responses in phonotaxis tests. The seven standardized call variables were the independent variables and the number of female responses in either the recognition or discrimination tests was the dependent variable. We then asked how well those variables predicted conspecific identity.

Final frequency was the only variable that entered into the equation for recognition and explains 28% of the variation in these female responses ($r^2 = 0.28$). The shapes of the rise, fall, and frequency entered into the multiple regression equation for the discrimination responses, and together they explain 62% of the variation ($r^2 = 0.62$). How well do these call variables discriminate conspecific signals from the heterospecific/ancestral ones we used in this study? Only final frequency was identified as influencing female responses in the recognition experiments. A discriminant function analysis cannot be performed using only one variable. Thus we added a second variable to the analysis: fall time. This variable had the lowest partial correlation coefficient in the multiple regression analysis and thus should least influence the ability of final frequency to correctly classify conspecific versus heterospecific. These variables had a classification accuracy of 70%. The three variables that were weighted most heavily in the discrimination experiments, the shapes of rise, fall, and frequency, correctly classified 85% of the calls as conspecific or heterospecific.

These results show that in recognition tests females do not attend to variables that resulted in accurate discrimination between conspecific and heterospecific/ancestral calls tested. In discrimination tests they weighted more heavily a suite of features that gave better species discrimination than in recognition tests (85% versus 70%), but were less accurate than the discrimination provided by the three best features for discriminating species (85% versus 96%).

Phylogenetic Divergence and Feature Weighting

Past history can influence how receivers respond to signal variation (Phelps, this volume and references therein). Thus we asked what call features are phylogenetically informative—those that best explain the amount of evolution between heterospecifics—and if these call features figure prominently in explaining female phonotactic responses. It is possible that the auditory system of these frogs are most sensitive to the phylogenetically informative signal features. Our measure of amount of evolution was derived from divergence of DNA sequences, which was the dependent variable in the stepwise multiple regression analyses (Ryan and Rand 1995, 1999b; Cannatella et al. 1998). We used only the heterospecific call variables and female responses and not the ancestors in the stepwise multiple regression analysis. Rise time, fall time, frequency-sweep shape, and fall shape entered into the equation, in that order, and predicted 98% of the variation in phylogenetic distance among species ($r^2 = 0.98$). Rise time and fall time, which were not weighted heavily in either the recognition or discrimination experiments, explained 66% of the variation ($r^2 = 0.66$) in a separate multiple regression analysis, as did frequency-sweep shape and fall shape ($r^2 = 0.66$), which are two of the three variables that were weighted heavily in the discrimination experiments.

The one variable that was weighted by the frogs in the recognition responses, final frequency, only explained 16% of the variation in phylogenetic distance ($r^2 = 0.16$), whereas the three variables that were heavily weighted in the discrimination experiments explained 80% of the variation in phylogenetic distance ($r^2 = 0.80$). These results suggest that females weight phylogenetically informative call features in discrimination experiments but not in recognition experiments.

Feature Invariance and Feature Weighting

A third factor that has been offered to explain feature weighting is the feature invariance hypothesis. This argument is that animals should weight most heavily those signal properties that vary little within the species regardless of how these signals compare with heterospecific signals (Nelson and Marler 1990; cf. Gerhardt 1991). In the sample of túngara frog calls used to estimate the mean, we find a pattern of variation among call variables that seems rather typical for frogs: spectral variables that are primarily determined by morphology have lower coefficients of variation than temporal properties that are usually under behavioral–physiological control (Ryan 1988; Cocroft and Ryan 1995). The coefficients of variation for the call variables analyzed are: initial frequency, 4.5%; final frequency, 11.3%; rise time, 67.1%; fall time, 20.1%; frequency-sweep shape, 21.3%; rise shape, 51.2%; fall shape, 46.2%. The variable that is weighted most heavily in the recognition responses has the second lowest coefficient of variation, whereas the three shape variables, which are weighted most heavily in discrimination responses, have three of the four highest coefficients of variation. Thus feature invariance predicts responses in recognition but not in discrimination experiments.

How Are Recognition and Discrimination Related?

It is usually assumed that the two types of experiments we use to assess phonotactic responses, recognition and discrimination, measure a similar phenomenon, but merely in different contexts. For example female choice tests usually use one but not both measures as assessments of female preferences. On the one extreme, recognition and discrimination might be two measures of the same phenomenon, as seems to be implicitly assumed, with the results of one perfectly predicting the results of the other. Alternatively these might be unrelated phenomena having only the bioassay in common but not being governed by any of the same rules or perceptual processes to evaluate the signals. We determined to what degree recognition and discrimination responses differ in feature weighting.

The responses of females to the same stimuli in the recognition and discrimination experiments were significantly correlated with one another ($r = 0.67$, $p < 0.001$). The amount of variation in female responses explained by this correlation is 45%, leaving 55% of the variation unexplained. The unexplained variance might be due to stochastic variation in two similar phenomena or to the fact that discrimination and recognition are two different phenomena. It is possible that there is a stronger relationship between the responses in the discrimination and recognition experiments, but that the relationship in not linear; transforming the data by logarithm, however, reduced the correlation slightly ($r = 0.64$).

Female responses in recognition and discrimination experiments were not influenced in the same way by overall similarity of the test call to the túngara frog call and phylogenetic distance of the test species/ancestor to túngara frogs. Differences among calls are not a good predictor of evolutionary relationships (Cannatella et al. 1998); the correlation between call similarity and DNA sequence divergence explains only 28% of the variation in these data sets. An earlier study with a smaller data set (Ryan and Rand 1995) had shown that call similarity and phylogenetic distance might differently influence female responses in recognition and discrimination experiments—discrimination was more influenced by the phylogenetic distance between the call being compared with the túngara frog call, whereas recognition was more influenced by the similarity of the test call to the túngara frog call. In the expanded data set analyzed here, phylogenetic distance and similarity still differently influenced responses but in the opposite way. A forward stepwise multiple regression analysis showed that similarity alone explained 51% of the variation in discrimination responses, adding phylogeny explained a total of 67% of the variation. Alternatively phylogeny explained 40% of the variation in recognition responses, whereas adding similarity to the equation explained a total of 50% of the variation.

There were also differences in how responses in recognition and discrimination experiments scaled to independent variables, and these differences might be related to perceptual factors. For example response variation might be linear in some tasks but nonlinear (e.g., logarithmic) in others. It is also possible that females compare the test call with an internal representation of the conspecific call in recognition experiments, in which comparison between two simultaneous stimuli is not an option. Therefore we transformed both the female response data and calls as a means of uncovering such perceptual processes. The responses, the dependent variables, were log-transformed, and the standardized call variables, the independent variables, were coded by their absolute difference to the túngara frog call variables (|túngara call variable − test call variable|). If these transformations influenced the explanation of responses in discrimination and recognition experiments differently, it might indicate a difference in underlying perceptual processes.

These data manipulations only decreased the variance explained in discrimination responses. The unmanipulated data in the multiple regression explained 67% of the variation, as reported above. When the responses were log-transformed the variation explained fell to 35%, and was 45% when the call variables were also coded relative to the túngara frog call variables. Alternatively these data manipulations had a profound effect on the variance explained by the multiple regression in the analysis of the recognition responses. The unmanipulated data, reported above, explained 28% of the variation, and only final frequency contributed significantly to explaining variation in female responses in a stepwise multiple regression. Log-transforming the response data increased the variation explained to 63%, with frequency-sweep shape now contributing to the call variables significantly explaining variation in preferences. Coding the call variables to the túngara frog call, in addition, further increased the variation explained to 77%, in which final frequency, frequency-sweep shape, and total duration became important predictors. Although these data transformations do not allow any insights into how the frogs are processing information in the discrimination and recognition experiments, it is yet another suggestion that the responses elicited in recognition and discrimination experiments might have been motivated by different weightings of factors.

Discussion

There are three general findings from the results of these phonotaxis experiments with túngara frogs. First, females often recognize other signals besides the conspecific signal as appropriate. Second, females do not weight signal features in a manner optimal for classifying conspecifics versus heterospecifics; in discrimination tests they weight heavily phylogenetically informative features and in recognition tests features with less intraspecific variation. Third, the two types of experiments we used to assess female phonotaxis preference, recognition and discrimination, are not highly correlated with one another and appear to involve different sets of feature weighting.

Classification of Conspecific versus Heterospecific

One of the general conclusions of animal communication is that females usually show a strong attraction for conspecific mate recognition signals relative to heterospecific signals, although exceptions to this trend are known (e.g., in katydids, Morris and Fullard 1983; swordtails, Ryan and Wagner 1987). Our results, however, show that a majority of non-conspecific (heterospecific and ancestor) calls are recognized as signaling an appropriate mate, and a substantial proportion of these calls are not significantly discriminated from the conspecific; this is unusual if not unprecedented. Considering only the seven heterospecific calls, females always preferred the conspecific call to those in discrimination tests, and in three of seven cases the heterospecific signals elicited significant phonotaxis. There are several possible explanations for these results.

It is possible that túngara frogs are unusual among species so far tested in possessing a quite permissive mate-recognition system. We doubt this is the case; nothing about the biology of these animals suggests that they are special, and other species of anurans, for example, will approach calls that are not conspecific even if they are sympatric (Oldham and Gerhardt 1975; Backwell and Jennions 1993; Gerhardt and Schwartz 1996).

We are sure that the experimental methodology we use, rather than the taxon we study, is responsible for revealing substantial evidence for attraction to heterospecific/ancestral signals. Most studies of mate recognition present pairs of stimuli simultaneously and the results are interpreted as females exhibiting a preference for one stimulus relative to the other or no preference (reviewed in Andersson 1994; Hauser 1996; Bradbury and Vehrencamp 1998). Although these studies reveal which stimulus is preferred, they tell us little about the unpreferred stimulus. Even in studies in which stimuli are presented separately, the strength of responses to two or several stimuli are usually compared to determine if one signal is preferred over another (Searcy and Marler 1981; Wagner 1998). Most studies have not compared the response in single-stimulus experiments with a null hypothesis of no response.

These methodological issues, however, cannot explain the lack of statistically significant discrimination between conspecific and some of the ancestral calls. This discrepancy with the bulk of previous studies on mate recognition is probably due to the types of signals we are testing in the phonotaxis experiments. Female túngara frogs always preferred a conspecific to a heterospecific call, but failed to discriminate in favor of the conspecific in 39% of the comparisons with 28 ancestral calls. The ancestral calls that were not significantly discriminated against were estimates of calls at the ancestral nodes closest to the túngara frogs: nodes c, d, e, and the root (Figures 8.3 and 8.5). Females discriminated against the call estimates at nodes a, b, and f. Thus it was the calls more similar to the túngara frog call that were not significantly discriminated against.

The ancestral calls were all significantly different from the túngara frog call. All of the values for their call variables were outside of the 95% confidence intervals of the analogous call variables for the túngara frog calls that were used to estimate the mean. It would be important, however, to know how the ancestral calls compare with a much larger sample of calls from the population of the túngara frog females tested. We tentatively conclude that these ancestral calls that were not significantly discriminated against are different from the conspecific call but, as seen in Figure 8.5, tend to be more similar to the túngara frog call than calls of the heterospecifics. We are probably testing signals more similar to the conspecific signal than other studies of species recognition.

Our study also differs from others in that our choice of test stimuli is guided by evolution rather than ecology. As noted above, we have not tested túngara frogs in response to calls of sympatric species. These calls are quite different from the túngara frog call, and we would expect near-perfect conspecific discrimination and little heterospecific recognition, but this needs to be tested. Because calls of more closely related species will tend to be more similar than calls of more distantly related species, especially over large phylogenetic distances, our phylogenetic approach might be more likely to test responses to stimuli that are more similar to conspecific signals. Furthermore ancestral signals might not only be more similar to conspecific signals, but are more likely to have been the types of sounds that past receivers have had to decode. Thus our choice of using a phylogenetic approach to these studies (intentionally) biases us toward choosing signals that play to these response biases (Phelps and Ryan 1998, 2000; Phelps, this volume).

A more general conclusion from all of these experiments is that the female's perceptual map of what it considers an acceptable stimulus is not restricted to the conspecific signal. In fact some signals that are outside the conspecific range are not even discriminated from the conspecific call. This is one more piece of evidence against the notion that properties of the signal and receiver need be tightly matched. These results also suggest that interactions with other species are not necessary for effective conspecific versus heterospecific recognition. Female túngara frogs discriminated against all of the heterospecifics in favor of the conspecific call, and only some but not all of the heterospecifics elicited recognition from females. This is consistent with Paterson's (1985) notion that self recognition might suffice for species recognition, but might not be considered strong support for it, as we now address.

We suggest that the túngara frog females' positive responses to ancestral calls results from a combined process of generalization and relaxed selection. Females use signal processing strategies that are under strong selection to guide them to conspecific and only conspecific signals. The precision of the strategies required for such a task is context dependent. The strategy will, of course, depend on the properties of the conspecific signals but need not be restricted by the bounds on these properties. If the signal of interest has the lowest frequency to be encountered by a species, then an open-ended preference for low frequency will be as effective in identifying conspecifics as a preference for the precise frequency of the signal. But the degree of generalization, or over-generalization, that will be transparent to selection depends on the types of errors that could be made. Generalization might be favored by selection because it reduces costly neural computations (e.g., Bernays 1998), as long as it does not lead to errors in identification that reduce fitness, as would be true for responses to allopatric or ancestral species. Selection will add precision when necessary. Until necessary, however, generalization can be an important response bias that could influence future signal evolution.

Feature Weighting

One of the early findings of ethology was that animals are not equally sensitive to all stimuli in their environments. Instead they filter out much of the potential stimulation they encounter and are more influenced by some stimuli than others (Tinbergen 1953). Signals are usually a composite of multivariate traits, and we know that all features of the signals are not attended to equally. Nelson and Marler (1990) review discussions of the forces that might shape feature weighting and contrast two hypotheses, the sound environment hypothesis and the feature invariance hypothesis. The sound environment hypothesis states that features that best discriminate between sympatric and synchronic conspecifics and heterospecifics are those features that should be weighted most heavily. The feature invariance hypothesis, alternatively, states that the features that vary least within the species should be weighted most heavily. In Nelson and Marler's study these two hypotheses predict primacy for some of the same features, but the results tend to favor the sound environment hypothesis.

Our analysis parallels Nelson and Marler's except that we are contrasting species along a phylogenetic rather than an ecological axis; because most of the species we are testing are not sympatric with the túngara frog, we are not confounding these axes. Our hypothesis can be thought of as a phylogenetic influence hypothesis. The rationale for a study of feature weighting along a phylogenetic axis is not as immediately clear as for an investigation along an ecological axis. Our assumption for testing feature weighting in a phylogenetic context is that the evolutionary divergence of signal preferences within a group might match those features of the signal that most diverge. Thus females should weight those features that have diverged the most and thus best predict the dichotomous classification of conspecific/heterospecific when females are making the same choice. The same emphasis on phylogenetic influences on communication evolution makes another prediction: the call variables used by females should be those that are the best predictors of phylogenetic distance.

Our results show that females do not weight most heavily those features of the signal that would be ideal for classifying species as conspecific versus heterospecific. The principal component factor that best explains call variation among extant species does not explain a significant amount of variation in either recognition or discrimination experiments. The set of correlated variables that best separates the species in multivariate call space appears to be ignored by the females.

Our analysis of feature weighting of individual call variables leads us to the same conclusions. Using all of the call variables results in 100% accurate classification, whereas using three variables can result in 96% accuracy. Túngara frog females only weight heavily three call variables in discrimination. However the three call variables they use are not the three that would result in optimal discrimination for that number of variables. Nevertheless, with the three variables the females use they would achieve 85% accuracy. The only variable heavily weighted in the recognition experiments gives only 28% accuracy; if the females were constrained for some reason to relying on only one variable, the most effective variable by itself would result in 70% accuracy.

Another view of phylogenetic influence on feature weighting, however, is less utilitarian. It does not assume that females have been selected to make the conspecific/heterospecific-ancestor discriminations we tested. Instead this view suggests that the features weighted most heavily by these kinds of frogs will be biased toward the acoustic features that are most likely to diverge. Four call variables explain almost all of the variation in phylogenetic distance, or evolutionary divergence, among the eight heterospecifics ($r^2 = 0.98$). Two of these call variables are not weighted heavily in female responses but two are. Each of the pairs of variables by themselves explains two-thirds of the variation in evolutionary divergence ($r^2 = 0.66$). Furthermore the three variables weighted most heavily in discrimination explained 85% of the variation in evolutionary divergence ($r^2 = 0.85$). The single call feature heavily weighted in recognition is not a good predictor of evolutionary divergence ($r^2 = 0.16$).

Our results suggest that there is a substantial phylogenetic influence on feature weighting in túngara frogs. This influence is not revealed in strategies that females use to discriminate against heterospecific/ancestral calls with which they have no current experience, in contrast to Nelson and Marler's (1990) result suggesting a significant influence of ecological signaling interactions on feature weighting strategies in some song birds. Instead it appears that there might be a very basic correlation between the properties of the signal and the receiver in that the receiver is more sensitive to signal features that are more likely to evolve. This is in contrast to the feature invariance hypothesis that suggests that feature weighting strategies are self referential and are biased to features that are less variable within the conspecific. We found no evidence for such an effect in this study when females were confronted with a discrimination task, but feature invariance appears to be relevant to recognition.

Recognition versus Discrimination

Studies of both species recognition and sexual selection by female choice have used laboratory experiments to ascertain female preferences based on male trait variation; this is especially true for studies in the acoustic domain (reviewed in Ryan and Keddy-Hector 1992; Andersson 1994). In many studies of anurans as well as some in insects and birds, females are given a choice between a pair of sounds; a choice of one stimulus to exclusion of the other is interpreted as females having a preference within the confines of the laboratory setting and the female's current physiological condition (Wagner 1998). The strength of the discrimination between the two stimuli can be used as a measure of the strength of the preference. These experiments do not, however, tell us anything about the saliency of the unpreferred signal—is it merely not as attractive as the alternative; is it recognized as a signal that should be avoided; or is it not recognized as a signal and relegated by the auditory system to the general category of "noise"?

Other studies of mate preference use what we refer to as recognition experiments, and the strength of the response to single stimuli are compared among stimuli to ascertain female preference. In studies of *Drosophila,* flies are put together in pairs and mating speed is measured (Coyne and Orr 1989; Boake et al. 1997); in crickets the female's duration and acceleration in response to a single sound is measured (Wagner 1998), and in birds the number of courtship solicitation displays to a single acoustic stimulus is measured (Searcy and Yasakawa 1996; Stoddard 1996). If a stimulus elicits a strong response from a female we assume that the signal is recognized as being appropriate. Weaker responses, however, are usually not compared against a null hypothesis to test for recognition. The goal of these experiments is usually to compare responses across experiments to assess variation in the strength of preference relative to the stimuli being compared.

Both of these experimental approaches reveal important information about how females respond to stimulus variation, and the tacit assumption has been that these are different measures of the same phenomenon, female signal preference. Each might be more appropriate for different questions. Wagner (1998), for example, argued that a single-stimulus experiment with a continuous measure of the female's response is a better metric for measuring the strength of preference than a dichotomous choice among two stimuli. This comparison, however, suggests that these are measures of the same phenomenon. As far as we know, no one has investigated the possibility that recognition and discrimination are different phenomena, a caution also raised by Wagner (1998). If they are not, then serious consideration must be given to the appropriateness of the type of experiment for the hypothesis being tested, the validity of comparing results from single-stimulus and two-stimulus experiments, and, more importantly, what might be the differences between these phenomena at the perceptual level.

Our comparisons of recognition and discrimination experiments in túngara frogs suggest that these two responses might operate differently in some ways. All 36 of the heterospecific/ancestral calls we used were tested for both recognition versus white noise and discrimination versus the túngara frog call. The responses to the same stimuli were significantly and positively correlated: the more responses to the heterospecific/ancestral call versus noise, the more responses to the same call when it was paired with the túngara frog call. The variability in one response predicted by the other is 45%; the unexplained variance could be stochastic or could indicate different sets of feature weightings in the processes of recognition and discrimination. Also there is no overlap in the call variables that explains the variation in female response in the two types of experiments in a stepwise multiple regression. In the discrimination experiments the shapes of the frequency sweep and the rise and fall times explain about two-thirds of the variation in responses. In contrast in the recognition experiments final frequency is the only call variable that contributes significantly to explaining the variation in female responses, and at that it explains only about a quarter of the variation.

It is especially interesting how transformation of the data influences the degree to which call variables predict recognition versus discrimination responses. Our data transformations only decreased the variance explained by the discrimination experiments. The variation explained in the recognition experiments increased from one-quarter to three-quarters when the responses were transformed by logarithm and the call variables were coded by their relative differences to the túngara frog call. Concomitant with the increased variance explained is the recruitment of additional call variables in explaining the variance. But of the three that significantly explain recognition responses using the transformed and coded data, only one of those, the shape of the frequency sweep, is shared with the three call variables that best predict the discrimination responses.

The above results suggest that females might be attending to different cues in recognition and discrimination experiments, and their responses might be scaled to stimulus variation differently. Log-transforming only the responses in the recognition experiments substantially increases the predictability of the regression equation. This might be expected if recognition tended to be a stepwise function, and we might expect this to be more likely of recognition than discrimination. Coding the variables relative to the túngara frog call increases the predictability of the regression equation, and this might indicate the nature of an internal template to which females are comparing the call in the recognition experiments. The lack of such an effect in the discrimination experiments might be expected if the túngara frog call is being presented to the female as one of the alternatives.

Some caveats need to be considered. It is possible that the lack of correspondence between recognition and discrimination results might derive from sampling error. Also many of our tentative conclusions are based on results of multiple regression analyses. Interpretation of such results should be made with caution. Ideally these results should be used to design experiments that test the importance of different variables in discrimination and recognition tests. If the phonotactic decisions being made in recognition and discrimination experiments are governed by different sets of feature weightings, however, we feel that we have revealed important context-dependent responses that have repercussions for methodological approaches to studies of acoustic preferences, and suggest a more complicated interaction of perceptual processes than previously appreciated.

Conclusions

The main conclusion of this study is that signals and receivers are not deconstructed and reconstructed with each speciation event. Instead they are jury-rigged innovations of ancestral conditions. Thus the kinds of recognition tasks that challenged the brains of ancestors will influence the manner in which brains of extant species solve analogous tasks. This is why responses of female túngara frogs to a variety of heterospecific and ancestral calls can be predicted by phylogenetic distance independent of any similarity of the test call to the conspecific call. Thus understanding the mechanisms underlying mate preferences requires a historical as well as a functional analysis.

Acknowledgments

We thank C. Gerhardt, N. Kime, and W. Wagner for comments on the manuscript, the Smithsonian Institution's Scholarly Studies Fund and the National Science Foundation for financial support, and the Smithsonian Tropical Research Institute for logistical support. We are especially grateful to C. Silva, S. Yoon, J. Ellingson, F. Bolaños, K. Mills, G. Alvarado, P. Monsivais, M. Dantzker, D. Lombeida, G. Rosenthal, M. Sasa, S. Rodriguez, Z. Tarano, M. Bridarolli, L. Dries, M. Kapfer, N. Kime, A. Angulo, L. Ferrari, M. Dillon, J. Fiaño, and S. Phelps for assistance in conducting phonotaxis experiments over the last decade.

References

Andersson, M. 1994. Sexual Selection. Princeton University Press, Princeton, NJ.

Backwell, P. R. Y., and M. D. Jennions. 1993. Mate choice in the Neotropical frog, *Hyla ebraccata:* Sexual selection, mate recognition and signal selection. Animal Behaviour 45:1248–1250.

Bernays, E. A. 1998. The value of being a response specialist: Behavioral support for a neural hypothesis. American Naturalist 151:451–464.

Boake C. R. B., M. P. Deangelis, and D. K. Andreadis. 1997. Is sexual selection and species recognition a continuum? Mating behavior of the stalk-eyed fly *Drosophila heteroneura*. Proceedings of the National Academy of Sciences USA 94:12442–12445.

Bradbury, J. W., and S. L. Vehrencamp. 1998. Principles of Animal Communication. Sinauer, Sunderland, MA.

Cannatella, D.C., and W. E. Duellman. 1984. Leptodactylid frogs of the *Physalaemus pustulosus* species group. Copeia 1984:902–921.

Cannatella, D.C., D. M. Hillis, P. Chippindale, L. Weigt, A. S. Rand, and M. J. Ryan. 1998. Phylogeny of frogs of the *Physalaemus pustulosus* species group, with an examination of data incongruence. Systematic Biology 47:311–335.

Cocroft, R. C., and M. J. Ryan. 1995. Patterns of mating call evolution in toads and chorus frogs. Animal Behaviour 49:283–303.

Coyne, J. A., and H. A. Orr. 1989. Patterns of speciation in *Drosophila*. Evolution 43:362–381.

Dobzhansky, T. 1940. Speciation as a stage in evolutionary divergence. American Naturalist 74:312–321.

Doherty J. A., and D. J. Howard. 1996. Lack of preference for conspecific calling songs in female crickets. Animal Behaviour 51:981–989.

Fisher, R. A. 1930. The Genetical Theory of Evolution. Clarendon Press, Oxford.

Gerhardt, H. C. 1988. Acoustic properties used in call recognition by frogs and toads. Pp. 455–483. *In* B. Fritzsch, M. J. Ryan, W. Wilczynski, T. Hetherington, and W. Walkowiak (eds.), The Evolution of the Amphibian Auditory System. John Wiley and Sons, New York.

Gerhardt, H. C. 1991. Female mate choice in treefrogs: Static and dynamic acoustic criteria. Animal Behaviour 42:615–635

Gerhardt, H. C. 1994. The evolution of vocalization in frogs and toads. Annual Review of Ecology and Systematics 25:293–324

Gerhardt, H. C., and Schwartz, J. J. 1996. Interspecific interactions in anuran courtship. Pp. 603–632. *In* H. Heatwole, and B. K. Sullivan (eds.), Amphibian Biology, vol. 2 Social Behavior, Surrey Beatty and Sons, New York.

Harnad, S. 1987. Categorical Perception: The Groundwork of Cognition. Cambridge University Press, Cambridge.

Hauser, M. D. 1996. The Evolution of Animal Communication. MIT Press, Cambridge, MA.

Kirkpatrick, M., and M. J. Ryan. 1991. The paradox of the lek and the evolution of mating preferences. Nature 350:33–38.

Martens, J. 1996. Vocalizations and speciation in palearctic birds. Pp. 221–240. *In* D. E. Kroodsma, and E. H. Miller (eds.), Ecology and Evolution of Acoustic Communication in Birds. Cornell University Press, Ithaca, NY.

Mayr, E. 1942. Systematics and the Origin of Species. Columbia University Press, New York.

Morris G. K., and J. H. Fullard. 1983. Random noise and conspecific discrimination in *Conocephalus* (Orthoptera; Tetigonnidae). Pp.73–96. *In* D. G. Gwyne, and G. K. Morris (eds.), Orthopteran Mating Systems: Sexual Competition in a Diverse Group of Insects, Westview Press, Boulder, CO.

Nelson, D. A., and P. Marler. 1990. The perception of bird song and an ecological concept of signal space. Pp. 443–478. *In* E. C. Stebbins, and M. A. Berkeley (eds.), Comparative Perception. Wiley, New York.

Oldham, R. S., and H. C. Gerhardt. 1975. Behavioral isolating mechanisms of the treefrogs *Hyla cinerea* and *H. gratiosa*. Copeia 1975:223–231.

Passmore, N. I. 1981. The relevance of the specific mate recognition concept to anuran reproductive biology. Monitore Zoologia Italia 6:93–108.

Paterson, H. E. H. 1985. The recognition concept of species. Transvaal Museum Monograph 4:21–29.

Pfennig K. S. 1998. The evolution of mate choice and the potential for conflict between species and mate quality recognition. Proceedings of the Royal Society London B 265:1743–1748.

Phelps, S. M., and M. J. Ryan. 1998. Neural networks predict response biases in female túngara frogs. Proceeding of the Royal Society London B 265:279–285.

Phelps, S. M., and M. J. Ryan. 2000. History influences signal recognition: Neural network models of túngara frogs. Proceedings of the Royal Society of London B 267:1633–1639.

Ryan, M. J. 1988. Constraints and patterns in the evolution of anuran acoustic communication. Pp. 637–677. *In* B. Fritzsch, M. J. Ryan, W. Wilczynski, W. Walkowiak, and T. Hetherington, (eds.), The Evolution of the Amphibian Auditory System. John Wiley and Sons, New York.

Ryan, M. J. 1991. Sexual selection and communication in frogs: Some recent advances. Trends in Ecology and Evolution 6:351–354.

Ryan, M. J., and A. Keddy-Hector. 1992. Directional patterns of female mate choice and the role of sensory biases. American Naturalist 139:S4–S35.

Ryan, M. J., and A. S. Rand. 1993a. Phylogenetic patterns of behavioral mate recognition systems in the *Physalaemus pustulosus* species group (Anura: Leptodactylidae): The role of ancestral and derived characters and sensory exploitation. Linnean Society Symposium Series 14:251–267.

Ryan, M. J., and A. S. Rand. 1993b. Sexual selection and signal evolution: The ghost of biases past. Philosophical Transactions of the Royal Society B 340:187–195.

Ryan, M. J., and A. S. Rand. 1993c. Species recognition and sexual selection as a unitary problem in animal communication. Evolution 47:647–657.

Ryan, M. J., and A. S. Rand. 1995. Female responses to ancestral advertisement calls in the túngara frog. Science 269:390–392.

Ryan, M. J., and A. S. Rand. 1999a. Phylogenetic inference and the evolution of communication in túngara frogs. Pp. 535–557. *In* M. Hauser and M. Konishi (eds.), The Design of Animal Communication. MIT Press, Cambridge, MA.

Ryan, M. J., and A. S. Rand. 1999b. Phylogenetic influence on mating call preferences in female túngara frogs (*Physalaemus pustulosus*). Animal Behaviour. 57:945–956.

Ryan, M. J., A. S. Rand, and L. A. Weigt. 1996. Allozyme and advertisement call variation in the túngara frog, *Physalaemus pustulosus*. Evolution 50:2435–2453.

Ryan, M. J., and W. E. Wagner Jr. 1987. Asymmetries in mating preferences between species: Female swordtails prefer heterospecific mates. Science 236:595–597.

Searcy, W. A., and P. Marler. 1981 A test for responsiveness in song structure and programming in female sparrows. Science 213:926–958.

Searcy, W. A., and K. Yasakawa. 1996. Song and female choice. Pp. 454–473. *In* D. E. Kroodsma and E. H. Miller (eds.), Ecology and Evolution of Acoustic Communication in Birds. Cornell University Press, Ithaca, New York.

Stoddard, P. K. 1996. Vocal recognition of neighbors by territorial passerines. Pp. 356–376. *In* D. E. Kroodsma and E. H. Miller (eds.), Ecology and Evolution of Acoustic Communication in Birds. Cornell University Press, Ithaca, NY.

Tinbergen, N. 1953. Social Behaviour in Animals. Butler and Tannet, London.

Veech, J. A., J. H. Benedix Jr., and D. J. Howard. 1996. Lack of calling song displacement between two closely related ground crickets. Evolution 50:1982–1989.

Wagner, W. E., Jr. 1998. Measuring female mating preferences. Animal Behaviour 55:1029–1042.

West-Eberhard, M. J. 1979. Sexual selection, social competition, and evolution. Proceedings of the American Philosophical Society 123:222–234.

Wilczynski, W., A. S. Rand, and M. J. Ryan. 1995. The processing of spectral cues by the call analysis system of the túngara frog, *Physalaemus pustulosus*. Animal Behaviour 49:911–929.

9

MURRAY J. LITTLEJOHN

Patterns of Differentiation in Temporal Properties of Acoustic Signals of Anurans

Introduction

The long-range acoustic repertoire of anuran amphibians (sensu Littlejohn 1977) includes two common signals. The conspicuous and most studied of these is the advertisement call (Wells 1977), which is produced by breeding males to attract appropriate mates and to repel calling males. The other signal, thus far studied less than the advertisement call, is emitted by a male when the advertisement calls of neighbors are received at a level above some critical threshold of intensity. In such a situation it is presumed that there is competition between conspecific males for mates, and between males of the same and different species for acoustic space (= active space, Brenowitz 1982) and calling sites (Littlejohn 1977). The terms *territorial call* (Bogert 1960), *encounter call* (McDiarmid and Adler 1974), and *aggressive call* (Wells and Greer 1981) have been applied to this type of signal. Of these, encounter call is preferred, however, because it is based only on the context and, as with the neutral term *advertisement call,* does not invoke any particular function. Although there are other long-range signals within the acoustic repertoires of some species of anuran, such as *Hyla regilla* (Snyder and Jameson 1965; Allan 1973; Whitney 1981) and *Crinia* (= *Ranidella*) *parinsignifera* (Hawe 1970; Littlejohn et al. 1985), their functions have not yet been established.

As is evident from a perusal of many of the papers cited in this chapter, the terminology associated with description of bioacoustic signals presently is lacking in standardization, especially for the temporal structure of such signals, and for envelope shape in particular.

Effective comparisons based on tables and graphs have been made of the advertisement calls of closely related taxa (e.g., Blair 1958; Littlejohn 1959, 1969) and of the advertisement calls and encounter calls within species (e.g., Schwartz and Wells 1984; Littlejohn et al. 1985). Multivariate applications to the assessment of geographic variation in advertisement calls (e.g., Snyder and Jamieson 1965) and the establishment of phylogenetic patterns (e.g., Cocroft and Ryan 1995) have also been developed. A standard measure or coefficient is needed, however, to facilitate the search for general principles of differentiation in key attributes of signals. One such attempt is that of Littlejohn (1999) through the formulation of a univariate coefficient of call difference (discussed below). Restriction of comparisons to taxa that have acoustic signals with similar or overlapping frequency bandwidths enables the emphasis to be placed on the time domain. This focus on temporal differentiation thus allows consideration of the most basic of coding systems that can distinguish the components of repertoires of pairs of closely related syntopic and synchronously breeding taxa. The ap-

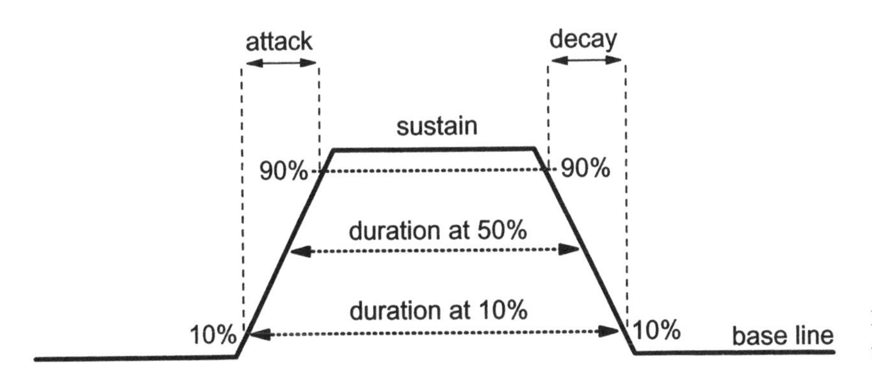

Figure 9.1. Terminology associated with the structure of a monopolar pulse.

proach also avoids the anatomical and physiological complications of production and reception of calls that can be associated with spectral differentiation of systems of acoustic communication.

Accordingly the aims of this chapter are to (1) review the terminology used for the description of bioacoustic signals; (2) consider the nature and possible functional significance of envelope shape; (3) investigate temporal features of the advertisement call and the encounter call within the acoustic repertoire of a taxon; (4) compare the temporal features of advertisement calls—and of encounter calls (where described)—of pairs of closely related species that breed at the same time and place; and (5) review the patterns of diversification of the repertoire, the nature and elaboration of temporal codes, and their significance in mate choice and speciation.

Structure of Bioacoustic Signals

For initial description and measurement, a bioacoustic signal can be viewed in the time domain as (1) a linear bipolar (i.e., with both positive and negative excursions) display of amplitude (usually as voltage), which is termed a waveform (previously oscillogram); or (2) a unipolar relative logarithmic display of power as decibels (dB) against time (time-series energy plot of the amplitude envelope). The signal may also be transformed into the frequency domain as (1) a linear display of frequency against time, with an indication of relative amplitude (logarithmic, dB) in gradations of color or density (audiospectrogram, sonagram or sonogram); or (2) a power spectrum with amplitude (logarithmic, dB) on the vertical axis, and frequency (usually linear) on the horizontal axis.

The basic element of a bioacoustic signal is a tone-burst consisting of at least several cycles of a sinusoidal fundamental frequency, which by analogy with amplitude-modulated radio transmissions can also be termed the carrier frequency. This signal may be an impulse or damped oscillation such as a transient "click" (perhaps of fewer than 20 cycles) or a sustained longer tone. The dominant frequency is that which contains the greatest amount of energy in a power spectrum, and may be the fundamental frequency or one of its harmonics. In other cases much of the energy is contained within one or several sidebands that result from amplitude modulation of a carrier frequency (Watkins 1967; Greenewalt 1968; Jackson 1996), or in harmonics of the fundamental (carrier) frequency. The value of the carrier frequency can usually be determined by inspection of the primary periodicity of an expanded waveform.

Although difficult to define precisely (Gottlieb 1958; Broughton 1963), the term *pulse* is generally applied to the short signals (e.g., Bennet-Clark 1989; Littlejohn and Wright 1997). The shape of the waveform of a pulse, pulse train, or tone-burst is termed its *envelope*, and the bipolar waveform is generally symmetrical about the zero axis or baseline. Thus measurements can usually be made on either side of the zero level; even so, in case there is asymmetry, protocols should be established for consistency. In conventional electronic terminology, derived primarily from a consideration of monopolar direct current pulses (e.g., Gottlieb 1958), the 10% level of attack and the 10% level of decay, relative to the maximum level of amplitude (= 100%), are used as the criteria for measuring the shape and duration of a signal (Figure 9.1). The rise time, or duration of the attack or leading edge of a signal (either a single pulse or a sustained sinusoid), is measured from the 10% to the 90% levels of amplitude, and the fall time (or decay) from the 90% to the 10% levels on the trailing edge (Figure 9.1). The duration of a pulse, note, or call may then be defined as the interval between the 10% levels at the leading and trailing edges of a waveform, or across the 50% level of amplitude, and the value chosen should be stated (Gottlieb 1958). The rise times of pulses that have very abrupt attacks and relatively low carrier frequencies cannot always be measured precisely because only three or four cycles of the fundamental frequency may be present, and so must be estimated or some other measure used—such as the number of cycles of the carrier from the baseline to the peak of amplitude (Littlejohn and Wright 1997). Likewise the

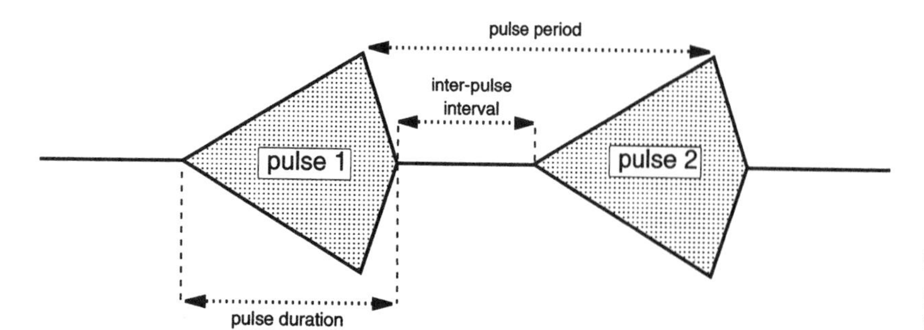

Figure 9.2. Terminology associated with pulse trains. Note that pulse duration has been set at the baseline for this illustration.

envelopes of many acoustic signals have extended and very gradual increases in amplitude, or ramps, so that the use of conventional rise times could result in the exclusion of a biologically significant proportion of the duration of such signals. Clearly other criteria must be applied in these cases; but it is crucial that they always be precisely defined and protocols adopted and consistently used.

Pulses may be repeated at regular intervals to result in a periodic or quasi-periodic pulse train with a pulse repetition rate (or pulse rate) that is the reciprocal of the pulse period (pulse duration plus interpulse interval) (Figure 9.2). The ratio of the pulse duration to the pulse period constitutes the duty cycle. The length (duration) of a pulse train can be arbitrarily decided by the presence of an interval (period of silence) that is at least twice as long as the longest interpulse interval in the pulse train.

The envelope of a sustained tone may be amplitude modulated at a rate (or frequency) much lower than the fundamental frequency, on a scale of 0% (no modulation) to 100% (full modulation) where the amplitude is reduced to that of the baseline. By convention, the level of amplitude modulation is expressed as a percentage (% AM) of an unmodulated carrier (x). Because this value is generally unknown in bioacoustic signals, the level of amplitude modulation can be calculated from the ratio of the sum and difference of the maximum (y) and minimum (z) values of amplitude of a waveform as: $\% \text{AM} = [(y - z) \div (y + z)] \times 100$ (Greenewalt 1968; American Radio Relay League 1978; Gerhardt and Davis 1988). The outcome of 100% amplitude modulation of a carrier frequency is an envelope pattern that is similar to that of a primary pulse train formed by the repeated regular production of pulses, but with a different basis for its formation. Greenewalt (1968) has provided detailed explanations for the several ways in which amplitude modulation can occur in acoustic communication and has discussed the process for birds. Bennet-Clark (1989, 1995) reviewed sound production and amplitude modulation in terrestrial insects. Martin (1972) and Gans (1973) described the possible mechanisms of amplitude modulation for call production in anurans. Distinct pulse trains in anuran calls can originate in two ways: (1) as a consequence of amplitude modulation of a carrier by the active or passive action of secondary membranes (e.g., arytenoid cartilages) in the larynx, and (2) as the successive production of discrete pulses by the active contraction of muscles of the body wall. Active processes are under neuromuscular control and are correlated with the temperature of the source. If the structure and mechanism of production of a pulse train can be determined, then this information could be useful in the establishment of phylogenic patterns—which may also be reflected in the structure of the larynx (Martin 1972).

The simplest basic or primary structure of an anuran acoustic signal (call) is a single sustained sequence of sine waves (tone-burst), a single pulse, or a pulse train that is separated from the next train by an interval of silence much longer than the interpulse interval of the signal (Figure 9.3). It is here suggested that an arbitrary definition of the call and the setting of its limits could be based on the intercall interval being at least of the same duration as the pulse, pulse train, or tone-burst that constitutes the call. A secondary structure to the call may be present when there are groups of pulse trains or tone-bursts, each separated by intervals shorter than the pulse train, followed by a longer interval, here arbitrarily defined as greater than twice the longest interval between the primary units (Figure 9.3). Where such a secondary structure is present, the primary units are termed *notes* (Littlejohn 1965). Monophasic calls consist of several notes of similar structure, and diphasic calls have two distinct types of note (Figure 9.3) (Littlejohn 1977). Michelsen (1985) and Bennet-Clark (1989) described comparable sets of terminology for the sounds of insects.

Naturally produced and broadcast bioacoustic signals are usually recorded for analysis under conditions of relatively high levels of background noise. Thus the beginning and the end of a waveform cannot be objectively ascertained because one or both may be concealed within the baseline level of noise. Even in recordings obtained under virtually noise-free conditions, the beginnings of positive linear and exponential ramps and the ends of negative exponential curves are by definition indeterminate. The maximum amplitude is, however, often present as a clear peak that can be used as a reference point for the measurement of inter-

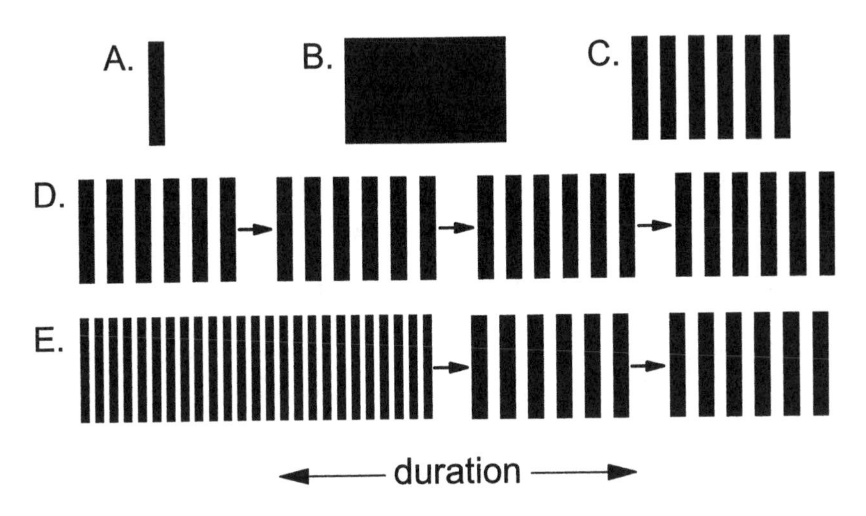

Figure 9.3. Types of call based on temporal patterns: (A) Short single note or pulse. (B) Long unpulsed single note. (C) Single periodic pulse train. (D) Monophasic call consisting of a group of similar pulse trains. (E) Diphasic call consisting of two distinct types of pulse train.

pulse intervals and the consequent calculation of pulse rate (Littlejohn and Wright 1997).

Production and Reception of Pulses

The degree of tuning of a resonant system or a filter is expressed by the symbol Q (quality factor) which is usually determined from the ratio of the center (dominant) frequency to the bandwidth at the 50% power level, which is at –3 dB (i.e., Q_{3dB}) (Greenewalt 1968; Bennet-Clark and Young 1992; Bennet-Clark 1999). Another value of Q, defined as the ratio of the center frequency to the bandwidth at –10 dB (i.e., Q_{10dB}), has been used in neurobiology, especially in describing tuning curves of receptors. Thus the best excitatory frequency (BEF) is the frequency at which a unit has its lowest threshold, and the sharpness of tuning is measured as the BEF divided by the bandwidth of the tuning curve at 10 dB above the threshold (e.g., Rose and Capranica 1984). Bennet-Clark (1999) critically evaluated the two measures of sharpness of tuning (Q_{3dB}, Q_{10dB}) and pointed out that "the two terms have been used in separate contexts and they measure different things" (p. 351). Given the origin of Q in electronics and physical acoustics (Bennet-Clark 1999) and its fundamental derivation, Q_{3dB} is preferred in considerations of resonant systems and descriptions of frequency spectra. The advertisement calls of many species are composed of pulses with very short rise times of about 1 ms; for example, *Crinia glauerti* and *Crinia signifera* (Littlejohn and Wright 1997) and *Acris crepitans* (e.g., Figures 6 and 7 of Nevo and Capranica 1985; N. M. Kime, personal communication). Such calls generate high but brief peak sound-pressure levels and may increase the maximum peak signal strength, while maintaining a low average power for small frogs with small driving muscles in their body walls (Loftus-Hills and Littlejohn 1971). The bandwidths of such signals are wide, especially in the attack phase when the amplitude reaches its maximum, because as the rise time decreases the leading edge approaches that of a square wave—which consists of the sum of a fundamental and its odd harmonics. There is thus a tradeoff between a greater range of effective propagation obtained with instantaneously high sound-pressure levels (Loftus-Hills and Littlejohn 1971) and the efficiency of transmission and reception that can be achieved with narrow bandwidths.

Bennet-Clark and Young (1992) have considered the level of tuning in acoustic systems of cicadas (Insecta: Homoptera: Cicadidae) with respect to the quality factor (Q_{3dB}) of a resonant system and the role of resonators in the production and reception of the clicks in songs of adult males. For a received signal in a resonant system of a female of the cicada *Cyclochila australasiae* to reach within 1 dB of its steady-state level, the driving pulse must persist for 5.5 cycles, where $Q_{3dB} = 8$. Thus "in a well-adapted song pulse, one might expect a build-up over about five cycles, then a plateau lasting five cycles (or 1.2 ms) and then a decay over five cycles. At a carrier frequency of 4.3 kHz, this gives a total pulse duration of 3.5 ms, or a pulse rate of 280 Hz" (Bennet-Clark and Young 1992, p. 152). For estimated values of Q_{3dB} of 7.4 (female) and 8.0 (male), and center (carrier) frequencies of 3700 (female) and 4000 Hz (male) in the auditory tuning curves of *Acris crepitans* (from an inspection of Figure 4B of Wilczynski et al. 1992), the minimum buildup of a received pulse is calculated at about 4 ms, and the duration of an ideal pulse would be about 12 ms. Inspection of figures of Nevo and Capranica (1985), however, indicates that rise times of pulses are about 1 ms and pulse durations are about 5–7 ms in the advertisement calls of *A. crepitans*. These estimates are supported by N. M. Kime (personal communication) with data (adjusted to 20°C) from a population of *A. crepitans* in central Texas (Gill Ranch, Austin), in which the mean rise time is 1.2 ms and the mean pulse duration is 6.2 ms (range, 3.9–7.7 ms). In *Crinia signifera* the minimum rise time of a pulse in the advertisement call is less than 2 ms and involves five cycles of the carrier frequency (Littlejohn and Wright 1997).

Gerhardt and Doherty (1988) made a detailed comparison of the duty cycles and shapes of pulses in advertisement calls of *Hyla chrysoscelis* and *H. versicolor.* No absolute measurements of durations and rise times are provided. The duty cycles (as pulse duration divided by pulse period) are similar, with means of 0.44 to 0.59 being contiguous at < 15°C and overlapping extensively at 20 and 24°C. From an inspection of Figure 1 of Gerhardt and Doherty (1988), pulse durations are estimated to be 14 ms for *H. chrysoscelis* and 30 ms for *H. versicolor,* and the relatively long rise times to be 4 ms and 20 ms, respectively.

Another problem to be considered for species with pulsatile advertisement calls, especially those with short rise times, is whether the tuning curves and thresholds of best frequencies are the same when stimuli with pulse characteristics similar to those of conspecific advertisement calls are used as stimuli, rather than sustained tones (e.g., 705 ms, Loftus-Hills 1973; 300 ms, Wilczynski et al. 1992). Another possible process that may allow for greater efficiency of transduction of pulsatile calls at the receiver is "impulsive excitation" (Fletcher 1992), which can occur with the arrival of a broadband pulse with a short rise time at the surface of a resonant receptor, and results in a much longer period of oscillation in the receiver at its tuned frequency. The time constants for temporal integration (Narins 1992) in the auditory receptors of species with advertisement calls composed of short pulses would also be of interest.

Accurate measurement of sound-pressure levels of short calls or where there are sharp attacks in pulsatile calls (i.e., transient signals with rise times of a few milliseconds), and the different rates and levels of amplitude modulation in longer calls and notes, can present problems. Only the peak sound-pressure level (PSPL) obtained by a detector and indicator system of a precision sound-level meter with an integration time/time constant of < 50 μs (Standards Australia 1990; Johnson et al. 1991) is meaningful for such signals. Standard integration times of the other settings on sound-level meters are far too long for calls of most anurans, namely, "slow" = 1000 ms, "fast" = 125 ms, "impulse" = 35 ms (Standards Australia 1990; Yeager and Marsh 1991). Furthermore the duration of the signal being measured should exceed the integration time by a factor of 3–4 to allow accurate measurement (Prestwich et al. 1989; Yeager and Marsh 1991). Thus for the fast setting the shortest unmodulated signal that can be accurately measured would have a duration of at least 375 ms. A wide-band linear frequency response (flat weighting) is recommended for measurement of the peak sound-pressure level to allow for the presence of high frequency components. The effect of introduction of a bandpass filter on the indicated peak sound-pressure level should be carefully determined by comparison with the value of the unfiltered signal. Because they only had access to a precision sound-level meter with slow and fast RMS (root mean square) ballistics (Brüel and Kjaer type 2203), Loftus-Hills and Littlejohn (1971) used an alternative means for determining peak values for sound-pressure levels in the field. A microphone was placed 50 cm from the subject and 15–20 cm above the substrate. The amplified signal from the microphone was then displayed directly on the screen of an oscilloscope and the amplitude measured. A free-field calibration was then obtained for a sustained sine wave of a frequency similar to the dominant frequency of the call replayed through a loudspeaker, and the amplitude was adjusted to match that of the call (previously measured). The RMS sound-pressure level of the sine wave was then measured at a standard distance to obtain the value of the sustained long tone that was equivalent to the maximum RMS value of the call. As pointed out by Gerhardt (1975), 3 dB should be added to these RMS-equivalent peak values to obtain true peak values. Littlejohn (1998) has also discussed historical aspects of techniques for determination of the peak sound-pressure levels of calls.

Temporal Features and Determination of Distance and Direction

Patterns of radiation of advertisement calls for most of the species so far measured are close to omnidirectional (Gerhardt 1975; Narins and Hurley 1982; Robertson 1984; Prestwich et al. 1989), as may be expected where the vocal sac, as a monopole source, is a pulsating sphere with dimensions much smaller than the wavelength of the carrier frequency. Location of the source of advertisement calls by the receiver is based on an internally acoustically coupled biaural system that operates on the principle of an asymmetrical pressure difference (Eggermont 1988). The discovery of an additional pulmonary pathway (Narins et al. 1988) for sound transmission to the inner ear has complicated the original simple model. Ehret et al. (1990) noted that the pulmonary pathway favors transmission of lower frequencies, especially around 1 kHz, and so could be of significance for reception and localization of certain sounds, such as the Co note in the diphasic advertisement call of *Eleutherodactylus coqui.*

Because of spherical divergence, the sound-pressure level of a signal decreases by 6 dB for each doubling of the distance between the source and the receiver. The sound-pressure level can be further reduced by absorption, reverberation, and boundary effects (excess attenuation). As a general fundamental neurobiological process, determination of the level or strength of the received signal—and of the associated distance between receiver and source—is based on a positive relationship between the intensity of a

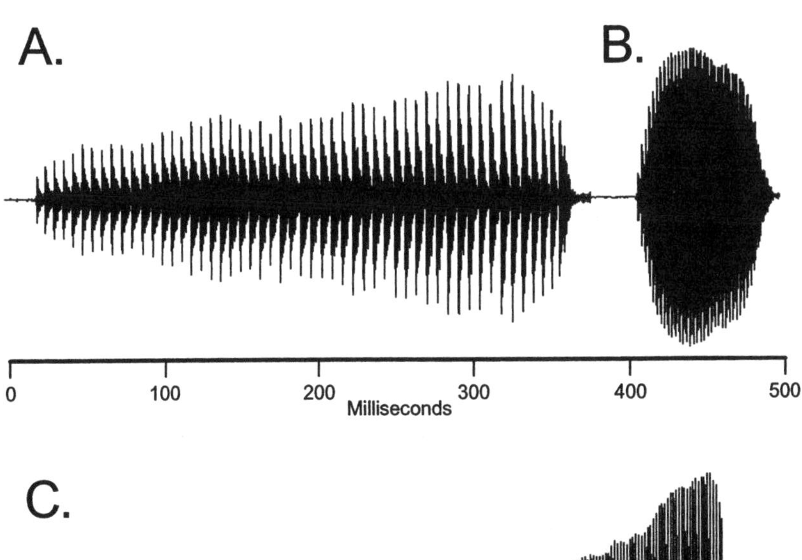

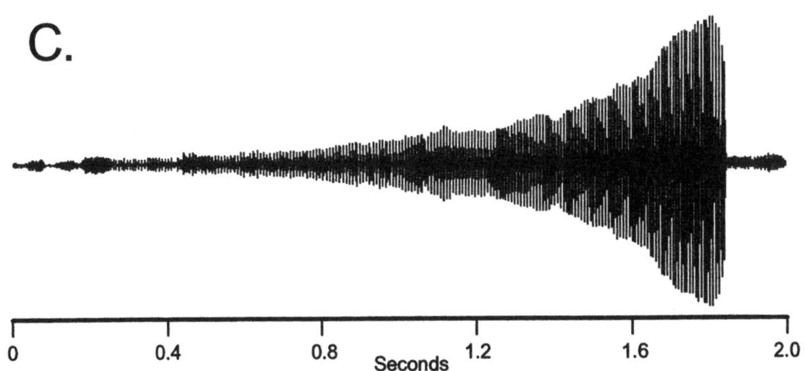

Figure 9.4. Waveforms of the three components in the acoustic repertoire of a male of *Geocrinia victoriana*, recorded in the Otway State Forest at a wet-bulb air temperature of 12.2°C. Upper panel (Reference: reel PB 83/2–4): notes from a naturally produced advertisement call: (A) introductory note, (B) one of a series of repeated notes. Lower panel (Reference: reel PB 83/3–4): (C) long call (= encounter call) evoked during a field playback experiment.

stimulus and the rate of production of action potentials, and different thresholds of sensitivity for different sensory neurons. Another possible mechanism for indicating the distance between a receiver and a source could be based on the ramps of amplitude in envelopes that are found in the long-range acoustic signals of many taxa. When referring to sawtooth pulses (which have linear ramps) in auditory pattern recognition in an acridid grasshopper, von Helversen (1993, p. 638) noted that "the pattern changes with increasing distance. Pulses are perceived to be shorter (reduced by that amount remaining below threshold) while the intervals appear enlarged." Furthermore "steepness [of a pulse] as such is perceived differently by a receiver nearby or far away." When considering pulse trains it can be added that, as well as the change in the detected envelope shape and duration of a signal noted by von Helversen (1993), the number of pulses detected would increase with the improvement of signal-to-noise ratio as the distance between source and receiver decreased, thus providing another possible code for proximity. Such a system could operate at the three levels of structural complexity—pulse, note, or call. The detection of such temporal changes with each successive movement of approach by a positively phonotactic receiver would thus provide a greater "dynamic range" in that information could still be obtained even though there was saturation of the sensory neurons. Envelopes with gradual rise times but abrupt fall times would provide clear timing reference points for call, note, and pulse repetition rates. The production of pulses with sharp attacks or clear peaks of amplitude, and of envelopes of long tones and trains of notes with abrupt terminal reductions in amplitude would allow assessment of rates from the relatively constant interpulse intervals or internote intervals, even though the perceived duration of the signal was increasing as the distance between the receiver and the source decreased. Thus the values of these temporal properties of a signal would be maintained along a gradient of increasing sound-pressure level and improving signal-to-noise ratio. Possible neural mechanisms for the processing of information provided by ramps with different rise times are discussed by von Helversen (1993).

Although both linear and exponential ramps of amplitude are commonplace in the calls of anurans, their presence and possible functions have not been the subject of discussion in earlier publications. Ramps can be identified clearly in waveforms as protracted rise times, and their presence can be implied from audiospectrograms by the gradation of increasing intensity and breadth of the marking at the leading edges of signals. For example in *Geocrinia victoriana* (Figure 9.4) there are pronounced ramps in introductory notes of advertisement calls and in long calls (= encounter calls) where the dynamic range can exceed 25 dB. There are also pronounced ramps in the notes of the advertisement calls of *Litoria ewingii* and *L. verreauxii*, with amplitude ranges of 15 dB and 23 dB,

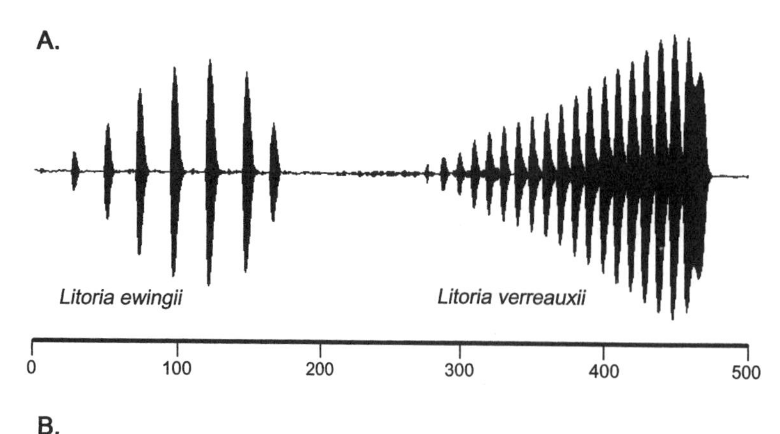

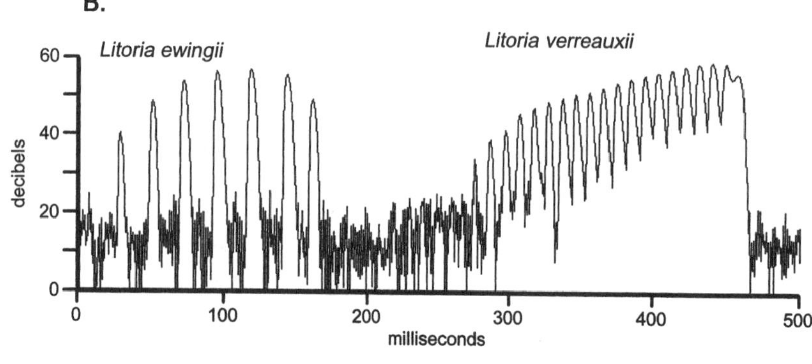

Figure 9.5. Single note from near the middle of the advertisement call of a male *Litoria ewingii* (Reference: reel R464–9, recorded at a wet-bulb air temperature of 9.3°C; left), and of a male of *L. verreauxii* (Reference: reel R464–6, recorded at a wet-bulb air temperature of 9.0°C; right) from a syntopic chorus at Wallagaraugh Retreat. (A) Waveforms. (B) Displays of relative amplitude with time.

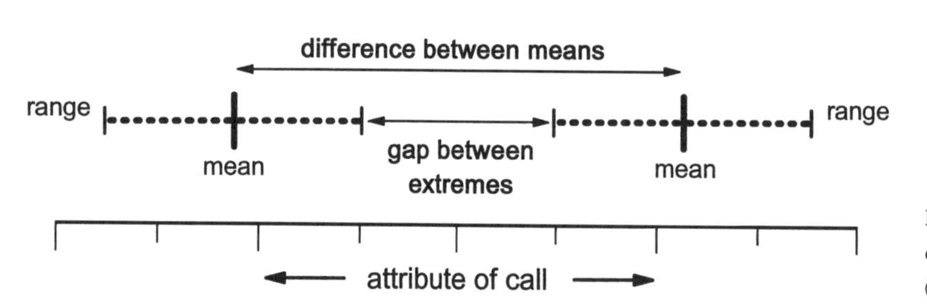

Figure 9.6. Derivation of values used for calculating the coefficient of call difference (CCD).

respectively (Figure 9.5), and in encounter calls of the latter species (Figure 5.2 of Harrison 1987).

Comparisons of Signals within Repertoires and Between Species

In the comparison of acoustic signals, both within and between taxa, attention is generally directed at the most distinctive or key attributes that do not overlap in ranges of variation and can thus provide the basis for coding the specificity of a signal. Measures of central tendency (mean, median, mode) generally are robust, especially where sample sizes are large; and Littlejohn (1965, 1969) noted that mean values of pulse rates in the advertisement calls of several closely related pairs of anuran species differed by a factor of about two or more. Published ranges of variation, however, may (1) include exceptional outliers, (2) be based on different sampling protocols (e.g., use of the average of values from several calls of each individual or of one value from one call of each individual), or (3) depend on the type of temperature measured (e.g., cloacal, dry-bulb air, or wet-bulb air). To facilitate the comparison of advertisement calls of pairs of closely related anuran taxa, a coefficient of call difference was developed by Littlejohn (1999) as the ratio of the gap between the extreme values at the converging ranges of variation to the difference between means of a key attribute of a signal (Figure 9.6).

As an extension of this approach to the differentiation of acoustic signals, I will first compare the key factors in the advertisement calls and encounter calls within species to obtain a measure of the elaboration of repertoire and then explore further the differentiation in pulse rates in the advertisement calls of pairs of taxa. Cases where two types of signals were produced by subjects during playback experiments with a conspecific advertisement call as the stimulus (*Crinia* [= *Ranidella*] *parinsignifera;* Hawe 1970; Littlejohn et al. 1985), or where the responses are complex and highly variable (*Rana berlandieri;* Gambs and Littlejohn 1979), are not discussed in this account. Thus only those taxa with signals that are of relatively simple structure and differentiated mainly in one or two attributes will be considered at this early stage of development of the coefficient of call difference. Accordingly a univariate approach is adopted, primarily to explore the simplest patterns of temporal differentiation.

Comparisons of Advertisement Calls and Encounter Calls in Single-Species Systems

Crinia signifera

As indicated in the brief overview by Littlejohn and Wright (1997), the acoustic repertoire of this widely distributed southeastern Australian myobatrachid species has been the subject of several detailed studies. Two discrete and distinctive types of call were identified: advertisement call and encounter call. Straughan and Main (1966) established the mate-attracting function of the advertisement call, and the territorial function of the encounter call was verified through field playback experiments of advertisement calls by Hawe (1970) and Littlejohn et al. (1985). Audiospectrograms of the two types of call are provided by Littlejohn et al. (1985), and waveforms by Littlejohn and Wright (1997).

Both types of call consist of short quasi-periodic pulse trains. The calls differ markedly in duration and pulse rate, with the advertisement call being shorter and of lower pulse rate than the encounter call (Table 9.1). Dominant frequencies overlap extensively and range from 2611 to 3600 Hz (means: advertisement call, 3145 Hz; encounter call, 3200 Hz) for calls of five subjects studied in the Willowmavin area (37° 16′S, 144° 54′E), about 60 km north of Melbourne, Victoria, by Littlejohn et al. (1985). Because of the small size of the samples, and the narrow range of effective temperatures (sensu Littlejohn and Wright 1997—wet bulb air or surface water, depending on the calling site) of 10.7 to 13.2°C (mean, 12.0°C), the original, uncorrected data are considered. Means and ranges for durations (from peak of first pulse to peak of last pulse) and pulse rates of advertisement calls and encounter calls of these five subjects are presented in Table 9.1. The coefficient of call difference is 0.65 for duration and 0.34 for pulse rate (Table 9.1).

Table 9.1 Values of two attributes and calculation of coefficients of call difference (CCD) of advertisement and encounter calls of *Crinia* (= *Ranidella*) *signifera* from Willowmavin, Victoria, Australia

Attribute	Duration (ms)	Pulse Rate (pulses/sec)
Advertisement call (mean, range)	83 (56–108)	41 (31–58)
Encounter call (mean, range)	192 (179–210)	88 (74–100)
Ratio of means	2.31	2.15
Difference in means ($\Delta\bar{x}$)	109	47
Gap in ranges (gap)	71	16
CCD (gap/$\Delta\bar{x}$)	0.65	0.34

Data are derived from 1 call of each type for 5 individuals.

Wet-bulb air temperatures at the calling sites were 10.7–13.2°C (mean, 12.0°C); values are not adjusted for temperature.

See Littlejohn et al. (1985) for further details.

Geocrinia victoriana

The acoustic repertoire of this southeastern Australian myobatrachid species was described by Littlejohn and Harrison (1985) on the basis of field playback experiments in an area of the Otway State Forest (38° 28′S, 143° 31′E), 16 km southwest of Colac in western Victoria. The data presented in Table 9.2 were derived from the laboratory records for 10 of the subjects in the playback experiments in Table 2 of Littlejohn and Harrison (1985). The subjects were at their natural calling sites and were presented with either the complete call or an introductory note at peak sound-pressure levels above about 110 dB at the subject as stimuli.

Audiospectrograms of the three types of note were provided by Littlejohn and Harrison (1985), and waveforms are presented in Figure 9.4. The advertisement call of this species is strongly diphasic (Figure 9.4), and consists of one (or rarely two or three) longer introductory notes of low pulse rate followed by a series of shorter repeated notes of higher pulse rate (Table 9.2). After presentation of one or the other of the above stimuli, the calling patterns of the subjects were altered—notes of structure similar to the introductory note but of longer duration and lower pulse rate, termed *long calls,* were emitted (Table 9.2), and the number of repeated notes was greatly reduced or production of them ceased (Littlejohn and Harrison 1985).

The following points emerge from an inspection of Table 9.2. The ranges of durations of introductory notes and long calls are separated by a small gap, and durations of repeated notes are much shorter than those of the other two types of note. Pulse rates of introductory notes and long calls overlap extensively, but those of the repeated notes are separated from the others by a considerable gap, and values are about three times higher. The means for dominant frequencies of the three types of note are very similar and ranges of variation overlap extensively (2480 and 3067 Hz). The coefficient of variation is highest (41.8%) for the durations of the long calls and very low (5.0–5.6%) for dominant frequencies of all three classes of note and the pulse rates of the repeated notes

Table 9.2 Values for three components of the acoustic repertoire of *Geocrinia victoriana* from a population in the Otway State Forest, Victoria, Australia

	Duration (ms)		Pulse Rate (pulses/sec)		Dominant Frequency (Hz)	
	Mean (range)	CV %	Mean (range)	CV %	Mean (range)	CV %
Introductory note of advertisement call[a]	376 (309–483)	16.5	144 (118–162)	9.9	2734 (2592–3067)	5.6
Repeated note of advertisement call[a]	76 (57–86)	10.8	439 (392–475)	5.0	2749 (2592–3046)	5.3
Long call[b] (= encounter call)	1076 (522–2110)	41.8	115 (81–146)	15.3	2710 (2480–2913)	5.0

Source: From original data of Littlejohn and Harrison (1985).

Wet-bulb air temperatures at calling sites were 9.7–12.3°C (mean, 11.5°C); values not adjusted for temperature.

[a] From 1 call each of 10 naturally calling males.

[b] Signals produced during field playback experiments to the same 10 individuals with stimulus of either a complete advertisement call (to 5 males) or an introductory note (to 5 males).

CV = coefficient of variation.

(5.0%) (Table 9.2). Thus the long calls are the most variable and the repeated notes the least variable in the two temporal attributes. Based on the appropriate values given in Table 9.2, the coefficient of call difference for duration is 0.06 for introductory notes and long calls, 0.74 for introductory notes and repeated notes, and 0.44 for the long calls and repeated notes. For pulse rates, the coefficient of call difference is 0.78 for introductory and repeated notes and 0.76 for long calls and repeated notes. Although the gap in ranges of variation of durations is small (39 ms) for each subject in the experiments that provided the data for Table 9.2, the duration of the long call is greater than that of the introductory note. Given the experimental context in which it was evoked, the long call may be considered as an encounter call. But as suggested by Littlejohn and Harrison (1985), the transition from diphasic advertisement calls to monophasic long calls is probably gradual rather than discrete.

Differentiation in Key Temporal Factors of Calls in Two-Species Systems

Litoria ewingii and *L. verreauxii*

These widely distributed southeastern and eastern Australian hylids have an extensive and linear zone of sympatry in coastal regions (Littlejohn 1965, 1982, 1999). The advertisement calls are similar, each consisting of a regular series of pulse-modulated notes. Littlejohn (1965) separated the samples from the elongated zone of sympatry, which has a length of about 800 km, into western and eastern regions and then considered pooled data (note duration, pulses per note, pulse rate, and dominant frequency) for each region, corrected to 10°C where appropriate. There is pronounced geographical variation in note structure across the zone of sympatry, especially in pulse rate (both species) and depth of amplitude modulation (*L. verreauxii*) (Littlejohn 1965; Littlejohn and Watson, unpublished data). Of particular interest is the progressive reduction from east to west through the zone of sympatry in depth of amplitude modulation in notes of advertisement calls of *L. verreauxii,* and the associated increased variation and loss of pulse modulation as a temporal coding pattern in the calls of some individuals (Littlejohn 1965; Gerhardt and Davis 1988).

The spectral structure of the advertisement call of each species consists of a clear carrier frequency together with the sidebands from the amplitude modulation (Littlejohn et al. 1993; Littlejohn and Watson, unpublished data). The carrier frequencies are similar in eastern sympatry (*L. ewingii* mean, 2307 Hz; *L. verreauxii* mean, 2252 Hz), and ranges of variation are coincidental (Littlejohn 1965). Ranges of variation of carrier frequencies overlap slightly in western sympatry, and the means differ by 568 Hz (*L. ewingii* mean, 2622 Hz; *L. verreauxii* mean, 2054 Hz; Littlejohn 1965).

Littlejohn and Loftus-Hills (1968) and Loftus-Hills and Littlejohn (1971) established that pulse rate is sufficient as the key factor distinguishing the advertisement calls of sympatric populations and is the basis for discrimination and mate choice by females in the western sympatry. Harrison (1987) investigated the structure of advertisement calls and encounter calls of both species in western sympatry at one locality near Melbourne, Victoria (Yan Yean, 37° 33′S, 145° 08′E). He also corrected the data to 10°C, where appropriate, before making comparisons, but did not include ranges of variation in his tables. The means obtained by Harrison (1987) for the advertisement calls of each species at 10°C are close to those obtained by Littlejohn (1965) for western sympatry, with those for pulse rates differing by less than three pulses per second and dominant frequencies by less than 60 Hz ($n = 28$ for *L. ewingii* and $n = 16$ for *L. verreauxii*). Gerhardt and Davis (1988) made a detailed study of variation in pulse rates and amplitude modulation in several western sympatric populations of *L. verreauxii;* they concluded that, whereas pulse rates were stable in recordings obtained from the same individual at different times, the depths of amplitude modulation differed, suggesting a high degree of intra-individual variation in this temporal attribute. Gerhardt and Davis (1988) obtained a mean value ($n = 24$) of 127.0 pulses per second at 10°C for their sample of *L. verreauxii* from western sympatry, which is also close to the previously cited values; however, ranges of variation were not included.

Littlejohn (1999) used the data for pulse rates from the original files and Table 1 of Littlejohn (1965) for the combined samples of advertisement calls obtained in western and in eastern sympatry by Littlejohn (1965) to calculate coefficients of call difference. For pulse rates corrected to an effective temperature of 10°C, the coefficient of call difference for western sympatry was 0.50, and that for eastern sympatry was 0.26. Because the latter value was the lowest obtained for several species, Littlejohn (1999) considered that it might be explained as a consequence of combining several samples from a wide geographic area. The limitations of the earlier technology and methodology may also have contributed by increasing the variances of both samples, with resultant expansions in ranges of variation. Accordingly pulse rates of advertisement calls of the two species were determined from recordings obtained with better equipment for temperature measurement (Takara D611 thermistor thermometers), recording (Beyer M 88 cardioid dynamic microphones, Nagra III, 4.2, 4-S tape recorders), and analysis (Kay Elemetrics DSP-5500 digital audiospectrograph) than that available to Littlejohn (1965). To avoid the possibility of increased variation that could arise from the combination of recordings from several populations within an area (Littlejohn 1999), samples from two syntopic sites separated by about 420 km were analyzed: one from western sympatry near Melbourne, Victoria (Yan Yean), and the other from eastern sympatry in East Gippsland, Victoria (Wallagaraugh Retreat, 37° 27′S, 149° 41′E) (Littlejohn and Watson, unpublished data). After exclusion of values for calls of two putative hybrids from Yan Yean, largely on the basis of intermediate pulse rates (Littlejohn and Watson, unpublished data), these values were then used to provide means and ranges adjusted to different temperatures. Statistical calculations were carried out with SYSTAT for Windows (Releases 7.0, 8.0, SPSS Inc., Chicago).

Linear regressions of pulse rates against effective temperatures indicated that the slopes differ significantly between species at each locality ($p < 0.001$), and diverge with increasing temperature. For each species the slopes also differ significantly between localities ($p < 0.001$). Coefficients of determination (r^2) ranged from 0.90 to 0.98 for both species and both localities. After logarithmic transformation (base 10) of the data, the recalculated linear regression coefficients did not differ significantly between localities for each species or between species. The probabilities of significance of the interaction terms from analyses of covariance are *L. ewingii*, $p = 0.515$ ($n = 44$); *L. verreauxii*, $p = 0.427$ ($n = 61$), and between species for both localities combined, $p = 0.595$ ($n = 105$). As the established effective temperature for comparison of pulse rates in advertisement calls of the two taxa with untransformed values is 10°C (Littlejohn 1965; Gerhardt and Davis 1988), normal and log-transformed values that have been corrected to this temperature are also provided (Table 9.3). To indicate the effect of a higher temperature on the coefficient of call difference, both untransformed and transformed values of pulse rate were also corrected to 15°C (Table 9.3), an effective temperature well within the ranges of recording at both sites. For comparisons between localities, both untransformed and transformed values were corrected to the grand mean temperature for species and localities of 11.85°C (Table 9.3). Coefficients of call difference were also calculated from the untransformed and log-transformed means and ranges at the three temperatures (Table 9.3).

As discussed earlier, exceptional values (outliers) would have a great effect on the determination of the gap between the ranges for a key attribute of acoustic signals being compared. Because there are no assumptions about the distribution, and the median and interquartile range are not sensitive to extreme values as are estimations based on normal statistics, box plots (McGill et al. 1978; Frigge et al. 1989) provide an alternative method for the comparison of acoustic signals. This procedure may thus provide a more robust basis for determination of a coefficient of call difference. Accordingly the data for log-transformed pulse rates, adjusted to the grand mean of effective temperature (11.85°C), for the samples of *L. ewingii* and *L. verreauxii* from Yan Yean and Wallagaraugh Retreat were also analyzed by these exploratory procedures (Figure 9.7). Although upper values for both samples of advertisement calls of *L. ewingii* fall within the fences (sensu Frigge et al.1989), there is one extreme value (outlier) on the lower side in each sample of *L. verreauxii* (Figure 9.7). By excluding these values the gap is increased from 43.3 to 49.5 for the sample from Yan Yean and from 45.5 to 51.2 for that from Wallagaraugh

Table 9.3 Determination of coefficients of call difference (CCD) for pulse rates at three temperatures (see text for explanation) for samples of advertisement calls from two syntopic populations of *Litoria ewingii* and *L. verreauxii* in south-central and southeastern Victoria, Australia

Locality Treatment	Effective Temperature (°C)	Mean Pulse Rate (Pulses/sec) *L. verreauxii*	*L. ewingii*	Ratio of Means	Difference in Means (Δx̄)	Minimum Pulse Rate *L. verreauxii* (y)	Maximum Pulse Rate *L. ewingii* (z)	Gap (y − z)	CCD (gap/Δx̄)
Yan Yean[a]									
Normal (linear)	10.00	131.81	66.07	2.00	65.74	108.96	73.18	35.78	0.54
	11.85	154.35	76.47	2.02	77.88	131.50	83.59	47.92	0.62
	15.00	192.69	94.17	2.05	98.52	169.84	101.28	68.56	0.70
Transformed (antilog)	10.00	129.73	64.05	2.03	65.68	107.66	71.09	36.57	0.56
	11.85	148.70	72.92	2.04	75.78	123.40	80.06	43.33	0.57
	15.00	187.52	92.58	2.03	94.94	155.62	102.76	52.86	0.56
Wallagaraugh Retreat[b]									
Normal (linear)	10.00	109.91	54.83	2.00	55.08	90.82	58.33	32.49	0.59
	11.85	127.15	62.82	2.02	64.33	108.06	66.32	41.74	0.65
	15.00	156.47	76.41	2.05	80.06	137.38	79.91	57.47	0.72
Transformed (antilog)	10.00	109.96	53.21	2.07	56.76	94.15	56.93	37.21	0.66
	11.85	125.50	60.73	2.07	64.78	107.45	64.98	42.47	0.66
	15.00	157.13	76.03	2.07	81.10	134.53	81.35	53.17	0.66

Sample sizes:

[a]*L. ewingii, n* = 28; *L. verreauxii, n* = 38.

[b]*L. ewingii, n* = 16; *L. verreauxii, n* = 23.

Retreat. The difference in the medians is smaller than that for the means for the sample from Yan Yean (78.58 versus 74.70), and greater for the sample from Wallagaraugh Retreat (64.78 versus 66.65). The coefficient of call difference, based on exclusion of outliers, for the gap and the median rather than the mean is higher for both localities, increasing from 0.56 to 0.66 for Yan Yean and from 0.66 to 0.77 for Wallagaraugh Retreat. Clearly there is scope for further consideration and development of this coefficient. The similarity of the values of coefficient of call difference for Yan Yean and Wallagaraugh Retreat (e.g., 0.54 and 0.59 for untransformed values adjusted to 10°C; Table 9.3) supports the suggestion that the low value of the coefficient of call difference for eastern sympatry (0.26 at 10°C) calculated by Littlejohn (1999) is a consequence of combining samples from several localities within the region and using a common regression coefficient, together with the greater variation resulting from earlier procedures of recording, analysis, and measurement of effective temperature.

Harrison (1987) evoked encounter calls from males of both species by field playback experiments in which conspecific and heterospecific advertisement calls were used as stimuli. In contrast to the clear and regular note structure of the advertisement calls, the encounter calls were highly variable in temporal structure. Even so, Harrison (1987) obtained values for durations and pulse rates of encounter calls for comparison with advertisement calls at 10°C and then calculated coefficients of variation (Table 9.4). As ranges of variation were not included by Harrison (1987), coefficients of call difference could not be calculated. For both species the durations and pulse rates are lower, and the coefficients of variation are higher, in the encounter calls than in the conspecific advertisement calls. In playback experiments with conspecific advertisement calls as the stimulus and presented at peak sound-pressure levels above a mean threshold of about 101 dB, 2 of 16 males of *L. ewingii* tested continued to produce advertisement calls; the others changed over to encounter calls. Of 16 males of *L. verreauxii* presented with the conspecific advertisement call above the mean threshold of peak sound-pressure level of about 104 dB, 5 produced both types of call, 10 produced encounter calls, and 1 became silent. When presented with advertisement calls of *L. verreauxii* above a critical threshold of sound-pressure level, only 2 of 13 males of *L. ewingii* tested responded with encounter calls, 8 continued to produce advertisement calls, and 3 ceased calling. Six of the seven males of *L. verreauxii* tested, however, responded to playback of advertisement calls of *L. ewingii* by producing only encounter calls—in much the same way as they would have responded to conspecific advertisement calls; the other subject continued to produce advertisement calls. Thus there is an asymmetry of response to the presentation of heterospecific ad-

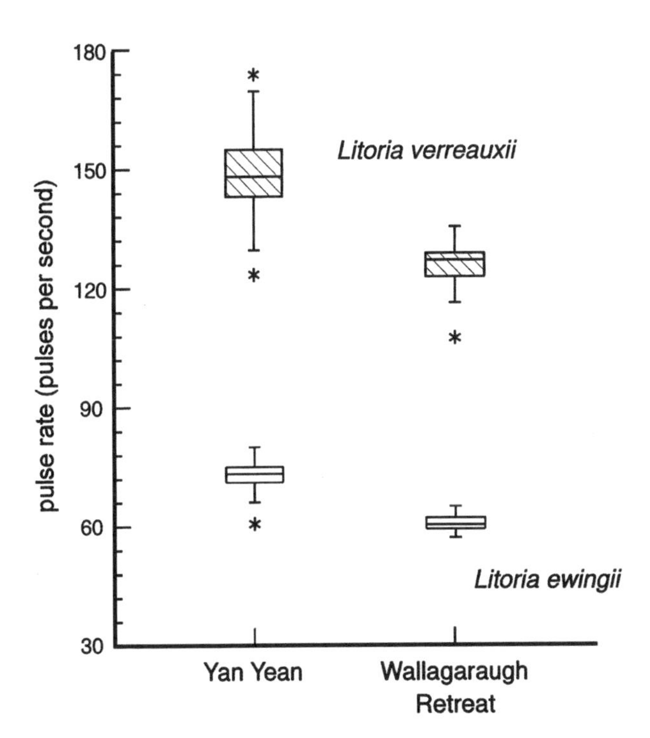

Figure 9.7. Box plots (Frigge et al. 1989) of log-transformed (base 10) pulse rates (adjusted to 11.85°C) of notes in advertisement calls of *Litoria ewingii* (lower, open boxes) and *L. verreauxii* (upper, hatched boxes) from Yan Yean and Wallagaraugh Retreat. See text and Table 9.3 for details. The boxes indicate the interquartile ranges and the horizontal line inside each box indicates the median. The limits for recognition of exceptional values as outliers (*) are set beyond the ends of the boxes (hinges) at 1.5 × the interquartile range, and the whiskers are extended to the smallest or largest value, respectively, within these limits.

vertisement calls. This asymmetry in the response by males of the deeply sympatric populations of *L. verreauxii* may reflect a long history of interactions with males of *L. ewingii*, or a reduced influence of gene flow because of the remoteness of conspecific allopatric populations. Conversely the low level of response in shallowly sympatric males of *L. ewingii* may reflect a much shorter historical period of interaction or a greater influence of gene flow from the closer allopatric conspecific stocks.

Hyla chrysoscelis and *H. versicolor*

The two closely related hylids of eastern North America, *Hyla chrysoscelis* and *H. versicolor*, have been the subjects of considerable and significant research in the areas of speciation and sound communication since the pioneering studies of Blair (1958), Ralin (1968), and Gerhardt (1974). The following description is based on the waveforms of the representative call of each species presented in Figure 1 of Gerhardt (1982). The advertisement calls are very similar in duration and envelope shape, and each consists of a single pulse-train that is fully amplitude modulated. There are more pulses in the call of *H. chrysoscelis* (41) than in that of *H. versicolor* (17). Maximum amplitude is reached at about the sixth pulse in the call of *H. chrysoscelis* and at about the fourth pulse in *H. versicolor*; thereafter the other pulses in each call are of nearly uniform amplitude. The positive ramps appear too short to be of significance in coding for distance, and may be a product of the process of initiation of calling. Pulse shapes differ between species, with the duration being about

Table 9.4 Mean values (adjusted to 10°C) of duration and pulse rate of notes in advertisement calls and encounter calls of *Litoria ewingii* and *L. verreauxii* from Yan Yean, Victoria

Call Attribute / Call Type	*L. ewingii* Mean (S_{yx})	*L. ewingii* CV(%)	*L. verreauxii* Mean (S_{yx})	*L. verreauxii* CV(%)	Ratio of Interspecific Means[a]
Note duration (ms)					
Advertisement	179.1 (28.7)	16.0	214.2 (20.0)	9.3	1.20
Encounter	109.1 (31.1)	28.5	91.4 (12.6)	13.8	1.19
Ratio of intraspecific means[a]	1.64	—	2.34	—	—
Pulse rate (pulses/sec)					
Advertisement	65.4 (4.4)	6.7	135.5 (10.3)	7.6	2.07
Encounter	46.3 (11.5)	24.8	87.0 (17.6)	20.2	1.88
Ratio of intraspecific means[a]	1.41	—	1.56	—	—

Source: Data from Harrison (1987).

S_{yx} = standard deviation from regression; CV = coefficient of variation.

Sample sizes: *L. ewingii:* advertisement calls = 28, encounter calls = 22; *L. verreauxii:* advertisement calls = 16, encounter calls = 14.

[a] Ratios are calculated as larger value divided by smaller value.

Table 9.5 Pulse rates (pulses per second), adjusted to 21°C (southeastern Texas) or 20°C (other localities), in advertisement calls of *Hyla chrysoscelis* and *H. versicolor*

	H. chrysoscelis			*H. versicolor*						
Area	*n*	Mean (SD)	Min. (*y*)	*n*	Mean (SD)	Max. (*z*)	Ratio of Means	Difference in Means ($\Delta\bar{x}$)	Gap ($y-z$)	CCD (gap/$\Delta\bar{x}$)
SE Texas[a]*	65	46.7 (2.7)	41.1	80	25.3 (1.8)	28.8	1.85	21.4	12.3	0.57
Louisiana[b]	43	44.4 (2.7)	38	12	22.9 (1.6)	26	1.94	21.5	12	0.56
Missouri[b]	63	49.2 (1.7)	45	50	20.9 (1.4)	25	2.35	28.3	20	0.71
Maryland[b]	45	38.7 (2.1)	32	39	19.2 (1.5)	25	2.02	19.5	7	0.36
Michigan[c]	10	53.2	40.7	9	23.3	26.0	2.28	29.9	14.7	0.49

Sources: [a]Ralin (1977); [b]Gerhardt (1994); [c]Bogart and Jaslow (1979).

*Sympatric localities 5–7 of Ralin (1977).

n = sample size; CCD = coefficient of call difference, sensu Littlejohn (1999); SD = standard deviation.

twice and the rise time about four times as long in the call of *H. versicolor;* decay times, however, appear similar (estimated from Figure 1 of Gerhardt and Doherty 1988).

Although values for dominant frequencies of the advertisement calls of the two species from syntopic synchronous choruses apparently have not been clearly stated in publications, it is generally assumed that spectral structure is similar (Gerhardt et al. 1996). Blair (1958) gave the mean values for five pooled regional samples from two transects: 2683, 2643, and 2650 Hz for "fast trillers" (*H. chrysoscelis*), and 2435 and 2407 Hz for "slow trillers" (*H. versicolor*). Bogart and Jaslow (1979) provided data for dominant frequencies in small samples of the two species (esophageal temperatures ranged from 20.2 to 26.0°C), each combined from several localities in Michigan, but made no clear statement about syntopy. Values are *H. chrysoscelis* (n = 10) mean, 2.52 kHz, range, 2.17–3.00 kHz; *H. versicolor* (n = 9) mean, 2.29 kHz, range, 2.13–2.55 kHz (Bogart and Jaslow 1979). Mable and Bogart (1991), in a consideration of the call structure of triploid hybrids, provided mean values for upper dominant frequencies at 22°C (site of temperature measurement, sample sizes, and ranges of variation were not stated) from calls of *H. chrysoscelis* from Wisconsin (2.796 kHz) and of *H. versicolor* from Ontario (2.002 kHz). Gerhardt et al. (1994), when describing the structure of advertisement calls of putative hybrids, gave values (means ?) for dominant frequencies of calls of representatives of the parental species (body temperature range, 19.7–21.4°C) from West Virginia (*H. chrysoscelis:* 2.212 kHz, *H. versicolor:* 1.850 kHz) and Missouri (*H. chrysoscelis:* 2.460 kHz, *H. versicolor:* 2.450 kHz). Means and ranges (in parentheses) of dominant frequencies for two large pooled samples of *H. chrysoscelis* are 2.45 kHz (2.05–2.80) in Georgia and South Carolina (n = 46), and 2.57 kHz (2.18–3.10) for Texas (n = 45) (Gerhardt 1974). The ranges given by Gerhardt (1974) encompass the values for the means of Blair (1958) for both species, confirming that the advertisement calls are of similar spectral structure.

For recordings obtained from the same areas and corrected to a common temperature of either 21 or 20°C, respectively, the ranges of variation for duration overlap broadly (Ralin 1977; Gerhardt 1994); those for pulse rate, however, are separated by a distinct gap (Ralin 1968; 1977; Zweifel 1970; Gerhardt 1978, 1982, 1994). Ralin (1977) pooled data for each species from localities in south-central and southeastern Texas for linear regression analyses. The slopes for the regressions against air temperature were highly significant ($p < 0.005$) for both species for duration: *H. chrysoscelis:* regression coefficient, $b = -0.02$ ($n = 158$), *H. versicolor:* $b = -0.04$ ($n = 97$); and pulse rate: *H. chrysoscelis:* $b = +2.36$ ($n = 158$), *H. versicolor:* $b = +1.19$ ($n = 97$). Coefficients of determination (r^2) were not given. Gerhardt (1978, 1982) explored the association of pulse rate with body temperature in pooled samples of each species from Missouri. The following comments are based on the scattergram and the linear regressions in Figure 2 of Gerhardt (1982). The slopes were highly significant ($p < 0.001$) for both species, and the following regression coefficients were obtained: *H. chrysoscelis:* $b = +3.56$, $r^2 = 0.96$ ($n = 48$); *H. versicolor:* $b = +1.18$, $r^2 = 0.86$ ($n = 81$). Means and converging extreme values for pulse rates of pooled samples from south-central and southeastern Texas, corrected to 21°C using the above regression coefficients (Ralin 1977), are presented in Table 9.5. Values for means and converging extremes for pooled sympatric samples from three areas in the eastern United States (Louisiana, Missouri, and Maryland), corrected to 20°C using the appropriate regression coefficient within the range of 1.95–2.8 for *H. chrysoscelis* and 0.99–1.03 for *H. versicolor* (Gerhardt 1994), are presented in Table 9.5. Similar values for pulse rates (at 20°C) in advertisement calls of the two species from Michigan (Bogart and Jaslow 1979) are also included in Table 9.5. Ratios of means and coefficients of call

Table 9.6 Attributes of Type II calls (encounter calls) and advertisement calls of *Hyla chrysoscelis* and *H. versicolor* from south-central and southeastern Texas and adjacent states

Species Call Type	Locality	Temperature (°C)	*n*	Duration Mean (Range) (s)	Pulse Rate Mean (Range) (pulses/sec)	Dominant Frequency Mean (Range) (kHz)
H. chrysoscelis						
Type II	13 km east of Bastrop, Texas	19.0–19.8	13	0.17[a] (0.14–0.25)	— (53–69)	1.912[a]
Advertisement	South-central and southeast Texas[b]	21[b]	65[b] 45[c]	0.57 (0.36–0.88)[b]	46.7 (41.1–56.5)[b]	2.57 (2.18–3.10)[c]
H. versicolor						
Type II	Lexington, Texas	18.5	6	0.21 (0.14–0.24)	— (50–63)	2.100
Advertisement	Eastern Texas,[b] Texas and adjacent states[d]	21[b]	80[b] 33[d]	0.64 (0.28–1.22)[b]	25.3 (21.3–28.8)[b]	2.435[d]

Sources: Unless indicated, data for Type II calls from Table 1 of Pierce and Ralin (1972) on the basis of locality and temperature.

[a]Average calculated from Table 1 of Pierce and Ralin (1972); no range given.

[b]From sympatric localities 5–7 in Table 4 of Ralin (1977).

[c]From Table 1 of Gerhardt (1974); no temperature data given.

[d]From Table 3 of Blair (1958) for "southern slow trillers" (= *H. versicolor*); no temperature data given.

difference were then calculated from these values (Table 9.5). Ratios of means are close to 2 and range from 1.85 in Texas to 2.35 in Missouri, and coefficients of call difference range from 0.36 in Maryland to 0.71 in Missouri. Even though the data for Texas were obtained by a different investigator and are corrected to air temperature (Ralin 1977), the derived ratios are very similar to those for the adjacent area, Louisiana, corrected to body temperature (Gerhardt 1994).

Pierce and Ralin (1972) described "Type II" calls of *Hyla chrysoscelis* and *H. versicolor* from south-central and southeastern Texas, which were presumed to be territorial calls (= encounter calls) on the basis of being vocal responses to the production of "a high intensity human imitation of a mating call nearby, while the frog is giving its mating call" (p. 330). Means were given for dominant frequencies, and ranges and means for duration, but only ranges for pulse rate. As there is no provision for temperature correction, only the values from recordings obtained at similar temperatures at two localities are considered here (Table 9.6). The values for these calls are compared with those for advertisement calls obtained by Blair (1958), Gerhardt (1974), and Ralin (1977) (Table 9.5). For each species the durations of the encounter calls are much shorter than those of the advertisement calls, and dominant frequencies of encounter calls are lower than those of advertisement calls (Table 9.6). Although the pulse rates of encounter calls and advertisement calls overlap slightly for *H. chrysoscelis*, those for *H. versicolor* are separated by a gap of 21 pulses per second (Table 9.6).

When the encounter calls of each species are compared, the ranges of variation for both durations and pulse rates are very similar, and the means for dominant frequency differ by less than 200 Hz (Table 9.6). Of particular note is the similarity in pulse rates of encounter calls (Table 9.6). Such values are consistent with the postulated sympatric origin of the tetraploid species *H. versicolor* from the diploid species *H. chrysoscelis* through autopolyploidy (Gerhardt et al.1994; Ptacek et al. 1994). The advertisement calls presumably have diverged, or any original differences associated with the phenomenon of polyploidy have been maintained, to allow efficient specificity of mate choice because the triploid F1 hybrids are sterile and there is thus an absolute cost in terms of reproductive success (Gerhardt et al. 1994). In contrast the encounter calls may have retained the original structure of the derived tetraploid or have subsequently converged due to natural selection acting in the context of competition for similar calling stations.

Intraspecific and Interspecific Differentiation of Repertoire: An Overview

The aim of this section is to explore the patterns of differentiation in key attributes of signals at the intraspecific and interspecific levels. Estimations of the degrees of difference between pairs of signals at these two levels may provide some insight into underlying mechanisms of sensory discrimination. Three aspects will be considered: the ratio of means, the size of the gap between converging extremes of ranges of variation, and the coefficient of call difference (described earlier) for key attributes of acoustic signals. As there are no really adequate and appropriate sets of data with sufficient cases for both advertisement calls and encounter calls, a quantitatively robust analysis is not yet possible. Even so the provisional and partial analysis of a few cases may provide guidelines and priorities for future research in this area.

The ratio for the means of durations of the advertisement calls and the encounter calls in the repertoire of *Crinia signifera* (2.31, Table 9.1), and those for the durations and pulse rates of the introductory notes (4.95) and repeated notes (3.05) in the diphasic advertisement calls of *Geocrinia victoriana* (Table 9.2), are higher than those in key attributes of the advertisement calls of the two pairs of syntopic taxa: 2.00–2.07 for pulse rates in *Litoria ewingii* and *L. verreauxii* (Table 9.3) and 1.85–2.35 in *Hyla chrysoscelis* and *H. versicolor* (Table 9.5). For *L. ewingii,* however, the ratios of means for durations and pulse rates of notes in advertisement calls and encounter calls at 10°C are lower than between signals within conspecific repertoires: 1.64 and 1.41, respectively (Table 9.4). The ratio for pulse rate (1.56) is also low in *L. verreauxii,* but that for duration (2.34) is higher (Table 9.4). Gaps for durations and pulse rates are 71 ms and 16 pulses per second for advertisement calls and encounter calls, respectively, in *C. signifera* (Table 9.1) and 223 ms and 230 pulses per second for introductory and repeated notes in *G. victoriana* (Table 9.2). For the two interspecific comparisons of pulse rates of advertisement calls, gaps range from 42 to 48 pulses per second at 11.85°C for *L. ewingii* and *L. verreauxii* (Table 9.3) and 7 to 20 pulses per second for *H. chrysoscelis* and *H. versicolor* at 20–21°C (Table 9.5). Intraspecific coefficients of call difference for pulse rate range from 0.34 in *C. signifera* (Table 9.1) to 0.78 in *G. victoriana* (computed from the values presented in Table 9.2). Levels of differentiation in key attributes of advertisement calls of the syntopic closely related taxa have already been addressed. For comparison, as discussed earlier, a minimal value of 0.5 for the coefficient of call difference would appear typical of the key attribute in the advertisement calls of syntopic pairs of taxa.

Although of primary concern in interactions between competing males, a basic and general mechanism of central discrimination might already exist for the intraspecific repertoire that could also be applied to interspecific discrimination (Littlejohn 1999). Such a mechanism for discrimination could also operate in the process of mate choice by females in normal sexual selection. The presence of graded encounter calls, as in *Geocrinia victoriana* (Littlejohn and Harrison 1985) and in *Hyla ebraccata* (Schwartz and Wells 1984), leads to the suggestion that there can be alternative underlying acoustic mechanisms and strategies for territorial behavior.

The greater similarity of encounter calls of each of the two syntopic pairs of taxa discussed earlier (Tables 9.4 and 9.6), when compared with their advertisement calls, indicates that there has been either convergence or conservation of a basic ancestral structure, and emphasizes the significance of interspecific territoriality. Furthermore, for *Litoria ewingii* and *L. verreauxii* the equivalent components in the encounter call are much more variable than that in the advertisement call (Table 9.4). A comparable situation, with greater similarity of encounter calls when compared with advertisement calls, has been described by Schwartz and Wells (1984) for three closely related syntopic species of *Hyla* in Panama.

Origin, Extent, and Maintenance of Gaps

As indicated earlier there is a distinct gap in the ranges of variation of at least one key temporal attribute of mate-attracting signals of members of different syntopic, synchronously breeding species. The origin, extent, and maintenance of such gaps in the primary feature or features of the signals, and the ranges of variation within the mate-attracting signals of each genetic system (i.e., species), are thus of considerable significance to understanding the process of mate choice in taxonomically complex reproductive environments (Littlejohn 1999). In a two-species situation, where there is a cost in terms of reduced reproductive success for both of the mating heterospecific individuals (Littlejohn 1993, 1999), but particularly for females with a greater investment in gametes, conspecific signals on the converging (ipsilateral) extreme of variation may be avoided because of the increased possibility of making mistakes in choice of an appropriate mate. Processes of sexual selection within each taxon would also be expected to tend toward normal (i.e., unskewed), possibly

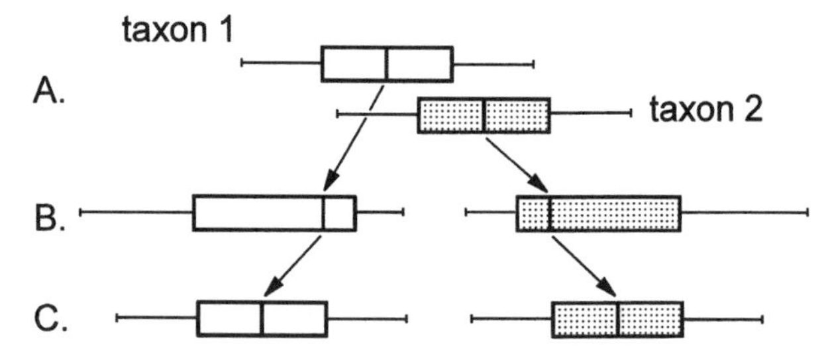

Figure 9.8. Hypothetical box plots (see legend to Figure 9.7 for explanation) depicting postulated changes in variation and distribution of a key attribute in the mate-attracting signals (e.g., pulse rate in advertisement calls of anurans) of two closely related taxa. (A) Overlapping unskewed distributions in disjunct allopatry before contact is made. (B) Skewed distributions associated with reproductive character displacement through reinforcing selection during initial syntopic interactions associated with synchronous breeding, after contact. (C) Restoration of unskewed distributions during a later equilibrium in syntopy.

Table 9.7 Variation in pulse rate (pulses per second), adjusted to 11.85°C, in syntopic populations of *Litoria ewingii* and *L. verreauxii*.

Species	Locality	*n*	CV[a] (%)	Skewness[a]	SES[a]	Skewness/SES[b]	Kurtosis[a]	SEK[a]	Kurtosis/SEK[b]
L. ewingii	Yan Yean	28	5.9	−0.681	0.441	1.544	1.233	0.858	1.437
	Wallagaraugh Retreat	16	3.8	0.437	0.564	0.775	−0.348	1.091	0.319
L. verreauxii	Yan Yean	38	6.7	0.193	0.383	0.504	0.927	0.750	1.236
	Wallagaraugh Retreat	23	4.7	−1.299	0.481	2.70	2.964	0.935	3.170
	Wallagaraugh Retreat (no outlier)	22	3.6	−0.295	0.491	0.600	0.163	0.953	0.171

Data were adjusted to the appropriate temperatures after log-transformation (base 10) then restored to normal values.

[a]Calculated with SYSTAT 8.0.

[b]Considered significant if the ratio of the absolute value to its standard error is greater than 2.

CV = coefficient of variation; SES = standard error of skewness; SEK = standard error of kurtosis.

leptokurtic, distributions at equilibrium. Thus a break and the associated gap are developed in the postulated initial continuum of variation, and the gap is then maintained by natural selection for the same reasons. Skewed distributions with longer tails on the contralateral (diverging) sides and truncation on the ipsilateral sides may be expected from the action of reinforcing selection during the early stages in the development of syntopic sympatry; but subsequently, as an equilibrium is approached, the normalizing processes of sexual selection should prevail (Figure 9.8). There is no significant skewness or kurtosis in pulse rates for the samples of *Litoria ewingii* from Yan Yean and Wallagaraugh Retreat, or of *L. verreauxii* from Yan Yean (Table 9.7). Because of an outlier in the sample of *L. verreauxii* from Wallagaraugh Retreat (Figure 9.7), however, the distribution is significantly skewed on the low side, and is also significantly leptokurtic (with longer tails than for a normal distribution). Recalculation after removal of the outlier resulted in nonsignificant values for skewness and kurtosis (Table 9.7) and emphasizes the need to develop a coefficient of call difference that is less sensitive to such values.

Conclusions

The temporal structure of the two main components in the long-range acoustic repertoire of anuran amphibians, the advertisement call and the encounter call, is considered both within species and between pairs of closely related synchronously breeding syntopic taxa. Restriction to the time domain allows focus on the basic coding patterns in key attributes that characterize these acoustic signals in species with calls that have similar spectral properties.

The terminology and criteria for describing temporal attributes at different time scales are reviewed, and the application of more rigorous standard definitions derived from physics and electronics is recommended as replacements for the variety of definitions currently in use in anuran bioacoustics.

The acoustic aspects of production, transmission, and reception of transient and pulsatile signals, especially resonance and the measurement of relevant sound-pressure levels of such signals, are explored. It is recommended that maximum (peak) values should be obtained as the relevant measurement in bioacoustic studies of such signals, and that stimuli of appropriate temporal structure should be used in determination of neurophysiological and behavioral auditory thresholds.

The significance of ramps of amplitude in the envelopes of long calls and call notes, and their possible role in distance ranging—as a change in the perceived duration to the receiver—is considered. Because envelope ramps are of common occurrence in the calls of anurans, this virtually unexplored aspect of temporal structure warrants further study.

The utility and limitations of the coefficient of call difference, a standardized univariate measure of the difference in the key factor that differentiates signals (e.g., pulse rate), based on the difference between means and the extent of the gap in ranges of variation, are further considered.

The patterns of intraspecific differentiation and extent of variation in temporal structure of advertisement calls and encounter calls are explored, and the key factor or code that may account for the specificity of each type of signal is identified. Because of the small amount of information that is available, and often obtained incidentally or in other contexts, further studies that directly address such differentiation are strongly recommended.

The extent of difference in the key temporal aspects of advertisement calls of closely related syntopic, synchronously breeding taxa is explored and the coefficient of call difference is calculated. These range from 0.56 to 0.66 (10–15°C, log-transformed) for pulse rates in samples of *Litoria ewingii* and *L. verreauxii* from two localities in southeastern Australia.

The nature of the gap in the ranges of variation of the key attribute of advertisement calls of closely related syntopic species and methodological problems associated with its derivation (e.g., individual variation, presence of hybrids, effect of temperature, stability of recording system) are considered.

Acknowledgments

The field and laboratory studies on *Litoria ewingii* and *L. verreauxii* were supported by grants from the Australian Research Council and funding was provided by the University of Melbourne. Patsy Littlejohn assisted with the field work at Wallagaraugh Retreat and Yan Yean. The investigations at Wallagaraugh Retreat were made under Research Permit RP-93-004 issued by the Department of Conservation and Environment (now Natural Resources and Environment), Victoria. John Wright carried out many of the acoustic analyses of the resulting tape recordings and also assisted with the preliminary statistical evaluations of the data. Nicole Kime kindly provided the values for calls of *Acris crepitans* from Gil Ranch, Austin. Access to the study sites at Yan Yean and at Wallagaraugh Retreat was kindly given by the Melbourne and Metropolitan Board of Works (now Melbourne Water) and Mr. John Riley, respectively.

References

Allan, D. M. 1973. Some relationships of vocalization to behavior in the Pacific treefrog, *Hyla regilla*. Herpetologica 29:366–371.

American Radio Relay League. 1978. The Radio Amateur's Handbook. 55th ed. American Radio Relay League, Newington, CN.

Bennet-Clark, H. C. 1989. Songs and the physics of sound production. Pp. 227–261. *In* F. Huber, T. E. Moore and W. Loher (eds.), Cricket Behavior and Neurobiology. Comstock Publishing Associates, Cornell University Press, Ithaca, NY.

Bennet-Clark, H. C. 1995. Insect sound production: Transduction mechanisms and impedance matching. Pp. 199–218. *In* C. P. Ellington and T. J. Pedley (eds.), Biological Fluid Dynamics. Symposia of the Society for Experimental Biology, No. 49. Company of Biologists, Cambridge.

Bennet-Clark, H. C. 1999. Which Qs to choose: Questions of quality in bioacoustics? Bioacoustics 9:351–359.

Bennet-Clark, H. C., and D. Young, 1992. A model of the mechanism of sound production in cicadas. Journal of Experimental Biology 173:123–153.

Blair, W. F. 1958. Mating call in the speciation of anuran amphibians. American Naturalist 92:27–51.

Bogart, J. P., and A. P. Jaslow. 1979. Distribution and call parameters of *Hyla chrysoscelis* and *Hyla versicolor* in Michigan. Life Sciences Contributions, Royal Ontario Museum No. 117. The Royal Ontario Museum, Toronto.

Bogert, C. M. 1960. The influence of sound on the behavior of amphibians and reptiles. Pp.137–320. *In* W. E. Lanyon, and W. N. Tavolga (eds.), Animal Sounds and Communication. American Institute of Biological Sciences, Washington, DC.

Brenowitz, E. A. 1982. The active space of red-winged blackbird song. Journal of Comparative Physiology 147:511–522.

Broughton, W. B. 1963. Method in bio-acoustic terminology. Pp. 3–24. *In* R.-G. Busnel (ed.), Acoustic Behaviour of Animals. Elsevier, Amsterdam.

Cocroft, R. B., and M. J. Ryan. 1995. Patterns of advertisement call evolution in toads and chorus frogs. Animal Behaviour 49:283–303.

Eggermont, J. J. 1988. Mechanisms of sound localization in anurans. Pp. 307–336. *In* B. Fritzsch, M. J. Ryan, W. Wilczynski, T. E. Hetherington, and W. Walkowiak (eds.), The Evolution of the Amphibian Auditory System. John Wiley and Sons, New York.

Ehret, G., J. Tautz, and B. Schmitz. 1990. Hearing through the lungs: Lung-eardrum transmission of sound in the frog *Eleutherodactylus coqui*. Naturwissenschaften 77:192–194.

Fletcher, N.H. 1992. Acoustic Systems in Biology. Oxford University Press, New York.

Frigge, M., D.C. Hoaglin, and B. Iglewicz. 1989. Some implementations of the boxplot. American Statistician 43:50–54.

Gambs, R. D., and M. J. Littlejohn. 1979. Acoustic behavior of males of the Rio Grande leopard frog (*Rana berlandieri*): An experimental analysis through field playback trials. Copeia 1979:643–650.

Gans, C. 1973. Sound production in the Salientia: Mechanism and evolution of the emitter. American Zoologist 13:1179–1194.

Gerhardt, H. C. 1974. Mating call differences between eastern and western populations of the treefrog *Hyla chrysoscelis*. Copeia 1974:534–536.

Gerhardt, H. C. 1975. Sound pressure levels and radiation patterns of the vocalizations of some North American frogs and toads. Journal of Comparative Physiology 102:1–12.

Gerhardt, H. C. 1978. Temperature coupling in the vocal communication system of the gray tree frog, *Hyla versicolor*. Science 199:992–994.

Gerhardt, H. C. 1982. Sound pattern recognition in some North American treefrogs (Anura: Hylidae): Implications for mate choice. American Zoologist 22:581–595.

Gerhardt, H. C. 1994. Reproductive character displacement of female mate choice in the grey treefrog, *Hyla chrysoscelis*. Animal Behaviour 47:959–969.

Gerhardt, H. C., and M. S. Davis. 1988. Variation in the coding of species identity in the advertisement calls of *Litoria verreauxi* (Anura: Hylidae). Evolution 42:556–563.

Gerhardt, H. C., and J. A. Doherty. 1988. Acoustic communication in the gray treefrog, *Hyla versicolor*: Evolutionary and neurobiological implications. Journal of Comparative Physiology A 162:261–278.

Gerhardt, H. C., M. L. Dyson, and S. D. Tanner. 1996. Dynamic properties of the advertisement calls of gray tree frogs: Patterns of variability and female choice. Behavioral Ecology 7:7–18.

Gerhardt, H. C., M. B. Ptacek, L. Barnett, and K. G. Torke. 1994. Hybridization in the diploid-tetraploid treefrogs *Hyla chrysoscelis* and *Hyla versicolor*. Copeia 1994:51–59.

Gottlieb, I. 1958. Basic Pulses. John F. Rider, New York.

Greenewalt, C. H. 1968. Bird Song: Acoustics and Physiology. Smithsonian Institution Press, Washington, DC.

Harrison, P. A. 1987. Vocal behaviour in the south-eastern Australian tree frogs, *Litoria ewingi* and *L. verreauxi* (Anura: Hylidae). M.Sc. thesis, Department of Zoology, University of Melbourne, Parkville.

Hawe, S. M. 1970. Calling behaviour and territoriality in males of two species of *Crinia* (Anura: Leptodactylidae). B.Sc.(Hon) thesis, Department of Zoology, University of Melbourne, Parkville.

Jackson, L. 1996. Sidebands: Artefacts or facts? Bioacoustics 7:163–164.

Johnson, D. L., A. H. Marsh, and C. M. Harris. 1991. Acoustical measurement instruments. Pp. 5.1–5.21. *In* C. M. Harris (ed.), Handbook of Acoustical Measurements and Noise Control, 3rd ed. McGraw-Hill, New York.

Littlejohn, M. J. 1959. Call differentiation in a complex of seven species of *Crinia* (Anura, Leptodactylidae). Evolution 13:452–468.

Littlejohn, M. J. 1965. Premating isolation in the *Hyla ewingi* complex (Anura: Hylidae). Evolution 19:234–243.

Littlejohn, M. J. 1969. The systematic significance of isolating mechanisms. Pp. 459–482. *In* National Academy of Sciences (eds.), Systematic Biology. Proceedings of an International Conference. National Academy of Sciences Washington, DC.

Littlejohn, M. J. 1977. Long-range acoustic communication in anurans: An integrated and evolutionary approach. Pp. 263–294. *In* D. H. Taylor and S. I. Guttman (eds.), The Reproductive Biology of Amphibians. Plenum Press, New York.

Littlejohn, M. J. 1982. *Litoria ewingi* in Australia: A consideration of indigenous populations, and their interactions with two closely related species. Pp. 113–135. *In* D. G. Newman (ed.), New Zealand Herpetology. Proceedings of a Symposium held at the Victoria University of Wellington 29–31 January 1980. New Zealand Wildlife Service, Wellington.

Littlejohn, M. J. 1993. Homogamy and speciation: A reappraisal. Pp. 135–165. *In* D. Futuyma and J. Antonovics (eds.), Oxford Surveys in Evolutionary Biology, Vol 9. Oxford University Press, New York.

Littlejohn, M. J. 1998. Personal perspectives: Historical aspects of recording and analysis in anuran bioacoustics: 1954–1997. Bioacoustics 9:69–80.

Littlejohn, M. J. 1999. Variation in advertisement calls of anurans across zonal interactions. The evolution and breakdown of homogamy. Pp. 209–233. *In* S. A. Foster and J. A. Endler (eds.), Geographic Variation in Behavior: Perspectives on Evolutionary Mechanisms. Oxford University Press, New York.

Littlejohn, M. J., and P. A. Harrison 1985. The functional significance of the diphasic advertisement call of *Geocrinia victoriana* (Anura: Leptodactylidae). Behavioural Ecology and Sociobiology 16:363–373.

Littlejohn, M. J., P. A. Harrison, and R. C. Mac Nally. 1985. Interspecific acoustic interactions in sympatric populations of *Ranidella signifera* and *R. parinsignifera* (Anura: Leptodactylidae). Pp. 287–296. *In* G. Grigg, R. Shine, and H. Ehmann (eds.), Biology of Australasian Frogs and Reptiles. Royal Zoological Society of New South Wales, Sydney, and Surrey Beatty and Sons, Chipping Norton.

Littlejohn, M. J., and J. J. Loftus-Hills. 1968. An experimental evaluation of premating isolation in the *Hyla ewingi* complex (Anura: Hylidae). Evolution 22:659–663.

Littlejohn, M. J., G. F. Watson, and J. R. Wright. 1993. Structure of advertisement call of *Litoria ewingi* (Anura: Hylidae) introduced into New Zealand from Tasmania. Copeia 1993:60–67.

Littlejohn, M. J., and J. R. Wright. 1997. Structure of the acoustic signals of *Crinia glauerti* (Anura: Myobatrachidae) from southwestern Australia, and comparison with those of *C. signifera* from South Australia. Transactions of the Royal Society of South Australia 121:103–117.

Loftus-Hills, J. J. 1973. Comparative aspects of auditory function in Australian anurans. Australian Journal of Zoology 21:353–367.

Loftus-Hills, J. J., and M. J. Littlejohn. 1971. Mating-call sound intensities of anuran amphibians. Journal of the Acoustical Society of America 49:1327–1329.

Mable, B. K., and J. P. Bogart., 1991. Call analysis of triploid hybrids resulting from diploid-tetraploid species crosses of hylid tree frogs. Bioacoustics 3:111–119.

Martin, W. F. 1972. Evolution of vocalization in the genus *Bufo*. Pp. 279–309. *In* W. F. Blair (ed.), Evolution in the Genus *Bufo*. University of Texas Press, Austin.

McDiarmid, R. W., and K. Adler. 1974. Notes on territorial and vocal behavior of neotropical frogs of the genus *Centrolenella*. Herpetologica 30:75–78.

McGill, R., J. W. Tukey, and W. A. Larsen. 1978. Variations of box plots. American Statistician 32:12–16.

Michelsen, A. 1985. Environmental aspects of sound communication in insects. Pp. 1–9. *In* K. Kalmring, and N. Elsner (eds.), Acoustic and Vibrational Communication in Insects. Verlag Paul Parey, Berlin.

Narins, P. M. 1992. Evolution of anuran chorus behavior: Neural and behavioral constraints. American Naturalist 139(Supp):S90-S104.

Narins, P. M., G. Ehret, and J. Tautz. 1988. Accessory pathway for sound transfer in a neotropical frog. Proceedings of the National Academy of Sciences, USA 85:1508–1512.

Narins, P. M., and D. D. Hurley. 1982. The relationship between call intensity and function in the Puerto Rican coqui (Anura: Leptodactylidae). Herpetologica 38:287–295.

Nevo, E., and R. R. Capranica. 1985. Evolutionary origin of ethological reproductive isolation in cricket frogs, *Acris*. Pp. 147–214. *In* M. K. Hecht, B. Wallace, and G. T. Prance (eds.), Evolutionary Biology, Volume 19. Plenum Press, New York.

Pierce, J. R., and D. B. Ralin. 1972. Vocalizations and behavior of the males of three species in the *Hyla versicolor* complex. Herpetologica 28:329–337.

Prestwich, K. N., K. E. Brugger, and M. Topping. 1989. Energy and communication in three species of hylid frogs: Power input, power output and efficiency. Journal of Experimental Biology 144:53–80.

Ptacek, M. B., H. C. Gerhardt, and R. D. Sage. 1994. Speciation by polyploidy in treefrogs: Multiple origins of the tetraploid, *Hyla versicolor*. Evolution 48:898–908.

Ralin, D. B. 1968. Ecological and reproductive differentiation in the cryptic species of the *Hyla versicolor* complex (Hylidae). Southwestern Naturalist 13:283–299.

Ralin, D. B. 1977. Evolutionary aspects of mating call variation in a diploid-tetraploid species complex of treefrogs (Anura). Evolution 31:721–736.

Robertson, J. G. M. 1984. Acoustic spacing by breeding males of *Uperoleia rugosa* (Anura: Leptodactylidae). Zeitschrift für Tierpsychologie 64:283–297.

Rose, G. J., and R. R. Capranica. 1984. Processing amplitude-modulated sounds by the auditory midbrain of two species of toads: Matched temporal filters. Journal of Comparative Physiology A 154:211–219.

Schwartz, J. J., and K. D. Wells 1984. Interspecific acoustic interactions of the neotropical treefrog *Hyla ebraccata*. Behavioural Ecology and Sociobiology 14:211–224.

Snyder, W. F., and D. L. Jameson. 1965. Multivariate geographic variation of mating call in populations of the Pacific tree frog (*Hyla regilla*). Copeia 1965:129–142.

Standards Australia. 1990. Australian Standard AS 1259.1-1990. Sound

Level Meters. Part 1: Non-integrating. Standards Australia, North Sydney.

Straughan, I. R., and A. R. Main. 1966. Speciation and polymorphism in the genus *Crinia* Tschudi (Anura, Leptodactylidae) in Queensland. Proceedings of the Royal Society of Queensland 78:11–28.

von Helversen, D. 1993. 'Absolute steepness' of ramps as an essential cue for auditory pattern recognition by a grasshopper (Orthoptera; Acrididae; *Chorthippus biguttulus* L.). Journal of Comparative Physiology A 172:633–639.

Watkins, W. A. 1967. The harmonic interval: Fact or artifact in spectral analysis of pulse trains. Pp.15–42. *In* W. N. Tavolga (ed.), Marine Bio-acoustics, Vol 2. Pergamon Press, Oxford.

Wells, K. D. 1977. The social behaviour of anuran amphibians. Animal Behaviour 25:666–693.

Wells, K. D., and B. J. Greer. 1981. Vocal responses to conspecific calls in a neotropical hylid frog, *Hyla ebraccata*. Copeia 1981:615–624.

Whitney, C. L. 1981. The monophasic call of *Hyla regilla* (Anura: Hylidae). Copeia 1981:230–233.

Wilczynski, W., A. C. Keddy-Hector, and M. J. Ryan. 1992. Call patterns and basilar papilla tuning in cricket frogs. 1. Differences among populations and between sexes. Brain Behaviour Evolution 39:229–237.

Yeager, D. M., and A. H. Marsh. 1991. Sound levels and their measurement. Pp. 11.1–11.18. *In* C. M. Harris (ed.), Handbook of Acoustical Measurements and Noise Control. 3rd ed. McGraw-Hill, New York.

Zweifel, R. G. 1970. Distribution and mating call of the treefrog, *Hyla chrysoscelis,* at the northeastern edge of its range. Chesapeake Science 11:94–97.

10

WALTER HÖDL AND ADOLFO AMÉZQUITA

Visual Signaling in Anuran Amphibians

Introduction

Acoustic communication plays a fundamental role in anuran reproduction and thus is involved in evolutionary processes such as mate recognition, reproductive isolation, speciation, and character displacement (Wells 1977a, 1977b, 1988; Rand 1988; Gerhardt and Schwartz 1995; Halliday and Tejedo 1995; Sullivan et al. 1995). Visual cues, however, have been thought to function only during close-range interactions (Wells 1977c; Duellman and Trueb 1986). Visual signaling is predicted to be predominantly employed by diurnal species at sites with an unobstructed view (Endler 1992). Diurnality, however, is not common for the majority of frog species. Thus vocalizations, which are highly efficient for communicating at night or in dense vegetation, are by far the best studied anuran signals (Duellman and Trueb 1986; Fritzsch et al. 1988; Hödl 1996). Nevertheless there are anecdotal reports of apparent visual signaling in intra- or intersexual interactions, or both (for summary see Lindquist and Hetherington 1996). Although the function of visual cues has only occasionally been tested experimentally (Emlen 1968; Lindquist and Hetherington 1996) the reports suggest that visual signaling is a significant mode of communication in at least a few anuran species.

In both classic and recent reviews on anuran communication, social behavior, or natural history, visual signaling was either not considered or was treated as a minor subject (Wells 1977a, 1977b; Arak 1983; Duellman and Trueb 1986; Rand 1988; Halliday and Tejedo 1995; Stebbins and Cohen 1995; Sullivan et al. 1995). The most detailed review of the subject is now more than 20 years old (Wells 1977b). Nevertheless some authors have discussed the possible evolutionary link between visual signaling and the reproductive ecology of species, such as reproduction associated with streams (Heyer et al. 1990; Lindquist and Hetherington 1996, 1998; Hödl et al. 1997; Haddad and Giaretta 1999) or reproduction within feeding territories (Wells 1977c).

Our aim in this review is (1) to propose a classification of reported behavioral patterns of visual signaling in frogs; (2) to describe the diversity of visual signals among living anuran taxa; and (3) to apply a comparative approach to exploring any associations between the diversity of visual signals and the ecological conditions in which they may have evolved.

Visual Signals and Communication

A communication process involves a transfer of information from a sender to a receiver by means of specifically designed signals. Signals are traits that have evolved specifically to manipulate the receiver's behavior (Harper 1991; Krebs and Davies 1993; Bradbury and Vehrencamp 1998). Signaling may occur at both the intra- and interspecific level (Bradbury and Vehrencamp 1998).

Because of the complexity of visual stimuli and signal processing and the difficulty of quantifying what animals actually perceive, the identification and analysis of visual signals is problematic (Bradbury and Vehrencamp 1998). In addition to their information content, features of ritualized signals generally involve redundancy, conspicuousness, stereotypy, and alerting components (Hailman 1977; Harper 1991). In general, communication systems that increase the signal-to-noise ratio will be favored by natural selection (Endler 1992; Alcock 1998). Conspicuousness in visual signals is achieved by enhancing the contrast between the signals and the brightness, color, spatial pattern, or movement of their background (Hailman 1977; Endler 1992; Bradbury and Vehrencamp 1998). Most reflected light signals involve muscular movement to change body shapes, to perform stereotyped displays or gestures, or to orient or position the individual in space. Often signal-generating movements are accentuated by strikingly colored or structured parts of the body (Bradbury and Vehrencamp 1998).

We exclusively define a visual signal if it is reported or personally observed that the behavioral event (1) provides a visual cue during an intra- or interspecific interaction, (2) is redundant, conspicuous, and stereotypical, and (3) most likely provokes an immediate response by the receiver that benefits the sender. The presence of alerting components in visual signals is difficult to prove (Lindquist and Hetherington 1998).

Any visually perceivable state (sensu Lehner 1979; Martin and Bateson 1993), such as persistent or temporarily limited sexual dimorphism, is excluded. Thus seasonal variation in color (as in breeding males of the European brown frog, *Rana arvalis wolterstorffi*, Mildner and Hafner 1990) or shape (e.g., hypertrophied forelimb musculature and nuptial excrescences in sexually mature males; Duellman and Trueb 1986) is not included in our analyses. We only considered changes of body coloration as visual signals if they occur within short time intervals and immediately affect the behavior of a conspecific receiver. Many postures and, in general, behavioral states can be used by conspecifics or heterospecifics as cues that provide information about the status of a sender. However they barely satisfy our definition of visual signals—that the behavioral change provoked in the receiver by the posture most probably benefits the sender. The same reasoning applies to many locomotory movements.

Consequently we do not consider visual cues that obviously originate as epiphenomena to be ritualized visual signals, for example, the inflation of the gular sac during vocalization and the vibration of flanks during ventilation (Rand 1988; Zimmermann and Zimmermann 1988). However we do not ignore the fact that the accompanying movements may provoke changes in the behavior of potential receivers hearing a call or looking at another individual. In the dendrobatid frog *Colostethus palmatus*, the presence of frog dummies and simultaneous playback of advertisement calls elicits in females low postures associated with courtship that are intensified when the dummy's gular sac is vibrated; however, gular sac vibrations without simultaneous playback of advertisement calls elicits predatory attacks (Lüddecke 1999). A serious reason to ignore epiphenomena is in many cases the trouble of separating, either experimentally or by observation, the concomitant effects of the sound or the presence of the individual. We also consider as epiphenomena actions that individuals perform to improve their visual or acoustic field or to prepare themselves for another behavioral pattern. For example, in some contexts "upright posture" or "body elevation" are followed by jumping in *C. palmatus* and may constitute orientation or preparatory movements (Lüddecke 1974, 1999). The stereotyped movements at underwater calling sites in pipid frogs (Rabb and Rabb 1960, 1963) have been classified as "postural or other visual displays" (Wells 1977b). However we do not include these behavioral traits in our considerations since it is not known whether these movements are performed to produce mechanical cues that propagate through water or whether they represent visual signals themselves.

Additionally we do not classify as signals visual cues that are permanently "on" (Bradbury and Vehrencamp 1998 and references therein), such as coloration or size, even though we cannot rule out the fact that some states provide substantial information in communication processes. Examples include aposematic body coloration in dendrobatid frogs, thought to indicate toxicity to potential predators (Myers and Daly 1976; Myers et al. 1978); sexual dimorphism in color patterns in *Atelopus* spp. (Lötters 1996) and *Bufo* spp. (Duellman and Trueb 1986), considered to play a role in intersexual communication; and intermale differences in the amount of ventral pigmentation, correlated with differences in reproductive success (Burrowes 1997).

Deimatic Behavior and Visual Signaling of Anurans during Interspecific Communication

Signals Addressed toward a Potential Predator

Some anuran species perform a deimatic behavior consisting of intimidating postures or actions when caught by a pursuing predator (Edmunds 1981). The unken reflex (*Un-*

ken is the German term for the genus *Bombina*), a common behavioral response to perceived threat in the discoglossid species *Bombina bombina* and *B. variegata*, is characterized primarily by lifting all four legs and arching the back, drawing attention to the bright ventral surface. Accompanying this posture and subsequent immobility is a slowing down of respiratory movements and an increase in skin secretion (Hinsche 1926; Noble 1955; Bajger 1980). The term *unken reflex* has been used broadly to describe a wide range of similar defensive postures restricted to species with ventral warning coloration (for anurans, see Haberl and Wilkinson 1997; for salamanders, see Brodie 1978; Brodie et al. 1979). Other visual defense actions in anurans include eyespot display, body inflation, hind parts elevation, and display of noxious glands or bright glandular secretions (Sazima and Caramaschi 1986; Martins 1989; Williams et al. 2000).

Arm waving in the harlequin frog, *Atelopus zeteki*, is elicited by the presence of human observers and continued by the frog approaching the intruder (Lindquist and Hetherington 1996). Movement of appendages (in lizards) has been hypothesized to serve as a strategy that by alerting potential predators and stimulating their movements would facilitate their detection by the signaling individual (Magnusson 1996). However this explanation would apply if the movements were performed both in the presence and absence of potential predators and were frequently repeated, a condition not met during the arm-waving displays of *A. zeteki*.

When picked up or tapped on the snout the horned frog, *Hemiphractus fasciatus* (*Cerathyla panamensis* auct.), widely opens the mouth and sometimes slightly arches its body by throwing the head up and back while exhibiting a bright yellowish orange tongue. Such behaviors are believed to function as a defensive display in which the unusual bright coloration of the tongue undoubtedly forms an integral part of the signal (Myers 1966).

Signals Addressed toward Potential Prey

Individuals of another horned frog, *Ceratophrys ornata*, confronted with potential prey perform pedal luring displays by twitching and curling the middle three toes of one or both hind feet. The movement can be gradually intensified when the potential prey remains visible and moving. In the extreme case one or both hind feet are raised out of the water and held almost vertically, with the soles of the feet and the toes directed laterally. Viewed from the front, the toes appear wriggling directly above the head. Pedal luring is not performed in front of all types of prey, but most likely in front of those animals that might be attracted to the display (Radcliffe et al. 1986). A similar response has been observed in some individuals of *C. calcarata* in the presence of a small leptodactylid frog of the genus *Pleurodema*. In this case not only the fourth and fifth toes of a hind foot were curled and undulated, but the foot itself was lifted and rotated. As in *C. ornata*, individuals do not signal to all types of prey. For example, the presence of crickets failed to elicit pedal luring (Murphy 1976).

Visual Signaling in Anurans during Intraspecific Interactions

Using our criteria for ritualized visual signals we chose from the literature and then classified, named, and described all behavioral patterns we considered to be visual signaling displays performed by anuran species (Table 10.1). Since the diversity of signals that an animal emits can be exaggerated from the human point of view (Harper 1991), we follow a conservative approach in our classification process. Thus we group all those behavioral patterns involving similar body parts and providing a similar visual impression into a single visual display category, assuming that potential receivers are approximately at the same level and in front of the sender. This approach neither assumes nor suggests behavioral homologies between the species performing similar displays. We name visual displays by describing both the involved part of the body and its main action (e.g., toe trembling). After noting the main components of the visual display we describe what we consider to be secondary interspecific variations. Since the communicative role of visual displays has rarely been tested we only provide information about the general context in which they are performed. Finally Figures 10.1 to 10.4 depict foot-flagging, the most conspicuous anuran visual display.

Context and Interindividual Distances

Visual displays have been observed during both agonistic and courtship interactions (Zimmermann and Zimmermann 1988; Narvaes 1997). Nonetheless when foot-flagging or arm-waving displays are considered, most of the reports indicate agonistic contexts. Such displays are usually performed after conspecific males approach or vocalize close to a signaling individual (Lindquist and Hetherington 1996, 1998; Hödl et al. 1997; Haddad and Giaretta 1999).

Visual displays are triggered by approaches of conspecific individuals (Brattstrom and Yarnell 1968; Durant and Dole 1975; Kluge 1981; Crump 1988; Richards and James 1992; Pombal et al. 1994; Hödl et al. 1997), by self reflections in a mirror (Pombal et al. 1994; Lindquist and Hetherington 1998), by dummies (Emlen 1968; Lüddecke 1999), or by visual signals performed by a conspecific individual (Pombal

Table 10.1 Anuran visual display patterns used during intraspecific communication (see Table 10.2 for information on the species performing the displays)

Visual Signals Repertoire

Limbs

1. **Toe trembling.** Twitching, vibrating, or wiggling the toes, without otherwise moving the leg. Toes may be moved sequentially, in wave like fashion (Lindquist and Hetherington 1996; Narvaes 1997; Amézquita and Hödl, unpublished manuscript), or without an apparent order (Brattstrom and Yarnell 1968; Zimmermann and Zimmermann 1988). The signal can be accentuated by the whitish upper surface of the toes (Heyer et al. 1990; Hödl et al. 1997; Narvaes 1997), and can be performed as part of the arm-waving and hind-foot-raising displays (Lindquist and Hetherington 1996). Toe trembling is performed after the sighting of potential rivals or partners (Zimmermann and Zimmermann 1988) or during agonistic interactions (Brattstrom and Yarnell 1968; Heyer et al. 1990; Hödl et al. 1997; Narvaes 1997; Haddad and Giaretta 1999, Amézquita and Hödl, unpublished data).
2. **Hind foot lifting.** Raising one hind foot dorsally, without extending the leg, and putting it back on the ground. The motion may (Lindquist and Hetherington 1996) or may not (Weygoldt and Carvalho e Silva 1992) be repeated before the limb is set back on the ground. The lifted foot may simultaneously perform toe trembling (Lindquist and Hetherington 1996). It has been reported to occur only during agonistic encounters.
3. **Arm waving.** Lifting an arm and waving it up and down in an arc above or in front of the head. The movements may (Lindquist and Hetherington 1996) or may not (Crump 1988; Richards and James 1992; Pombal et al. 1994; Amézquita and Hödl, unpublished manuscript) be performed in a temporal pattern. It may also be performed while the animal is walking and the lifted hand may perform simultaneous toe trembling (Lindquist and Hetherington 1996). Arm waving has been observed during agonistic encounters.
4. **Limb shaking.** Rapid up-and-down movements of a hand or foot. Limb shaking differs from hind foot lifting and arm waving in the very high speed of the movement during limb shaking, which probably causes a different visual impression to the receiver. The movement can be repeated several times before the leg returns to the resting position (Lescure and Bechter 1982). Limb shaking is commonly performed in the course of both courtship and agonistic encounters (Lüddecke 1974; Lescure and Bechter 1982; Zimmermann and Zimmermann 1988; Zimmermann 1990).
5. **Wiping.** Moving jerkily the hand or foot upon the ground, without lifting it. Wiping differs from limb shaking in that the hand or foot remains in contact with the substrate and differs from leg stretching in that the limb is not fully extended. It is performed during advertisement displays (Zimmermann and Zimmermann 1988).
6. **Leg stretching.** Stretching a single leg or both hind legs rapidly at the substrate level. The leg may (Hödl et al. 1997; Narvaes 1997; Haddad and Giaretta 1999) or may not (Winter and MacDonald 1986; Wevers 1988; Richards and James 1992; Weygoldt and Carvalho e Silva 1992) remain extended for some time. Leg stretching occurs in agonistic contexts.
7. **Foot flagging** (Figures 10.1–10.4). Raising one or both hind legs, extending it/them slowly out and back in an arc above the substrate level, and returning it/them to the body side. At maximum extension feet can be vibrated (Richards and James 1992) or toes can be outstretched and spread, or curled revealing contrasting (Harding 1982; Davison 1984; Heyer et al. 1990; Malkmus 1992; Hödl et al. 1997) or unconspicuous (Richards and James 1992) toes or webbing. Both hind legs can be simultaneously (Figures 10.2, 10.3) or alternately (Malkmus 1989, 1992; Heyer et al. 1990; Richards and James 1992; Narvaes 1997; Pavan et al. 2000) extended, in the latter case with no evidence of regular alternation (Heyer et al. 1990; Hödl et al. 1997; Narvaes 1997; Haddad and Giaretta 1999). It is performed during both male advertisement displays and agonistic encounters (Hödl et al. 1997, Pavan et al. 2000).

Body—Stationary

8. **Color changing.** Changing the color of the whole body (Wells 1980b), or at least the dorsal surface and the throat (Lüddecke 1974, 1976), to a near black tone. It occurs within 1–30 minutes after the beginning of calling activity. The reverse change, from black to brown coloration, may occur within similar time intervals after the male ceases calling (but see Lüddecke 1999). Color changing affects the course of interindividual encounters, leading to either agonistic or courtship interactions (Wells 1980b).
9. **Body lowering.** Lowering either the anterior part of the body or the whole body, pressing it against the substrate. Limbs may remain pressed to the sides of the body (Brattstrom and Yarnell 1968; Zimmermann and Zimmermann 1988) or lie outstretched (Jaslow 1979). It may be performed as part of circling (Silverstone 1973; Zimmermann 1990) and during both courtship and agonistic interactions (Zimmermann 1990). At least in the latter case it is believed to signal submission or lack of aggressive intention (Wells 1977a; Jaslow 1979; Zimmermann and Zimmermann 1988).
10. **Upright posture.** Extending the angled arms and raising the anterior part of the body (Zimmermann and Zimmermann 1988). It may occur merely as a posture (Zimmermann 1990) or precede walking toward an intruder or displaying the throat (Wells 1980b). In shallow water it is also performed through the inflation of the body (Emlen 1968; Wells 1977c). When the extension of the arms was associated with the facilitation of calling activity (Pombal et al. 1994), we did not consider the posture as a visual display. Upright posture occurs after the sighting of potential rivals or partners, and thus in both agonistic and courtship contexts.
11. **Throat display.** Pulsation of the throat without vocalizing after adopting an upright posture. The vocal sac usually contrasts with the background because of the brilliant yellow coloration (Dole and Durant 1974; Wells 1980b; Juncá 1998). Throats of males in other species are also colored and probably contrasting (Lüddecke 1974; Wells 1980a; Hödl 1991), but we do not consider them as visual displays because the extensions of the vocal sac occur only during vocalization.

Table 10.1 Continued

12. **Body raising.** Elevating the body by extending all four legs. Sometimes the individual stands on the toes of the rear feet (Durant and Dole 1975). It may occur as a posture (Wells 1980a) or to precede, and perhaps represent, an initial step of body jerking (Zimmermann 1990). Some individuals slowly approach the receiver while maintaining this body posture, that is, without flexing the limbs ("strutting" in Zimmermann and Zimmermann 1988). It is performed at the initial stages of both courtship and agonistic interactions. (For body raising during complex agonistic interactions see Figure 2 in Haddad and Giaretta 1999.)
13. **Body inflation.** Increasing the apparent body size by pumping air into the body. It is commonly performed after body raising, during both agonistic and courtship interactions (Zimmermann and Zimmermann 1988; Zimmermann 1990). Some ranids also inflate their bodies to adopt upright postures (Emlen 1968; Wells 1978), but we do not treat it as a visual display in this case because the visibility (to conspecifics) of the trunk partially submerged in shallow water is in doubt (Emlen 1968).
14. **Two-legged pushups.** Moving the anterior part of the body up and down through jerky and repeated extensions of the forelegs. The fore part of the body may remain higher (Winter and MacDonald 1986; Zimmermann and Zimmermann 1988; Zimmermann 1990) or lower than the horizontal axis ("nodding" in Zimmermann and Zimmermann 1988) during the movements. It is performed at least during advertisement displays.
15. **Body jerking.** Performing jerky movements with the body without lifting either hands or feet off the ground. The movements can be performed forward and backward in a repeated manner ("bockeln" in Zimmermann and Zimmermann 1988) or without an apparent pattern (Weygoldt and Carvalho e Silva 1992). Body jerking is performed during agonistic encounters or as part of advertisement displays.
16. **Back raising.** Elevating the posterior part of the body through the simultaneous extension of the rear legs. It may occur as a series of movements and be reinforced by a particular coloration of the dorsal posterior end of the body (Brattstrom and Yarnell 1968). It has been reported during both agonistic (Brattstrom and Yarnell 1968) and courtship (Narvaes 1997) interactions.

Body—Non-stationary

17. **Running-jumping display.** Running quickly back and forth or sideways along the substrate or a calling perch. It may be accompanied by raising of front feet off the ground or by "four-feet" jumps (Dole and Durant 1974; Wells 1980b). These extremely conspicuous movements are performed at the initial stages of courtship interactions (Wells 1980b; Juncá et al. 1994).
18. **Circling.** Moving around another individual or simply pivoting around its own axis. It is mostly performed as a discontinuous movement and combined with body lowering (Silverstone 1973; Zimmermann 1990). Circling occurs during courtship interactions (Silverstone 1973; Wells 1980a; Zimmermann and Zimmermann 1988; Zimmermann 1990).

et al. 1994; Narvaes 1997; Amézquita and Hödl, unpublished data). Individuals of the Brazilian torrent frog, *Hylodes asper*, provided with a waist band and tied onto a fishing hook immediately elicited foot flagging in vocalizing conspecific males when brought into their visual fields (Hödl, unpublished data). The playback of conspecific calls has elicited both acoustic and visual signaling in the harlequin frog, *Atelopus zeteki*, although the initial response of males perceiving a rival acoustically is to vocalize (Lindquist and Hetherington 1996). In both *Hylodes asper* and the Amazonian tree frog, *Hyla parviceps*, visual signaling occurs only after the sighting of intruders; the playback of conspecific advertisement calls provokes reorientation and sometimes displacement toward the sound source, but fails to elicit visual signaling (Hödl, Amézquita, unpublished data).

Visual signals are reported to be performed mainly at interindividual distances of less than 50 cm (mean = 22.8 cm, data from 7 species reported in McDiarmid and Adler 1974; Durant and Dole 1975; Davison 1984; Malkmus 1989; Richards and James 1992; Amézquita and Hödl, unpublished data). Visual communication at interindividual distances up to a few meters is exclusively reported for foot-flagging species (for *Hylodes dactylocinus* see Narvaes 1997, Pavan et al. 2000; for the Brazilian torrent frog, *Hylodes asper*, see Hödl et al. 1997; and for the rock skipper, *Staurois latopalmatus*, see R. Zeiner, unpublished data). Variation in the distance to potential receivers may exist within species when either individuals or different visual displays are compared. For example, in *Hyla parviceps*, foot flagging is performed at distances of 1 to 74 cm (mean = 19.7 cm, n = 22 from 8 individuals) from the receiver, and arm waving at 5 to 10 cm (mean = 7.5 cm, n = 6 from 4 individuals) (Amézquita and Hödl, unpublished data).

Elicitation of visual signaling suggests that visual signals represent a facultative communicative mode whose use depends on the visibility of potential intruders. Once the first visual displays are performed the outcome of the interaction depends largely on the reaction of the potential intruder. If the intruder calls, approaches further, or displays visually, the initial display is usually followed by further displays or by physical combat (Weygoldt and Carvalho e Silva 1992; Pombal et al. 1994; Hödl et al. 1997; Narvaes 1997; Amézquita and Hödl, unpublished manuscript). Initial displays or combats usually end when the intruder retreats (Brattstrom and Yarnell 1968; Kluge 1981; Richards and James 1992; Weygoldt and Carvalho e Silva 1992). On the other hand, if a potential intruder performs some form of body lowering (Brattstrom and Yarnell 1968; Emlen 1968;

Figure 10.1. Foot-flagging and vocalizing male of the Brazilian torrent frog, *Hylodes asper* (Leptodactylidae, SVL = 41 mm), drawn by H. C. Grillitsch after a photograph taken by W. Hödl at Picinguaba, São Paulo, Brazil. The first field notes on the spectacular limb-flagging behavior in anurans were made in this species by Stanley Rand at Boracéia, São Paulo, Brazil, in 1963 (published by Heyer et al. 1990). (Graphics courtesy of H. C. Grillitsch)

Zimmermann 1990) the attack will be inhibited and a courtship sequence may follow.

Visual signals are commonly associated with movement and sometimes function as alerting signals that call the attention of the receiver before the emission of the main message (Endler 1992). We suspect that toe trembling, upright posture, and body raising most likely have an alerting function because they are commonly followed by or performed in conjunction with other visual or acoustic displays (Wells 1980b; Zimmermann 1990; Lindquist and Hetherington 1996; Hödl et al. 1997). In spite of this, they are treated as visual displays themselves since they may elicit attention in the receiver, increasing its probability of detecting the following signal. This reaction most probably benefits the sender.

The Distribution of Visual Signaling among Anuran Taxa

The Database

To investigate the patterns of visual signaling among anuran taxa, we listed—based on the behavioral categories described in Table 10.1—the species and their visual displays known to occur during intraspecific communication (Table 10.2). Probably most if not all dendrobatid frogs communicate visually, but the behavior of only a small subset of species has been adequately studied. We include this subset of species, although the amount of knowledge on the behavioral repertoire is also highly variable among them. Where available the sex of the individual performing the visual display is given (Table 10.2).

Diversity of Visual Signaling Behaviors

The diversity of behavioral patterns related to visual communication (i.e., the repertoire of visual displays) varies enormously among anuran taxa (see Table 10.2). Studies on dendrobatid species reveal the most impressive variety of visual displays, which are coupled with acoustic signals during both courtship and agonistic interactions (Lüddecke 1974, 1999; Wells 1980b; Zimmermann and Zimmermann 1988; Zimmermann 1990). The extraordinarily complex visual communication apparent in dendrobatids may be partially explained by the fact that these species are relatively easy to observe and breed in captivity, and are thus attractive for both professional researchers and hobbyists (Schmidt and Henkel 1995). It is therefore not surprising that their behavior is fairly well known. At the other extreme, reports on the behavior of species that breed in streams remain highly anecdotal (Harding 1982; Winter and MacDonald 1986; Heyer et al. 1990; Richards and James 1992), which is partially a consequence of the difficulties imposed by the habitat and by their complicated breeding behavior (Narvaes 1997).

Figure 10.2. Foot-flagging male of the rock skipper, *Staurois latopalmatus* (Ranidae, SVL = 42 mm), drawn by H. C. Grillitsch after a video taken by R. Zeiner at Danum Valley, Sabah. Individuals of *S. latopalmatus* performed visual signals by raising the left foot, the right foot, or both feet simultaneously in irregular alternation. Of 125 flaggings during 283 minutes of observation of at least 11 individuals, 43 signals were given with the left foot, 38 with the right foot, and 44 with both feet simultaneously (W. Hödl and eight Austrian field course students, unpublished observations at waterfall near Danum Valley Field Center, 7.2.97–11.2.97, between 10:30 and 18:00). (Graphics courtesy of H. C. Grillitsch)

Species in 7 of 20 "families" (clades following Ford and Cannatella 1993) within Neobatrachia are reported to display visually, and most of the species perform more than one category of visual display during communication (Table 10.2). As indicated above, further studies on the behavior of the less-known species will probably reveal a greater variety of visual displays than those reported here.

Visual Display Repertoires

Behavioral patterns seem to be more similar within rather than between families, and particularly within rather than between genera (Table 10.2). Although ascertaining behavioral homologies is difficult given the present state of knowledge, some species in phylogenetically distant clades present very similar visual displays. The most striking examples of similarities involve up to four anuran families: the foot-flagging displays performed by *Hylodes asper* (Leptodactylidae), *Taudactylus eungellensis* (Myobatrachidae), *Litoria genimaculata* (Hylidae), and *Staurois parvus* (Ranidae); the arm-waving displays performed by *Brachycephalus ephippium* (Brachycephalidae), *Atelopus zeteki* (Bufonidae), *Crossodactylus gaudichaudii* (Leptodactylidae), and *Hyla parviceps* (Hylidae); and the leg-stretching displays performed by *Hylodes asper, Taudactylus eungellensis, Litoria rheocola,* and *Epipedobates parvulus* (Dendrobatidae).

Likewise the number of different visual displays performed by individuals of one species (hereafter, the "species' visual signaling repertoire") appears also to vary both between species and between families. To group species according to their similarities in repertoire diversity, we used hierarchical cluster analyses. Each of the 18 categories of visual displays was considered a binary variable and its presence (1) or absence (0) in a species' repertoire was correspondingly coded. We then used the within-group linkage and the Euclidean Gamma Index (Hand 1981; SPSS 6.1.1) to estimate the degree of similarity between species' behavioral repertoires.

We contrasted graphically the similarity in the diversity of species' repertoires with the phylogenetic relatedness at the family level (Figure 10.5). Repertoires of species in the family Dendrobatidae were clearly associated with cluster 2, but repertoires of species in other families were grouped less distinctively among the clusters. Cluster analyses also revealed between-group differences in diversity of the visual signaling repertoires. The median number of visual displays performed by species increased progressively from cluster 1A to cluster 2B (Figure 10.6). Box plots also revealed outliers, that is, species that performed more (cluster 1A and

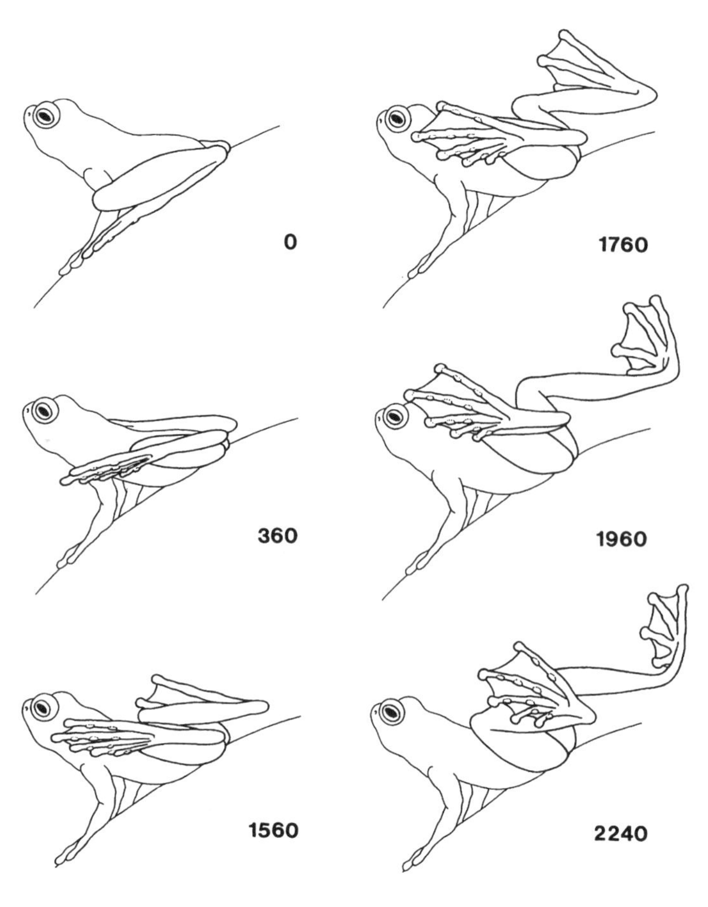

Figure 10.3. Complete foot-flagging sequence of *Staurois latopalmatus* (SVL = 42 mm) drawn by H. C. Grillitsch after a video taken by R. Zeiner at Danum Valley, Sabah. Time is given in milliseconds after the onset of the behavior. Note the toe spreading, exposing the whitish interdigital webbing during the flagging phase. (Graphics courtesy of H. C. Grillitsch)

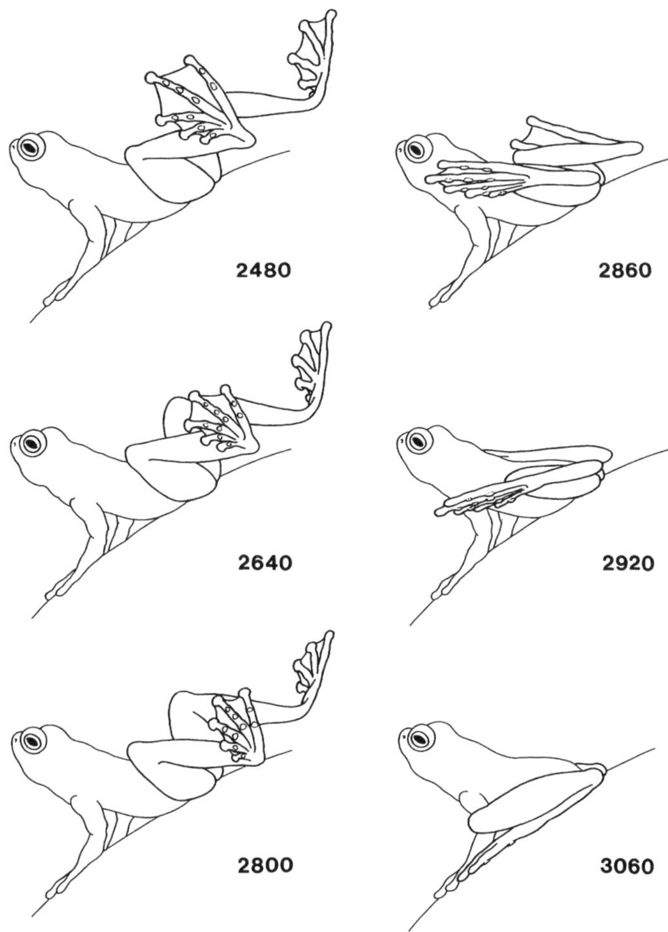

Figure 10.4. Foot-flagging male *Hyla parviceps* (Hylidae) (SVL = 19 mm) during call interval drawn by H. C. Grillitsch after a photograph taken by W. Hödl at Surumoni River near La Esmeralda, Amazonas, Venezuela. (Graphics courtesy of H. C. Grillitsch)

2A) or fewer (cluster 2B) displays than expected according to the within-cluster distribution. Thus the most parsimonious sequence of clusters, from less to more diverse, is 1A-1B-2A-2B (Figures 10.5 and 10.6). On the other hand, distances between clusters 2A and 2B (within the family Dendrobatidae) are larger than distances between 1A and 1B (each of them including species in different families). These results support the idea that diversity of visual displays is higher in dendrobatids than in any other family. Most of the dendrobatid species in cluster 2B were also grouped by Zimmermann and Zimmermann (1988) for presenting the highest diversity in their behavioral repertoire.

The patterns revealed by cluster analysis partly result from some species lacking many visual displays and performing few displays that may be similar between them. Some clusters are easily identifiable in these terms. About half of the species in cluster 1A perform arm waving and have very few other visual displays. Likewise cluster 1B is distinguished by the presence of foot flagging.

Whereas multiple occurrence of visual signaling among anuran families suggests that similar evolutionary pressures have led to convergent behavior, cluster analysis suggests some common patterns at the family level. The study of the evolution of visual signaling must involve both the phylogenetic effects and the more recent effects of natural selection (Brooks and McLennan 1991; Bradbury and Vehrencamp 1998), although homologies of behavioral patterns are difficult to ascertain. Therefore our analysis aims only at contrasting the diversity of visual signaling repertoires with the ecological conditions in which they could evolve. Although we cannot test whether foot flagging represents convergent evolution in species breeding at noisy streams, we can look for associations between the diversity of visual signaling repertoires and breeding under these environmental conditions.

Hypotheses on the Evolution of Visual Displaying in Anurans

The evolution of signals as phenotypic traits results from interactions among at least three sets of factors: (1) the restrictions imposed by other phenotypic attributes of the individual or species, (2) the intrinsic properties of signals, and (3) the effect of ecological factors. We discuss each of these aspects, emphasizing the differences between acoustic and visual signals.

The Origin of Signals

Signals are believed to have evolved from pre-existing cues or from other signals, and comparative studies indicate four main sources of raw material for the evolution of ritualized visual signals (Harper 1991): intention movements, motivational conflict, autonomic responses, and protective responses. In this sense the visual displays we report here could originate from other activities performed in either intra- or interspecific interactions.

Table 10.2 Distribution of visual signaling behaviors among anuran species (see description of behavioral patterns in Table 10.1)

	Behavioral Pattern[a]																					
	Appendages							Body—Stationary									Body—Non-stationary					
Species	Toe trembling 1	Hind foot lifting 2	Arm waving 3	Limb shaking 4	Wiping 5	Leg stretching 6	Foot flagging 7	Color changing 8	Body lowering 9	Upright posture 10	Throat display 11	Body raising 12	Body inflation 13	Two-legged pushups 14	Body jerking 15	Back raising 16	Running-jumping display 17	Circling 18	Ecology[b]			References[c]
Brachycephalidae																						
Brachycephalus ephippium			M																FF	A	D	1
Bufonidae																						
Atelopus limosus			M																SS	C	D	2
A. varius			B																FS	A	D	3
A. zeteki			M																FS	A	D	4
A. chiriquiensis			M						M										FS	A	D	4, 5
Leptodactylidae																						
Leptodactylus melanonotus	M								M							M			PO	C	N	6; 44
Crossodactylus gaudichaudii		M	M			M									M				FS	C	D	7
Hylodes asper	M			m[d]		M	M					m[d]							FS	C	D	8, 46, 47
H. dactylocinus	M			M		M	M									M			FS	C	D	9, 48
Myobatrachidae																						
Taudactylus eungellensis						M	M							M					FS	C	D	10
Hylidae																						
Litoria genimaculata							M												SS	C	N	11; 38
L. nannotis			M				M												FS	C	N	11; 38
L. rheocola						M													FS	C	N	11; 38
L. fallax							M												EP	C	N	11; 39
Hyla parviceps	M		B				M				M								EP	C	N	12
H. rosenbergi												M							SS	C	N	13
Phyllomedusa sauvagii							M												EP	C	N	14
Centrolenidae																						
Hyalinobatrachium fleischmanni															M				SS[e]	C	N	37
Dendrobatidae																						
Colostethus trinitatis	U							M		B	F						M		SS	C	D	15, 25, 26, 36
C. collaris	U							M		U	F	B		B	M		M		SS	C	D	16, 26, 35
C. inguinalis	U								F	U		B						M	SS	C	D	17, 18, 26
C. palmatus				M				M	F	B		B							SS	C	D	19, 20, 26

C. marchesianus									F	F	F							B	FF	C	D	45
C. stepheni									F			M					M	B	FF	C	D	21, 45
Epipedobates parvulus/bilinguis	U					M				U									FF	A	D	23, 26; 40
E. pictus	U			U					U	U					U				FF	C	D	26; 40
E. pulchripectus	U									U									FF	A	D	26; 40
E. tricolor	U			B	F				F	M		M	M					F	FF	A	D	22, 26; 40
E. anthonyi/tricolor	U			U					U	U									FF	A	D	26; 40
E. boulengeri	U			U					U	U					U				FF	A	D	26; 40
E. silverstonei	U									U									FF	A	D	26; 41
E. bassleri	U			U	U				U	U				U					FF	A	D	26; 41
E. trivittatus	U									U									FF	A	D	26; 41
E. femoralis	U								U	U								U	FF	C	D	26; 40, 41
Phyllobates terribilis	U			B	B				M	M		M	M	U				F	FF	A	D	22, 26; 41
P. lugubris	U			U						U								U	FF	A	D	26; 41
P. vittatus	U			U	U				U	U								U	FF	A	D	26; 40, 41
Dendrobates auratus	U			U					U	U				U				U	FF	A	D	26; 41
D. truncatus	U			U					U	U				U				U	FF	A	D	26; 41, 42
D. leucomelas	U			U	U					U									FF	A	D	26; 41, 42
D. tinctorius	U			U	U					U				U				U	FF	A	D	26; 41, 42
D. azureus	U				U					U								U	FF	A	D	26; 41, 42
D. fantasticus	U									U		U							FF	A	D	26; 41
D. reticulatus	U			U	U				U	U		U	U	U				U	FF	A	D	26; 41
D. variabilis	U			U	U					U		U	U	U					FF	A	D	26; 41
D. imitator	U			U	U					U		U	U	U				U	FF	A	D	26; 41
D. quinquevittatus complex	U			U	U					U		U	U	U				U	FF	A	D	26; 41, 42
D. granuliferus	U			M	U					M						F			FF	A	D	26, 28; 41, 42
D. pumilio	U			U						U		M	U				M		FF	A	D	26, 27; 41, 42
D. speciosus	U			U	U				U	U									FF	A	D	26; 41, 42
D. lehmanni	U			U	U				U	U		U	U	U				U	FF	A	D	26; 41
D. histrionicus	U			B	B				B	M		M	U	B				F	FF	A	D	22, 24, 26
Ranidae																						
Staurois parvus							M												FS	C	D	29, 30, 31
S. latopalmatus							M												FS	C	D	32
Rana clamitans										M									PO	C	N	33
R. catesbeiana										M									PO	C	N	34; 43

Reference codes: 1. Pombal et al. (1994); 2. Ibáñez et al. (1995); 3. Crump (1988); 4. Lindquist and Hetherington (1996); 5. Jaslow (1979); 6. Brattstrom and Yarnell (1968); 7. Weygoldt and Carvalho e Silva (1992); 8. Heyer et al. (1990); 9. Narvaes (1997); 10. Winter and MacDonald (1986); 11. Richards and James (1992); 12. Amézquita and Hödl (unpubl. manusc.); 13. Kluge (1981); 14. Halloy (unpubl. manusc.); 15. Wells (1980b); 16. Dole and Durant (1974); 17. Duellman (1966); 18. Wells (1980a); 19. Lüddecke (1974); 20. Lüddecke (1976); 21. Juncá et al. (1994); 22. Zimmermann (1990); 23. Wevers (1988); 24. Silverstone (1973); 25. Test (1954); 26. Zimmermann and Zimmermann (1988); 27. Forester et al. (1993); 28. Crump (1972); 29. Harding (1982); 30. Malkmus (1989); 31. Malkmus (1992); 32. Davison (1984); 33. Wells (1978); 34. Emlen (1968); 35. Durant and Dole (1975); 36. Sexton (1960); 37. McDiarmid and Adler (1974); 38. Frith and Frith (1987); 39. Robinson (1993); 40. Silverstone (1976); 41. Schmidt and Henkel (1995); 42. Silverstone (1975); 43. Wiewandt (1969); 44. Gregory (1983); 45. Juncá (1998); 46. Hödl et al. (1997); 47. Haddad and Giaretta (1999); 48. Pavan et al. (2000).

[a]Codes for behavior: M = performed by males; F = females; B = or both; U = unspecified by the author.

[b]Codes for ecological traits: PO = ponds; SS = slow-flowing streams; FS = fast-flowing streams and waterfalls; EP = elevated perches; FF = forest floor; C = cryptic; A = aposematic; N = nocturnal; D = diurnal.

[c]References after semicolon provide ecological data but do not report on visual displays.

[d]The "m" in *Hylodes asper* was added during the editing phase and was therefore not included in the grouping and phylogenetic analyses.

[e]This species displays at elevated perches (EP) above streams (SS).

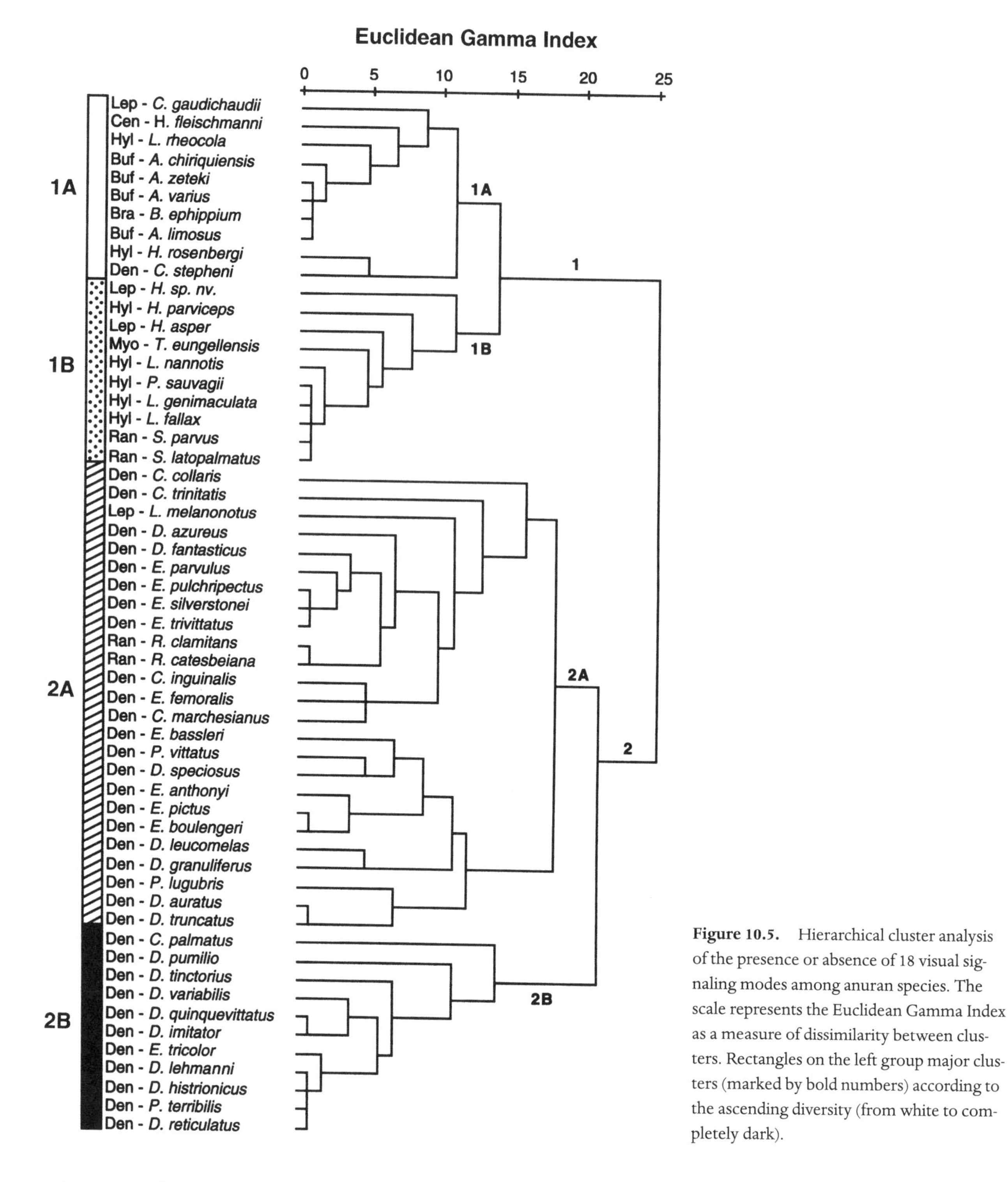

Figure 10.5. Hierarchical cluster analysis of the presence or absence of 18 visual signaling modes among anuran species. The scale represents the Euclidean Gamma Index as a measure of dissimilarity between clusters. Rectangles on the left group major clusters (marked by bold numbers) according to the ascending diversity (from white to completely dark).

Pombal et al. (1994) suggested that arm waving in the pumpkin toadlet, *Brachycephalus ephippium*, is derived from cleaning behavior because of the similarities in both movements. Alternatively, arm waving and hind-feet lifting may well have evolved from displacement movements related to intrasexual competition for access to mates (Woodward 1982). Likewise visual displays involving either stationary or nonstationary body movements (e.g., body jerking, two-legged pushing up, running-jumping display) may have appeared from either intention movements or motivational conflict. Foot-flagging displays are commonly performed during physical combat and in these instances they are easily confounded by the observer with kicking motions (Amézquita and Hödl, unpublished observations in *Hyla parviceps*). Foot-flagging may represent the ritualization of an aggressive behavior (attack) now performed before phys-

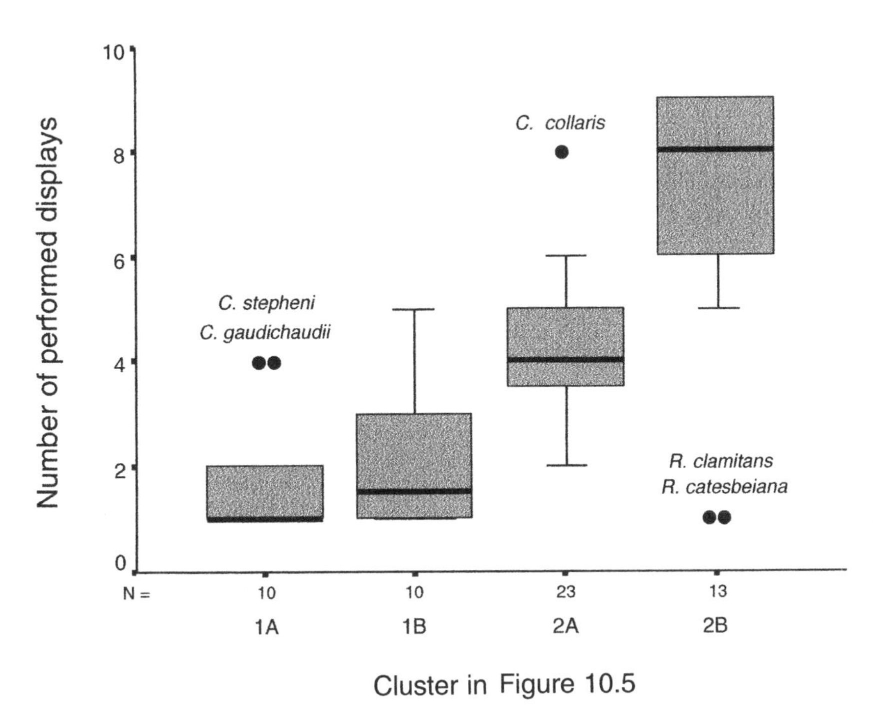

Figure 10.6. Distribution of repertoire diversity within and between clusters of anurans species (see Figure 10.5) that display visually. Bars indicate median values, upper and lower ends of the boxes denote upper and lower quartiles; points represent outliers from each distribution.

ical contact with the rival. In any case ritualization should lead to signals becoming highly stereotyped, repetitive, and exaggerated, thus improving their communicative role (Krebs and Davies 1993).

The ritualization of signals can also be initiated or greatly accelerated if perceptual systems in the species have inherent biases (Ryan and Keddy-Hector 1992; Arak and Enquist 1993), for example, if potential receivers are particularly sensitive to some movements in specific environments. Fleishman (1992) suggested that in anoline lizards selection on the visual system to detect insect or prey movements has biased males toward making the kinds of visual displays that they do. In a similar way females of some *Colostethus* species might be sensitive to objects that look and move like a gular sac because it is important to identify calling males during initial stages of courtship (Lüddecke 1999). This sensory bias may have facilitated the evolution of throat displaying as a signal, even in female–female agonistic interactions, given that a female throat differs from that of a male. Such a difference is found in *Colostethus trinitatis* and *C. collaris;* both species perform throat displays during agonistic interactions.

Visual versus Acoustic Signals

Signals themselves possess characteristics that impose restrictions on the communication system (Alcock 1998). Visual signals differ from auditory signals, the most prominent anuran signals, in many important aspects (Krebs and Davies 1993), to be discussed below.

Locatability

Although visual signals are similar to acoustic signals in their rate of change, a property that may affect the amount of information transmitted (Endler 1992), visual signals are more easily located than acoustic signals. Location is probably one of the most important aspects of the information an individual transmits in advertisement, courtship, and agonistic contexts. Females approaching sound sources that are broadcast from elevated perches show greater jump-error angles than when they approach at the two-dimensional ground level (data reviewed in Klump 1995). Furthermore female *Hyperolius marmoratus* made extensive use of visual cues when approaching elevated sound sources, and both lateral head scanning (before jumps) and vertical changes in head orientation were frequent (Passmore et al. 1984). Though we are not aware of reports on visual signaling in the species above, this evidence suggests that visual cues enhance the ability of females to locate males. In this way the sender's fitness might be increased in visual-signaling males in those species where mating or defense of territories occur on perches above ground level, such as *Hyla parviceps, Phyllomedusa sauvagii, Litoria fallax,* and *Hyalinobatrachium fleischmanni.*

Diurnality

Visual signals can be transmitted over relatively short distances and are often obscured by obstacles in the environment. These problems impose distance and light restrictions on the communication process (Harper 1991). Data on the distance between sender and potential receiver of visual signals support this prediction (see "Context and Interindividual Distances" above); with the exception of the foot-flagging species (Hödl, personal observations) almost all reported visual displays were performed when individuals were less than 50 cm from one another. The distance constraint can be overcome if different types of signals are used facultatively, that is, acoustic communication may prevail for long-distance interactions, whereas the spatial component added by visual communication becomes useful at medium or short ranges (Endler 1992; Lindquist and Hetherington 1996). Actually all species reported here also make use of vocalizations during intraspecific communication. The alternative (short-distance) role of visual signals would explain why they also are used by species that are active at low light levels and that occur at sites with high vegetation density (McDiarmid and Adler 1974; Richards and James 1992; Amézquita and Hödl, unpublished data), even though visual communication is predicted to be inefficient under these conditions (Harper 1991; Endler 1992; but see Buchanan 1998 for low-illumination prey detection in tree frogs).

Predation Pressure

Visual signals can make the sender vulnerable to localization by predators, but this problem can be partly overcome when the signals are performed rapidly (Harper 1991; Bradbury and Vehrencamp 1998) or are highly directional. Because cryptic coloration is the first line of defense against predation for most animals, selection may favor short-range distinctiveness for communication with conspecifics and long-range crypsis for protection against predators. A compromise between conspicuousness and crypsis is dynamic coloration (Butcher and Rohwer 1989): an individual may, for example, maintain colored or structured body parts hidden and flash them during visual displays by briefly uncovering them (Bradbury and Vehrencamp 1998). In some species that perform foot-flagging displays, brilliantly colored webbing is visible only during the last part of the display (Figure 10.3, Table 10.1). Likewise the brilliant orange spots on the limbs of *Hyla parviceps* are visible only during foot-flagging displays (Figure 10.4; Amézquita and Hödl, unpublished manuscript). Bright coloration may serve two purposes: to increase the signal-to-noise ratio and/or to indicate some quality of the sender. The latter possibility has not been considered for flashing coloration in anuran amphibians. On the other hand extended appendages do not necessarily have to be colored since the movement itself contrasts against the stationary background (Bradbury and Vehrencamp 1998).

The evolution of visual signals can also be favored when predators are deterred from attacking by strategies such as toxicity and aposematism, as has been proposed for some dendrobatid, brachycephalid, and bufonid frogs (Duellman and Trueb 1986; Pombal et al. 1994; Lötters 1996). Aposematism may also favor visual signaling because the aposematic coloration increases the conspicuousness of individual frogs to both conspecifics and heterospecifics within the habitat (Pombal et al. 1994).

Ambient Noise

Noise can affect the evolution of a signal (Harper 1991). Visual signaling may represent an alternative or complementary form of communication for those species that display and breed at sites with high ambient noise levels. Such an explanation has been suggested for the evolution of visual signaling in species that live near waterfalls or torrential streams (Heyer et al. 1990; Lindquist and Hetherington 1996; Hödl et al. 1997; Narvaes 1997; Haddad and Giaretta 1999; Figures 10.1–10.3). Several frog species (for birds, see also Martens and Geduldig 1988) that breed in the midst of high ambient noise generated by rushing water utter high-pitched calls concentrated within a narrow frequency band (Dubois and Martens 1984). The higher the frequency, the better the contrast to the low-frequency-dominated noise produced by the turbulent waters (see Figure 10.7; Hödl, unpublished data). However, perhaps to compensate for the increased sound attenuation at high frequencies, visual signaling (i.e., leg-waving) has evolved in these frogs.

Acoustically streams do not represent a homogeneous habitat. According to published descriptions and our own experience, we found that of the species reported to communicate visually and breed at streams (Table 10.2), only some of them seem to do it under conditions of high background noise. The intensity and quality of noise varies in relation to characteristics of the streams and the season of the year. In addition some species avoid breeding in streams when or where water flow is rapid (Wells 1980a), whereas others seem to prefer rushing water (Malkmus 1989; Heyer et al. 1990; Richards and James 1992; Hödl et al. 1997).

Nevertheless high-pitched calls are also emitted by males of other species that inhabit apparently less noisy streams (e.g., *Colostethus trinitatis*, Wells 1980b). Low noise levels may represent a selective force for many of the stream-breeding *Atelopus* species, since their vocalizations are particularly weak (Jaslow 1979; Cocroft et al. 1990; Lötters 1996). We thus considered in Table 10.2 two kinds of stream conditions according to the most probable prevalent condition during the displaying and breeding activity of the

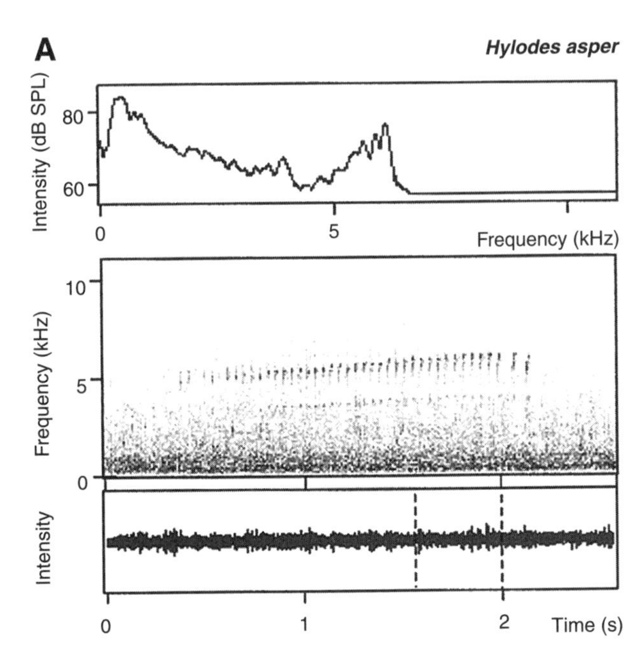

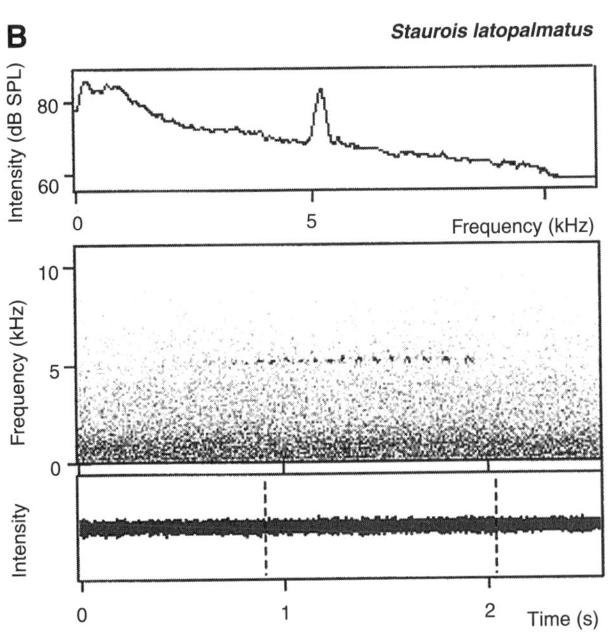

Figure 10.7. Oscillograms (below), spectrograms, and spectral section displays (above) taken at the marked period of advertisement calls of (A) the Brazilian torrent frog, *Hylodes asper* (Picinguaba, São Paulo, Brazil; 24°C air temperature). The sound intensity at 1 m from the calling individual and the rushing water measured 86 dB (flat, re 20 μPa, impulse time constant). Note that the maximum energy of the high-frequency call is below the ambient noise level, which drops as frequency increases. (B) The rock skipper, *Staurois latopalmatus* (Danum Valley, Sabah, 25°C air temperature). The sound intensity at 1 m from the calling individual and 70 cm from the rushing water measured 87 dB (flat, re 20 μPa, impulse time constant). Note that the maximum energy of the call exceeds that of the constant ambient noise level within its frequency range by less than 15 dB. In both analyses Canary sampling rate was 22 kHz, filter bandwidth 266, 58 Hz, frame length 23.22 ms, time 2.902 ms.

species: waterfalls and fast-flowing streams (FS) and low-forest, slow-flowing streams (SS).

An interesting potential counterexample to the relationship between high ambient noise level and visual signaling is given by the Bornean frog *Rana blythi:* this apparently mute species builds its nests within gravel bars of streams. Its behavior was thoroughly studied by Emerson and Inger (1992) and they did not report on any kind of visual display.

Ecology and Phylogeny in the Evolution of Visual Signaling

We have discussed arguments leading to predictions about ecological conditions that may have favored the evolution of visual displaying in anurans, namely displaying at elevated perches, diurnality, aposematism, and displaying at continuous high ambient noise levels. We collected ecological information from the literature about these factors in the species reported to perform visual displays (Table 10.2). We then mapped these ecological traits onto a phylogenetic tree depicting the relationships among the main clades of neobatrachian anurans. The tree was modified from Ford and Cannatella (1993) to include those clades reported to display visually. Additional information on the phylogenetic relationships within Dendrobatidae was obtained from Myers et al. (1991) and Toft (1995). We also separated closely related clades when they differed in some ecological or behavioral characteristic.

Ecological information was coded as discrete characters for three variables: display site (pond or swamp, elevated perch, stream, waterfall or noisy stream, forest floor), time of activity (nocturnal, diurnal), and the most probable antipredator strategy (crypsis, aposematism) according to dorsal coloration. As a working definition we classified as aposematic those species whose dorsal color is mostly dominated by bright colors, excluding all shades of green, brown, and black. For the ancestral conditions we assumed nocturnal activity, reproduction at ponds, and crypsis (Duellman and Trueb 1986).

The comparative method has been neglected in elucidating the ecological correlates of signal diversity (Harper 1991, but see Ryan 1988; Cocroft and Ryan 1995). We do not intend to compare the ecological attributes of species that perform visual displays with those of species not reported to display visually. There is of course not enough information and probably too much variation in ecological traits within the taxonomic (family) level that we use to represent nonvisually displaying clades. Instead our aim here is to look for patterns of association between the ecological attributes of the clades and the diversity of behavioral repertoires. We estimated diversity of behavioral repertoires from the cluster analysis performed in the section "Visual Displaying Repertoires" (Figure 10.5). Clusters 1A-1B-2A-2B are ordered from lowest to highest level of diversity and are represented by gradually darker rectangles in Figures 10.5 and 10.8. Species recognized as outliers in Figure 10.6 were reassigned to the cluster whose median number of performed displays was more similar. In this way *Colostethus stepheni* and *Crossodactylus gaudichaudii* were considered better represented as part of cluster 2A, *Colostethus collaris* was reassigned to cluster 2B, and *Rana clamitans* and *R. catesbeiana* were reassigned to cluster 1A.

The observed patterns suggest that visual communication may have evolved in anuran amphibians under several ecological conditions (Figure 10.8). Species that breed near waterfalls and fast-flowing parts of streams, mainly diurnal, in the families "Hylodinae" (*Hylodes*), Myobatrachidae (*Taudactylus*), Hylidae (*Litoria*), and Ranidae (*Staurois*) constitute a likely convergent group. They perform visual displays by hind-leg extension or flagging that apparently serve for communication at "long" distances (see "Context and Interindividual Distances"). These species differ from *Atelopus* spp. by using posterior instead of anterior limbs and their group is mostly represented by cluster 1B in Figure 10.5. The genera less consistent with this pattern are *Litoria* and *Phyllomedusa,* as they include nocturnal species, three of them apparently not breeding near waterfalls. This result, however, strongly supports a former explanation for the evolution of visual signaling. According to Heyer et al. (1990) high levels of background noise may have favored the evolution of visual signaling.

Visual signaling in diurnal and highly aposematic species of the families Brachycephalidae (*Brachycephalus*) and Bufonidae (*Atelopus*) may also have evolved under similar ecological conditions. Their behavioral repertoires are of similar diversity (see within cluster 1A in Figure 10.5) and involve arm waving and, in only two species of *Atelopus,* some additional displays. Unfortunately information on the (diurnal) behavior of several cryptically colored *Atelopus* species and the two non-aposematic brachycephalid species are lacking (McDiarmid 1971; Pombal et al. 1994; Lötters 1996). Further research on these species should provide an interesting phylogenetic contrast to test at a finer level the possible evolutionary link between their ecological attributes and visual signaling (Pombal et al. 1994). Although we classified here *Atelopus* species as displaying beside streams and waterfalls, some species not included in this study display and mate at forested sites, sometimes far from the streams where oviposition occurs (Dole and Durant 1974; Lötters 1996).

The diurnal, forest-floor inhabitant species in the family Dendrobatidae (*Colostethus, Epipedobates, Phyllobates,* and *Dendrobates*) perform many different displays, grouped as behavioral repertoires in clusters 2A and 2B (Figure 10.5). Some displays are widespread among the species, such as toe trembling, limb shaking, body lowering, and upright posture. In cluster 2B most species that are included additionally perform body raising, body inflation, two-legged push-ups, and circling. The shift to diurnality seems to be an ancestral condition for almost all dendrobatid taxa (but see Myers et al. 1991), making it difficult to assess its potential relevance for within-family differences in repertoire diversity. The association between aposematism and repertoire diversity also remains unclear, since most of cryptic *Colostethus* species perform as many visual displays as the aposematic *Dendrobates* in cluster 2A (Figure 10.5). Nevertheless some well-studied species in the cluster with highest diversity of visual signals (2B) are recognized as highly poisonous (Myers and Daly 1976; Myers et al. 1978).

The evolution of visual signaling in dendrobatids may also be associated with displaying and mating on the forest floor. As already proposed by Wells (1977a), the fact that both males and females live all year in the same areas and

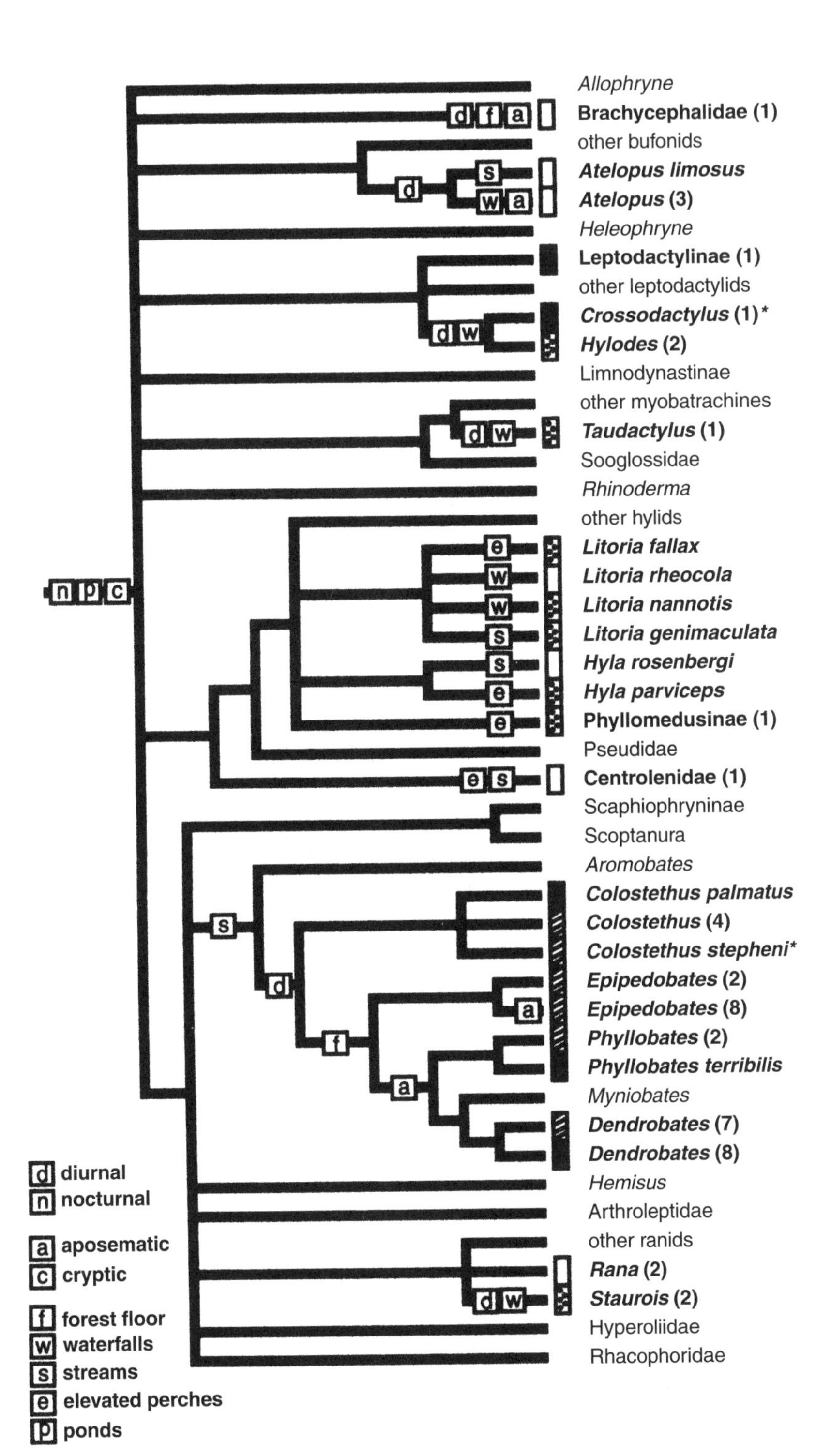

Figure 10.8. Mapping of ecological traits in anuran taxa that perform visual signals (bold letters). Rectangles indicate repertoire clusters resulting from the analysis depicted in Figure 10.5. Shading (from white to black) reflects increasing repertoire diversity. Numbers within parentheses give the number of species reported to display visually within the taxon. The phylogenetic tree was modified from Ford and Cannatella (1993), Myers et al. (1991), and Toft (1995). Fine captions indicate taxa in which visual signaling has not been reported to date. (Drawing by A. Amézquita)

compete for similar resources makes it important to recognize the sexual receptivity of potential partners. Individuals should therefore signal whether they are ready to mate or to defend a territory. This reasoning also applies to those *Colostethus* species in which both males and females establish territories close to streams (Lüddecke 1974, 1999; Durant and Dole 1975; Wells 1980b).

Explanations for the evolution of visual signaling in species not included in the groups mentioned above must surely wait until there is more information. Though no clear pattern was detected on the phylogenetic tree (Figure 10.8), we believe that displaying at elevated perches may have favored the evolution of visual signals that facilitate location and sex recognition in some hylid and centrolenid species.

The possibility also remains that some species perform visual signals as an ancestral condition: they evolved visual signals because they bred in abiotically noisy environments in the past. This explanation could particularly apply to *Hyla parviceps,* which presents one of the most diverse visual signaling repertoires reported here. Weygoldt (1986) studied a population of *H. parviceps* calling from perches 1 to 2 m above creeks. Our study on visual signaling by *H. parviceps* was performed at pools along a lowland stream that actually formed the pools by flooding once the rainy season advanced. Even though the environment was not as noisy as at waterfalls, Weygoldt's and our observations suggest that this species or its recent ancestor may have been a stream-breeding species in the past.

Conclusions

Modes of visual signaling are diverse and are widespread among and have evolved independently in several anuran families. Unfortunately data on receivers' responses to visual signals are scarce. Research in this area should expand the number of known signals and provide valuable information on their communicative role. The combination of ecological attributes associated with the diversity of visual signaling repertoires seems to vary between anuran taxa. Ecological patterns suggest the shift to diurnality to be the most ancestral change that facilitated the evolution of visual signaling. The visual signal repertoires are more complex in species that breed at noisy streams and even more in species that breed at terrestrial sites also used as feeding grounds. We strongly suggest that further comparative analyses be made at a finer resolution in particularly interesting groups, namely elosiine frogs (*Hylodes* and *Crossodactylus*) and *Atelopus* and *Staurois* species, to provide stronger tests that will lead to a clearer picture of the evolution of visual communication in anuran amphibians.

Acknowledgments

This research was funded by the Austrian Science Foundation (FWF) grant to WH (P 11565-Bio). B. Rojas provided invaluable help in collecting and organizing the reviewed bibliographic material. Field and filming assistance was provided by the following: G. M. Accacio, P. C. Fernandes, L. B. Ladeira, P. H. Lara, P. Narvaes, D. Pavan, E. Reisenberger, M. T. Rodrigues, L. C. Schiesari, G. Skuk, M. A. P. Valle (Brazil); A. Franz, U. Fraunschiel, H. Gasser, B. Gottsberger, G. Grabenweger, M. Hable, K. Humer, R. Jehle, R. Zeiner (Sabah, Borneo); and N. Ellinger (Venezuela). We highly appreciated comments on the manuscript by N. M. Kime, P. M. Narins, M. J. Ryan, R. Wassersug, K. D. Wells, and two anonymous referees. H. C. Grillitsch drew the illustrations. WH is especially indebted to M. Ryan for the invitation to participate in this symposium.

References

Alcock, J. 1998. Animal Behavior. Sinauer Associates, Sunderland, MA.

Arak, A. 1983. Male-male competition and mate choice in the anuran amphibians. Pp. 181–209. *In* P. Bateson (ed.), Mate Choice. Cambridge University Press, Cambridge.

Arak, A., and M. Enquist. 1993. Hidden preferences and the evolution of signals. Philosophical Transactions of the Royal Society of London B 340:207–213.

Bajger, J. 1980. Diversity of defensive responses in populations of fire toads (*Bombina bombina* and *Bombina variegata*). Herpetologica 36:133–137.

Bradbury, J. W., and S. L. Vehrencamp. 1998. Principles of Animal Communication. Sinauer Associates, Sunderland, MA.

Brattstrom, B. H., and R. M. Yarnell. 1968. Aggressive behavior in two species of leptodactylid frogs. Herpetologica 24:222–228.

Brodie, E. D., Jr. 1978. Biting and vocalization as antipredator mechanisms in terrestrial salamanders. Copeia 1978:127–129.

Brodie, E. D., Jr., R. T. Nowak, and W. R. Harvey. 1979. The effectiveness of antipredator secretions and behaviour of selected salamanders against shrews. Copeia 1979:270–274.

Brooks, D. R., and D. A. McLennan. 1991. Phylogeny, Ecology, and Behavior: A Research Program in Comparative Biology. The University of Chicago Press, Chicago.

Buchanan, B. W. 1998. Low-illumination prey detection by squirrel treefrogs. Journal of Herpetology 32:270–274.

Burrowes, P. A. 1997. The reproductive biology and population genetics of the Puerto Rican cave-dwelling frog *Eleutherodactylus cooki*. Ph.D. thesis, University of Kansas.

Butcher, G. S., and S. Rohwer. 1989. The evolution of conspicuous and distinctive coloration for communication in birds. Pp. 51–108. *In* D. M. Power (ed.), Current Ornithology, Vol. 6. Plenum Press, London.

Cocroft, R. B., R. W. McDiarmid, A. P. Jaslow, and P. M. Ruiz-Carranza. 1990. Vocalizations of eight species of *Atelopus* (Anura: Bufonidae) with comments on communication in the genus. Copeia 1990:631–643.

Cocroft, R. B., and M. J. Ryan. 1995. Patterns of advertisement call evolution in toads and chorus frogs. Animal Behaviour 49:283–303.

Crump, M. L. 1972. Territoriality and mating behavior in *Dendrobates granuliferus* (Anura: Dendrobatidae). Herpetologica 28:195–198.

Crump, M. L. 1988. Aggression in harlequin frogs: Male-male competition and a possible conflict of interest between the sexes. Animal Behaviour 36:1064–1077.

Davison, G. W. H. 1984. Foot-flagging display in Bornean frogs. The Sarawak Museum Journal 33:177–178.

Dole, J. W., and P. Durant. 1974. Courtship behavior in *Colostethus collaris* (Dendrobatidae). Copeia 1974:988–990.

Dubois, A., and J. Martens. 1984. A case of possible convergence between frogs and a bird in Himalayan torrents. Journal of Ornithology 125:455–463.

Duellman, W. E. 1966. Aggressive behavior in dendrobatid frogs. Herpetologica 22:217–221.

Duellman, W. E., and L. Trueb. 1986. Biology of Amphibians. McGraw-Hill, New York.

Durant, P., and J. W. Dole. 1975. Aggressive behavior in *Colostethus* (=*Prostherapis*) *collaris* (Anura: Dendrobatidae). Herpetologica 31:23–26.

Edmunds, M. 1981. Defensive behavior. Pp. 121–129. *In* D. McFarland (ed.), The Oxford Companion to Animal Behavior. Oxford University Press, Oxford.

Emerson, S. B., and R. F. Inger. 1992. The comparative ecology of voiced and voiceless Bornean frogs. Journal of Herpetology 26:482–490.

Emlen, S. T. 1968. Territoriality in the bullfrog, *Rana catesbeiana*. Copeia 1968:240–243.

Endler, J. A. 1992. Signals, signal conditions, and the direction of evolution. American Naturalist 139:S125–S153.

Fleishman, L. J. 1992. The influence of the sensory system and the environment on motion patterns in the visual displays of anoline lizards and other vertebrates. American Naturalist 139:S36–S61.

Ford, L. S., and D.C. Cannatella. 1993. The major clades of frogs. Herpetological Monographs 7:94–117.

Forester, D.C., J. Cover, and A. Wisnieski. 1993. The influence of time of residency on the tenacity of territorial defense by the dart-poison frog *Dendrobates pumilio*. Herpetologica 49:94–99.

Frith, C. B., and D. W. Frith. 1987. Australian tropical reptiles and frogs. Frith and Frith Books, Malanda, Australia.

Fritzsch, B., M. J. Ryan, W. Wilczynski, T. E. Hetherington, and W. Walkowiak (eds.) 1988. The Evolution of the Amphibian Auditory System. John Wiley and Sons, New York.

Gerhardt, H. C., and J. J. Schwartz. 1995. Interspecific interactions and species recognition. Pp. 603–632. *In* H. Heatwole, and B. K. Sullivan (eds.), Amphibian Biology. Vol. 2. Social Behaviour. Surrey Beatty and Sons, Chipping Norton, Australia.

Gregory, P. T. 1983. Habitat structure affects diel activity pattern in the Neotropical frog *Leptodactylus melanonotus*. Journal of Herpetology 17:179–181.

Haberl, W., and J. W. Wilkinson. 1997. A note on the unkenreflex and similar defense postures in *Rana temporaria* (Anura, Amphibia). British Herpetological Society Bulletin 61:16–20.

Haddad, C. F. B., and A. A. Giaretta.1999. Visual and acoustic communication in the Brazilian torrent frog, *Hylodes asper* (Anura: Leptodactylidae). Herpetologica 55:324–333.

Hailman, J. P. 1977. Optical Signals: Animal Communication and Light. Indiana University Press, Bloomington, IN.

Halliday, T. R. and M. Tejedo. 1995. Intrasexual selection and alternative mating behaviour. Pp. 419–468. *In* H. Heatwole, and B. K. Sullivan (eds.), Amphibian Biology. Vol. 2. Social behaviour. Surrey Beatty and Sons, Chipping Norton, Australia.

Hand, D. J. 1981. Discrimination and classification. John Wiley and Sons, New York.

Harding, K. A. 1982. Courtship display in a Bornean frog. Proceedings of the Biological Society of Washington 95:621–624.

Harper, D. G. C. 1991. Communication. Pp. 374–397. *In* J. R. Krebs, and N. B. Davies (eds.), Behavioural Ecology: An Evolutionary Approach. Blackwell Scientific Publications, Oxford.

Heyer, R. W., A. S. Rand, C. A. Gonçalves da Cruz, O. L. Peixoto, and C. E. Nelson. 1990. Frogs of Boraceia. Arq. Zool. S. Paulo 31:231–410.

Hinsche, G. 1926. Vergleichende Untersuchungen zum sogenannten Unkenreflex. Biol. Zentralbl. Leipzig 46:296–305.

Hödl, W. 1991. Calling behavior and spectral stratification in dart-poison frogs (Dendrobatidae). Scientific film, Ctf 2444 ÖWF Vienna, Austria.

Hödl, W. 1996. Wie verständigen sich Frösche? Pp.53–70. *In* Hödl, W., and G. Aubrecht (eds.), Frösche Kröten Unken—aus der Welt der Amphibien. Stapfia 47.

Hödl, W., M. T. Rodrigues, G. M. Accacio, P. H. Lara, L. Schiesari, and G. Skuk 1997. Foot-flagging display in the Brazilian stream-breeding frog *Hylodes asper* (Leptodactylidae). Scientific film, Ctf 2703 ÖWF Vienna, Austria.

Ibáñez, R. D., C. A. Jaramillo, and F. A. Solís. 1995. Una nueva especie de *Atelopus* (Amphibia: Bufonidae) de Panama. Carib. J. Sci. 31:57–64.

Jaslow, A. P. 1979. Vocalization and aggression in *Atelopus chiriquiensis* (Amphibia, Anura, Bufonidae). Journal of Herpetology 13:141–145.

Juncá, F. A. 1998. Reproductive biology of *Colostethus stepheni* and *Colostethus marchesianus* (Dendrobatidae), with the description of a new anuran mating behavior. Herpetologica 54:377–387.

Juncá, F. A., R. Altig, and C. Gascon. 1994. Breeding biology of *Colostethus stepheni*, a dendrobatid frog with a nontransported nidicolous tadpole. Copeia 1994:747–750.

Kluge, A. G. 1981. The life history, social organization, and parental behavior of *Hyla rosenbergi*, a nest-building gladiator frog. Miscellaneous Publications, Museum Zoology, University of Michigan 160: 1–170.

Klump, G. M. 1995. Studying sound localization in frogs with behavioral methods. Pp. 221–233. *In* G. M. Klump, R. J. Dooling, R. R. Fay, and W. C. Stebbins (eds.), Methods in Comparative Psychoacoustics. Birkhäuser Verlag, Basel.

Krebs, J. R., and N. B. Davies. 1993. An Introduction to Behavioural Ecology. Blackwell Scientific Publications, London.

Lehner, P. N. 1979. Handbook of Ethological Methods. Garland STPM, New York.

Lescure, J. and R. Bechter. 1982. Le comportement de reproduction en captivité et le polymorphisme de *Dendrobates quinquevittatus* Steindacher (Amphibia, Anura, Dendrobatidae). Rev. Fr. Aquariol. 8:107–118.

Lindquist, E. D., and T. E. Hetherington. 1996. Field studies on visual and acoustic signaling in the "earless" Panamanian golden frog, *Atelopus zeteki*. Journal of Herpetology 30:347–354.

Lindquist, E. D., and T. E. Hetherington. 1998. Semaphoring in an earless frog: The origin of a novel visual signal. Animal Cognition 1:83–87.

Lötters, S. 1996. The neotropical toad genus *Atelopus*. M. Vences and F. Glaw Verlags, Cologne, Germany.

Lüddecke, H. 1974. Ethologische Untersuchungen zur Fortpflanzung von *Phyllobates palmatus* (Amphibia, Ranidae). Ph.D. dissertation, Johannes Gutenberg-Universität, Mainz, Germany.

Lüddecke, H. 1976. Einige Ergebnisse aus Feldbeobachtungen an Phyllobates palmatus (Amphibia, Ranidae) in Kolumbien. Mitt. Inst. Colombo-Alemán Invest. Cient. 8:157–163.

Lüddecke, H. 1999. Behavioural aspects of the reproductive biology of the Andean frog *Colostethus palmatus* (Amphibia, Dendrobatidae). Rev. Acad. Colomb. Cient. 23:303–316.

Magnusson, W. E. 1996. Tail and hand waves: A come-on for predators? Herpetol. Rev. 27:60.

Malkmus, R. 1989. Herpetologische Beobachtungen am Mount Kinabalu, Nord Borneo II. Mitt. Zool. Mus. Berl. 65:179–200.

Malkmus, R. 1992. Herpetologische Beobachtungen am Mount Kinabalu, Nord-Borneo. III. Mitt. Zool. Mus. Berl. 68:101–138.

Martens, J., and G. Geduldig. 1988. Acoustic adaptations of birds living close to Himalayan torrents. Proc. Int. 100 DOG Meeting: Current Topics in Avian Biology, Bonn 123–131.

Martin, P., and P. Bateson. 1993. Measuring Animal Behaviour: An Introductory Guide. Cambridge University Press, Cambridge.

Martins, M. 1989. Deimatic behavior in *Pleurodema brachyops*. Journal of Herpetology 23:305–307.

McDiarmid, R. W. 1971. Comparative morphology and evolution of frogs of the neotropical genera *Atelopus, Dendrophryniscus, Melanophryniscus,* and *Oreophrynella*. Museum of Natural History, Los Angeles County, Science Bulletin 1–66.

McDiarmid, R. W., and K. Adler. 1974. Notes on territorial and vocal behavior of neotropical frogs of the genus *Centrolenella*. Herpetologica 30:75–78.

Mildner, P., and F. Hafner. 1990. Die Amphibien Kärntens. Carinthia II 180:55–121.

Murphy, J. B. 1976. Pedal luring in the leptodactylid frogs, *Ceratophrys calcarata* Boulenger. Herpetologica 32:339–341.

Myers, C. W. 1966. The distribution and behavior of a tropical horned frog, *Cerathyla panamensis* Stejneger. Herpetologica 22:68–71.

Myers, C. W., and J. W. Daly. 1976. Preliminary evaluation of skin toxins and vocalizations in taxonomic and evolutionary studies of poison-dart frogs (Dendrobatidae). Bulletin, American Museum of Natural History 157:177–262.

Myers, C. W., J. W. Daly, and B. Malkin. 1978. A dangerously toxic new frog (*Phyllobates*) used by Emberá indians of western Colombia, with discussion of blowgun fabrication and dart poisoning. Bulletin, American Museum of Natural History 161:309–365.

Myers, C. W., A. Paolillo, and J. W. Daly. 1991. Discovery of a defensively malodorous and nocturnal frog in the family Dendrobatidae: Phylogenetic significance of a new genus and species from the Venezuelan Andes. American Museum Novitates 3002:1–33.

Narvaes, P. 1997. Comportamento territorial e reprodutivo de uma nova espécie de *Hylodes* (Amphibia, Anura, Leptodactylidae) da mata Atlántica do sudeste do Brasil. MSc Tesis, Universidade de São Paulo, São Paulo.

Noble, G. K. 1955. The Biology of the Amphibia. Dover Publications, New York.

Passmore, N. I., R. R. Capranica, S. R. Telford, and P. J. Bishop. 1984. Phonotaxis in the painted reed frog (*Hyperolius marmoratus*): The localization of elevated sound sources. Journal of Comparative Physiology A. 154:189–197.

Pavan, D., P. Narvaes, and M. T. Rodrigues. 2000. A new species of leptodactylid frog from the Atlantic forest of southeastern Brazil with notes on the status and on the speciation of the *Hylodes* species groups. Pap. Avul. Zool. S. Paulo 41:405–423.

Pombal, J. P. J., I. Sazima, and C. F. B. Haddad. 1994. Breeding behavior of the pumpkin toadlet, *Brachycephalus ephippium* (Brachycephalidae). Journal of Herpetology 28:516–519.

Rabb, G. B., and M. S. Rabb. 1960. On the mating and egg-laying behavior of the Surinam toad, *Pipa pipa*. Copeia 1960:271–276.

Rabb, G. B., and M. S. Rabb. 1963. Additional observations on breeding behavior of the Surinam toad, *Pipa pipa*. Copeia 1963:636–642.

Radcliffe, C. W., D. Chiszar, K. Estep, J. B. Murphy, and H. M. Smith. 1986. Observations on pedal luring and pedal movements in leptodactylid frogs. Journal of Herpetology 20:300–306.

Rand, A. S. 1988. An overview of anuran acoustic communication. Pp. 415–431. *In* B. Fritzsch, M. J. Ryan, W. Wilczynski, T. E. Hetherington, and W. Walkowiak (eds), The Evolution of the Amphibian Auditory System. John Wiley and Sons, New York.

Richards, S. J., and C. James. 1992. Foot-flagging displays of some Australian frogs. Mem. Queensland Museum 32:302.

Robinson, M. 1993. A field guide to frogs of Australia. Australian Museum/Reed, Sydney.

Ryan, M. J. 1988. Constraints and patterns in the evolution of anuran acoustic communication. Pp. 637–675. *In* B. Fritzsch, M. J. Ryan, W. Wilczynski, T. E. Hetherington, and W. Walkowiak (eds.), The Evolution of the Amphibian Auditory System. John Wiley and Sons, New York.

Ryan, M. J., and A. Keddy-Hector. 1992. Directional patterns of female mate choice and the role of sensory biases. American Naturalist 139:S4–S35.

Sazima, I., and U. Caramaschi. 1986. Descrição de *Physalaemus deimaticus,* sp. nv., e observações sobre comportamento deimatico em *P. nattereri* (Steindn.) Anura, Leptodactylidae. Rev. Biol. 13:91–101.

Schmidt, W., and F. W. Henkel. 1995. Pfeilgiftfrösche im Terrarium. Landbuch-Verlag. Hannover, Germany.

Sexton, O. J. 1960. Some aspects of the behavior and of the territory of a dendrobatid frog, *Protherapis trinitatis*. Ecology 41:107–115.

Silverstone, P. A. 1973. Observations on the behavior and ecology of a Colombian poison-arrow frog, Kõkoé-pá (*Dendrobates histrionicus* Berthold). Herpetologica 29:295–301.

Silverstone, P. A. 1975. A revision of the poison-arrow frogs of the genus *Dendrobates* Wagler. Natural History Museum, Los Angeles County, Science Bulletin 21:1–55.

Silverstone, P. A. 1976. A revision of the poison-arrow frogs of the genus *Phyllobates* Bibron in Sagra (family Dendrobatidae). Natural History Museum, Los Angeles County, Science Bulletin 27:1–53.

Stebbins, R. C., and N. W. Cohen. 1995. A Natural History of Amphibians. Princeton University Press, Princeton, New Jersey.

Sullivan, B. K., M. J. Ryan, and P. A. Verrell. 1995. Female choice and mating system structure. Pp. 469–517. *In* H. Heatwole, and B. K. Sullivan (eds.), Amphibian Biology. Vol. 2. Social Behaviour. Surrey Beatty and Sons, Chipping Norton, Australia.

Test, F. H. 1954. Social aggressiveness in an amphibian. Science 120:140–141.

Toft, C. A. 1995. Evolution of diet specialization in poison-dart frogs (Dendrobatidae). Herpetologica 51:202–216.

Wells, K. D. 1977a. The courtship of frogs. Pp. 233–262. *In* D. H. Taylor, and S. I. Guttman (eds.), The Reproductive Biology of Amphibians. Plenum Publishing.

Wells, K. D. 1977b. The social behaviour of anuran amphibians. Animal Behaviour 25:666–693.

Wells, K. D. 1977c. Territoriality and male mating success in the green frog (*Rana clamitans*). Ecology 58:750–762.

Wells, K. D. 1978. Territoriality in the green frog (*Rana clamitans*): Vocalizations and agonistic behaviour. Animal Behaviour 26:1051–1063

Wells, K. D. 1980a. Behavioral ecology and social organization of a dendrobatid frog (*Colostethus inguinalis*). Behavioral Ecology and Sociobiology 6:199–209.

Wells, K. D. 1980b. Social behavior and communication of a dendrobatid frog (*Colostethus trinitatis*). Herpetologica 36:189–199.

Wells, K. D. 1988. The effect of social interactions on anuran vocal behavior. Pp. 433–454. *In* B. Fritzsch, M. J. Ryan, W. Wilczynski, T. E. Hetherington, and W. Walkowiak (eds), The Evolution of the Amphibian Auditory System. John Wiley and Sons, New York.

Wevers, E. 1988. Enige opmerkingen over de pijlgifkikker *Dendrobates parvulus*. Lacerta 46:51–53.

Weygoldt, P. 1986. Beobachtungen zur Ökologie und Biologie von Fröschen an einem neotropischen Bergbach. Zool. Jb. Syst. 113:429–454.

Weygoldt, P., and S. P. de Carvalho e Silva. 1992. Mating and oviposition in the hylodine frog *Crossodactylus gaudichaudii* (Anura: Leptodactylidae). Amph.-Rept. 13:35–45.

Wiewandt, T. A. 1969. Vocalization, aggressive behavior, and territoriality in the bullfrog, *Rana catesbeiana*. Copeia 1969:276–285.

Williams, C. R., E. D. Brodic Jr., M. Tyler, and S. J. Walker. 2000. Antipredator mechanisms of Australian frogs. Journal of Herpetology 34:431–433.

Winter, J., and K. McDonald. 1986. Eungella: The land of the cloud. Australian Natural History 22:39–43.

Woodward, B. D. 1982. Sexual selection and nonrandom mating patterns in desert anurans (*Bufo woodhousei, Scaphiopus couchi, S. multiplicatus* and *S. bombifrons*). Copeia 1982:351–355.

Zimmermann, E. 1990. Behavioral signals and reproduction modes in the neotropical frog family Dendrobatidae. Pp. 61–73. *In* W. Hanke (ed.), Biology and Physiology of Amphibians. Fortschritte der Zoologie. Vol. 38. Gustav Fischer Verlag, Stuttgart, Germany.

Zimmermann, H., and E. Zimmermann. 1988. Etho-Taxonomie und zoogeographische Artengruppenbildung bei Pfeilgiftfröschen (Anura: Dendrobatidae). Salamandra 24:125–160.

PART FOUR

Neural Processing

11

ELIOT A. BRENOWITZ, GARY J. ROSE, AND TODD ALDER

The Neuroethology of Acoustic Communication in Pacific Treefrogs

Introduction

Pacific treefrogs (*Hyla regilla*, also sometimes referred to as *Pseudacris regilla*) are widely distributed throughout the western part of North America. They have a prolonged breeding season that can extend from February to July. The duration of the breeding season appears to be influenced by rainfall patterns because these frogs breed in ephemeral ponds throughout much of their range. Each night during the breeding season males aggregate at ponds, begin calling at sunset, and may continue until dawn, depending on the water temperature and wind conditions. They adopt calling positions at the water's edge or on vegetation floating on shallow water. Females visit the chorus in relatively low numbers on any given night to select mates. When a pair has entered amplexus, the female moves into the water with the male clinging to her, and she lays her eggs on aquatic vegetation.

We have studied the vocal communication system of Pacific treefrogs over the past decade. Anurans, in general, provide an excellent model system to study the proximate and ultimate bases of behavior. Calling behavior in frogs and toads also allows us to investigate the role of communication in reproduction. In this chapter we will discuss a number of factors that make *Hyla regilla* of particular interest.

Anurans have traditionally been regarded as having rather stereotyped behavior patterns. This view stems from a number of observations, including the difficulty of training anurans to operant tasks (e.g., Megela-Simmons et al. 1985), "the widespread view that amphibians are of inferior learning ability" (e.g., Thorpe 1963), the prevalence of innate releasing mechanisms and fixed action patterns in anuran behavior (e.g., Ewert 1980), and the occurrence of pronounced stimulus filtering in the anuran visual and auditory peripheries (e.g., Lettvin et al. 1959; Capranica 1976). In this chapter we demonstrate that, contrary to this traditional view of anurans as being inflexible, male Pacific treefrogs show extreme plasticity in sensory aspects of their vocal communication behavior. We suggest that this plasticity may have evolved in response to the preference of conspecific females for one type of call, highly variable patterns of intermale spacing in choruses within and between nights, and the limited availability of suitable breeding habitat in the geographic range occupied by this species. In addition we will discuss evidence that this plasticity arises in the central auditory nervous system. We will conclude by considering specializations in the auditory system for processing different call types that have essentially identical spectral composition but very different temporal structures.

Vocal Communication System

Male Pacific treefrogs produce two different types of advertisement calls (monophasic and diphasic) and an encounter call (Allen 1973) (Figure 11.1). These call types are highly similar spectrally, with dominant energy at about 2 kHz and a fundamental frequency at about 1 kHz, but the advertisement and encounter calls differ in a key temporal feature, the pulse repetition rate (Allen 1973; Straughan 1975; Rose and Brenowitz 1997). Males produce primarily diphasic advertisement calls throughout the night. These signals attract females to choruses (Whitney and Krebs 1975a). The diphasic advertisement call also influences intermale spacing in a chorus (Allen 1973; Whitney and Krebs 1975b; Awbrey 1978). The function of the monophasic advertisement call is not yet clear. A male produces this call at a high rate when approached by females. The male continues to produce the call as he clasps the female in amplexus and stops only when his vocal sac deflates against her back. Monophasic calling can be induced in a male by making gentle vibrations on the ground near him or by separating an amplexed pair (Brenowitz and Rose, unpublished observations). The encounter call is considered to be an aggressive signal that functions in establishing spacing between calling males (Awbrey 1978; Whitney 1980). It is heard most often early in the night when males first enter the chorus and establish their calling positions. During the rest of the night males will switch from producing the advertisement call to the encounter call after hearing either the advertisement or encounter calls of an intruding male (Awbrey 1978; Whitney 1980; Brenowitz 1989; Rose and Brenowitz 1991, 1997; Brenowitz and Rose 1994).

Calling, Plasticity, and Intermale Spacing

Within the choruses, males space nonrandomly and are more clumped than one would predict stochastically (Whitney and Krebs 1975b). Vocalizations play an important role

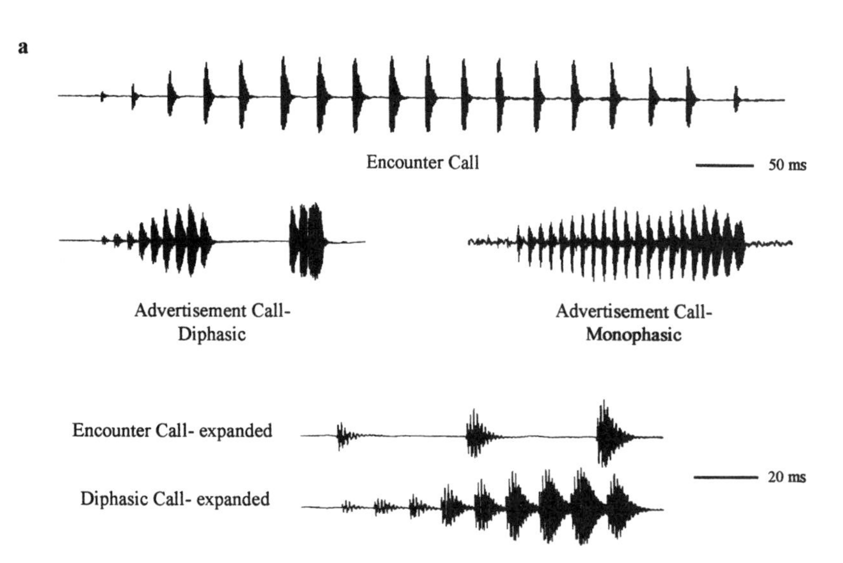

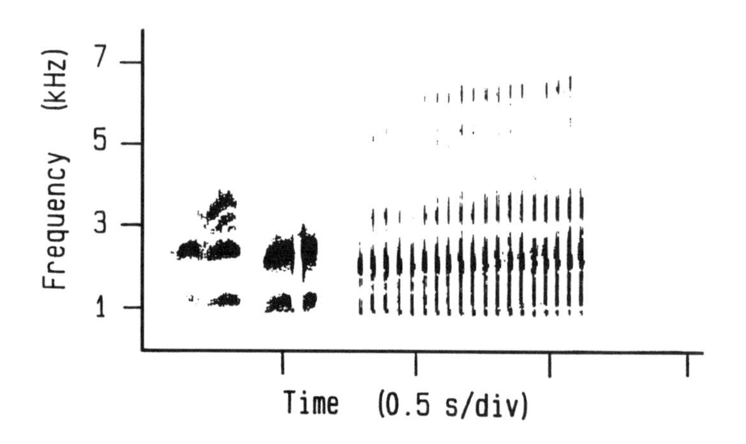

Figure 11.1. (a) Oscillograms of encounter and advertisement calls of *Hyla regilla*. (b) Sound spectrograms of, left to right, monophasic advertisement call, diphasic advertisement call, and aggressive encounter call.

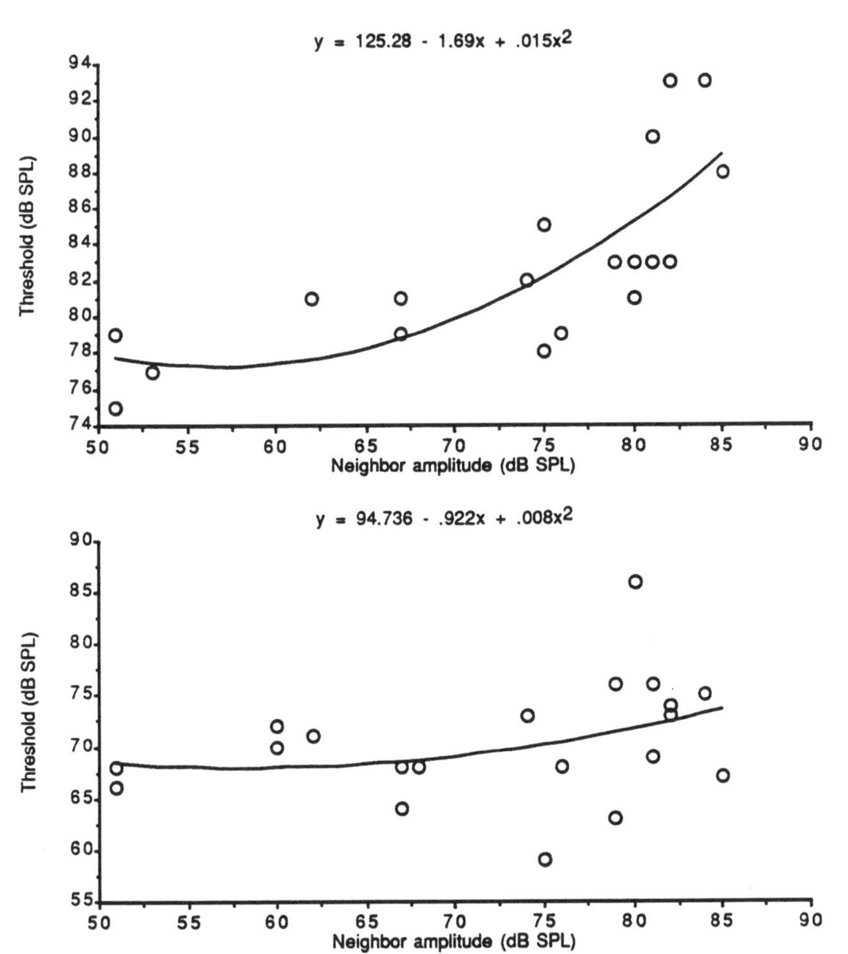

Figure 11.2. Aggressive thresholds of males to playback of the diphasic advertisement call versus the root mean square (RMS) sound-pressure level (SPL) of the loudest calls of the nearest neighboring frog. The regression line was derived from the equation at the top. There is a significant relationship ($r^2 = 0.58, p < 0.001$) between aggressive thresholds and the neighbors' call amplitude.

Figure 11.3. Aggressive thresholds of males to playback of the encounter call versus RMS sound-pressure level (SPL) of the loudest diphasic advertisement calls of the nearest neighboring frog. The equation used to calculate the regression line is shown at the top. Aggressive thresholds for the encounter call are not significantly related to the amplitude of neighbors' advertisement calls ($r^2 = 0.11, p = 0.367$).

in establishing the spatial distribution of males in a chorus (Whitney and Krebs 1975b; Awbrey 1978; Whitney 1980; Brenowitz 1989; Rose and Brenowitz 1991, 1997; Brenowitz and Rose 1994). The amplitude of neighbors' calls at the location of a resident male serves as the proximate cue in regulating the spacing of males in a chorus (Brenowitz 1989). We refer to the lowest amplitude at which a male gives encounter calls in response to the calls of another frog as his aggressive threshold for that call. Aggressive thresholds to the diphasic advertisement call are positively correlated with the maximum amplitude of the advertisement calls of a male's nearest neighbor (Figure 11.2; Rose and Brenowitz 1991). Aggressive thresholds to the encounter call are considerably lower than those to the diphasic call, and are not correlated with the amplitude of neighbors' calls (Figure 11.3). These observations are consistent with the aggressive function of the encounter call.

Two hypotheses can be proposed to explain the positive correlation between aggressive thresholds to diphasic advertisement calls and neighbor call amplitudes. (1) Each male has a fixed aggressive threshold and actively adjusts his spacing from other males in a chorus so that the amplitude of his neighbors' calls falls below his threshold. (2) A male has a plastic aggressive threshold that is modified on a short-term basis according to the amplitude of his neighbors' calls. We conducted two experiments to test these hypotheses (Brenowitz and Rose 1994).

In the first study we identified pairs or groups of closely spaced (i.e., within 5 m) frogs and removed all but one frog. The aggressive threshold to diphasic calls of this individual was measured before and 15 minutes after removal of his neighbors. The prediction of the fixed threshold hypothesis is that this manipulation will not alter the male's aggressive threshold. The plastic threshold hypothesis, on the other hand, predicts that his behavioral threshold will decrease after the neighbors are removed. We found that when his neighbors were removed, a focal frog responded aggressively to playback of the diphasic advertisement call at an amplitude about half as great as that required to evoke such a response only 15 minutes earlier. The mean preremoval aggressive threshold was 93.3 dB sound pressure level (SPL, re: 20 μPa). By 15 minutes after removal of all calling neighbors within 5 m, the mean postremoval threshold was 87.0 dB SPL, a decrease in the mean aggressive threshold of 6.3 dB.

In the second study we located a male that had no calling neighbors within 1 m and measured his aggressive threshold to diphasic calls. We then played the diphasic call at an amplitude 4 dB above the male's initial threshold level for 5 minutes and counted the number of encounter and advertisement calls given over this period. During this 5-minute period, males accommodated to the playback. At the end of the 5-minute period, we measured the male's post-accommodation aggressive threshold by raising the amplitude of the playback in 4 dB increments until the focal frog once again gave at least one encounter call. The fixed threshold hypothesis again predicts that this manipulation will not alter the male's aggressive threshold. The plastic threshold hypothesis, however, predicts that the behavioral threshold will increase with this manipulation. We observed that frogs accommodated to presentation of the diphasic advertisement call at a suprathreshold amplitude such that the mean amplitude necessary to evoke an aggressive response increased almost fourfold. The mean pre-accommodation aggressive threshold to playback of the diphasic call was 79.3 dB. The mean post-accommodation aggressive threshold was 90.2 dB. We also measured the behavioral thresholds of several of these frogs after 15 minutes of silence following the end of the accommodation playback. Thresholds were elevated by the playback and then returned to near their original levels within 15 minutes after the end of the playback. The mean aggressive thresholds for this subgroup rose from a pre-accommodation level of 80.5 dB to 90.9 dB at the end of the 5-minute accommodation playback, and then decreased to 83.7 dB after 15 minutes.

In this second study, male *Hyla regilla* accommodated to playback of the diphasic call at the higher amplitude quite rapidly (Figure 11.4). During the first 15 seconds of the 5-minute accommodation playback, a mean of 82% of the calls given by the test frogs were aggressive encounter calls. By 45 seconds, only about 50% of the calls given were encounter calls. After 2.5 minutes, the proportion of encounter calls dropped to only about 10% and remained at or below that level for the remainder of the 5-minute playback. By the end of the playback, most males produced exclusively advertisement calls. Only a few males continued to produce encounter calls occasionally.

These results indicate that there is pronounced plasticity in the aggressive thresholds of male Pacific treefrogs to the diphasic advertisement calls. This plasticity explains the observed correlation between a male's aggressive threshold to the diphasic advertisement call and the amplitude of his neighbors' calls at his position. When a resident male first adopts a stable calling position within a chorus early in the evening, his aggressive threshold to the advertisement calls of his neighbors is low and he responds aggressively to the

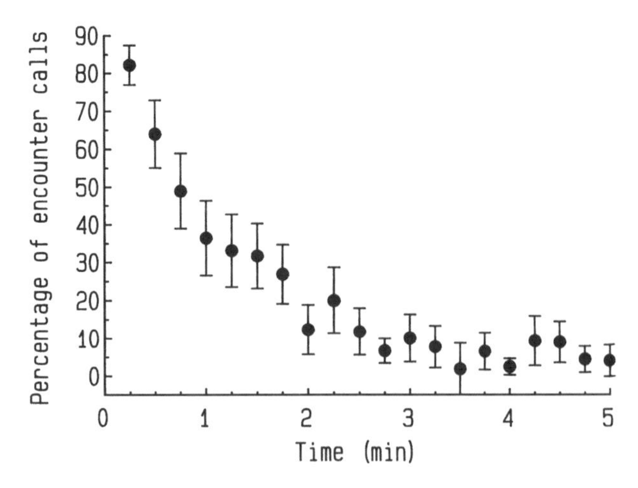

Figure 11.4. Time course of accommodation to playback of the diphasic advertisement call at an amplitude 4 dB above the initial aggressive threshold to this call. The percentage of encounter calls produced ($\bar{x} \pm$ SEM) at 15-second intervals throughout the 5-minute playback period is shown.

calls of nearby frogs. As neighboring males settle into regular calling, the resident male accommodates to their calls so that his aggressive threshold now lies slightly above the maximum amplitude of their calls, as measured at the resident's position. Thus a resident male's recent experience with the calls of his nearest neighbor determines his aggressive threshold. Consistent with this conclusion is the observation that when there are extended silent intervals on the order of 5 to 10 minutes between bouts of calling, as occurs on cold nights, males often initially respond to the advertisement calls of their neighbors by producing encounter calls. These silent intervals are comparable to the situation in our removal study described above; in the absence of calls from close neighbors, a male becomes "disaccommodated" to their calls and his aggressive threshold decreases. After a few minutes of sustained calling, however, a resident male rapidly accommodates again to his neighbors' calls and he switches from encounter to advertisement calling.

Given that the correlation between a frog's aggressive threshold to the advertisement call and the amplitude of his nearest neighbor's advertisement calls results from plasticity of these thresholds, then perhaps the lack of such a correlation for the encounter call indicates a lack of plasticity of their behavioral thresholds to this call. We tested this hypothesis by determining whether frogs accommodated to playback of the encounter call at amplitudes above their aggressive threshold to this signal (Rose and Brenowitz 1997). Males rapidly accommodated to playback of encounter calls 4 dB above their aggressive thresholds for this call. Only 4 of 27 frogs continued to give encounter calls after 2 minutes of playback. When we played the encounter call at 8 dB above the aggressive threshold, males accommodated more slowly and sometimes moved away from the speaker. Smaller males accommodated more rapidly than did larger frogs, perhaps reflecting the fact that smaller males are more likely to be defeated should an encounter escalate to physical combat.

These results indicate that the aggressive threshold to the encounter call is plastic in the same manner as the threshold to the advertisement call. How can we explain the fact that these encounter thresholds do not correlate with the amplitude of the neighbors' calls at a resident male's position? This lack of correlation results from the pattern of calling in a chorus during a given night. Encounter calls are only produced at a high rate early in the evening, when males first enter the chorus. By the time of full darkness males with stable neighbors have accommodated to each others' calls and produce predominantly advertisement calls during the rest of the night. During most of the night, therefore, a resident male rarely hears encounter calls from his nearest neighbors and does not have an opportunity to accommodate to them. His aggressive threshold to encounter calls consequently remains low compared with his threshold to the advertisement call, which he hears almost continuously from his neighbors. Consistent with this scenario is our observation that when neighboring males exchange encounter calls, they usually accommodate and return to producing advertisement calls within 5 to 10 minutes.

Our removal and accommodation studies indicate that the aggressive thresholds to both the diphasic advertisement and encounter calls are very plastic, undergoing pronounced changes over short time intervals as a function of a resident male's recent history of vocal interactions with neighboring males. These results combined with our observations of male behavior in choruses suggest that these frogs pursue the following strategy. After a male has accommodated to the calls of his nearest neighbors early in the evening, he produces mainly diphasic advertisement calls. If the calls of an intruder reach the resident male at an amplitude above his current aggressive threshold, the resident male will switch to producing encounter calls. The encounter call usually deters the intruder from approaching closer to the resident frog (Allen 1973; Awbrey 1978; Whitney 1980). If the intruder maintains his position while giving either advertisement or encounter calls, the resident will rapidly accommodate to the intruder's calls and return to producing diphasic advertisement calls. If, however, the intruder approaches the resident or selects a site close to the resident and begins calling, his calls will arrive at the resident's ears at amplitudes well above his aggressive threshold. In such cases the resident may not accommodate to the intruder's calls, or do so only slowly. Consistent with this suggestion is our observation that males took longer to accommodate to encounter calls played 8 dB above their aggressive threshold than they did when these calls were played 4 dB above the threshold (Rose and Brenowitz 1997). Instead, in this situation, both males may be "locked" into producing encounter calls and they may escalate to physical combat. An aggressive interaction only rarely escalates to this level, however.

Adaptive Value of Plasticity

The above scenario implies that there is a cost to producing the encounter call. One possible cost is that the time spent in aggressive interactions with other frogs might decrease a male's ability to attract females. In several congeneric species of hylid frogs, it has been shown that aggressive encounter calls are less attractive to females than are advertisement calls (Oldham and Gerhardt 1975; Schwartz 1986; Wells and Bard 1987). To determine whether female Pacific treefrogs prefer the diphasic advertisement call over the encounter call, we presented these calls in alternation from separate speakers and scored their choices; phonotactic responses were used as a measure of "choice" (Brenowitz and Rose 1999). We only considered females to have chosen a signal in this experiment if their responses met four criteria: (1) they had to maintain a neutral position near the point at which they were released, and make head and body movements typical of phonotactic orientation (Gerhardt 1994) for at least five calls before beginning to move toward a speaker; (2) they had to approach within 10 cm of the speaker or contact it; (3) they had to approach the speaker within 5 minutes of the start of the playback; and (4) a female had to move within 10 cm of the other speaker when the same signal that they chose in the first test was presented from that speaker in a second test. This speaker "reversal" criterion is not commonly used in phonotactic studies, but we view it as essential to demonstrating that an individual female actually has chosen a signal rather than a specific location in a test arena. Of 20 females tested, 12 met our criteria for demonstrating a phonotactic response, and each approached the speaker playing the advertisement call rather than the encounter call. None preferred the encounter call. The approach of all 12 females to the advertisement call might reflect a failure to recognize the encounter call, rather than a preference for the advertisement call per se. To test this hypothesis, we played the encounter call by itself to five females. Two females approached to within 10 cm of the speaker from which the call was played. This preliminary observation suggests that the encounter call is recognized as being a relevant signal. Therefore selective orientation to diphasic calls in two speaker playbacks likely represents a true preference for the advertisement call, rather than a failure to recognize the encounter call. This strong female preference for the advertisement call supports the hypothesis that males reduce their chance of attracting a female by engaging in prolonged bouts of encounter calling.

With this information on female choice we can suggest an adaptive value to the plasticity of aggressive thresholds that we observe in males. We propose that this plasticity allows males to balance the benefits and costs of aggression and to maximize the chance of attracting a female by adjusting their aggressive behavior to changes in male spacing within choruses. Male spacing may change during any single night as males enter, leave, and move about within the chorus. Between nights, the absolute number of males, and consequently the spacing between males, in the chorus can change as a function of weather; there are fewer males on cold, windy nights and more males on warm, calm nights. Males in a chorus appear to establish a stable distribution each evening, despite these variations in male density (Gerhardt et al. 1989). As a consequence of such changes in spacing, a resident frog with a stable calling position may hear the calls of other males at a range of amplitudes. If a male's aggressive threshold was fixed at a low level, then on nights when the density of calling males is high he might be obligated to engage in prolonged encounter calling. Given female preference for the advertisement call, engaging in encounter calling to repel closely spaced neighbors most of the night would result in a male greatly reducing his chance of attracting females. If thresholds were fixed at high levels, nearby intruders would be tolerated even when male density is low. A resident male's fitness might be compromised by such an intruder intercepting a female or by masking detection of the resident's calls (Telford 1985; Schwartz and Gerhardt 1989). Accommodation permits a male to minimize these costs of aggressive calling by adjusting his aggressive threshold to changes in the detected amplitude of his neighbors' calls that accompany changes in male spacing. Plasticity therefore allows males to make the best of what might be an unavoidably suboptimal situation and maximize the time spent in advertisement calling to attract females. A similar tradeoff between costs and benefits of aggressive signaling may occur in other species of hylid frogs in which females prefer advertisement calls to encounter calls (Oldham and Gerhardt 1975; Schwartz 1986, 1987; Wells and Bard 1987).

Discrete Accommodation and the Two-Channel Hypothesis

As described above, males accommodate to playback of either diphasic or encounter calls at amplitudes that exceed their aggressive threshold for that call type. Yet in stable choruses only aggressive thresholds of males for the diphasic advertisement call are correlated with the maximum amplitude of neighbor's calls. These findings suggested that exposure to diphasic calls at suprathreshold amplitudes may raise a male's aggressive threshold for this call without elevating his threshold for the encounter call. Conversely exposure to encounter calls at suprathreshold amplitudes might not elevate a male's aggressive threshold to the diphasic call. The hypothesis that accommodation to one call type does not elevate a male's aggressive threshold for the other call type was tested in two "cross accommodation" experiments (Brenowitz and Rose 1994; Rose and Brenowitz 1997). Both studies supported this hypothesis.

In the first study the aggressive thresholds of a focal male to the diphasic and encounter calls were measured. The diphasic call alone was then played to the male at an amplitude 4 dB above his aggressive threshold for that call type for 5 minutes, and then the amplitude was increased by another 4 dB for 5 minutes. Subsequently the male's aggressive thresholds for the diphasic and encounter calls were remeasured and compared with their values before the accommodation period. In this experiment the mean aggressive threshold to the diphasic call increased from a pre-accommodation level of 81.4 dB to a post-accommodation level of 88.4 dB. The aggressive threshold to the encounter call, however, increased only from a pre-accommodation mean of 74.8 dB to a post-accommodation mean of 76.7 dB.

The second study was conducted in the same way, except that we played the encounter call alone at suprathreshold amplitudes. In this case the median aggressive threshold to the encounter call increased from a pre-accommodation level of 68 dB to a post-accommodation level of 83 dB. The aggressive threshold to the diphasic call increased only from a pre-accommodation median of 78 dB to a post-accommodation median of 80 dB.

The behavioral implication of these results is that males may accommodate to the diphasic calls of a neighbor while maintaining a low threshold for responding aggressively to the encounter calls of subsequent intruders. Once the intruding male has resumed producing diphasic calls, the resident male can quickly accommodate to these calls and resume advertisement calling. Presumably it is unlikely during natural vocal interactions that a male's aggressive threshold for the encounter call would be raised without concomitant elevation of the threshold for the diphasic call. Otherwise in a natural interaction males would never resume exclusive advertisement calling.

The lack of cross accommodation also suggests that two independent "channels" exist for processing the advertisement and encounter calls. These two call types are highly similar spectrally, and selective accommodation therefore cannot be explained simply by a process of sensory adaptation in the peripheral auditory system. Nor can a process of generalized change in aggressive motivation be a tenable mechanism. In both of these cases, accommodation to each call type would induce a parallel increase in aggressive thresholds for the other call type. Instead selective accommodation suggests that two independent channels exist centrally for processing these two call types. Because these calls primarily differ in their pulse repetition rate, it is likely that these channels consist of neurons that are tuned to each of these temporal patterns.

Neural Processing

How might selective responses to pulse repetition rates characteristic of advertisement or encounter calls be generated? First let us consider what is known about processing amplitude-modulated sounds in the anuran auditory system. Previous neurophysiological work has shown that there is a transformation in the manner that amplitude modulation (AM) is represented in the auditory systems of anurans that produce AM calls (Rose and Capranica 1983, 1985). Primary eighth nerve fibers code the rate of sinusoidal amplitude modulation (SAM) in the periodicity of their discharges. The mean spike rate of these fibers, however, is independent of the AM rate, provided that spectral properties of the stimulus are held constant across AM rates. In the torus semicircularis, however, the firing rate of most neurons is strongly dependent on the AM rate (Rose and Capranica 1983, 1985; Walkowiak 1988; Gooler and Feng 1992). With AM tones as a stimulus, approximately two-thirds of the cells respond best over a narrow range of modulation rates (i.e., are "AM-tuned"; Alder and Rose 1998). This transformation from periodicity to rate coding is of particular interest with regard to the abilities of anurans to discriminate between different patterns of amplitude modulation. The distribution of AM tuning values for toral neurons is species-specific and nicely related to the range of pulse repetition rates observed in their calls (Rose and Capranica 1984; Rose et al. 1985).

Might neurons that are tuned to particular rates of SAM form the substrates of selective filters for the advertisement or encounter calls of *H. regilla*? The answer to this question depends on the mechanisms that underlie the AM selectivity of auditory neurons. One hypothesis attempts to account for band-pass AM selectivity in terms of a neuron's underlying sensitivity to stimulus rise time and duration. Individual "pulses" (cycles of AM) decrease in rise time and duration as the rate of SAM is increased. Band-pass selectivity could result if neurons in the torus, or cells upstream from them, were excited best for particular combinations of these two variables. Such a mechanism, however, would be insufficient for generating selective filters for the temporal pulse patterns that distinguish the advertisement and encounter calls of *H. regilla*. This is because these calls differ in pulse repetition rate but not in the rise time or duration of individual pulses.

Alternatively, tuning to particular rates of SAM could result from processes that control the phasic responses of neurons and their recovery times. As the rate of SAM is decreased, "pulse" number per unit time decreases. The decline in response of band-pass cells as the AM rate is decreased below their best rate would be expected for neurons that respond phasically to pulses; fewer spikes will be elicited, in part, because fewer pulses are present as the rate of AM is decreased. Following a cell's phasic response to a pulse, a "recovery time" might be required before a response to a subsequent stimulus pulse can be generated. As interpulse interval decreases below an optimal level with increased pulse repetition rate, neural spike rate might also decrease. According to this hypothesis, neurons might show tuning to particular rates of SAM, but not discriminate in their response magnitude between calls that vary only in pulse repetition rate (i.e., constant pulse structure and number). Indeed neurons have been recorded in the torus that show sharp tuning to the rate of SAM, but are low-pass for stimuli in which only pulse repetition rate is varied (Alder and Rose 1998). Clearly additional mechanisms of temporal selectivity must be postulated to account for *H. regilla*'s ability to discriminate between its encounter and advertisement calls.

Recently neurons have been recorded in the torus semicircularis of *H. regilla* that respond selectively to pulse repetition rates characteristic of advertisement calls (Alder and Rose 1998). Representative recordings from a single cell are shown in Figure 11.5. The selectivity of this neuron type was largely independent of whether the AM was sinusoidal or of more natural shape. Cells of this type failed to respond to tone-bursts.

The band-pass tuning to AM rate could result from a process wherein neural activities that stem from individual pulses are combined when in an appropriate temporal pattern. To test this hypothesis, the number of pulses per stimulus presentation was varied (Figure 11.5). In this case, at least eight pulses, presented at approximately 80 pulses per second, were required to elicit spiking. The long response latency of this neuron, therefore, resulted primarily from an integration process with a time constant that is longer than the time required to conduct signals to this area of the brain. The minimum number of pulses, delivered at the optimal pulse repetition rate, that was sufficient for eliciting at least

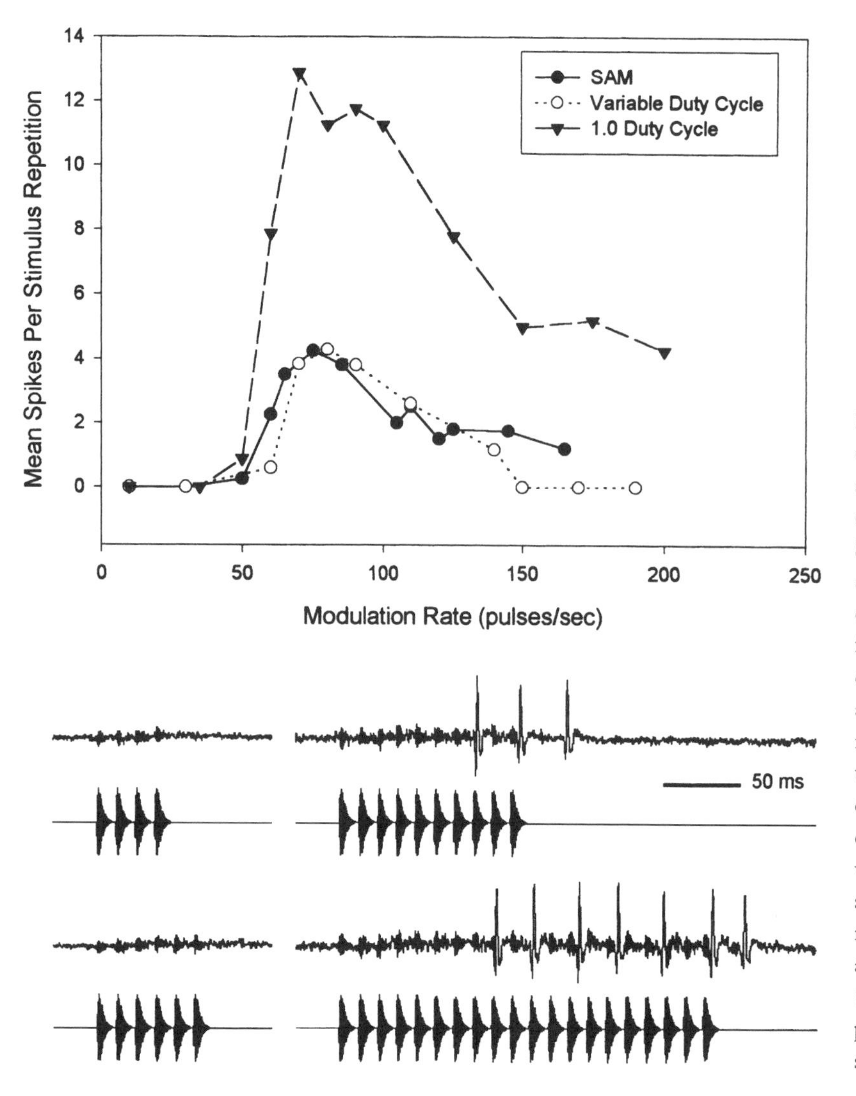

Figure 11.5. Pulse integrating properties of a neuron in the torus of *Hyla regilla*. Response level versus the rate of modulation of the amplitude of a 1.2-kHz carrier signal. Signal amplitude was modulated sinusoidally (●), or such that the relative rise and fall characteristics of pulses resembled those found in natural calls (▼ and ○). In the latter case, pulse duration was either held constant at 10 ms (○), or varied with pulse repetition rate to maintain a duty cycle = 1.0 (▼). Stimulus duration was 400 ms, except when pulse duration and number (15 pulses) were held constant. In the latter case stimulus duration varied with pulse repetition rate. Traces show recordings from this neuron in response to stimuli that consisted of 4, 6, 10, and 20 pulses per presentation. Pulses were 12.5 ms in duration. A faint microphonic potential, reflecting the stimulus, can be seen in these recordings.

one spike per stimulus presentation (pulse train) ranged from 4 to 15 (median, 8.5, $n = 18$) for this group of neurons.

Interestingly the minimum number of pulses required for eliciting responses from these cells was almost entirely independent of stimulus intensity. If the responses of neurons reflected the integration of stimulus intensity that was presented in a temporally appropriate pattern, then as the stimulus intensity was increased, fewer pulses should have been required to elicit spiking. These auditory neurons, therefore, integrate information concerning the number of pulses that are presented at the optimal rate, not simply stimulus intensity that is presented in a temporally appropriate pattern.

Significance and Future Directions

We have studied accommodation behaviors in *H. regilla* and their adaptive significance. We have also used these behaviors to understand further the proximate mechanisms for detection and discrimination of encounter and advertisement calls. Parallel neurophysiological investigations have increased our understanding of the mechanisms that generate selective filters for these calls. These findings are relevant to a number of issues in behavioral ecology and neurobiology and have raised additional interesting questions that can be addressed in future experiments.

Evolution of Plasticity

Our studies of Pacific treefrogs show that males are extremely plastic in their aggressive behavior. An interesting question is why they evolved this behavioral plasticity. Alternative mechanisms might have evolved that would allow males to minimize the time spent in encounter calling. When challenged by a nearby intruder, for example, a male could simply move away until the intruder's calls fall in amplitude below his aggressive threshold. This option might not have been available to Pacific treefrogs, however, because of limited breeding sites available to them. Rainfall is limited throughout much of this species' range in the western North America, and the ponds required for egg laying and tadpole development are therefore often small, ephemeral, and limited in number in any given area. This combination of factors may severely constrain the ability of a resident male to respond to intrusion by moving away. The density of males calling around a small pond can become very high on warm nights and it might be impossible for a male with a fixed low aggressive threshold to find a suitable site at the water's edge where the amplitude of other frog's calls fall below his threshold. Also calling sites may differ in their sound transmission qualities and in the protection they provide from attacks by giant water bugs (*Lethocerus americanus*), which is a significant predator of male Pacific treefrogs (Rose and Brenowitz, unpublished observations). This limitation of available breeding habitat, combined with the strong female preference for advertisement calls and highly variable intermale spacing patterns in choruses within and between nights, may have exerted strong selection on males to evolve plastic aggressive thresholds. This hypothesis could be tested through comparative studies of different closely related species. One could search for plasticity of aggressive thresholds in other species with limited breeding habitat in comparison with species such as the grey treefrog that occupy geographic locations with greater rainfall and less-limited breeding sites. The prediction would be that plasticity should be common to species with limited breeding habitat and less common or absent from species in which breeding sites are less constrained. Plasticity might also be absent in species in which females do not prefer advertisement calls over encounter calls.

Moving in response to the calls of an intruder might also be disadvantageous if a female samples the calls of a given male for a prolonged time before choosing to mate with him, as seems to occur in *H. regilla*. By changing his position, a male may sacrifice his investment in attracting a particular female. In this case accommodating to the calls of an intruder would be preferable to abandoning a calling site. This factor may also reinforce selection for plasticity.

Temporal Feature Recognition and Sensory Exploitation Models of Sexual Selection

Cross-accommodation studies indicated that two independent channels exist, at some level of the central auditory system, for processing the advertisement and encounter calls. Because these calls differ primarily in pulse repetition rate, mechanisms of temporal pattern analysis appear to underlie the generation of these selective filters. This analysis cannot be based on temporal parameters such as pulse duration and rise / fall times because pulses in the two call types are of similar duration and shape. These findings and considerations have stimulated a rethinking of how temporal patterns of amplitude modulations are represented in the brain and how selective temporal filters are generated. The lack of cross-accommodation suggests that these filters are remarkably selective. Our recent neurophysiological work shows that tuning to pulse repetition rates characteristic of the advertisement calls is, in part, due to long-term integration processes. Cells such as the one shown in Figure 11.5 fail to respond unless a threshold number of pulses, presented at the optimal rate, are delivered. This threshold appears to be virtually independent of stimulus intensity. The responses of these cells, therefore, are not simply reflective of stimulus energy that is presented in a temporally appropriate manner. At lower pulse repetition rates, these neurons do not respond, regardless of the number and amplitude of pulses in the stimulus (Alder and Rose 1998). Future experiments will test whether a single inappropriate interpulse interval "resets" the integration process, or whether additional "correct" intervals can compensate for the "incorrect" interval. The latter model is akin to a statistical decision-making process, where a certain probability of error is deemed acceptable.

Finally studies of neural mechanisms should, eventually, shed light on the forces that shape the evolution of behaviors. In the case of acoustic communication in anurans it is interesting, therefore, to speculate as to how long-term integration mechanisms in the auditory system might have influenced the evolution of anuran mating systems. The primary function of long-term integration in this system is to generate selective filters for the advertisement calls. Because of this mechanism, however, calls of longer duration, and therefore more correct consecutive interpulse intervals, should more effectively excite "pulse integrating" neurons than will calls of shorter duration. Males that produce calls of longer duration, therefore, should be preferred over males that produce shorter duration calls; this should be true even if males that produce shorter duration calls produce them more often. Klump and Gerhardt (1987), accordingly, have shown that female *H. versicolor* prefer calls of longer duration, in an energy-independent fashion. Also consistent with this notion is the observation that when a female is detected, male *H. regilla* shift from producing diphasic to monophasic calls. The latter calls have more pulses and more correct consecutive interpulse intervals.

This scenario is consistent with the "sensory exploitation" model of sexual selection (Ryan et al. 1990). The process of integrating information concerning the number of consecutive correct interpulse intervals most likely functions first as a mechanism for generating a filter that responds to pulse repetition rates characteristic of the advertisement calls, and not to that of the encounter call. The abilities of males and females to discriminate between advertisement and encounter calls depends on the selectivity of this filter. We propose that this integration process predisposes females to prefer males that produce calls of longer duration. Such a sensory "bias" could set in motion the forces of sexual selection.

Conclusions

Anurans traditionally have been regarded as having stereotyped behavior patterns. Recent studies of the Pacific treefrog and other species indicate, however, that there is considerable plasticity of both call production and behavioral thresholds. Intermale spacing within Pacific treefrog choruses is dynamic within and between nights, and a given resident male therefore hears the calls of other males at a variety of amplitudes. Rather than becoming "locked" into producing encounter calls when the density of males in a chorus is high, Pacific treefrog males have evolved plastic aggressive thresholds. This plasticity enables a male to rapidly adjust his threshold for responding aggressively to the calls of other males as a function of his recent experience with the calls of his nearest neighbor. Adjusting his aggressive threshold in this way allows a male to increase his chances of mating by minimizing the time that he spends producing encounter calls rather than advertisement calls; females preferentially approach males producing advertisement calls.

Males can adjust their aggressive thresholds to both the advertisement and encounter calls. In stable choruses, however, only the thresholds to advertisement calls are correlated with the amplitude of neighbors' calls. Playback studies indicate that the aggressive threshold to either call type can be raised without concomitantly raising the threshold to the other call type. This observation indicates that the advertisement and encounter calls are processed by separate channels within the central auditory system. These call types share similar frequency structures but differ in temporal structure. There are neurons in the midbrain torus semicircularis of Pacific treefrogs that respond selectively to the temporal structures characteristic of the advertisement or encounter call, but not both. These cells provide a neural substrate for the separate processing of the advertisement and encounter calls.

References

Alder, T. B., and G. J. Rose. 1998. Long-term temporal integration in the anuran auditory system. Nature Neuroscience 1:519–523.

Allen, D. M. 1973. Some relationships of vocalization to behavior in the Pacific treefrog, *Hyla regilla*. Herpetologica 29:366–371.

Awbrey, F. T. 1978. Social interaction among chorusing Pacific treefrogs, *Hyla regilla*. Copeia 1978:208–214.

Brenowitz, E. A. 1989. Neighbor call amplitude influences aggressive behavior and intermale spacing in choruses of the Pacific treefrog (*Hyla regilla*). Ethology 83:69–79.

Brenowitz, E. A., and G. J. Rose. 1994. Behavioural plasticity mediates aggression in choruses of the Pacific treefrog. Animal Behaviour 47:633–641.

Brenowitz, E. A., and G. J. Rose. 1999. Female choice and plasticity of male calling behavior in the Pacific treefrog. Animal Behaviour 57:1337–1342.

Capranica, R. R. 1976. Morphology and physiology of the auditory system. Pp. 551–575. *In* R. Llinas, and W. Precht. (eds.), Frog Neurobiology. Springer-Verlag, Berlin.

Ewert, J. P. 1980. Neuroethology. Springer-Verlag, Berlin.

Gerhardt, H. C. 1994. Reproductive character displacement of female mate choice in the grey treefrog, *Hyla chrysoscelis*. Animal Behaviour 47:959–969.

Gerhardt, H. C., B. Diekamp, and M. Ptacek. 1989. Inter-male spacing in choruses of the spring peeper *Pseudacris (Hyla) crucifer*. Animal Behaviour 38:1012–1024.

Gooler, D. M., and A. S. Feng. 1992. Temporal coding in the frog midbrain: The influence of duration and rise-fall time on the processing of complex amplitude-modulated stimuli. Journal of Neurophysiology 67:1–22.

Klump, G. M., and H. C. Gerhardt. 1987. Use of non-arbitrary acoustic criteria in mate choice by female grey treefrogs. Nature 326:286–288.

Lettvin, J. Y., H. R. Maturana, W. S. McCulloch, and W. H. Pitts. 1959. What the frog's eye tells the frog's brain. Proceedings of the Institute of Radio Engineering 47:1940–1951.

Megela-Simmons, A., C. F. Moss, and K. M. Daniel. 1985. Behavioral audiograms of the bullfrog (*Rana catesbeiana*) and the green tree frog (*Hyla cinerea*). Journal of the Acoustical Society of America 78:1236–1244.

Oldham, R. S., and H. C. Gerhardt. 1975. Behavioral isolating mechanisms of the treefrogs *Hyla cinerea* and *H. gratiosa*. Copeia 1975:223–231.

Rose, G. J., and E. A. Brenowitz. 1991. Aggressive thresholds of male Pacific treefrogs for advertisement calls vary with amplitude of neighbor's calls. Ethology 89:244–252.

Rose, G. J., and E. A. Brenowitz. 1997. Plasticity of aggressive thresholds in *Hyla regilla:* Discrete accommodation to encounter calls. Animal Behaviour. 53:353–361.

Rose, G. J., E. A. Brenowitz, and R. R. Capranica. 1985. Species specificity and temperature dependency of temporal processing by the auditory midbrain of two species of treefrogs. Journal of Comparative Physiology A 157:

Rose, G. J., and R. R. Capranica. 1983. Temporal selectivity in the central auditory system of the leopard frog. Science 219:1087–1089.

Rose, G. J., and R. R. Capranica. 1984. Processing amplitude-modulated sounds by the auditory midbrain of two species of toads: Matched temporal filters. Journal of Comparative Physiology A 154:211–219.

Rose, G. J., and R. R. Capranica. 1985. Sensitivity to amplitude-modulated sounds in the anuran auditory system. Journal of Neurophysiology 53:446–465.

Ryan, M. J., J. H. Fox, W. Wilczynski and A. S. Rand. 1990. Sexual selection for sensory exploitation in the frog, *Physalaemus pustulosus.* Nature 343:66–67.

Schwartz, J. J. 1986. Male calling behavior and female choice in the neotropical treefrog *Hyla microcephala.* Ethology 73:116–127.

Schwartz, J. J. 1987. The importance of spectral and temporal properties in species and call recognition in a neotropical treefrog with a complex vocal repertoire. Animal Behaviour 35:340–347.

Schwartz, J. J., and H. C. Gerhardt. 1989. Spatially mediated release from auditory masking in an anuran amphibian. Journal of Comparative Physiology A 166:37–41.

Straughan, I. R. 1975. An analysis of the mechanisms of mating call discrimination in the frogs *Hyla regilla* and *H. cadaverina.* Copeia 1975:415–424.

Telford, S. R. 1985. Mechanisms and evolution of inter-male spacing in the painted reedfrog, *Hyperolius marmoratus.* Animal Behaviour 33:1353–1361.

Thorpe, W. H. 1963. Learning and instinct in animals. Harvard University Press. Cambridge, MA.

Walkowiak, W. 1988. Central temporal encoding. Pp. 275–294. *In* B. Fritszch, W. Wilczynski, M. J. Ryan and W. Walkowiak (eds.), Evolution of the Amphibian Auditory System. Wiley, New York.

Wells, K. D., and K. M. Bard. 1987. Vocal communication in a neotropical treefrog, *Hyla ebraccata:* Responses of females to advertisement and aggressive calls. Behaviour 101:200–210.

Whitney, C. L. 1980. The role of the "encounter" call in spacing of Pacific treefrogs. Canadian Journal of Zoology 58:75–78.

Whitney, C. L., and J. R. Krebs. 1975a. Mate selection in Pacific treefrogs. Nature 255:325–326.

Whitney, C. L., and J. R. Krebs. 1975b. Spacing and calling in Pacific treefrogs. Canadian Journal of Zoology 53:1519–1527.

12

DARCY B. KELLEY, MARTHA L. TOBIAS,
AND SAM HORNG

Producing and Perceiving Frog Songs

Dissecting the Neural Bases for Vocal Behaviors in *Xenopus laevis*

Introduction

The vocalizations of frogs allow us to examine some important but unresolved issues in behavioral biology and neuroethology that include the bases for sexual attraction, categorization of social signals, and mechanisms of vocal communication. Our work has focused on the role of sex hormones in the development and adult function of vocal systems in South African clawed frogs (*Xenopus laevis*). We review here the mechanisms for song production, including the sexual differentiation of the vocal organ and the neural circuitry that drives it, and outline initial approaches to how attractive and repulsive vocalizations are perceived.

Vocal communication is used to establish territories, to repel rivals, and to attract potential mates. Species-specific advertisement calls of male frogs function in mate recognition by females, and the evolution of vocal differences is generally regarded as an important prespecies isolation mechanism. The classic (and current, Nei and Zhang 1998) view of speciation is that of geographical isolation of populations, the accumulation of genetic differences between them, and ultimately reproductive isolation. In an alternative view, "speciation can result from the introduction of a morphological, physiological or behavioral novelty that causes some individuals not to mate with one another, in spite of the fact that they could still do so, while permitting others to do so" (Schwartz 1999). In using this approach to think about the introduction of novelty in frog songs, an immediate problem is the requirement for matching of production and perception (reviewed in Doherty and Hoy 1985). Does genetic variation have to coordinate changes in song with preference for the change (or vice versa; genetic coupling) or can production and perception evolve independently (genetic independence with selection leading to coevolution)? Divergence may arise first from changes in male vocalizations or from female preferences for specific vocal features. The latter forms the basis for the sensory exploitation hypothesis advanced by Ryan, Rand, and their colleagues (e.g., Ryan 1985, 1998; Ryan et al. 1990). *X. laevis* has an unusual vocal characteristic: females produce a fertility advertisement call, rapping, which stimulates male calling whereas their unreceptive call, ticking, suppresses male calling (Tobias et al. 1998). The *X. laevis* vocal system thus allows us to determine whether the same factors thought to be important in male advertisement also operate in females. The effects of female calls on male vocal and approach behaviors allow us to use both to probe the perceptual basis of attraction.

Xenopus laevis as an Experimental System

X. laevis is a totally aquatic pipid species that is native to southern Africa. Calls are made and heard underwater, at night, in silt-filled ponds—circumstances that undoubtedly have contributed to its unpopularity for field studies. However, *X. laevis* has a rich intra- and intersexual communicatory repertoire; the ease with which it can be studied in the laboratory permits a number of mechanistic approaches that are more difficult in terrestrial frogs. At present we have a basic understanding of how calls are produced and how the sex of the caller determines its vocal behavior (see Kelley 1996 and Kelley and Tobias 1999 for detailed reviews of this topic). The recent discovery of rapping, a female fertility advertisement that acts as an acoustic aphrodisiac for males (Tobias et al. 1998), opens a new avenue for the exploration of the perceptual features of vocal communication between the sexes.

Male Advertisement and Aggressive Calls

X. laevis relies on acoustic communication to convey location and reproductive status. The male advertisement call consists of a biphasic trill with fast and slow phases (Figure 12.1). Each trill lasts about 1 second and consists of a series of rapid clicks. Each click is a brief and noisy sound burst with most of its energy between 1.7 and 2.2 kHz. The duration of the fast phase of the trill is approximately 250 ms with a click rate of 70 Hz, whereas the slow phase lasts approximately 750 ms with a click rate of 35 Hz. The amplitude of successive clicks within the fast trill phase increases throughout the trill; we call this characteristic feature amplitude modulation. The advertisement call is prolonged; an individual male will call for many hours. Advertisement calls are given both by isolated males and by males in the presence of conspecifics. Laboratory studies, including recordings from the laryngeal nerve of vocalizing animals (Yamaguchi and Kelley 1998), reveal that only males produce the advertisement call (Picker 1983; Wetzel and Kelley 1983). The calls can act as a positively phonotactic signal for a receptive female (Picker 1983; Tobias et al. 1998).

We conducted a series of field observations on vocal behaviors of *X. laevis* during their extended breeding season in 1995 and again in 1997. The observation sites were natural ponds in the vicinity of Cape Town, South Africa. The ponds contained very high population densities and were heavily silted with decomposing vegetation that formed the pond substrate. We could record vocal behaviors at night using a hydrophone but could not observe the frogs. In 1995 we sampled several ponds across the breeding season (July to December; the end of winter and the beginning of spring).

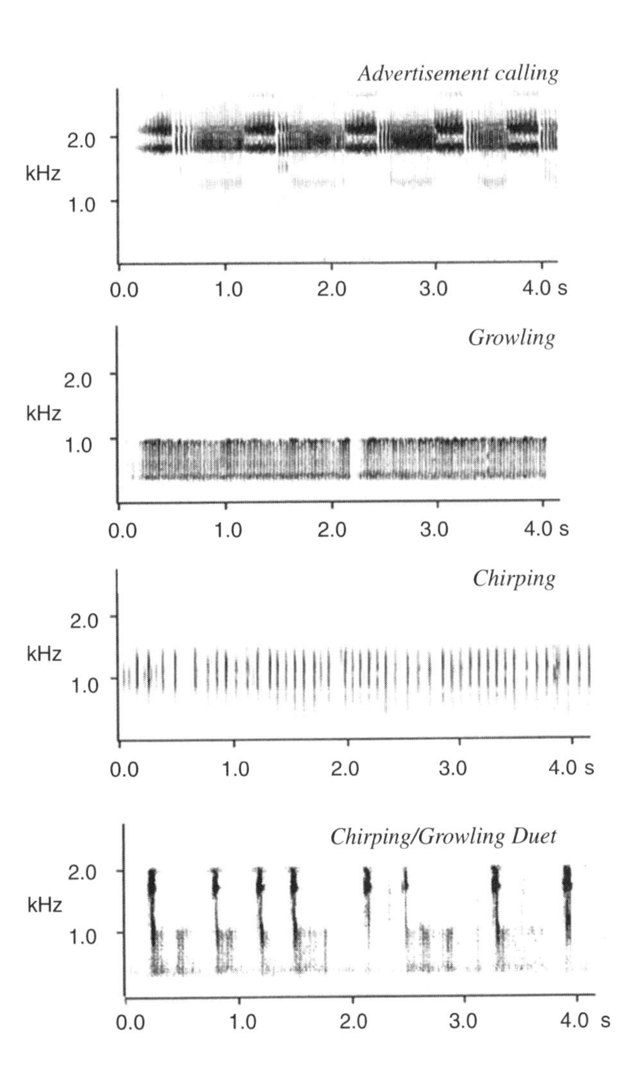

Figure 12.1. Sound spectrograms of male vocal behavior in *Xenopus laevis.* Advertisement calling is given by isolated male; growling functions as a male–male release call; chirping is given during male-male encounters. In these situations chirping and growling can alternate.

In 1997 we made repeated recordings at a single pond at the beginning of the breeding season (June to August).

Each pond contained many sexually receptive males (as judged by nuptial pad development, reviewed in Kelley 1996). Every male with well-developed nuptial pads called when removed from the pond. In the ponds themselves, however, only a subset of males called on a given night; on some nights, no males called. That relatively few males call in a pond where most males are capable of song suggests that calling is suppressed in many males. What is the source of this suppression? One possibility is the establishment of a calling dominance hierarchy due to male–male aggressive interactions. To explore this idea we have categorized the vocalizations given by pairs of males in a laboratory setting and have determined whether these vocal behaviors can be recorded in the Cape Town ponds, particularly at the beginning of the breeding season.

Two distinct vocal behaviors, growling and chirping, accompany male–male interactions (Figure 12.1). Growling is a rapid trill made up of low-frequency (< 1.0 kHz) clicks. Our observations and those of Picker (1980) suggest that growling functions as a release call when one male attempts to clasp another. Growling is thus probably functionally analogous to female ticking (see below) and some common acoustic features—such as the low (< 1.0 kHz) frequencies comprising individual clicks—could play a role in release from amplexus.

Chirping (Figure 12.1) is a brief trill made up of higher frequency clicks. Growling and chirping are produced together (Figure 12.1) by male–male pairs both during attempted amplexus and without clasp attempts. Both growling and chirping have been recorded from the natural ponds; the relationship of these male–male interactions to the determination of which male produces advertisement calls is a topic of current study in the laboratory.

Female Attractive and Repulsive Calls

Female *X. laevis* also use vocalizations to convey their state of sexual receptivity. A female about to oviposit produces a rapid (~ 12.5 Hz) series of clicks, rapping (Figure 12.2), in response to a calling male (Tobias et al. 1998). Rapping stimu-

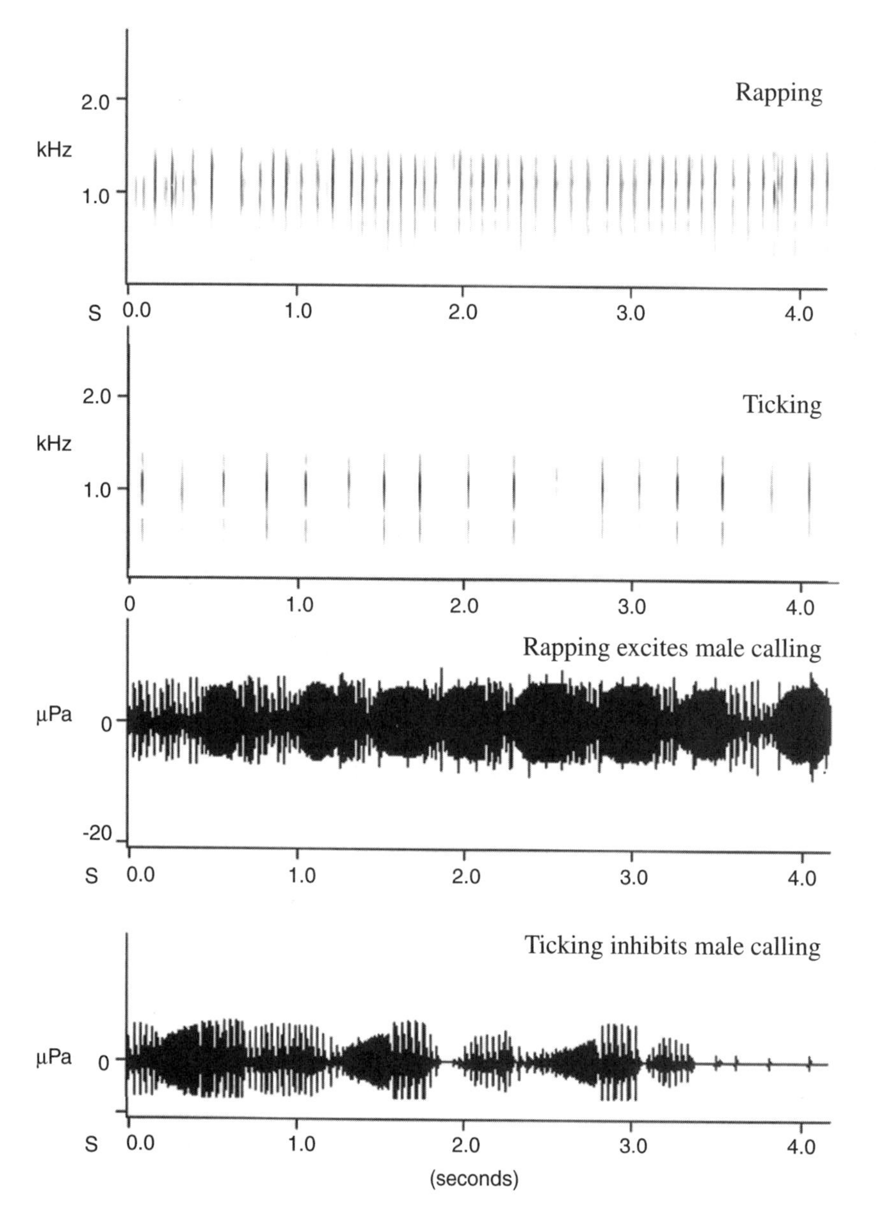

Figure 12.2. Female vocal behaviors and female–male duets. Rapping and ticking are sound spectrograms. Male's response to rapping and male's response to ticking are oscillograms.

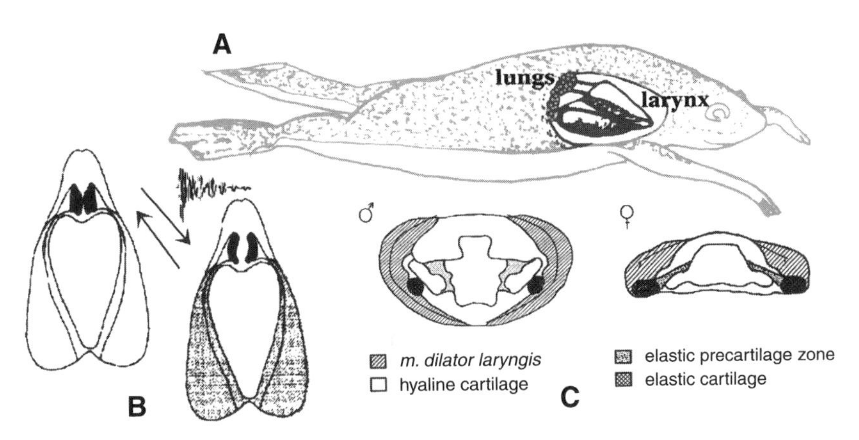

Figure 12.3. The larynx of *Xenopus laevis.* (A) The larynx is located in the body cavity, just dorsal to the heart. This vocal organ connects, via the glottis, with the buccal cavity anteriorly and, via the tracheae, with the lungs posteriorly. Modified from an original drawing by Julio Perez. (B) Clicks are produced when laryngeal muscles contract, pulling apart the arytenoid disks. (C) Transverse sections (shown in schematic diagram) through the adult male and female larynx. The male laryngeal muscles are more extensive and the interior lumen formed by hyaline cartilages more complex than those of the female. Elastic cartilage is absent in the adult female. (Drawings by D. Kelley and L. Fischer)

lates both the male's approach (phonotaxis) and his vocalizations. Males answer rapping by increasing the length and amplitude modulation of the fast trill phase of the advertisement call as well as decreasing the duration of the slow trill phase of the call; this alteration results in a distinct and intense answer call (Figure 12.2). As long as the female raps, the male answers. If rapping slows down, the male reverts to advertisement calling. Thus sexually receptive pairs duet before amplexus; if the sexes are prevented from initiating amplexus, duetting is prolonged.

Sexually unreceptive females produce a different call, ticking (Figure 12.2). Ticking inhibits male calling (Tobias et al. 1998). Ticking is a slow (~ 6 Hz) series of clicks, part of a suite of behaviors produced by sexually unreceptive females in response to attempted amplexus by males. Males initially respond to tapes of ticking with the answer call but then, in sharp contrast to the response to rapping, stop calling as long as ticking continues (Figure 12.2). Both ticking and rapping have been recorded from the natural ponds. Ticking and rapping are click trains; sound frequencies in female clicks are distinct from those of male clicks, permitting identification during duets.

Mechanisms of Vocal Production: The Larynx

In *X. laevis* all calls are made and heard underwater and one would thus expect to find some differences in sound production and perception relative to terrestrial anurans. The larynx, or vocal organ, has been modified to produce sounds using a "click" mechanism that does not require respiration. The larynx (Figure 12.3) is a box-like structure of muscle and cartilage that communicates with the buccal cavity, anteriorly, and with the lungs, posteriorly, via paired tracheae. The "skeleton" of the larynx is made up of several kinds of cartilage: the predominant form is hyaline cartilage, but the larynx also contains calcified cartilaginous rods, the thyrohyals, and sound-producing disks of arytenoid cartilage. As first described by Yager for *X. borealis* (Yager 1992), adult male *X. laevis* larynges also contain elastic cartilage, a type absent from the mature female (Fischer et al. 1995). The size and complexity of the hyaline cartilages differs markedly in the sexes and results in a more elaborate internal lumen in males than in females.

The cartilaginous skeleton is flanked by bipennate laryngeal dilator muscles that insert, at one end, onto a tendon attached to the arytenoid disks. In *X. borealis,* where the larynx is sufficiently transparent that disk movement can be observed, clicks are produced when the arytenoid disks open (Yager 1992); we assume that a similar mechanism operates in *X. laevis.* The force required for disk movement is supplied by contraction of the bipennate muscles in response to activity in the laryngeal nerve; provided that muscle activity generates supra-threshold force, each cycle of muscle contraction and relaxation is associated with the production of a single click.

This relatively simple mechanism of vocal production has been useful in analyzing the physiological properties of male and female larynges that are associated with their distinctive calls. The larynx can be removed and stimulated in vitro via the laryngeal nerves; muscle contraction and resultant tension on the tendon can be recorded as can the actual sounds produced (Tobias and Kelley 1987). Because the isolated larynx produces sounds that closely resemble actual vocalizations when stimulated via the nerves, respiration is clearly not required for sound production. However, the glottal nerve runs with the axons of laryngeal motor neurons, and trains of stimuli to the nerve result in glottal opening. It is possible that in the animal, access to the buccal cavity (or the lungs) is coordinated with vocalization and that glottal opening, for example, contributes to acoustic features of the call. The production of vocal patterns is widely believed to have arisen from the intrinsic rhythms of respiration. The aquatic habitat is a secondary adaptation (i.e., the pipid taxon is believed to have originated from terrestrial frogs) and it would be surprising if no remnant of respiratory control could be found in vocal output pathways.

One question is whether sexually differentiated features of the distinctive male and female calls can be tied to characteristics of the vocal organ. There are four basic differences between male and female vocal behaviors: the dominant frequency of each click, the rate at which clicks are given, the temporal pattern of the click trains, and the amplitude modulation of click trains. We do not know how the dominant frequency of each click is determined. Fast Fourier transform analyses reveal that clicks produced by isolated larynges contain more high-frequency components (> 4 kHz) than are found in actual clicks; some of the acoustic features of clicks may be shaped by resonance properties of extralaryngeal structures. In general the dominant frequency of clicks in male vocalizations is higher (~ 1.7 kHz) than that in female vocalizations (~ 1.2 kHz). The lumen of the hyaline cartilages is considerably more complex in males than in females and could contribute to differences in click frequencies. However the shape of the larynx does not impose an invariant constraint on click frequency. The clicks in growling (< 1 kHz) for example, produced only by males, are lower in frequency than the clicks in advertisement calls (~ 1.7 kHz).

Male and Female Laryngeal Muscle

The best understood laryngeal correspondence is between click rate and twitch characteristics of laryngeal muscle fibers. The fast trill portion of the advertisement call is a short (250 ms), rapid (70 Hz) click train and is produced by bipennate muscles that can contract and relax at rates up to 100 Hz. Female muscle tetanizes at rates as slow as 35 Hz and cannot contract and relax completely at rates faster than 25 Hz (Tobias and Kelley 1987). Since, as long as enough fibers are involved, each contraction of the bipennate muscle is accompanied by a click, the contractile properties of the muscle fibers themselves set limitations on the rapidity with which calls can be produced. Males can produce both fast and slow trills, whereas females can only produce slow trills.

What accounts for sex differences in the contraction rate of laryngeal muscle fibers? The entirely fast twitch muscle of males is the result of a masculine developmental program that relies on gonadal secretion of steroid hormones, particularly the androgens (reviewed in Sassoon et al. 1987; Kelley 1996). At the end of metamorphosis males and females have equivalent numbers of laryngeal muscle fibers and most are slow twitch. Males then proceed to add fibers at an average rate of approximately 150 a day until the adult complement is reached 6 months later (Marin et al. 1990). From this point on, muscle fibers start to convert from slow to fast twitch and it is not until the last slow twitch fibers change over, generally 10 to 12 months after metamorphosis is complete, that males can produce the fast trills of advertisement calls (Tobias et al. 1991a). Castration at metamorphosis or 6 months later halts, but does not reverse, conversion of slow to fast twitch fibers (Tobias et al. 1991b). The larynges of juvenile frogs, either males or females, can be converted to entirely fast twitch by the provision of a testis transplant or exogenous androgen that will also reverse the effects of castration at an early age (Tobias et al. 1991b; Watson et al. 1993).

The results described above suggest that androgen secretion from the gonads is required for the masculinization of muscle fiber type in the *X. laevis* larynx. Since it is clearly not adaptive to convert all nonvocal muscles to fast twitch we have asked what mechanisms confine the effects of gonadal androgen to the larynx. One possibility is that androgen sensitivity is limited to laryngeal muscle. Sensitivity to androgen is conferred by the expression of an intracellular protein, the androgen receptor, which binds the hormone; the receptor/hormone complex accumulates in the cell nucleus and influences function by regulating transcription of target genes. We have cloned the *X. laevis* androgen receptors and have analyzed their pattern of expression in larynx during development (Fischer et al. 1995). Laryngeal muscle and cartilage, particularly in juveniles, express extremely high levels of receptor, among the very highest levels of any vertebrate tissue. However, all muscle has some androgen receptor and thus could be influenced by circulating hormones.

Different myosin heavy chains (produced by separate genes) have different adenose triphosphatase (ATPase) activities that contribute substantially to differences in speed of muscle contraction and relaxation. Thus another possibility is that laryngeal muscle expresses unique, androgen-sensitive contractile proteins. We examined this possibility by screening a laryngeal cDNA library for myosin heavy chain (MHC) genes expressed predominantly in adults males (Catz et al. 1992). The screen revealed an abundant MHC that we have called laryngeal myosin heavy chain or LM. Using a probe containing untranslated 3′ sequence (important because this region is poorly conserved among myosin heavy chains), we can detect LM expression in all laryngeal muscle fibers of adult males, but in only some of females. LM expression is regulated by gonadal androgens; castration at the end of metamorphosis prevents the male-typical increase in LM expression, and LM expression is increased by exogenous androgen secretion in both sexes (Catz et al. 1995).

In vertebrates the MHC genes expressed by embryos are different from those expressed in newborns and in adults. In amino acid sequence, the portion of the LM gene that we have examined is more like embryonic fast myosins than those of neonates or adults. It thus seems likely that this tissue-specific, androgen-regulated gene arose from the duplication of—and then divergence from—an ancestral embryonic, fast MHC. That divergence probably included changes that permit regulation of the gene by several hormone systems. Recent studies by Chris Edwards in the laboratory (Edwards et al. 1999) suggest that the secretion of the pituitary hormone prolactin is required to establish the androgen sensitivity of LM expression. Male muscle fibers cannot become fast twitch in response to androgen without the normal developmental sequence of exposure to prolactin at the end of metamorphosis.

Temporal Pattern of Calling

The twitch characteristics of laryngeal muscle fibers set limits on the rate at which clicks can be produced. The temporal pattern of calling (including click rate) is generated by the central nervous system. This feature was first demonstrated by the vox in vitro experiments in which isolated larynges were driven to produce rapid trills (males: advertisement calling) or slow trills (females: ticking) by stimulating the laryngeal nerves at the behaviorally appropriate rates (35–70 Hz vs. 7 Hz); the rate of click production mirrored the rate of stimulation (Tobias and Kelley 1987).

What was not clear from these experiments was the actual temporal pattern produced by the central nervous system during vocal behavior. For example, the central nervous system might generate a single pattern conveying a sexually receptive state and an alternative pattern for the unreceptive state; the pattern for advertisement calling might be identical to that for rapping and that for growling be identical to ticking. The call produced would be determined by how faithfully the larynx could follow the brain; the twitch properties of laryngeal muscle fibers could, for example, filter the growling input into a ticking output or an advertisement calling input into rapping. To examine this question, Ayako Yamaguchi in the laboratory has recorded en passant from the laryngeal nerves of vocalizing male and female frogs (Yamaguchi and Kelley 2000). These studies reveal a unique pattern of nerve activity that corresponds 1:1 with the vocalizations produced by the animal. Thus sexually differentiated laryngeal muscle fiber characteristics do not shape a state-dependent neural output (receptive vs. unreceptive) into a sexually appropriate vocal behavior, but rather subserve the distinct demands of a varied and sexually dimorphic vocal repertoire.

Click Amplitude

A characteristic feature of the male's advertisement call is variation in the amplitude of individual clicks in the fast trill portion; clicks are initially soft and become progressively louder (amplitude modulation). An individual male can produce advertisement calls with both amplitude modulated and nonmodulated fast trills. The answer call (a modified advertisement call elicited by rapping), however, is always markedly amplitude modulated. How is the progressive increase in click loudness produced?

Recordings of electromyographic (EMG) activity from the vox in vitro preparation reveal that the loudness of individual clicks is directly proportional to EMG amplitude: the more muscle fibers that are active, the louder each click (Tobias and Kelley 1987). An increase in the number of contracting laryngeal muscle fibers reflects both a progressive increase in neurotransmitter release from individual synapses (facilitation, Tobias et al. 1995; Ruel et al. 1998) and a progressive increase in the activity of laryngeal motor neurons (recruitment, Yamaguchi and Kelley 2000). Thus the extent to which motor units (laryngeal motor neurons and the muscle fibers they innervate) are recruited and facilitated determines the progressive loudness of individual clicks in the fast portion of the advertisement call. Amplitude modulation, like the temporal pattern of vocal behaviors, is a function of the central nervous system, both recruitment of laryngeal motor neurons in the hindbrain and facilitation of the vocal neuromuscular synapse in the larynx.

A striking characteristic of the male vocal synapse is that a single action potential invading the presynaptic terminal releases only a small amount of neurotransmitter, an amount insufficient to depolarize the muscle fiber to the threshold level required to produce an action potential and a muscle contraction (Tobias and Kelley 1988). Action potentials in muscle require repetitive activity in the laryngeal nerve; the weak male synapse strengthens with use (facilitation). With weak vocal synapses, synaptic failures occur even during repetitive nerve activity; neuromuscular transmission is unreliable (Tobias and Kelley 1988). At most vertebrate neuromuscular synapses, each action potential in a motor axon releases sufficient (or more than sufficient) neurotransmitter to evoke a muscle fiber action potential. Even the laryngeal synapse of adult female *X. laevis* is stronger than that of males and synaptic failures are infrequent (Tobias and Kelley 1987, 1988; Tobias et al. 1998). Does improved vocal courtship compensate for the synaptic failures that characterize the weak vocal synapses of male *X. laevis?*

At the level of individual synapses, subthreshold neurotransmitter release permits facilitation, one of the cellular mechanisms responsible for the progressive amplitude increase of fast trills. If this feature—increasing click loudness—is important for attracting females (and/or discouraging male competitors) then it may reflect the process of sexual selection believed to shape characteristics of anuran advertisement calls (Wells 1977). Progressive increases in click amplitude accompany some advertisement calls. Preliminary experiments in the laboratory suggest that these calls can be more attractive to females than advertisement calls with constant amplitude fast trills (Tobias et al. 1991c). The male answer call is also highly amplitude modulated. We do not know whether answer calls are more attractive to females than advertisement calls. One possible scenario is that the answer call evoked from a male by a rapping female also attracts other gravid females, increasing the probability that a male will encounter an ovipositing female with which to mate.

The Localizability of Sounds

The first requirement of a mating system is locating a sexual partner; advertisement calls have to provide accurate location information. A parsimonious explanation for specific acoustic features of *Xenopus* calls (and those of the túngara frog; see Ryan and Rand, this volume) is that some are more easily localized than others. The localizability of natural sounds has been examined in a number of vertebrate species and several features of easily located sounds have been identified; louder sounds are easier to locate than softer ones, longer sounds are found more readily than shorter ones. Sounds such as clicks with abrupt onsets are more easily localized than sounds that gradually wax or wane in intensity. Repeated click trains with distinctive signatures (e.g., progressive increases in amplitude) may be more easily localized than single clicks or more monotonous trains (see Marler 1955; Erulkar 1972). The broad range of frequencies present in *X. laevis* clicks, their rapid onsets, the frequent repetition of clicks, the amplitude of advertisement and rapping, and the amplitude modulation of advertisement and answer calls all suggest an adaptation for easily localized sounds. How *X. laevis* actually localizes sounds, however, is not well understood (Elepfandt et al. 1995; Elepfandt 1996).

The auditory periphery of *X. laevis* includes the tympanic disk, amphibian and basilar papillae, and their associated nerve fibers (Figure 12.4; Wever 1985; Elepfandt 1996). The tympanic discs are oblongs of thin cartilage into which one end of the columella inserts, the other end being inserted into the oval window. The middle ear cavity is air filled and communicates with the contralateral cavity via a midline eustachian tube. The auditory sensitivity of the peripheral apparatus is well matched to the dominant frequencies of the clicks. Measurements of inner ear potential changes (reflecting activity of the papillae; Wever 1985) reveal maximum sensitivities at 1200 to 2500 Hz (Figure 12.4), a range that includes the dominant frequencies of clicks from most vocal behaviors. Behavioral tests (Figure 12.4; Elepfandt 1996) reveal maximum frequency discriminations in the range 1400 to 2500 Hz (2% limen), values that are in good agreement with the electrophysiological recordings of Wever. Biophysical studies reveal that the amplitude and phase of vibration of the tympanic disks varies with sound direction and frequency (reviewed in Elepfandt 1996). How these differences translate into the ability to localize sound sources is not yet clear, but could be augmented by interaural comparisons carried out within auditory regions of the central nervous system. Brain regions that participate specifically in sound localization are present in other vertebrates (the nucleus laminaris of birds for example; reviewed in Knudsen and Brainard 1995), but have not yet been identified in *X. laevis.*

The Anuran Auditory System: Sex Differences

How are anuran communication signals perceived? Answers to this question might shed some light on the preexisting preferences thought, in the sensory exploitation hypothesis (see Ryan and Rand, this volume), to shape vocal behaviors. For the most part, these issues have been examined in terms of the biases in the female's nervous system (assayed by phonotaxis to male advertisement calls). It is not clear, however, whether a specific auditory preference must represent a sexually dimorphic character; the auditory preference could be characteristic of both sexes in a particular species and exploited by both sexes, or it could be sex-specific and shape distinct features of male and female vocal behavior. In *X. laevis* we have the opportunity to examine the female's response to the male advertisement and answer

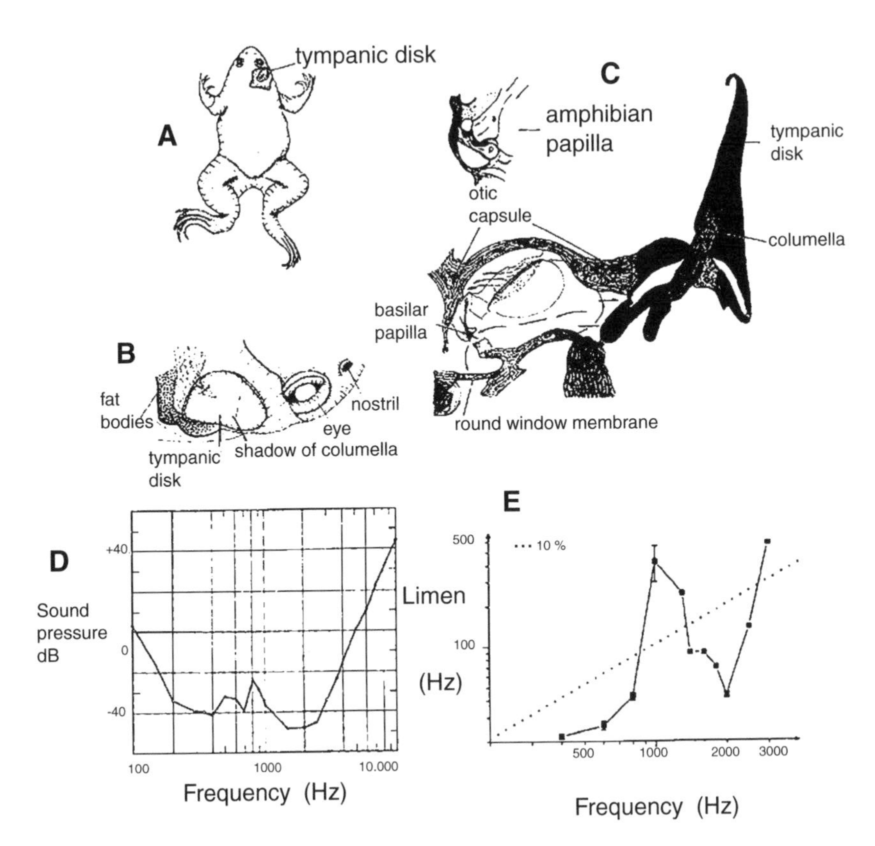

Figure 12.4. Some features of the *Xenopus laevis* auditory system. (A) Sounds are detected via the tympanic disks located under the skin, posterior to the eye. (B) The columella inserts into the tympanic disk. (C) The inner ear contains amphibian and basilar papillae innervated by processes from cells in the acoustic ganglion. (D) Auditory sensitivity determined using the "cochlear" microphonic. All of the above modified from Wever (1985). (E) Auditory discrimination determined using a behavioral choice task (modified from Elepfandt 1996). The sound frequencies to which adults are most sensitive match the dominant frequencies in vocal behaviors.

calls as well as the male's response to both repulsive and attractive female calls (ticking and rapping). Pacific treefrogs (see Brenowitz et al., this volume) present an opportunity to look at neural coding of salient auditory stimuli used in male–male competitions (which should also, provided that they influence reproductive opportunities, be subject to sexual selection).

Sound Frequency

The best evidence for sexually differentiated auditory processing comes from the work of Narins and Capranica (1976, 1980) on the Puerto Rican tree frog, *Eleutherodactylus coqui*. Males have a two-note call, Co and Qui; the first note is used in male–male territorial encounters whereas the second note is attractive to females. The Co note is a constant frequency (1.1–1.2 kHz) whereas the Qui note is an upwards frequency sweep (1.8–2.2 kHz). The anuran inner ear contains two papillae (see Figure 12.4), amphibian and basilar, tuned to different frequency ranges: low to mid (amphibian) and high (basilar). The papillae are innervated by axons of neurons that form the eighth nerve, and relative frequency sensitivity can be determined by an axon's best excitatory frequency (BEF)—the frequency most effective, at the lowest amplitude, in increasing the firing rate of eighth nerve axons.

The major result of these studies is the higher proportion (relative to the opposite sex) of neurons with BEFs in the Co range in males and in the Qui range in females. Further the sharpness of tuning of nerve axons with high frequency sensitivity (innervating the basilar papillae) was greater in males than in females; for low and midrange frequencies (amphibian papillae), results were the same in the sexes. Does this finding mean that females are deaf to Co and males to Qui? No. As a female approaches a male and the loudness of his call increases, her auditory system should respond to Co notes; only at relatively greater distances (and lower sound amplitudes) should the sex difference be apparent. A similar argument applies to a male approaching another male; responses to Qui should be robust at shorter distances.

Temporal Aspects of Sounds, Inter- and Intrasexual Communication

Ticking and rapping have very different effects on male vocal behaviors in *X. laevis* (inhibition and excitation, respectively) and this differential response has enabled us to design a series of experiments aimed at determining how these female vocalizations are distinguished. These vocalizations differ primarily in rate; the individual clicks in ticking and rapping are not distinguished, but the former is a

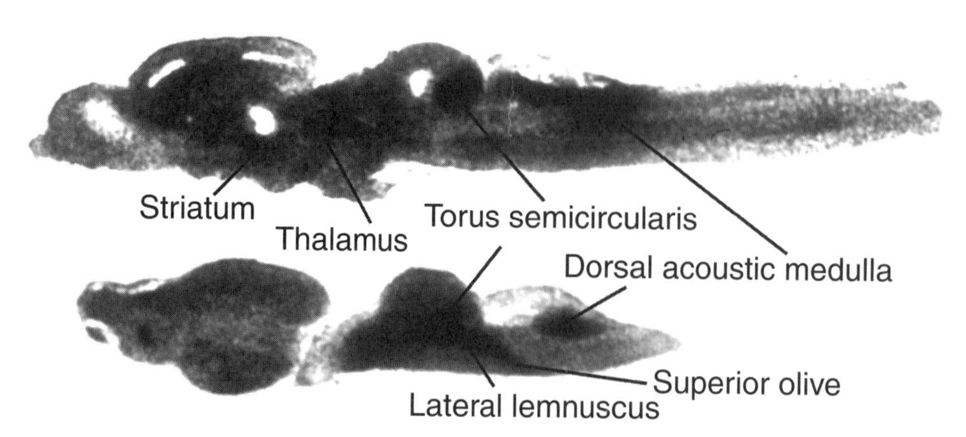

Figure 12.5. An autoradiogram of a saggital section of an adult male *Xenopus* brain and rostral spinal cord exposed to acoustic stimulation after injection with radioactive 2-deoxyglucose (see Kelley 1980 for experimental details). The terminals of acoustic fibers innervating the papillae terminate in the dorsal acoustic nucleus. These cells project to the torus semicircularis and thence to auditory nuclei of the thalamus.

slow (~ 4 Hz) whereas the latter is a more rapid (~12 Hz) trill. To determine how males label click trains we have constructed artificial trills at intervals intermediate between ticking and rapping and are examining the amount of time males spend calling in response to these artificial calls.

Of particular interest is whether the calls are labeled categorically (see Wyttenbach et al. 1996 for a discussion of this issue). For example, up to some boundary rate, males may respond to artificial calls as though to ticking and past this boundary, as though to rapping. Alternatively there may be a simple monotonic relationship between the rate of click production by the female and the male's vocal response. These outcomes have different implications for how female songs are decoded by the male's nervous system. In the first case, the nervous system initially sorts the click trains into one of two categories; vocalizations are stimulated if acoustic stimuli are categorized as rap-like and suppressed if categorized as tick-like. In the second, there is no initial categorization and vocalization is simply a direct function of the rate of stimulation. Our preliminary data suggest that the first scenario is correct and that the criteria for labeling are stringent.

When males hear rapping they can switch from advertisement to answer calling within a few hundred milliseconds. The differential response to the two vocal behaviors (ticking: suppression; rapping: stimulation) is slower, on the order of seconds, and could require processing in higher brain regions involving neurons that are tuned to the rate of stimulation by clicks. Studies by Rose and colleagues (Alder and Rose 1998 and reviewed by Brenowitz et al., this volume) reveal the presence of rate-selective neurons in the torus semicircularis (Figure 12.5), a midbrain region that processes temporal aspects of sound in many vertebrate species.

Male Pacific treefrogs defend calling territories and switch from a trill with a faster repetition rate (advertisement call) to one with a slower rate (encounter call) in response to a vocalizing intruder. Males discriminate between call-appropriate trill rates in playback experiments. Alder and Rose (1998) have recently described a population of rate-sensitive neurons in the torus of these frogs whose properties match call features. Neurons responsive to fast trill rates only start firing after the eighth click (multiple clicks are needed to determine rate) and the response continues after termination of the stimulus. These toral neurons have features that could be used as part of an encounter call detection system. Their presence in the anuran torus suggests that it is in this region that we may find portions of the neural machinery used in discriminating ticking from rapping in *X. laevis*. The torus of *X. laevis* (Figure 12.5) contains several candidate nuclei for participation in rate tuning including one, the laminar nucleus, that is a target for gonadal steroids (Kelley 1980). This system should provide a rich arena in which to investigate the neural processing of courtship signaling and whether it differs in the sexes.

Conclusions

Xenopus laevis live in dark, silt-filled ponds and call at night. Under these conditions, sound plays a preeminent role in social communication; *X. laevis* has a rich and complex repertoire of vocal signals. The two female-specific calls, rapping and ticking, have powerful but opposite effects on male calling. Rapping stimulates whereas ticking suppresses male songs. Both sexes can advertise their sexual readiness, females via rapping and ticking and males via their advertisement calls. Males also communicate vocally and some male–male duets appear to function in the establishment of vocal dominance. We are beginning to understand the cellular and molecular mechanisms that produce sexually differentiated vocal behaviors. Male- and female-specific developmental programs require secretion of sex-specific gonadal steroids that act on both the vocal organ and on the neural pathways for vocal production in the central nervous system. In the larynx, androgen controls muscle fiber addition and muscle fiber type. In the central nervous system, androgen prevents the death of developing vocal motor neurons. Estrogen controls synaptic strength. How vocal behaviors are perceived is less well understood, although the major brain pathways and some sex differences in auditory processing have been described in other anurans. Of particular interest is how the temporal characteristics of calls are coded for in the central nervous system. Understanding, for example, how the rapid trills of rapping and the slower trills of ticking, respectively, excite or suppress a calling male is a fascinating problem.

References

Alder, T., and G. J. Rose. 1998. Long term temporal integration in the anuran auditory system. Nature Neuroscience 1:519–523.

Andersson, M. 1994. Sexual Selection. Princeton University Press, Princeton, NJ.

Catz, D., L. Fischer, and D. B. Kelley. 1995. Androgen regulation of a laryngeal-specific myosin heavy chain isoform whose expression is sexually differentiated. Developmental Biology 171:448–457.

Catz, D., L. Fischer, M. Tobias, T. Moschella, and D. B. Kelley. 1992. Sexually dimorphic expression of a laryngeal-specific, androgen-regulated myosin heavy chain gene during *Xenopus laevis* development. Developmental Biology 154:366–376.

Darwin, C. 1871. The Descent of Man and Selection in Relation to Sex. J. Murray, London.

Doherty, J., and R. Hoy. 1985. Communication in insects III, The auditory behavior of crickets: Some views of genetic coupling, song recognition and predator selection. Quarterly Review of Biology 60:457–472.

Edwards, C. J., K. Yamamoto, S. Kikuyama, and D. B. Kelley. 1999. Prolactin opens the sensitive period for androgen regulation of a larynx-specific myosin heavy chain gene. Journal of Neurobiology 41:443–451.

Elepfandt, A. 1996. Underwater acoustics and hearing in the clawed frog, *Xenopus*. Pp. 177–193. *In* R. Tinsley and H. Kobel (eds.), The Biology of *Xenopus*. Oxford University Press, Oxford.

Elepfandt, A., A. Ringeis, and W. Fischer. 1995. Calling and territoriality in the clawed frog, *Xenopus laevis*. P. 343 (abstract). *In* M. Burrows, T. Matheson, P. Newland and H. Schuppe (eds.), Nervous Systems and Behaviour. Georg Thieme Verlag, Stuttgart.

Erulkar, S. D. 1972. Comparative aspects of spatial localization of sound. Physiological Reviews 52:237–326.

Fischer, L., D. Catz, and D. B. Kelley. 1995. Androgen-directed development of the *Xenopus laevis* larynx: Control of androgen receptor expression and tissue differentiation. Developmental Biology 170:115–126.

Kelley, D. B. 1980. Auditory and vocal nuclei of frog brain concentrate sex hormones. Science 207:553–555.

Kelley, D. B. 1996. Sexual differentiation in *Xenopus laevis*. Pp. 143–176. *In* R. Tinsley and H. Kobel (eds.), The Biology of *Xenopus*. Oxford University Press, Oxford.

Kelley, D. B., and M. L. Tobias. 1999. Vocal communication in *Xenopus laevis*. Pp. 9–35. In M. Hauser and M. Konishi (eds.), Neural Mechanisms of Communication. MIT Press, Cambridge, Massachusetts.

Knudsen, E., and M. Brainard. 1995. Creating a unified representation of visual and auditory space in the brain. Annual Review of Neuroscience 18:19–43.

Marin, M., M. Tobias, and D. B. Kelley. 1990. Hormone sensitive stages in the sexual differentiation of laryngeal muscle fiber number in *Xenopus laevis*. Development 110:703–71.

Marler, P. 1955. Characteristics of some animal calls. Nature 176:6–8.

Narins, P., and R. Capranica. 1976. Sexual differences in the auditory system of the tree frog, *Eleutherodactylus coqui*. Science 193:378–380.

Narins, P., and R. Capranica. 1980. Neural adaptations for processing the two-note call of the Puerto Rican tree frog, *Eleutherodacylus coqui*. Brain Behavior and Evolution 17:48–66.

Nei, M., and J. Zhang. 1998. Molecular origin of species. Science 282:1428–1429.

Picker, M. 1980. *Xenopus laevis* (Anura: Pipidae) mating systems: A preliminary synthesis with some data on the female phonoresponse. South African Journal of Zoology 15:150–158.

Picker, M. D. 1983. Hormonal induction of the aquatic phonotactic response of *Xenopus*. Behaviour 86:74–90.

Ruel, T., D. B. Kelley, and M. L. Tobias. 1998. Facilitation at the sexually differentiated laryngeal synapse of *Xenopus laevis*. Journal of Comparative Physiology 182:35–42.

Ryan, M. J. 1985. The Túngara Frog: A Study in Sexual Selection and Communication. University of Chicago Press, Chicago.

Ryan, M. J. 1998. Sexual selection, receiver biases and the evolution of sex differences. Science 281:1999–2003.

Ryan, M. J., J. H. Fox, W. Wilczynski, and A. S. Rand. 1990. Sexual selection for sensory exploitation in the frog *Physalaemus pustulosus*. Nature 343:66–67.

Sassoon, D., G. Gray, and D. B. Kelley. 1987. Androgen regulation of muscle fiber type in the sexually dimorphic larynx of *Xenopus laevis*. Journal of Neuroscience 7:3198–3206.

Schwartz, J. H. 1999. Homeobox genes, fossils and the origin of species. Anatomical Record (New Anat.) 257:15–31.

Tobias, M. L., and D. B. Kelley. 1987. Vocalizations of a sexually dimorphic isolated larynx: Peripheral constraints on behavioral expression. Journal of Neuroscience 7:3191–3197.

Tobias, M. L., and D. B. Kelley. 1988. Electrophysiology and dye coupling are sexually dimorphic characteristics of individual laryngeal muscle fibers in *Xenopus laevis*. Journal of Neuroscience 8:2422–2429.

Tobias, M. L., M. Marin, and D. B. Kelley. 1991a. Development of functional sex differences in the larynx of *Xenopus laevis*. Developmental Biology 147:251–259.

Tobias, M. L., M. Marin. and D. B. Kelley. 1991b. Temporal constraints on androgen directed laryngeal masculinization in *Xenopus laevis*. Developmental Biology 147:260–270.

Tobias, M. L., R. Bivens, S. Nowicke, and D. B. Kelley. 1991c. Amplitude modulation is an attractive feature of *X. laevis* song. Society for Neuroscience Abstracts 17:1403.

Tobias, M. L., D. B. Kelley, and M. Ellisman. 1995. A sex difference in synaptic efficacy at the laryngeal neuromuscular junction of *Xenopus laevis*. Journal of Neuroscience 15:1660–1668.

Tobias, M. L., S. Viswanathan, and D. B. Kelley. 1998. Rapping, a female receptive call, initiates male/female duets in the South

African clawed frog. Proceedings of the National Academy of Science USA 95:1870–1875.

Watson, J., J. Robertson, U. Sachdev, and D. B. Kelley. 1993. Laryngeal muscle and motor neuron plasticity in *Xenopus laevis:* Analysis of a sensitive period for testicular masculinization of a neuromuscular system. Journal of Neurobiology 24:1615–1625.

Wells, K. D. 1977. The social behavior of anuran amphibians. Animal Behaviour 17:388–404.

Wetzel, D., and D. B. Kelley. 1983. Androgen and gonadotropin control of the mate calls of male South African clawed frogs, *Xenopus laevis.* Hormones and Behavior 25:666–693.

Wever, G. E. 1985. The Amphibian Ear. Princeton University Press, Princeton, NJ.

Wyttenbach, R., M. May, and R. Hoy. 1996. Categorical perception of sound frequency by crickets. Science 273:1542–1543.

Yager, D. 1992. A unique sound production system in the pipid anuran *Xenopus borealis.* Zoological Journal of the Linnean Society 104:351–375.

Yamaguchi, A., and D. B. Kelley. 1998. Generating sexually differentiated vocal patterns: Nerve and laryngeal EMG recordings from singing male and female *Xenopus laevis.* Abstracts of the 5th International Congress of Neuroethology. P. 100.

Yamaguchi, A., and D. B. Kelley. 2000. Generating sexually different vocal patterns: Laryngeal nerve and EMG recordings from vocalizing male and female African clawed frogs (*Xenopus laevis*). Journal of Neuroscience 20:1559–1567.

13

STEVEN M. PHELPS

History's Lessons

A Neural Network Approach to Receiver Biases and the Evolution of Communication

Introduction

The last two decades have seen an enormous amount of work aimed at understanding the role the receiver plays in shaping the evolution of signal form. Strategic analyses have employed game theoretic models to predict the information that receivers entice signalers to reveal (Dawkins and Krebs 1978; Krebs and Dawkins 1984; Grafen 1990; Hauser 1996). Discussions of signal efficacy have added sensory and perceptual biases of the receiver (Basolo 1990; Ryan 1990; Guilford and Dawkins 1991; Endler 1992; Endler and Basolo 1998) to the traditional emphasis on environmental transmission of signals (Littlejohn 1977; Brenowitz 1982; Römer and Bailey 1986). Recent models based on psychophysical (Greenfield and Roizen 1993; Greenfield et al. 1997) and mnemonic (Leimar et al. 1986; Weary et al. 1993; MacDougall and Stamp Dawkins 1998; Yachi and Higashi 1998) processes yield insights into phenomena as diverse as chorusing behavior and the evolution of aposematism.

In collaboration with Mike Ryan, Stan Rand, and Walt Wilczynski, I have been using phonotaxis experiments and artificial neural network models to investigate how the evolutionary history of a receiver might shape its perceptual biases. Neural network models are computer simulations that perform computations analogous to those of real nervous systems. We use these models in a manner comparable to their use in cognitive psychology. Rather than explicitly modeling a particular nervous system we treat the networks as convenient empirical tools that share relevant attributes of living receivers. We manipulate the selection pressures acting on the networks and use the resulting data to formulate testable hypotheses about the evolutionary bases of receiver biases. Neural networks evolved under distinct selection regimes, for example, can be tested with novel stimuli; network responses can then be compared to female responses in conventional phonotaxis experiments.

The chapter begins with a brief review of neural network models and their application to studies of animal communication. This is followed by a description of the neural network approach we have developed, a review of the studies we have performed to date, and a discussion of the role that historically derived receiver biases may play in the evolution of signal form.

Artificial Neural Networks

Artificial neural networks, like nervous systems, can use incomplete information and generalize to novel stimuli (Rumelhart and McClelland 1986a). They have proved to be valuable predictors of behavioral responses (Rumelhart and

McClelland 1986b; Sejnowksi and Rosenberg 1987; Montague et al. 1995) and represent a scarcely tapped resource for students of animal communication. Artificial neural networks supersede the complexity allowed in analytic mathematical models, yet permit far more controlled manipulations than could reasonably be performed in real animals.

A neural network is a computer program that can be represented as a web of interacting computational units. A network or group of networks can be trained or "evolved" to recognize patterns in many domains—time-varying voltages, handwritten characters, or patterns of amino acids related to protein structure, for example—and have been used to gain insight into how nervous systems function. I will argue here that they are also excellent tools for the investigation of animal communication.

The simplest of the powerful architectures is known as a feed-forward network (Rumelhart and McClelland 1986a). This network comprises three groups, or layers, of neurons that are connected serially to one another: the input layer, the hidden layer, and the output layer. The activity of a neuron usually serves as an input to every neuron in a layer downstream.

The activity of a neuron in the input layer is often simply the value of one element in the input matrix—that is, the activity of the input layer may represent the unprocessed data presented to the networks. In anuran communication, for example, one datum may represent the pulse rate of a call, the intensity of sound in a particular range of frequencies, or any number of call descriptors. In contrast each neuron in the hidden layer associates a weight (some coefficient, perhaps between –1.0 and +1.0) with each of its inputs, multiplies this weight by the activity of the input neuron, and computes a weighted sum. These weights can be viewed as synaptic strengths, transmitter types, or the position of synapses on the target cell. Once a neuron has been excited or inhibited by its inputs, it must somehow transform those inputs into an activity that can serve as an output. The artificial neurons do this by entering the weighted sum of inputs into an "activation function." The activation function often takes one of two forms: (1) a step or threshold function that assigns an activation level of 1.0 to the neuron (maximally active) when the weighted sum exceeds some threshold and an activity level of 0.0 (inactive) when the weighted sum falls short of the threshold; or (2) a sigmoid function that assigns activity using a continuous function that approaches 0.0 when the input is very low and approaches 1.0 when the input is very high (Figure 13.1a). The activities of hidden layer neurons serve as input for the neuron or neurons of the output layer. The neurons of the output layer then calculate their activities, which define the response of the network to the original stimulus.

To summarize, a feed-forward network calculates a response to a given input in three steps. First, input layer neurons are activated by a stimulus. Second, neurons in the hidden layer weight the activities of neurons in the input layer to compute their own activation. Finally, neurons from the hidden layer send their outputs to the neuron or neurons of the output layer, thus determining the response of a network to a given stimulus (Figure 13.1b).

The neurons' weights ultimately determine the responses of the network as a whole. Algorithms that train networks consist of various ways to adjust these weights so that networks make desired responses to particular inputs. In one common training method, "back-propagation," the output of the network is compared with the desired output, and the weights that each neuron assigns to its inputs are adjusted to reduce the discrepancy between the desired output and the observed output (Haykin 1994). Over a number of iterations this algorithm causes the network to converge on some locally optimal algorithm for matching inputs to outputs. The network can then be tested with novel or incomplete inputs to see how it classifies various test stimuli.

A classic implementation of the back-propagation training procedure is exemplified by the problem of optical character recognition. In this problem, networks receive a matrix of inputs representing different pieces of a letter (see Figure 13.1b) and must produce a unique output for each letter. For example, if you were to train a network to distinguish between a T and a C, you would need an input layer large enough to house a complete letter, with enough resolution that the two letters produce different inputs. If the network had only one output neuron, the task might be to produce an output of 1.0 whenever viewing a T as an input and an output of 0.0 whenever receiving an input corresponding to a C. To make matters more difficult it might be desirable to attempt to produce networks that made these distinctions regardless of whether the letters were presented upright, sideways, upside down, or off-centered. Although this appears to be a trivial task for humans, it has been a benchmark test for the technologies of artificial intelligence. Using back propagation, a neural network with randomly assigned weights was trained to perform this discrimination; the trained network passed this test by making its hidden neurons into specialized T-detectors (Rumelhart, Hinton, and Williams 1986).

Computer scientists realized that another set of "machine learning" techniques, known as genetic algorithms, were applicable to neural network training. If one treats the weights that neurons place on their inputs as genes, and a list of weights as a genome, one can employ mutation, recombination, and selection to train networks to perform a particular task (reviewed in Mitchell 1996). Neural networks were introduced into the study of animal communication when Enquist and Arak (1993) realized that tasks like the

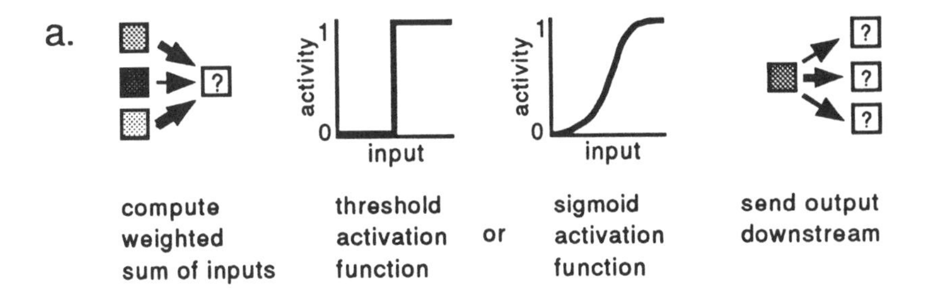

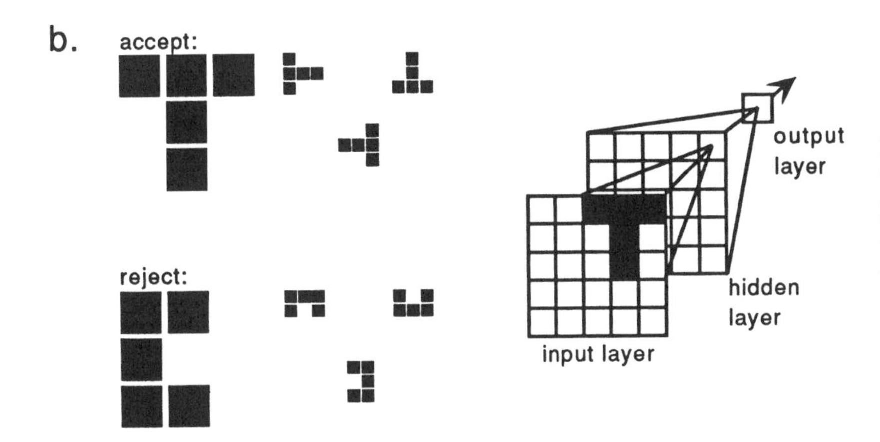

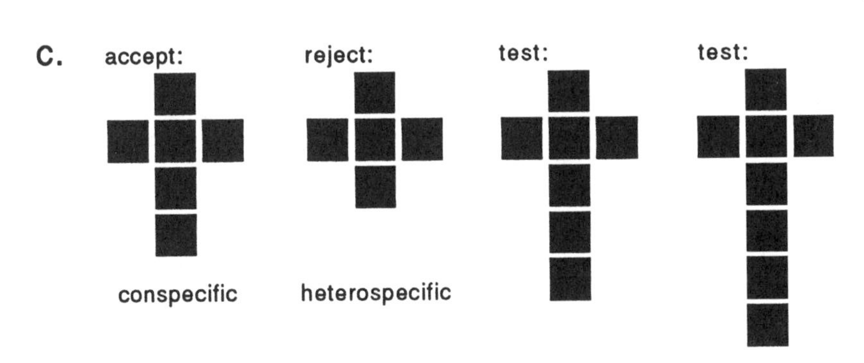

Figure 13.1. Artificial neural networks. (a) The left panel depicts neurons with differing levels of activity (gray scale) projecting to a target neuron. The target neuron (in white) weights these inputs differently (arrow thickness) and computes a weighted sum of its inputs. The center panels display two common activation functions: a threshold function and a sigmoid function. The target neuron uses one of these functions to compute its own activity based on its inputs. In the right panel, the neuron is now activated and projects to downstream neurons that weight its activity differently. (b) A feed-forward network trained to discriminate a T from a C when both letters vary in rotation and position (modified from Rumelhart, Hinton et al. 1986). (c) Stimuli used by Enquist and Arak (1993) using a feed-forward network similar to the one shown in panel b. Networks trained to discriminate between conspecific and heterospecific stimuli (left) exhibited a preference for novel stimuli with exaggerated traits (right).

T/C discrimination were basically problems in signal recognition. They used an architecture similar to the T/C feed-forward network, and replaced the back-propagation algorithm with a genetic algorithm. The authors were then able to manipulate selection pressures and observe the outcome on the neural network models.

Enquist and Arak (1993) selected networks to recognize a stimulus representing a conspecific long-tailed bird and reject a stimulus representing a heterospecific short-tailed bird (Figure 13.1c). They found that the networks exhibited an emergent preference for birds with still longer tails (but see Kamo et al. 1998). Subsequent work demonstrated that selecting networks to accommodate fluctuating asymmetry could result in networks with an emergent preference for symmetry (Enquist and Arak 1994; Johnstone 1994). A number of researchers suggest that females of various species prefer exaggerated or symmetric traits because such traits provide indicators of the genetic quality of mates (e.g., Zahavi 1975; Watson and Thornhill 1994). The significance of these network studies was twofold: They showed that networks could reproduce female preferences for exaggerated or symmetric patterns, and they demonstrated that they could do so without having been selected to cue in on the good genes these traits are often thought to signal.

Of course, a neural network model is not a brain, and the processing that occurs within the vertebrate retina is arguably more complex than a standard feed-forward network. The relative simplicity of neural networks has led many researchers to view the simulation results with caution. Admittedly, if one were primarily concerned with how the parietal cortex represents visual information or how midbrain neurons process frog calls, the neural networks that have been used in animal communication studies would be of little use. They are not designed to represent processing in a particular nervous system (but see Linsker 1986; Montague et al. 1995). In fact there is good reason for evolutionary biologists seeking general principles to avoid rooting their models in the particular neural architecture of a single species—the wiring of a nervous system is the adaptation responsible for the behavior under investigation. A detailed model of a nervous system presumes the very adaptation it aims to study. This circularity is an example of conflating the mechanistic and evolutionary causes of trait expression.

Perhaps an ideal compromise between neurobiological realism and evolutionary legitimacy would use neural networks that mimicked an architecture common to a clade of interest, allowing selection pressures to operate on finer details of the networks and then comparing the resulting networks to the nervous systems of the animals under study. To date no work of this sort has been done, and few communication systems seem tractable at this level (but see Heiligenberg 1991; Horseman and Huber 1994; Webb 1994; Webb and Hallam 1996).

An additional concern raised by researchers in animal communication is that the stimuli used to date do not adequately represent the multivariate complexity of most animal displays (Kirkpatrick and Rosenthal 1994; Dawkins and Guilford 1995). The simplicity of existing models is in many respects a major strength, and the use of actual animal signals is a daunting challenge to this simplicity. Using signals that more closely approximate natural displays has one chief advantage: it permits the comparison of network responses with the responses of living receivers. A study species can then be used as a yardstick for measuring the validity of simulation results.

Neural Networks Predict Receiver Biases

We began our studies by investigating whether neural networks could evolve to recognize a relatively simple mate-recognition signal, the call of the túngara frog, *Physalaemus pustulosus*. Previous work demonstrated that a female túngara frog responds to a conspecific call if and only if it contains the fundamental frequency sweep of the call (Ryan and Rand 1993; Wilczynski et al. 1995). In addition the wealth of information on the call preferences of túngara frog females enabled us to assess the external validity of the networks by comparing their responses to the responses of real females (Phelps and Ryan 1998).

There is at least one trait however that makes a frog call qualitatively different from stimuli presented in the neural network simulations mentioned above. Like all frog calls the túngara frog whine varies in time (see Ryan and Rand, this volume). The T of the character recognition example and the conspecific bird of the Enquist and Arak study are static signals. No time-dependent information is necessary for their recognition. To extract time-varying features from a stimulus, networks need to possess attributes that enable them to respond to current stimuli in a manner that depends upon the stimuli that preceded them. This can be accomplished by using what is known as a *recurrent neural network*. The recurrent neural network allows neurons to feedback on one another, to set up feature detectors that are activated by some precisely timed set of events.

The network architecture used in our studies (Figure 13.2a; Phelps and Ryan 1998) is a modified version of the simple recurrent network, or Elman net, a standard architecture for grammar-processing tasks in studies of human language abilities (Elman 1990). The input layer in our architecture consists of 15 neurons, each of which receives input from a distinct range of frequencies. These neurons project to the first of two hidden layers. The first hidden layer projects to neurons of the second hidden layer, which then feed back onto the neurons of the first hidden layer. The neurons of the first hidden layer have information from preceding time-steps relayed by the second hidden layer coinciding with information from current frequency inputs. This allows the extraction of the time-varying features necessary for call recognition.

A spectrogram was used to decompose the túngara frog whine into a matrix of numbers representing the relative amount of sound energy within a given range of frequencies and times. This matrix was then placed randomly in a time window large enough to house any of the calls of the *P. pustulosus* species group (Figure 13.2a). The activity of the output neuron at the end of the time window defined the response of a network to a given stimulus.

Once the network architecture has been chosen, it is possible to fully specify a network simply by making a list of the weights between all pairs of neurons. We represented this list as a bit-string that could then be subject to mutation, recombination, and selection (Figure 13.2b; see Smith et al. 1994 for code).

The evolutionary simulation begins by generating 100 networks at random and evaluating how well each network discriminates the túngara frog call from noise in the same amplitude envelope. Those networks that have an output nearest 1.0 in response to the túngara frog whine and near-

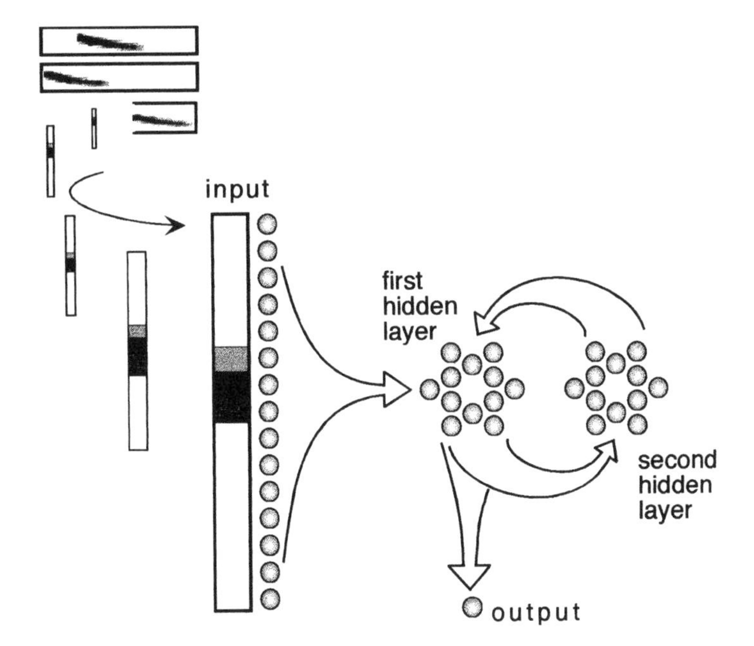

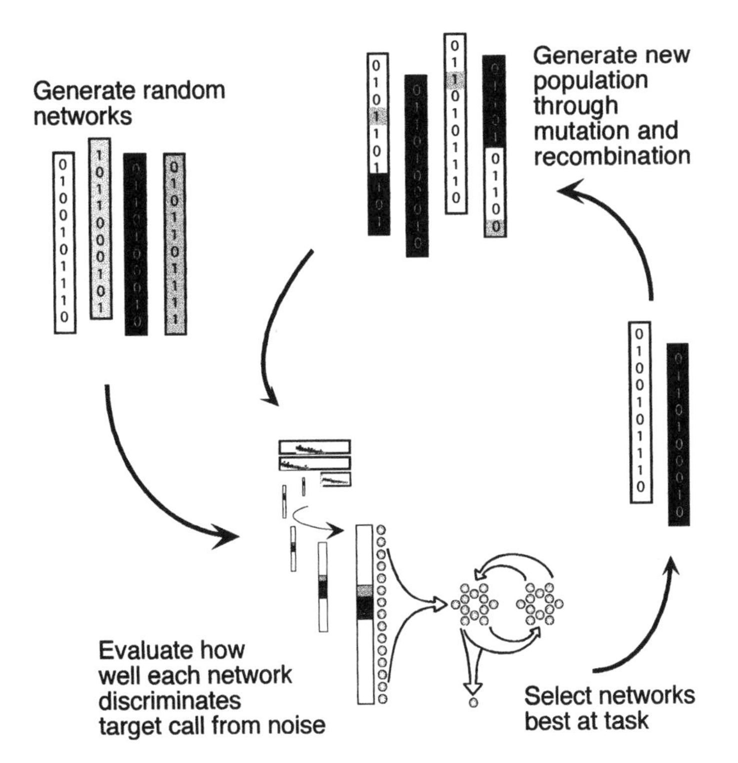

Figure 13.2. Evolving neural networks for call recognition. (a) The structure of a recurrent network that was trained to recognize the call (upper left) of the túngara frog. (b) A genetic algorithm used to train networks using mutation, recombination, and selection of networks.

est 0.0 in response to noise are assigned the highest fitness. The precise fitness function used is

$$F \equiv \sqrt{\sum_{i=1}^{n}(C_i - N_i)^2/N} + 0.01$$

where F is fitness, C_i is the response of a network to one of three examples of a particular call i, N_i is the response of networks to noise in a matching amplitude envelope, and n is the number of call examples (3) that were tested on the network.

The probability that a network would be represented in the next generation was proportional to its relative fitness. Because of this proportional selection scheme, known as roulette wheel selection, very minor differences in fitness early in a run can lead to disproportionately strong selection and subsequent loss of genetic diversity. Computer scientists refer to this as "premature convergence" (Mitchell 1996), and one of the simpler ways to deal with it is to add a small constant (0.01 in our fitness function) to the fitness of each network (Phelps and Ryan 1998).

Networks were chosen for the next generation at random, with a probability of being chosen proportional to a network's fitness. After 100 networks were selected, the networks were randomly paired to form 50 couples. Each couple had a fixed probability (0.5) of recombining to form two new networks. If the networks recombined, the "crossover" point was equally likely to occur at any point along the bit-string. Of the resulting 100 networks, each bit was mutated with a probability of 0.001. Mutation and recombination produced a daughter population with suitable variation for continued selection. The procedure continued until two consecutive generations met two criteria: the best network had a fitness of at least 0.90, and the average fitness of the population was at least 0.75.

We performed 20 runs of the simulation, and at the end of each run we recorded the weights of the best network in the population as a representative of the run. This yielded a total of 20 networks that we then tested on 34 novel stimuli representing the calls of various heterospecifics and the calls of reconstructed ancestors (for stimuli see Phelps and Ryan 1998). We used the fitness function defined above to describe how well a network discriminated the novel stimuli from noise. We then compared the network responses with responses of 20 females in phonotaxis experiments. Females were tested for phonotaxis using the recognition paradigm described by Ryan and Rand (this volume), in which a test call is played from one speaker and noise from another. The proportion of 20 females approaching the call was defined as the average female response to the stimulus.

We found that network responses were surprisingly strongly correlated with female responses ($r = 0.80$, $p < 0.001$; Phelps and Ryan 1998). Apparently selection for species recognition is sufficient to explain much of the variation in the preferences exhibited by real females. Changing the network architecture by reducing or enlarging the size of the hidden layers had no discernible effect on the predictions of networks, indicating that the predictions were robust to the manipulations of the architecture we investigated.

Historical Influences on Response Biases

Signals are multivariate in nature, and as such there are typically many strategies a receiver can use to decode a signal. Classic work from ethologists describes releasing stimuli—key components of a natural signal that are capable of eliciting a response from a receiver (Tinbergen 1951; Hauser 1996). What determines the particular features to which a receiver attends? Presumably females attend to features that convey relevant information (Grafen 1990) or features that are particularly easy to detect against environmental noise (Wiley 1983, 1994; Endler 1992), but are these really sufficient to explain receiver biases?

Given that species do not emerge spontaneously in the laboratories of systematists, one might suppose that the particular computational strategies that receivers use are molded from the strategies inherited from ancestral species. Students of animal behavior are familiar with Tinbergen's (1952) stickleback fish who performed threat displays at the sight of a postal truck. Perhaps fewer are aware of studies by McPhail (1969) on sticklebacks of the Olympic Peninsula. McPhail showed that females of an all-black species (*Gasterosteus aculeatus*) retain a strong preference for an apparently ancestral red-bellied display.

Similarly McLennan and Ryan (1997) recently found that female swordtails prefer the olfactory cues of more closely related species to those of more distantly related heterospecifics. Marler and Ryan (1997) demonstrated that females of the gynogenetic molly species *Poecilia formosa* retain a preference for large males exhibited by their two sexual ancestors *P. latipinna* and *P. mexicana,* despite the fact that the preferred males are heterospecifics who make no genetic or resource contribution to the *P. formosa* offspring. Ryan and Rand (1995, 1999, this volume) demonstrated that both call similarity and phylogenetic relatedness are significant predictors of female responses. These studies suggest that current female preferences may reflect signal recognition strategies of antecedent species.

Having demonstrated that neural network models could significantly predict the responses of female túngara frogs, we felt we could use them to assess the consequences of history on receiver responses. We manipulated the historical trajectories of various populations of networks and tested whether history influenced the particular decoding strategy used. As before we investigated the particular stimuli that elicit or fail to elicit responses from receivers to assess the underlying mechanisms of signal recognition.

We gave networks one of three history types: mimetic, mirrored, or random histories (Phelps and Ryan 2000). Mimetic histories were first trained to recognize the root call of the clade (after Ryan and Rand 1995, 1999; see Cannatella and Duellman 1984, and Cannatella et al. 1998 for phylogenetic hypotheses), but reject noise in a matching amplitude envelope. Once a population met predetermined criteria (the same as in the simulation described in the preceding section), networks were no longer selected to recognize the root call, and were instead selected to recognize the call at the next node along the phylogenetic trajectory that ultimately led to the túngara frog. Once this was performed to our criteria, networks were selected to recognize the immediate ancestor to the túngara frog, and finally the call of the túngara frog (Figures 13.3 and 13.4).

The random history consisted of three calls chosen at random from the extant and reconstructed ancestors of the *P. pustulosus* species group, followed by the call of the túngara frog. A different sequence of calls was chosen for each of 20 runs of this history type. Random histories, like the mimetic history, consisted of calls that were real or hypothesized to exist, and ended with selection for recognition of the túngara frog call. Unlike the mimetic history, the particular sequences that made up the random histories did not correspond to any hypothesized trajectory of túngara frog evolution.

Random histories were often acoustically more diverse than mimetic histories. To control for this diversity we used what we refer to as a mirrored history. Networks given a mirrored history evolved through a sequence of calls that matched the path length of each step in the mimetic history (Figure 13.3a), but the sequence originated from a different point in acoustic space. To perform this multidimensional rotation, the distance between an ancestral call and the túngara frog call was reversed in sign for 10 of 12 axes defined by a principal components analysis (PCA). (PC2 and PC4 were unmanipulated. Reversing the sign on all 12 components produced some call features that were physically impossible. Because the principal component axes are orthogonal, we were able to selectively manipulate some axes without breaking correlations normally observed among call parameters). We then solved for the original call variables and used this information to synthesize sounds corresponding to the ancestral states of mirrored history networks.

We found that all networks could evolve to recognize the call of the túngara frog. However networks with different histories displayed distinct patterns of responses to novel stimuli. The responses of mimetic history networks were significantly correlated with the responses of females in phonotaxis experiments ($r = 0.62, p < 0.01$), whereas networks with either random or mirrored histories were not (random $r = 0.20$, mirrored $r = 0.32, p > 0.05$). The correlation coefficient for the mimetic history networks was significantly better than either the random ($p < 0.01$) or the mirrored history networks ($p < 0.01$). The distinctions remain when the data are analyzed using likelihood ratio tests (Phelps and Ryan 2000), which do not make normality assumptions about the distribution of network or female responses. Apparently history does influence the response biases of neural network models: neural networks with histories lacking a natural referent exhibit the least external validity.

A more detailed analysis of network responses revealed preferences for calls resembling those of recent ancestors (Phelps and Ryan 1998). (Note I use the term *preference* to

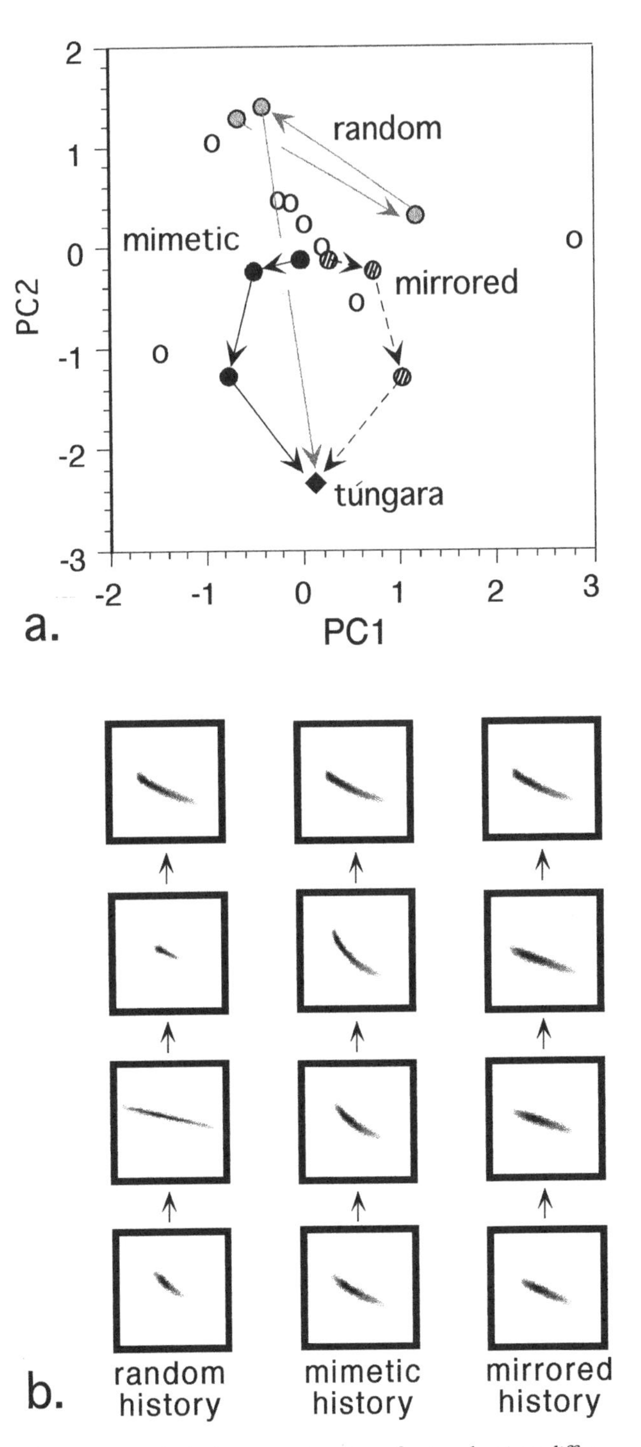

Figure 13.3. Evolutionary trajectories of networks given different histories. (a) Plot of calls in the first two principal components (PCs). Filled circles represent the trajectory taken by mimetic history networks, striped by mirrored history networks, and stippled by random history networks. The random history shown is one of 20 different random histories. Open circles represent other calls in the species group (see Figure 13.4), which were used in the random histories. (b) Sonograms of calls from each historical trajectory.

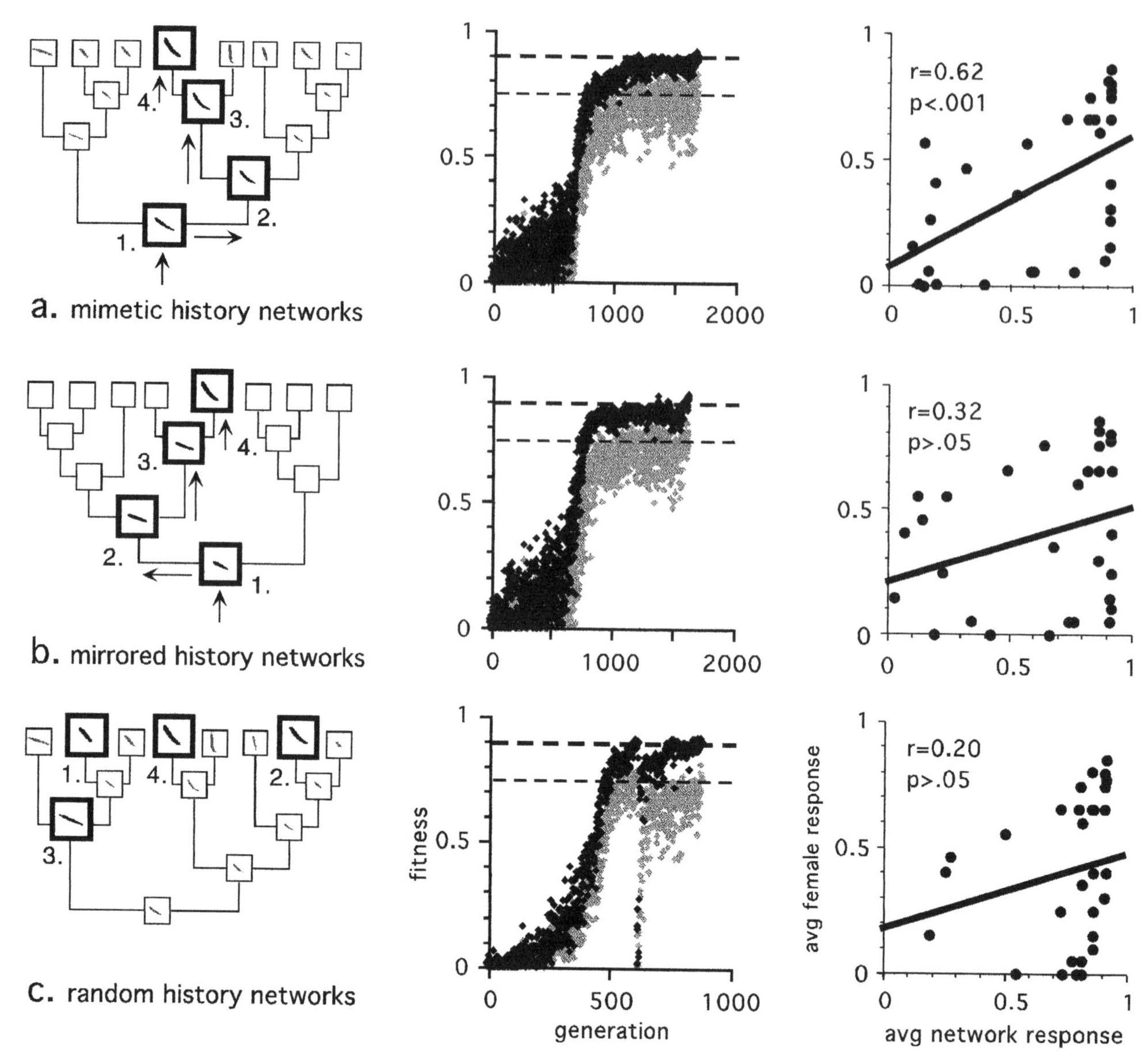

Figure 13.4. Evolution of networks along different historical trajectories. (a) The left panel represents the historical trajectory of mimetic history networks. The center panel illustrates changes in the fitness of a population of networks over one representative run. Black dots represent the fitness of the best network in a population, gray dots the average fitness of the population. The upper dashed line displays the criteria required of best networks (0.90) and the lower dashed line the criteria for average fitness (0.75; see text for explanation). The right panel displays the average network response (*x*-axis) and female response (*y*-axis) to each of 34 test stimuli. (b) Evolution and responses of neural networks provided with a mirrored history. (c) Evolution and responses of networks provided with a random history.

refer to an instance in which one stimulus is more likely to evoke a response than another stimulus, whether these responses are measured in one-choice or two-choice paradigms.) Preliminary data from additional phonotaxis experiments suggest that females share these preferences. In both recognition and discrimination tasks, females seem to prefer calls resembling the reconstruction of the most recent ancestor to nonancestral calls of matched similarity (Phelps, Ryan, and Rand, unpublished data). The possibility of detailing such vestigial preferences suggests an exciting direction for research into receiver biases that is intuitively obvious only after having performed these simulations.

Historical Effects on Receiver Permissiveness

One of the most widely documented examples of differential permissiveness among receivers is the sexually dimorphic "choosiness" of parents making unequal investments in their offspring (Trivers 1985). The primary resource provider—whether the resource is yolk volume or nest guarding—is the more selective mate. Another example occurs when sympatric populations of two or more species become less permissive than allopatric populations of the same species, presumably in response to the probability and cost of a heterospecific pairing (Fisher 1958; Gerhardt 1994; Márquez and Bosch 1997; Pfennig 1998; Rundle and Schluter 1998). These examples have much in common with discussions of receiver permissiveness in kin recognition (Sherman et al. 1997), prey detection (Dukas and Ellner 1993), and even generalization and categorization tasks (Mackintosh 1974; Ashby and Perrin 1988). All of them can be couched in terms of the tradeoffs that receivers make between the risks of inappropriate responses and missed detections. Substitute the term *false alarm* for *inappropriate response* and one finds that this compromise is ultimately equivalent to the tradeoffs predicted by signal detection theory (Wiley 1983, 1994). Despite its explanatory power, signal detection theory does not predict how selective a receiver ought to be when permissiveness does not impose costs, or when all levels of permissiveness within a range entail equivalent costs. For example the receiver that recognizes a supernormal stimulus constructed in the laboratory does not necessarily pay costs for those responses. Yet it is precisely this mismatch between the range of acceptable traits and existing variation that has been hypothesized to allow the evolution of novel signal forms (West-Eberhard 1979; Basolo 1990; Ryan 1990; Rosenthal and Evans 1998). A body of theory that addressed this disparity could prove extremely useful.

When comparing the responses of networks of different histories we noticed something surprising. Although mimetic history networks are best among the historically derived networks at predicting female responses, the mimetic, random, and mirrored histories are all worse than networks with no history at all. There seemed to be a paradox: If history is unimportant, why do mimetic history networks outperform mirrored and random history networks; and if history is important, why do mimetic history networks perform more poorly than ahistoric networks?

A closer examination of the disparities between network and female responses revealed that the networks of all types tended to overgeneralize. Permissive networks poorly predict female behavior. Random history networks, which have diverse histories, recognized most of the whines that were presented. This can be seen quite readily by noting the rightward skew in network responses in the rightmost panel of Figure 13.4c. Ahistoric networks also overgeneralize, but less. These differences occur despite the fact that all networks were selected to recognize the call of the túngara frog with equal accuracy. Perhaps receiver permissiveness is influenced by signal diversity in recent history.

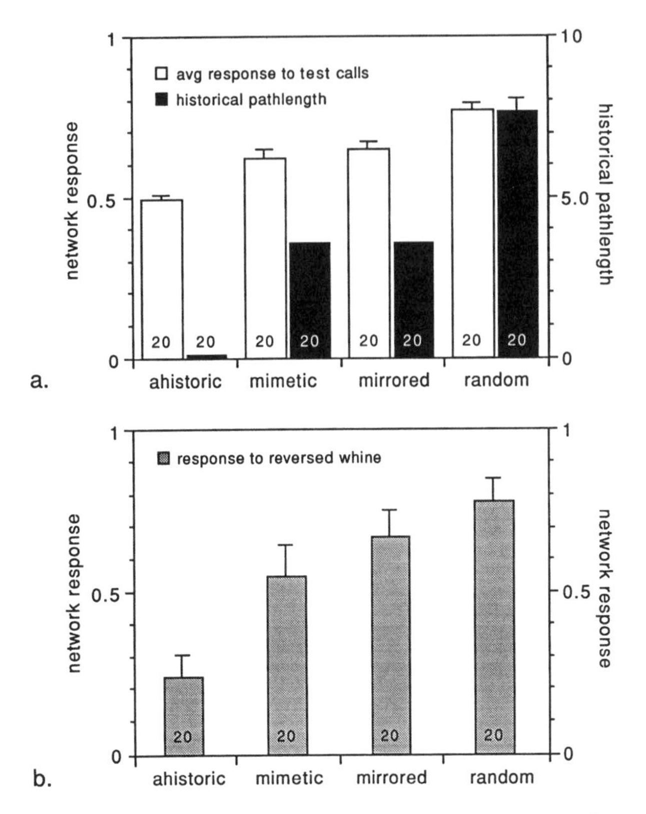

Figure 13.5. Receiver permissiveness in networks with varied histories. (a) The average response of networks to the 34 test stimuli as an index of receiver permissiveness. Networks with more diverse histories are more permissive. (b) Responses of networks to a reversed whine. Longer historical pathlengths produce networks that not only respond to more diverse descending frequency sweeps, but are more likely to respond to a novel ascending frequency sweep constructed by reversing the whine of *P. pustulosus*.

I estimated the permissiveness of the networks by measuring the average response of networks to the 34 whines that were used as test stimuli. I then measured the path length of the history by tracing the distance in three-dimensional acoustic space traversed by each history type (measured by PCA). Since the distance between any two calls in this acoustic space is a measure of dissimilarity, I take the cumulative distance that a history traverses as a measure of historical diversity. The two-dimensional plot of the history types shown in Figure 13.3a illustrates the approximate path lengths that were measured. These prove to be 0 for ahistoric networks, 3.49 for mimetic and mirrored history networks, and an average of 7.56 ($\pm$ 0.42) for random history networks (see Figure 13.5a). As predicted the path length of the history is significantly correlated with the permissiveness of networks (Kendall rank $\tau = 0.58$, corrected for ties, $p < 0.001$). Mimetic and mirrored history networks, which traversed identical path lengths through distinct regions of acoustic space, show no apparent differences in permissiveness ($p > 0.05$; two-tailed paired t-test), even though all other comparisons are significant ($p < 0.001$).

To determine whether this increase in permissiveness was simply due to similarities between the test stimuli and the ancestral signals assigned to networks, I examined network responses to a stimulus unlike any in the training set—a time-reversed whine of the túngara frog. There is again a correlation between network responsiveness and historical path length (Kendall rank $\tau = 0.42$, corrected for ties, $p < 0.001$). Networks with mirrored histories show responses similar to networks given mimetic histories ($p > 0.05$); all other comparisons (except mirrored versus random networks, $p > 0.05$) are statistically significant ($p < 0.05$).

If this permissiveness is truly the result of history, then within each historical simulation networks should become progressively more permissive over time. I tested this by looking at how well a population of networks in a run responded to a new target call the first time the target was presented. Before changing a target call, the average fitness of the population was required to reach 0.75, a level indicated by a dashed line in Figure 13.6. Since the fitness of the population is defined by how well the networks recognize the target call, I used the average fitness of the population immediately after changing the target as an estimate of how broadly the population generalized. A permissive population of receivers ought to generalize well, and so its fitness should remain near 0.75; a population of fastidious receivers should show a drop in fitness in response to a change in the target. If history increases permissiveness, the average fitness of the population should be lower immediately following the switch from call 1 to call 2 than it is following the switch from call 2 to call 3, or from call 3 to the túngara frog call. These data, taken from the 60 runs of the simulations and pooled across history types, are presented in Figure 13.6.

The permissiveness of populations does increase over time. The average fitness at the first switch (call 1 to call 2; $\mu = 0.52$, SE $= 0.03$) is significantly lower than it is at either the second (call 2 to call 3; $\mu = 0.62$, SE $= 0.03$, $p < 0.05$, two-tailed paired t-test) or the third (call 3 to the túngara frog call; $\mu = 0.69$, SE $= 0.02$, $p < 0.001$). The phenomenon is particularly salient in the case of the random histories, where the length of each step is greatest. The center panel of Figure 13.4c illustrates a fairly typical drop in population fitness at generation 620, when networks were switched from the first to the second call of this particular trajectory. Notice how relatively inconspicuous the declines are in subsequent steps of this run (switched at generations 853 and 854). These data consistently indicate that historical diversity may promote the evolution of receiver permissiveness.

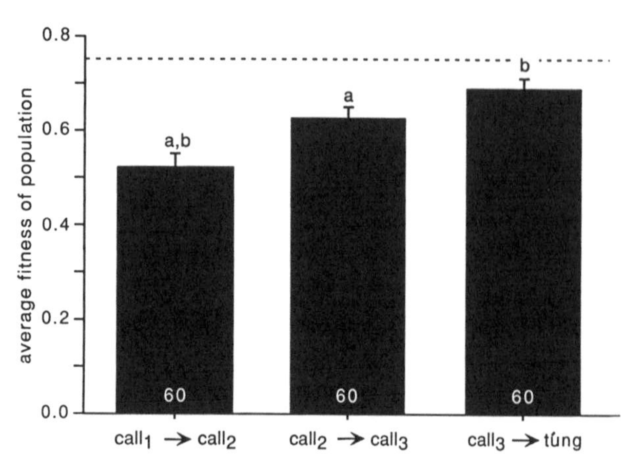

Figure 13.6. Receiver permissiveness increases over time. The ability of a receiver to generalize to a novel stimulus (measured by the average population fitness in response to a change in the target call) was measured at different stages of history. The dashed line represents the minimum average fitness of the population in the generation preceding the change in the target stimulus. When networks trained to Call 1 were switched to Call 2, there was a substantial drop in population fitness, indicating poor generalization and low permissiveness. The population fitness was significantly lower after this first switch than it was at either subsequent switch (Call 2 to Call 3, Call 3 to túngara frog call), demonstrating an increase in permissiveness as the diversity of past history increased. The numbers in each bar represent the sample size (the number of networks tested); "a" indicates bars that are significantly different from one another at $p < 0.05$; "b" indicates significance at $p < 0.001$.

Receiver History and Signal Evolution

The ability of historically derived receivers to generalize to novel stimuli has clear implications for the evolution of signal diversity. As other authors have noted, the capacity to respond to a novel signal facilitates changes in signal form and the emergence of new signal types (West-Eberhard 1979; Basolo 1990; Ryan 1990; Guilford and Dawkins 1991; Arak and Enquist 1993). Interestingly, although previous work has discussed how permissive receivers permit signal diversity, the neural network data suggest that historical diversity promotes receiver permissiveness. Both hypotheses predict a correlation between extant signal diversity and receiver permissiveness. Sensory exploitation and related models indicate that permissiveness predates signal diversity (Ryan 1990, 1998; Ryan et al. 1990; Endler and Basolo 1998), whereas the current analysis suggests that signal diversity may promote receiver permissiveness.

In one of the only studies to investigate correlations between receiver perceptual constraints and signal diversity, Ryan (1986) described a correlation between the length of the amphibian papilla (AP, which influences the range of frequencies perceivable by a frog) and the number of species in a clade sharing a particular AP morphology. He interpreted this correlation as indicating that increasing available signal space by increasing the size of the AP facilitated signal divergence; divergence in calls, a principal means of species recognition (Blair 1964), promoted speciation. The influence of historical diversity on network permissiveness suggests a new twist to this interpretation: perhaps selection for the recognition of novel signal forms—signals with a high dominant frequency, for example—drove the lengthening of the amphibian papilla. Historical diversity may produce receiver permissiveness, whereas such permissiveness facilitates further divergence. If both are at work in a given clade, then the positive feedback between signal divergence and receiver permissiveness will cause both to increase in time.

It is worth noting how distinct this putative historical influence is from the recent "chase-away" hypothesis of sexual selection, which predicts that female permissiveness declines over time (Holland and Rice 1998). In the chase-away scenario, mating with males imposes costs on females, so females are selected to become more resistant; males are selected to overcome female resistance, so their traits are elaborated, again at the expense of the most permissive females. A similar scenario is described in terms of female "coyness" and male "persuasion" by Dawkins and Krebs (1978) and in related terms by Wiley (1994) and Enquist and Arak (1998). Like most descriptions of receiver permissiveness, these scenarios share an emphasis on the costs of inappropriate receiver responses. In contrast the historical hypothesis forwarded in this chapter assumes that receiver permissiveness is not costly. Any costs of permissiveness limit the ability of history to produce broadly responsive receivers and the signal divergence that might result.

Contemporary comparative methods enable one to test these predictions by comparing the evolutionary order of permissiveness and signal innovation (Harvey and Pagel 1991; Martins 1996). It is less clear how one should compare permissiveness among species or populations. Perhaps the total range of perceivable signals in a given modality will be a useful index (as in Ryan 1986). If a "preference function" is defined as a density function describing the probability that a receiver will respond to signals varying in some dimension, the permissiveness in that dimension could also be estimated by the variance in accepted trait values. Because different species may exhibit distinct levels of motivation under common test circumstances, estimating the size of permissiveness may be particularly challenging (see Reeve 1989 for a related discussion on comparing "acceptance thresholds" of distinct populations of receivers).

Conclusions

We find that artificial neural network models selected to recognize the call of the túngara frog do a remarkably good job at predicting the responses of females to novel calls. Species recognition mechanisms appear to be sufficient to explain many of the preferences exhibited by real females.

Network preferences were sensitive to evolutionary history. Differences among the mean preferences of networks with varied histories reveal a deterministic component to this contingency: knowing the particular historical trajectory of the networks enables us to predict preferences of the descendent networks. For example network permissiveness and the ability to generalize to ancestral signals were reliably different among history treatments. However even networks resulting from replicates of the same historical sequence exhibit substantial variation in receiver biases.

The generalizations made by the different network types suggest that evolutionary history may make some novel signals more viable than others—not simply because of their ability to carry information or their ability to penetrate a noisy environment, but because once such signals reach a receiver the responses they evoke depend on the evolutionary history of the receiver's perceptual apparatus. This fits nicely with several studies suggesting that females exhibit ancestral preferences for male traits that have since been lost (McPhail 1969; Ryan and Wagner 1987; Marler and Ryan 1997; Ryan 1998). Preceding ideas about the origins of receiver biases do not predict the existence of these preferences, nor do they predict how these preferences persist in the absence of countervailing selection.

Artificial networks have been used to model processes of learning and generalization long before their application to animal communication (McClelland and Rumelhart 1985; Rumelhart and McClelland 1986a, 1986b). Based on the results of network studies, some authors have gone on to use a second type of generalization models—gradient interaction models—to investigate the origins of symmetry preferences (Enquist and Johnstone 1997). Perhaps similar models of generalization and extinction can be used to predict the persistence of vestigial preferences in extant taxa. One can imagine how such a model would make predictions regarding the relationship between branch lengths to ancestral nodes and the preference functions of extant taxa.

As mentioned in the introduction to this chapter, the models presented here are not designed to emulate the anuran auditory system. Nevertheless correlations between network and female responses prompt one to ask whether networks are recognizing the call of the túngara frog with strategies that have been proposed for the recognition of other frequency-modulated sounds. Of particular interest is whether neurons will exhibit any of the complex response properties of the torus semicircularis or auditory thalamus, regions that mediate call recognition in some species (Fuzessery and Feng 1982, 1983). Responses to frequency-modulated calls, for example, are thought to rely on neurons that respond selectively to combinations of frequencies present in the call (Fuzessery and Feng 1982, 1983). Discovering whether such neuron types are present, and how they are modified by history, may yield testable predictions regarding the neurophysiological substrates of vestigial preferences.

In short, neural network models enable us to conduct elaborate simulation experiments into the evolution of communication systems. They allow us to control selection pressures, perceptual capabilities, and genetic architectures, and to perform the replicates needed to discern evolutionary patterns. They are not adequate substitutes for neurophysiology, behavioral experiments, or analytic models. They are, however, useful tools that can help us formulate testable hypotheses regarding the behaviors of animals that interest us.

Acknowledgments

I would like to thank Stan Rand, Mike Ryan, and Walt Wilczynski for providing thoughtful discussion of this work and its presentation. Rand and Ryan provided the behavior data I present in this chapter, as well as financial support from NSF grant IBN 9316185. Additional support came from predoctoral and postdoctoral fellowships awarded by the Smithsonian Institution, and by a predoctoral fellowship from NIMH F31 MH11194. Lastly, I am indebted to Pete Hurd, Nicole Kime, Risto Miikkulainen, and Gil Rosenthal who reviewed this chapter and improved it immensely.

References

Arak, A., and M. Enquist. 1993. Hidden preferences and the evolution of signals. Philosophical Transactions of the Royal Society of London, series B 340:207–213.

Ashby, F. G., and N. A. Perrin. 1988. Toward a unified theory of similarity and recognition. Psychological Review 95:124–150.

Basolo, A. 1990. Female preference predates the evolution of the sword in swordtails. Science 250:808–810.

Blair, W. F. 1964. Isolating mechanisms and interspecies interactions in anuran amphibians. Quarterly Review of Biology 39:334–344.

Brenowitz, E. A. 1982. The active space of red-winged blackbird song. Journal of Comparative Physiology, series A 147:511–522.

Cannatella, D.C., and W. E. Duellman. 1984. Leptodactylid frogs of the *Physalaemus pustulosus* group. Copeia 1984:902–921.

Cannatella, D.C., D. M. Hillis, P. T. Chippindale, L. Weight, A. S. Rand, and M. J. Ryan. 1998. Phylogeny of frogs of the *Physalaemus pustulosus* species group, with an examination of data incongruence. Systematic Biology 47:311–335.

Dawkins, M. S., and T. Guilford. 1995. An exaggerated preference for simple neural network models of signal evolution? Proceedings of the Royal Society of London, B 261:357–360.

Dawkins, R., and J. R. Krebs. 1978. Animal signals: Information or manipulation? Pp. 282–309. In J. R. Krebs and N. B. Davies (eds.), Behavioural Ecology: An Evolutionary Approach. Blackwell Scientific, Oxford, UK.

Dukas, R., and S. Ellner. 1993. Information processing and prey detection. Ecology 74:1337–1346.

Elman, J. L. 1990. Finding structure in time. Cognitive Science 14:179–211.

Endler, J. A. 1992. Signals, signal conditions, and the direction of evolution. American Naturalist 139:S125-S153.

Endler, J. A., and A. L. Basolo. 1998. Sensory ecology, receiver biases and sexual selection. Trends in Ecology and Evolution 13:415–420.

Enquist, M., and A. Arak. 1993. Selection of exaggerated male traits by female aesthetic senses. Nature 361:446–448.

Enquist, M., and A. Arak. 1994. Symmetry, beauty and evolution. Nature 372:169–172.

Enquist, M., and A. Arak. 1998. Neural representation and the evolution of signal form. Pp. 21–88. *In* R. Dukas (ed.), Cognitive Ecology. University of Chicago Press, Chicago.

Enquist, M., and R. Johnstone. 1997. Generalisation and the evolution of symmetry preferences. Proceedings of the Royal Society, series B. 264:1345–1348.

Fisher, R. A. 1958. The Genetical Theory of Natural Selection. Dover Publications, New York.

Fuzessery, Z. M., and A. S. Feng. 1982. Frequency selectivity in the auditory midbrain: Single unit responses to single and multiple tone stimulation. Journal of Comparative Physiology, series A 146:471–484.

Fuzessery, Z. M., and A. S. Feng. 1983. Mating call selectivity in the thalamus and midbrain of the leopard frog (*Rana p. pipiens*): Single and multiunit analyses. Journal of Comparative Physiology, A 150:333–344.

Gerhardt, H. C. 1994. Reproductive character displacement of female mate choice in the grey treefrog, *Hyla chrysoscelis*. Animal Behaviour 47:959–969.

Grafen, A. 1990. Biological signals as handicaps. Journal of Theoretical Biology 144:517–546.

Greenfield, M. D., and I. Roizen. 1993. Katydid synchronous chorusing is an evolutionarily stable outcome of female choice. Nature 364:618–620.

Greenfield, M. D., M. K. Tourtellot, and W. A. Snedden. 1997. Precedence effects and the evolution of chorusing. Proceedings of the Royal Society of London, B 264:1057–1063.

Guilford, T., and M. S. Dawkins. 1991. Receiver psychology and the evolution of animal signals. Animal Behaviour 42:1–14.

Harvey, P. H., and M. D. Pagel. 1991. The Comparative Method in Evolutionary Biology. Oxford University Press, New York.

Hauser, M. D. 1996. The Evolution of Communication. MIT Press, Cambridge, MA.

Haykin, S. 1994. Neural Networks: A Comprehensive Foundation. Macmillan Press, New York.

Heiligenberg, W. 1991. Neural Nets in Electric Fish. MIT Press, Cambridge, MA.

Holland, B., and W. R. Rice. 1998. Chase-away sexual selection: Antagonistic seduction versus resistance. Evolution 52:1–7.

Horseman, G., and F. Huber. 1994. Sound localisation in crickets II: Modeling the role of a simple neural network in the prothoracic ganglion. Journal of Comparative Physiology, A 175:399–413.

Johnstone, R. A. 1994. Female preference for symmetrical males as a by-product of selection for mate recognition. Nature 372:172–175.

Kamo, M., T. Kubo, and Y. Iwasa. 1998. Neural network for female mate preference, trained by a genetic algorithm. Philosophical Transactions of the Royal Society, London, B 353:399–406.

Kirkpatrick, M., and G. Rosenthal. 1994. Symmetry without fear. Nature 372:134–135.

Krebs, J. R., and R. Dawkins. 1984. Animal signals: Mind reading and manipulation. Pp. 380–402. *In* J. R. Krebs and N. B. Davies (eds.), Behavioural Ecology: An Evolutionary Approach. Oxford University Press, Oxford, UK.

Leimar, O., M. Enquist, and B. Silen-Tullberg. 1986. Evolutionary stability of aposematic coloration and prey unprofitability: A theoretical analysis. American Naturalist 128:469–490.

Linsker, R. 1986. From basic network principles to neural architecture (series). Proceedings of the National Academy of Sciences, USA 83:7508–7512, 8390–8394, 8779–8783.

Littlejohn, M. J. 1977. Long range acoustic communication in anurans: An integrated evolutionary approach. Pp. 263–294. *In* D. H. Taylor and S. I. Guttman (eds.), The Reproductive Biology of Amphibians. Cambridge University Press, London, UK.

MacDougall, A., and M. Stamp Dawkins. 1998. Predator discrimination error and the benefits of Müllerian mimicry. Animal Behaviour 55:1281–1288.

Mackintosh, N.J. 1974. The Psychology of Animal Learning. Academic Press, New York.

Marler, C. A., and M. J. Ryan. 1997. Origin and evolution of a female mating preference. Evolution 51:1244–1248.

Martins, E. P. (ed.) 1996. Phylogenies and the Comparative Method in Animal Behavior. Oxford University Press, New York.

Márquez, R., and J. Bosch. 1997. Male advertisement call and female preference in sympatric and allopatric midwife toads. Animal Behaviour 54:1333–1345.

McClelland, J. L., and D. E. Rumelhart. 1985. Distributed memory and the representation of general and specific information. Journal of Experimental Psychology, Learning 114:159–188.

McLennan, D. A., and M. J. Ryan. 1997. Responses to conspecific and heterospecific olfactory cues in the swordtail *Xiphophorus cortezi*. Animal Behaviour 54:1077–1088.

McPhail, J. D. 1969. Predation and the evolution of a stickleback (*Gasterosteus*). Journal of the Fisheries Research Board, Canada 26:183–208.

Mitchell, M. 1996. An Introduction to Genetic Algorithms. MIT Press, Cambridge, MA.

Montague, P. R., P. Dayan, C. Person, and T. J. Sejnowksi. 1995. Bee foraging in uncertain environments using predictive Hebbian learning. Nature 377:725–728.

Pfennig, K. S. 1998. The evolution of mate choice and the potential for conflict between species and mate quality recognition. Proceedings of the Royal Society of London, series B 265:1743–1748.

Phelps, S. M., and M. J. Ryan. 1998. Neural networks predict the response biases of female túngara frogs. Proceedings of the Royal Society of London, series B 265:279–285.

Phelps, S. M., and M. J. Ryan. 2000. History influences signal recognition: Neural network models of túngara frogs. Proceedings of the Royal Society of London B 267:1633–1699.

Reeve, H. K. 1989. The evolution of conspecific acceptance thresholds. American Naturalist 133:407–435.

Römer, H., and W. J. Bailey. 1986. Insect hearing in the field: II. Male spacing behavior and correlated acoustic cues in the bushcricket *Mygalopsis marki*. Journal of Comparative Physiology, series A 159:627–638.

Rosenthal, G. G., and C. S. Evans. 1998. Female preference for swords in *Xiphophorus helleri* reflects a bias for large apparent size. Proceedings of the National Academy of Sciences, USA 95:4431–4436.

Rumelhart, D. E., G. E. Hinton, and R. J. Williams. 1986. Learning internal representations by error propagation. Pp. 318–362. *In* D. E. Rumelhart and J. L. McClelland (eds.), Parallel Distributed Processing: Explorations in the Microstructure of Cognition, Vol. 1, Foundations. MIT Press, Cambridge, MA.

Rumelhart, D. E., J. L. McClelland, and the PDP Research Group. 1986a. Parallel Distributed Processing: Explorations in the Microstructure of Cognition, Vol. 1, Foundations. MIT Press, Cambridge, MA.

Rumelhart, D. E., J. L. McClelland, and the PDP Research Group. 1986b. Parallel Distributed Processing: Explorations in the Microstructure of Cognition, Vol. 2, Psychological and Biological Models. MIT Press, Cambridge, MA.

Rundle, H. D., and D. Schluter. 1998. Reinforcement of stickleback mate preferences: Sympatry breeds contempt. Evolution 52:200–208.

Ryan, M. J. 1986. Neuroanatomy influences speciation rates among anurans. Proceedings of the National Academy of Sciences, USA 83:1379–1382.

Ryan, M. J. 1990. Sexual selection, sensory systems, and sensory exploitation. Oxford Surveys in Evolutionary Biology 7:157–195.

Ryan, M. J. 1998. Receiver biases, sexual selection and the evolution of sex differences. Science 281:1999–2003.

Ryan, M. J., J. H. Fox, W. Wilczynski, and A. S. Rand. 1990. Sexual selection for sensory exploitation in the frog *Physalaemus pustulosus*. Nature 343:66–67.

Ryan, M. J., and A. S. Rand. 1993. Species recognition and sexual selection as a unitary problem in animal communication. Evolution 47:647–657.

Ryan, M. J., and A. S. Rand. 1995. Female responses to ancestral advertisement calls in túngara frogs. Science 269:390–392.

Ryan, M. J., and A. S. Rand. 1999. Phylogenetic influence on mating call preferences in female túngara frogs. Animal Behaviour 57:945–957.

Ryan, M. J., and W. E. Wagner. 1987. Asymmetries in mating preferences between species: Female swordtails prefer heterospecific mates. Science 236:595–597.

Sejnowski, T. J., and C. R. Rosenberg. 1987. Parallel networks that learn to pronounce English text. Complex Systems 1:145–168.

Sherman, P. W., H. K. Reeve, and D. Pfennig. 1997. Recognition systems. Pp. 69–96. *In* J. R. Krebs and N. B. Davies (eds.), Behavioural Ecology: An Evolutionary Approach, 4th ed. Blackwell Scientific Publications, London, UK.

Smith, R. E., D. E. Goldberg, and J. A. Earickson. 1994. SGA-C: A C-language implementation of a simple genetic algorithm. Available from ftp://www.aic.nrl.navy.mil/pub/galist/src/sca-c.tar.z.

Tinbergen, N. 1951. The Study of Instinct. Oxford University Press, Oxford, UK.

Tinbergen, N. 1952. The curious behavior of the stickleback. Pp. 5–9. *In* Psychobiology, the Biological Bases of Behavior: Readings from Scientific American. Freeman and Company, San Francisco, CA.

Trivers, R. 1985. Social Evolution. Benjamin/Cummings, Menlo Park, CA.

Watson, P. J., and R. Thornhill. 1994. Fluctuating asymmetry and sexual selection. Trends in Ecology and Evolution 9:21–25.

Weary, D. M., T. C. Guilford, and R. G. Weisman. 1993. A product of discrimination learning may lead to female preferences for elaborate males. Evolution 47:333–336.

Webb, B. 1994. Robotic experiments in cricket phonotaxis. Pp. 45–54. *In* From Animals to Animats 3. Proceedings of the 3rd International Conference on Simulation of Adaptive Behavior. MIT Press, Cambridge, MA.

Webb, B., and J. Hallam. 1996. How to attract females: Further robotic experiments in cricket phonotaxis. Pp. 75–83. *In* From Animals to Animats 4. Proceedings of the 4th International Conference on Simulation of Adaptive Behavior. MIT Press, Cambridge, MA.

West-Eberhard, M. J. 1979. Sexual selection, social competition, and evolution. Proceedings of the American Philosophical Society 123:222–234.

Wilczynski, W., A. S. Rand, and M. J. Ryan. 1995. The processing of spectral cues by the call analysis system of the túngara frog. Animal Behaviour 49:911–929.

Wiley, R. H. 1983. The evolution of communication: Information and manipulation. Pp. 156–189. *In* Animal behavior, Vol. 2, Communication. Blackwell Scientific Publications, Oxford, UK.

Wiley, R. H. 1994. Errors, exaggeration and deception in animal communication. Pp. 157–189. In L. Real (ed.), Behavioral Mechanisms in Evolutionary Ecology. University of Chicago Press, Chicago.

Yachi, S., and M. Higashi. 1998. The evolution of warning signals. Nature 394:882–884.

Zahavi, A. 1975. Mate selection: A selection for a handicap. Journal of Theoretical Biology 53:205–214.

PART FIVE

Behavior and Evolution

14

JOSHUA J. SCHWARTZ

Call Monitoring and Interactive Playback Systems in the Study of Acoustic Interactions among Male Anurans

Introduction

Communication is a dynamic process in which a signal given by one individual can influence the behavior of another (Wiley 1994). Often communicative interactions involve the reciprocal exchange of signals as well as changes in the character of signals (Bradbury and Vehrencamp 1998). If many individuals are signaling within a shared active space, the details of the interactions may be quite complex. Additional complexities are imposed by changing social conditions, the presence of background noise, interference of signals, and the fact that the same signal may transmit different messages and have different meanings to different receivers. The acoustic environment in a chorus of frogs is characterized by just such complexities (Littlejohn 1977; Narins and Zelick 1988; Wells 1988; Gerhardt and Schwartz 1995). Nevertheless males successfully advertise information on their location, species identity, and perhaps even condition, not only to other males but also to gravid females that they are attempting to attract (Gerhardt 1994). How males communicate such information may involve the use of adjustments in call timing, elaboration of calls, exchange of graded signals, and selective attention to a subset of chorus members. Moreover vocal competition among males in choruses can be intense, and males need to use flexible calling strategies and favorably distinguish themselves from competitors while also maintaining sufficient reserves of energy to fuel signaling for those periods of time in which females are available (Schwartz et al. 1995).

Although courtship calls may be directed at an adjacent member of the opposite sex before pairing in some species (e.g., Given 1993; Bush 1997; Ovaska and Caldbeck 1997), advertisement calls are the principal signal type used by males to attract females in choruses of anurans (Wells 1977a, 1977b; Gerhardt 1994). In fact selection of a mate by females from among those conspecific males available on a given night is based largely on features of the advertisement call or the manner in which it is given. These features may be spectral (e.g., dominant frequency), temporal (e.g., call duration), or females may even evaluate such calling attributes as call rate and call complexity (Gerhardt 1994; Sullivan et al. 1995). If a male is to be reproductively successful, he must effectively transmit those attributes of his signals or signaling behavior relevant to decision-making by females, often in a noisy environment. He must also be aware of the calling behavior of his neighbors and respond rapidly to changes in their vocal performances to maintain his relative attractiveness. In addition males may not tolerate the loud calls of nearby individuals and respond to these with aggressive calls (Brenowitz and Rose 1994). Aggressive calls also may be exchanged during physical interactions (Wagner 1989a). Both advertisement calls and aggressive calls may be simple, consisting of a single note or notes of one type or complex, formed from notes of different form and even function (Narins and Capranica 1978; Littlejohn and Harrison 1985; Wells and Bard 1987; Backwell 1988; Jehle and Arak 1998). During vocal exchanges male frogs may alter call rate, complexity, duration, and intensity (Lopez et al. 1988; Wells 1988). There is evidence that males of a few species change their call frequency during interactions (Lopez et al. 1988; Wagner 1989a, 1989b; Grafe 1995; Bee and Perrill 1996; Howard and Young 1998; Bee et al. 1999; Given 1999), perhaps to deceive opponents during assessment of size or to increase their relative attractiveness to females.

How male anurans communicate in a chorus environment has been the focus of much research (Narins and Zelick 1988; Wells 1988) and is one focus of this chapter. I concentrate on recent studies of how males shift the timing of calls or call elements, adjust qualities of their advertisement calls, and use aggressive calls. I use primarily examples from those species that I have studied extensively: *Pseudacris crucifer, Hyla microcephala,* and *H. versicolor.* Interspecific interactions receive little attention in this chapter, and interested readers are referred to the review of Gerhardt and Schwartz (1995). A challenge for students of animal communication has been to develop effective techniques for data acquisition from interacting individuals and the presentation of signals in a way that elucidates how and why they are used. Accordingly a second focus of this chapter is how computer hardware has been and can be used to answer questions about communication in the complex acoustic environment of anuran choruses. I focus on systems that allow researchers to simultaneously monitor the vocal activity of more than two members of a chorus and interactive playback methods that facilitate subject-driven presentation of acoustic stimuli. I discuss how multichannel systems can reveal call-timing interactions that reduce acoustic interference with just a subset of chorus members, help us identify acoustic criteria used by females during mate choice, and illustrate how energetic constraints and male vocal competition may shape chorusing dynamics. I also describe how interactive playback can be used to dissect the details of call-timing behaviors in a way that may illuminate their underlying mechanisms and functions and enable researchers to set up vocal contests between a male and a simulated competitor in investigations of aggressive calling.

Computer-Based Monitoring and Playback Techniques

When I began research on anuran communication in 1980 I studied male–male acoustic interactions with playback experiments and by making stereo tape recordings of male–male vocal exchanges. In the playback tests males were presented with prerecorded calls of one type and their responses, along with the stimulus calls, were tape-recorded on a second machine. Analysis of the recordings was extremely tedious and involved listing each subject's call types and measuring the timing of notes on a storage oscilloscope. Since the late 1980s I have employed a computer-based system to conduct interactive playback experiments and monitor the vocal interactions among male frogs. With a computerized record of experiments, the analysis of data is relatively expeditious. Therefore within a field season I can design additional experiments or modify my testing protocol based on my results. Another important advantage of a computer-based approach is that one is not constrained by the unnatural form of a fixed sequence of tape-recorded playback stimuli. Rather, with computer-controlled playback, one has considerable flexibility in both the timing and the type of stimuli that are broadcast to the subjects. Furthermore, with computer-based monitoring, it is relatively easy to obtain a record of the calls given by many more males than just the two individuals that are possible with a stereo tape recorder. Because males within choruses often call relatively near one another, a multichannel system allows one to determine who is and who is not interacting vocally. The spatial structure of vocal interactions will shape the pattern of vocal competition among males and also has important consequences for acoustic interference within the aggregation.

Call-Monitoring Systems

Two general approaches have been used recently to monitor the vocal activity of more than two males in what McGregor and Dabelsteen (1996) have called a "communication network." In the first of these, individual microphones are placed near each male and an electronic interface transforms their analog signals before they are either tape-recorded or sampled by a computer. For example, in the system used by Brush and Narins (1989) to study the chorusing dynamics of *Eleutherodactylus coqui,* above-threshold sound production by each of up to eight subjects triggered a tone pulse at a unique frequency by a multichannel encoder. Recordings of these tones were decoded using bandpass filters and additional circuitry that produced a voltage change

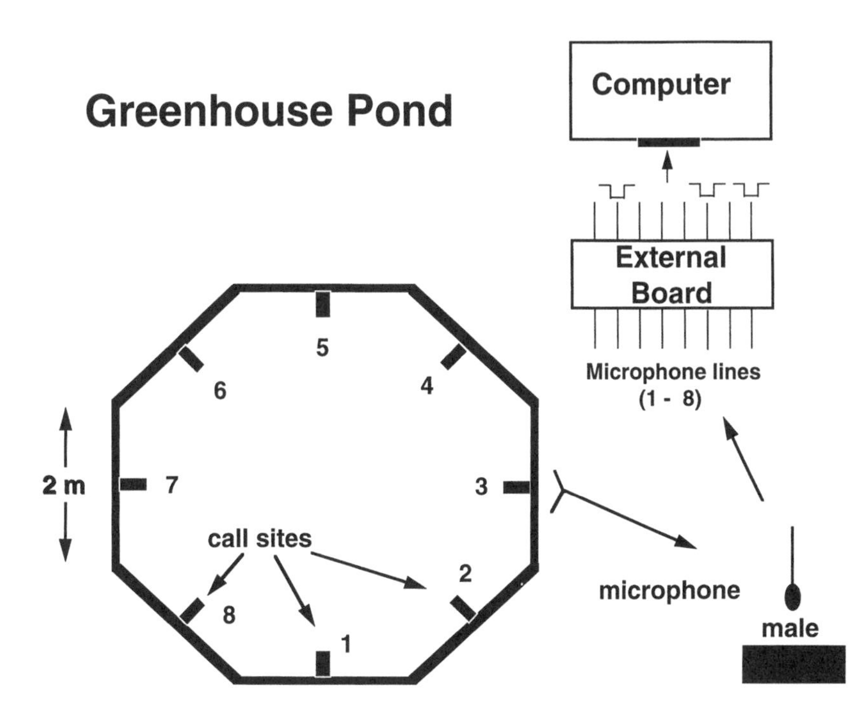

Figure 14.1. Schematic diagram of the artificial pond in the greenhouse as seen from above. The calls of up to eight males are monitored with a computer and an eight-channel custom-built interface board. A unidirectional microphone is suspended above each perch and a cable is run to the computer interface box outside the enclosure. If frogs call at positions 2, 6, and 8 in the pond, high to low voltage transitions, as shown, will be detected on the corresponding data lines of the computer's parallel port. The call monitoring system has also been used to monitor males in the field.

in response to an above-threshold signal at a particular frequency. These voltage changes were detected by a computer sampling the eight data lines of its parallel port and were stored as either zero or one. Therefore each of the eight-bit values in every byte of data indicated whether a male was calling at a particular location for any time sample.

The approach I have used with *Hyla microcephala* in the field (see Schwartz 1993 for additional details) and *H. versicolor* in an artificial pond is similar but avoids the frequency encoding and decoding steps used by Brush and Narins. With my system, up to eight microphones are connected to a battery-powered interface board that sends its output to a computer via the parallel port (Figure 14.1). The output for each of eight parallel channels of the board undergoes a voltage transition, of preset duration, in response to above-threshold input from a microphone. By adjusting the duration of the output pulse from the board and the sampling rate of the computer, it is possible to change the temporal resolution of the system. Thus calls, notes, or even individual pulses within calls may be discerned, and the specific frogs that are calling, or not calling, at a particular time are indicated in the pattern of zeros and ones in each byte of data. Because the interface board does not encode the spectral information in calls, the high sampling rate that would be needed to store such data is unnecessary. I have found that a sampling rate of 100 or 200 Hz is adequate to answer questions about call duration or note timing, and so it is possible to record many hours of data with relatively small amounts of computer memory.

Conceptually similar systems have been used in other studies of either anurans (Passmore et al. 1992) or chorusing insects (Bertram and Johnson 1998; Michael Greenfield, personal communication). A difference between these systems and my own is that the analog signals from each microphone are sampled sequentially rather than simultaneously and the data multiplexed in the computer record. Although the hardware used by the first two research groups cited limited their observations to a maximum of eight males, in principle an unlimited number of individuals could be monitored. Maintaining temporal resolution requires an increase in sampling rate with an increase in the number of data channels multiplexed. Moreover this method requires that each time sample be correctly identified as coming from a particular channel. If any glitches in computer timing occur they can cause a serious problem during analysis.

While working on the African painted reed frog (*Hyperolius marmoratus broadleyi*), Grafe (1997) used a second general approach originally developed to monitor the songs of cetaceans in the open sea (Clark 1989; Spiesberger and Fristrup 1990). He was able to use differences in the time of arrival of male calls at three microphones at the edge of a chorus to calculate both the location of each individual and its calling rate. McGregor and Dabelsteen (1996) and McGregor et al. (1997) discuss in more detail advantages and limitations of acoustic localization systems (ALS) as applied in studies of avian acoustics.

Interactive Playback

During interactive playback the broadcast of an experimental stimulus is triggered, after a preset delay, by the calls of the subject (Pina and Channing 1981). If an analog synthesizer is used the kinds of sounds produced can be seriously constrained by the electronics of the device, although careful design can allow for the creation of a variety of stimuli that can be used with different species (Narins and Capranica 1978; Narins 1982a; Schwartz 1987a, 1989). Moore et al. (1989) used a custom digital recording and playback device that allowed call-triggered playback of calls previously digitized and stored in RAM. Under circumstances requiring a clone of a natural stimulus call, this kind of system has some advantages over an analog synthesizer, but for greatest flexibility a PC-based approach should be used. The use of a computer (with sound hardware built in or added as a peripheral) essentially frees the researcher from the limitations on signal form imposed by an analog synthesizer, while allowing playback of digitized natural or synthetic calls if they are required.

Use of a computer also provides the researcher with many options for the delivery of stored sounds to the subject (Table 14.1). These sounds might be broadcast after a key is pressed or delivered automatically. Interactive playback experiments with birds have until now relied on the first form of delivery (reviewed in Dabelsteen and McGregor 1996). For example the "SingIt" program written by Bradbury and Vehrencamp (1994) responds to strikes of the keyboard or clicks of the mouse and can be run on a Macintosh PowerBook in the field. A system in which the computer responds automatically should be used for experiments in which very rapid or low-variability stimulus-response times are required. In a simple scheme for automatic stimulus delivery, one could have the computer respond after a fixed time delay following the detection of sound produced by the subject above a preset intensity. Following a programmed "dead" time, the computer could be triggered again. The interactive playback "effect module" (for SoundMaker software running on Apple computers) written by Alberto Riccio in collaboration with Rafael Márquez (Márquez et al. 1998) permits such tests. In a more complicated experiment the computer could identify different call or note types produced by a subject and then respond to each with a different kind of stimulus (e.g., Schwartz 1994). One could even have the computer digitize a subject's calls and echo these calls back to him. Unfortunately it would be difficult to design a software package that could accommodate the range of experiments that many different users might desire. Learning how to program the hardware of the computer used in the field will often be a necessary chore for those requiring a complex or idiosyncratic algorithm. This is the approach I have taken for interactive playback using Commodore Amiga computers.

Table 14.1 Some options for interactive playback

Trigger Reference	Response Delay	Response Type	
Call start	Fixed	Matching	Different
Call end	Variable	Single	Multiple
Specific call	Random	Discrete	Graded
Call element		Interrupting	Alternating
Alternate calls		Stored	Echo of subject
		Constant amplitude	Variable amplitude

Types of Vocal Interactions

Fine-Scale Call Timing Adjustments by Males

For communication to occur, information must be transmitted from a signaler in a form recognizable to a receiver. If abiotic noise or the signals of other organisms attempting to use the same communication channel obscure important elements of a call, information transfer will be impaired. One possible solution to this problem of interference is to move away from the noise source. Another is to call loudly. A caller may also make his signals redundant—that is, call a lot or use a call with copies of important elements. However anuran amphibians often call in choruses, and in dense aggregations of conspecifics this response may be inadequate. Not only are different males calling at similar frequencies, but also they typically elevate their rate of signal production (but see Sullivan and Leek 1986) because females may approach the most vigorous callers. In multispecies assemblages, heterospecific males may also pose a serious noise problem if the frequencies of their calls fall within the critical bands surrounding the frequencies of the conspecific call (Scharf 1970; Schwartz and Wells 1983a, 1983b).

Males of diverse taxa of chorusing organisms adjust the timing of calls or notes to those of other individuals (Klump and Gerhardt 1992; Greenfield 1997), and this behavior often mitigates the problems of masking and other forms of acoustic interference (e.g., Schwartz 1987a). Nevertheless there are different adaptive explanations for certain temporal adjustments (see below). Call-timing interactions in anurans have been studied extensively using descriptive and empirical approaches (for review see Klump and Gerhardt 1992). Although the calling of some species may have little or no effect on the calling of others (e.g., calls of *Rana catesbeiana* on *Hyla versicolor,* Schwartz et al., unpublished data), careful analysis of recordings can reveal nonrandom patterns of call or note timing among conspecific or heterospecific males. Moreover the temporal relationships can be influenced profoundly by the spatial relationships among callers (Schwartz 1993).

Male anurans may answer the individual vocalizations of conspecific or heterospecific males with short-latency responses or react with small shifts in the timing of their calls. Although this behavior can produce overlapping calls or notes as in *Hyla ebraccata* (Wells and Schwartz 1984a), *Smilisca sila* (Ryan 1986), or *Centrolenella granulosa* (Ibáñez 1993), the resulting "entrained" calls often fall in the intercall (or internote) intervals of the stimulating male and may produce a pattern of alternation. For example, Schwartz and Wells (1984a, 1984b, 1985) demonstrated, using playbacks of recorded vocalizations, that males of *H. ebraccata, H. phlebodes,* and *H. microcephala* made similar adjustments in the timing of their calls to calls of conspecifics and congeners. Males often answered stimulus calls within a few hundred milliseconds, producing a loose pattern of alternation. A particularly impressive study is that of Zelick and Narins (1985) on the Puerto-Rican species *Eleutherodactylus coqui* and it demonstrates the importance that gap detection may have on the relative call timing of interacting males. Zelick and Narins presented males with long and short tone-bursts in which they had placed 750 ms gaps in a pattern that should have been unpredictable to their subjects. These gaps were just long enough for a male to place a Co and a Qui note. Nevertheless male "coquis" had little difficulty with a task that proved beyond the abilities of human subjects: two-note calls were placed in the intervals without overlapping the tones. An additional experiment demonstrated that males could detect and respond to reductions in tone intensity at least as small as 4 dB SPL (Narins and Zelick 1988). Grafe (1996) has demonstrated that *Hyperolius marmoratus broadleyi* males are also quite adept at placing their calls within brief gaps during tone broadcasts.

My studies of *Hyla microcephala* demonstrated that detection of brief drops in background noise intensity may contribute to the ability of males to avoid acoustic interference (Schwartz 1993). Males of this species produce calls that may contain over 15 notes, and because males answer their neighbors with short-latency responses, calls of vocally interacting individuals often overlap. Inspection of stereo recordings from pairs of males revealed that in such instances acoustic interference rarely occurred because the notes of each male's calls alternated with those of the other (Schwartz and Wells 1985). Recordings also indicated that this note alternation is facilitated by an increase in internote spacing during interruptions. Similar behavior has been reported in *Smilisca sila* (Ibáñez 1991), although in this species males are not as effective in avoiding note interference as in *H. microcephala.*

I used an interactive playback experiment to investigate the phenomenon in *Hyla microcephala* in more detail. I programmed my computer to answer a male's introductory note with a synthetic note ranging from 20 to 1000 ms in duration (Schwartz 1991). During the experiment, the computer stored a record of the timing of the subject's notes. I found that males increased the spacing between their notes during interruptions of all durations. If the interruption was 400 ms or longer, however, this increase was often insufficient to prevent overlap with the stimulus (Figure 14.2). In interactions with conspecific males, the majority of interruptions (by the introductory and secondary notes of advertisement calls) would be less than 100 ms in duration. Therefore acoustic interference with close neighbors can often be avoided.

The increase in internote spacing during an interruption probably results from a temporary inhibition of note production whose strength decays with time. For two males to effectively alternate notes in overlapping multinote calls, they need also to respond rapidly following the termination of each interrupting note. A second experiment using interactive playback of interrupting 200 ms notes with centrally placed gaps of 10, 20, 30, or 40 ms indicated that gaps greater than 10 ms resulted in a disinhibition of note production (Schwartz 1993). Thus males often overlapped the second half of the split stimulus note with the second note of their own call.

Work on call timing in *Hyla microcephala* could benefit from additional research using interactive playback or randomly timed stimuli (Klump and Gerhardt 1992). For example, interrupting stimuli with varying time delays could be used to test elements of the model of call timing put forth by Moore et al. (1989) based on their study of *Leptodactylus albilabris.* The model is an elaboration of an earlier formulation of Narins (1982a) and posits that there are periods in a male's intercall interval during which acoustic stimulation may inhibit calling, stimulate calling to different degrees, or have no effect.

The problems of acoustic interference and call leadership are more difficult in a large chorus than when only one competing male is nearby. How do male frogs adjust the timing

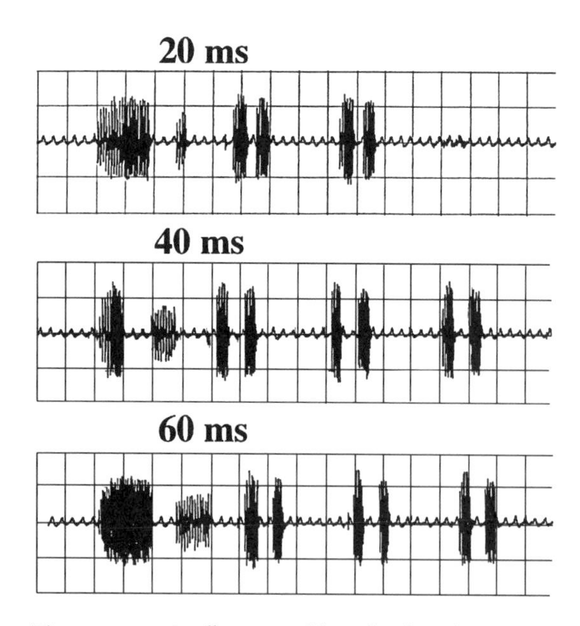

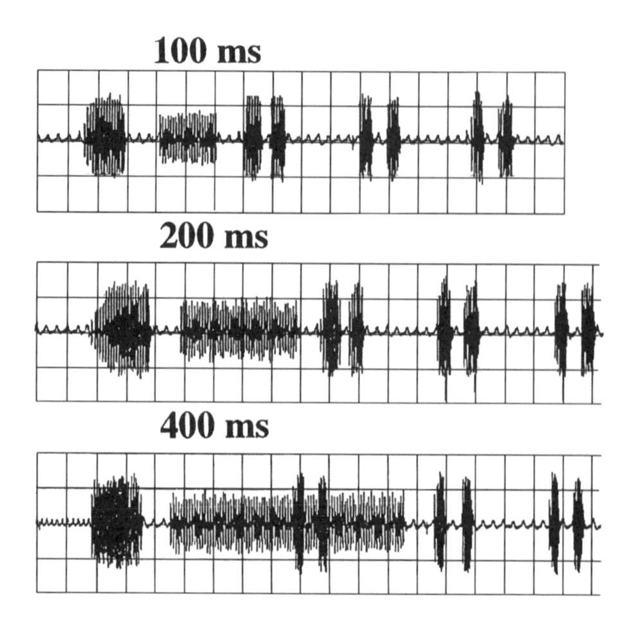

Figure 14.2. Oscillograms of the calls of a male *Hyla microcephala* in response to interruptions with computer-generated synthetic notes of 20–400 ms. The changes in note timing by the male accommodated interruptions below 400 ms in duration. Modified from Schwartz (1991).

of their calls in situations in which many other males are signaling? For example, do males exhibit selective attention by responding to just a subset of chorus members (Table 14.2)? Moreover, under conditions of increased acoustic complexity would these adjustments in timing result in enhanced attractiveness to females? I attempted to answer these questions using call monitoring, interactive playback, and female choice tests with *H. microcephala* (Schwartz 1993).

In four groups of 6, 5, 4, and 4 males, I found that males were most responsive to other chorus members whose calls were loudest at their position. These were typically, but not always, their nearest neighbors in the group. That is, males increased the spacing between their notes in response to interruptions by those individuals whose calls might be expected to pose the most serious interference problem. However males often failed to respond in this way to interrupting notes of certain males that should have been audible to them (Figure 14.3). For example, in Figure 14.3 one can see that male 4 responded when interrupted by male 3 but not by male 5—although the estimated intensity of the calls of male 5 was just 1 dB below those of male 3 (96 vs. 97 dB SPL at the position of male 4). The data suggested that males near the edge of a chorus were more likely to ignore the calls of other males than were more centrally located individuals.

When I used interactive playbacks to interrupt males with the simultaneous broadcast of a lower intensity long (200 ms) note and a higher intensity short (60 ms) note, I predicted that the longer note would be overlapped because of selective attention to the shorter note. This expectation was only partially met. In the experiment I varied the relative intensity of the two interrupting notes (from 0 to –15 dB). As the relative intensity of the longer note increased to approach that of the shorter note, there was a gradual increase in the spacing between the notes of the interrupted males. Therefore, the all-or-nothing character of responses that was observed in the natural chorus was not evident. One possible explanation for this difference is that directional cues were available to test subjects in the natural chorus that may have enabled them to identify neighboring callers as discrete sound sources at different positions. Because a single speaker was used to broadcast both interrupting notes in the playback experiment, only information on the intensities of notes was available to the test subjects. The data suggest that information on proximity of calling males augments information on intensity during calling in *H. microcephala.* This hypothesis could be tested in playback tests using two speakers at different distances from the test subjects.

Four-speaker choice tests with females of *H. microcephala* demonstrated that males who allow their notes to be overlapped by members of the chorus close to themselves might jeopardize their opportunities to attract a female. However there may not be much of a penalty for acoustic interference among most males whose calls differ by at least 6 dB at the position of a female (Schwartz 1993). Because calling males are usually ignorant of the position of females in or near the chorus, the best realizable strategy may be for males to time-shift notes in response to their loudest neighbors.

Selective attention by chorusing males of other taxa has also been reported. Individuals of the tarbush grasshopper,

Table 14.2 Hypotheses that address the form, proximate causes, or functional significance of aspects of anuran vocal behavior discussed in this chapter; the explanations are not necessarily mutually exclusive

Selective Attention

1. **Call Interference.** Call interference with a male's nearest or loudest neighbors jeopardizes that male's chances of attracting a mate more than interference with other males in the chorus. Therefore males selectively adjust the timing of their calls with respect to only the most potent sources of acoustic interference (Brush and Narins 1989; Schwartz, 1993).
2. **Threshold Shift.** A noise-induced change in auditory sensitivity reduces the number of callers a male can hear (Narins 1987; Schwartz and Gerhardt 1998).
3. **Call Leadership.** Selective attention facilitates adjustments in call-timing during contests among males for call leadership position that attracts females (Greenfield et al. 1997).
4. **Proximity-Based Call Elaboration.** Males selectively elaborate their calls with respect to their nearest neighbors in the chorus. This may be because males do not detect the calls of other males or because a female only evaluates a subset of chorus members in her immediate vicinity (this chapter).
5. **Attractiveness-Based Call Elaboration.** Males selectively elaborate their calls with respect to their most attractive audible neighbor in the chorus (this chapter).
6. **Context-Dependent Attention.** Males respond to different subsets of chorus members with respect to the timing and elaboration of their calls (this chapter).

Call Alternation

1. **Call Detection.** Alternation helps a male to either detect a neighbor's calls or assess their intensity (Lemon 1971; Passmore and Telford 1981; Schwartz 1987; Schwartz and Rand 1991; Narins 1992).
2. **Signal Degradation.** Males alternate because overlap can degrade fine temporal information within advertisement calls critical to call recognition or species discrimination by females (Littlejohn 1977; Wells and Schwartz 1984a; Schwartz 1987; Schwartz and Rand 1991).
3. **Localization.** Alternating calls are easier for females to localize than overlapped calls (Passmore and Telford 1981; Zelick and Narins 1985; Wells and Schwartz 1984a; Schwartz 1987a; Schwartz and Rand 1991; Grafe 1996).
4. **Epiphenomena of Signaling Interactions.** Alternation is an epiphenomenon of call-timing mechanisms that evolved in response to selection favoring males whose calls lead those of other males (Greenfield and Roizen 1993; Greenfield et al. 1997).

Cyclical Calling

1. **Fatigue.** Males must periodically cease calling because lactate accumulates in their calling muscles (Whitney and Krebs 1975; Pough et al. 1992).
2. **Female Preference.** Females discriminate in favor of males that call in bouts (Schwartz 1991).
3. **Noise.** Calls stimulate calling at low noise levels, but at high noise levels calling is inhibited (Schwartz 1991).
4. **Energetic Constraint.** Periodic pauses in calling conserve long-term energy reserves. This allows males to call for many hours each night at high short-term rates (Schwartz 1991; Schwartz et al. 1995).
5. **Predation Risk.** Periodic pauses in calling of the group make it more difficult for predators to find males within the group (Tuttle and Ryan 1982).
6. **Emergent Property.** Cyclical calling emerges as a mathematically predictable byproduct of mutual acoustic stimulation of chorus members (Cole and Cheshire 1996; Goodwin 1998).

Elaboration of Calls

1. **Female Attraction.** Males attempt to exceed the acoustic energy content of the advertisement calls of their competitors (via elevation of call rate, call intensity, call duration, or number of notes) because females are preferentially attracted to calls of relatively higher perceived energy (Ryan 1988; Ryan and Keddy-Hector 1992; Schwartz et al. 1995; Sullivan et al. 1995).
2. **Energy Conservation.** To avoid wasting energy males should modulate calling behavior in ways that reflect the level of competition in the chorus (Wells and Taigen 1986, 1989; Ryan 1988; Pough et al. 1992; Schwartz et al. 1995).
3. **Noise Level-Based Response.** Males gauge the intensity of competition for females indirectly and so elaborate their calls with respect to the overall background noise level of the chorus (this chapter).
4. **Focally Directed Responses.** Males elaborate their calls with respect to particular (one or more) chorus members (see "Selective Attention," this chapter).

Graded Aggressive Calls

1. **Bifunctionality.** Graded calls allow males to simultaneously vary the aggressive content of the call as well as its attractiveness to females (Wells and Schwartz 1984b; Wells and Bard 1987; Wagner 1989a; Grafe 1995).
2. **Motivation.** Longer calls signal an increased likelihood that the signaler will vigorously defend his calling site and even attack an opponent (Schwartz 1989, 1994; Wagner 1989a; Grafe 1995).
3. **Assessment.** Longer calls signal greater resource-holding potential (e.g., size, strength, energetic reserves) (Schwartz 1989, 1994; Wagner 1989a; Grafe 1995).
4. **Call Disruption.** Longer calls more effectively disrupt a rival's calling and so impair his ability to attract a female (this chapter).
5. **Behavioral Efference.** Aggressive calling increases the level of aggression in the caller. Therefore call duration of males escalates during an encounter through a process of positive feedback (Bond 1989a, 1989b).

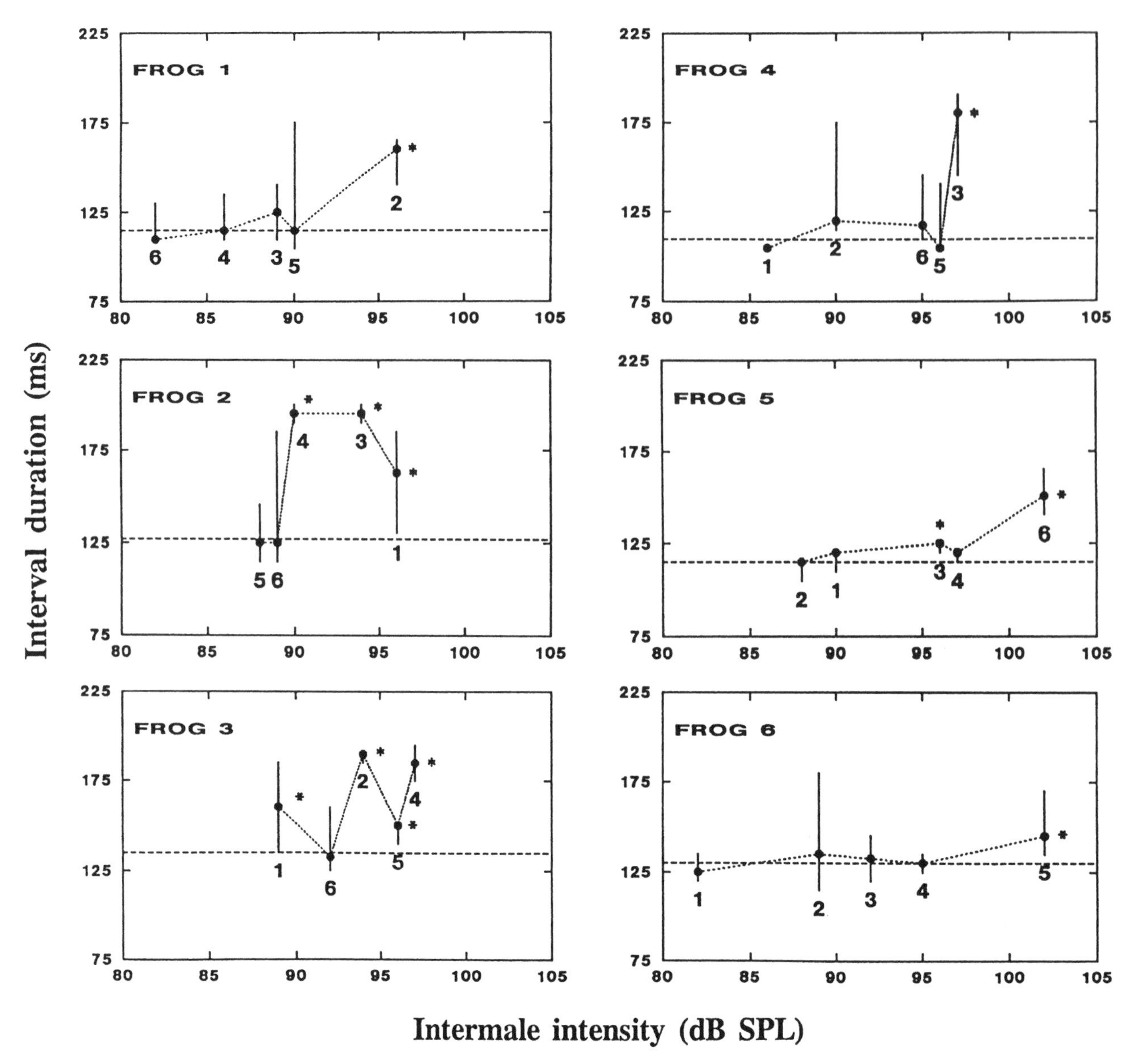

Figure 14.3. Durations (median values, approximate 95% confidence intervals) of the interrupted internote intervals of each of six males in a group of *Hyla microcephala* when interrupted by one other frog in this group. The number below each data point indicates the interrupting male. The estimated sound intensity (peak, dB SPL) of the notes for each interrupting male at the position of the interrupted male is shown along the abscissa. An asterisk indicates that the durations of interrupted internote intervals were significantly greater ($p < 0.05$, Wilcoxon two-sample test) than those internote intervals that were not interrupted (median duration indicated by the horizontal dashed line). The 95% confidence interval around the median duration of uninterrupted intervals (not shown) was 5 ms or less for all frogs except male 1 (10 ms). From Schwartz (1993).

Ligurotettix planum, attempt to shift their calls to a leading position with respect to their nearest neighbors, but not with respect to more distant members of the chorus (Minckley et al. 1995). Moreover this is a dynamic process whereby the sphere of acoustic interaction can change depending on the spacing and relative intensities of males in the chorus (Snedden et al. 1998). Especially interesting is the finding that the inhibitory-resetting of call timing modeled by Greenfield et al. (1997) is evolutionarily stable in choruses only if males exhibit selective attention. Brush and Narins (1989) used their eight-channel call-monitoring system to study call-timing interactions within groups of four and five males of *E. coqui.* Randomization of the call durations and intercall intervals enabled them to test whether patterns of acoustic interference among pairs of males differed significantly from those expected by chance. The analysis revealed that males typically avoided overlap with just two of their neighbors. Recently, Snedden and Rand (unpublished data) have discovered that males of the túngara frog, *Physalaemus pustulosus,* reset the timing of their "whines" with respect to only their nearest and loudest neighbors in the chorus.

With Bryant Buchanan and Carl Gerhardt, I have investigated call timing of *Hyla versicolor* following manipulations of chorus size. Up to eight males were allowed to call from equally spaced perches around the edge of an octagonal artificial pond in a greenhouse (Figure 14.1). As in *H. microcephala,* the percentage of a male's calling that was free from acoustic interference decreased as a function of chorus size (Figure 14.4A; Kruskal-Wallis test, $\chi^2 = 106.3$, $p < 0.0001$). However selective avoidance of interference with nearest neighbors appeared to be absent; rather, individuals were more likely to overlap an adjacent male in the pond than males separated by two or more positions (Figure 14.4B; Kruskal-Wallis test, $\chi^2 = 47.7$, $p < 0.0001$).

It is possible to test statistically for nonrandom calling interactions between pairs of male frogs in various ways. For example, Klump and Gerhardt (1992) describe approaches using distributions and plots of among- and within-male intercall intervals, distributions of relative phase angles, cross-interval histograms, and cross-correlation. Using an approach initially outlined by Popp (1989) and also used by Brush and Narins (1989) in which observed call durations and intercall intervals are randomized, we found that the males in choruses of eight frogs were more likely to overlap their closest neighbors than was expected by chance (Figure 14.4C). In fact just 11.25% of such pair-wise interactions showed significantly less than expected interference. Active avoidance of overlap occurred only when chorus size was reduced to two males and was significant in 70% of interactions. Therefore if large groups ($n \geq 8$) of *H. versicolor* call in close proximity (as in the artificial pond), on average more than half of the pulse trains produced by a male will be overlapped by the calling of other individuals. In a natural setting, males have considerable freedom to adjust their spacing and thus more easily avoid acoustic interference with callers they perceive as loud. Nevertheless it is obvious to anyone who has listened to males in a dense aggregation that overlap of calls occurs.

Given that call interference can impair a male's ability to attract a female, why do males of *H. versicolor* not exhibit selective attention to timing of calls? One explanation may be that the expectation for call overlap between any particular pair of males is already quite low. Even for adjacent males in our eight-male choruses, such overlap represented only about 11% of calling time (Figure 14.4B). Another factor is that females may not discriminate against males with overlapping calls if these males are well separated. This failure to discriminate against them may occur because directional cues provided by the female's auditory system reduce degradation of the inherent pulse structure within overlapping calls of the interacting males (Schwartz and Gerhardt 1995). Interestingly, in *H. microcephala,* a smaller species than *H. versicolor* that shows selective attention, I found angular separation of speakers did not change female discrimination against overlapping calls (Schwartz 1993).

Perhaps what we observe as "selective attention" may be a by-product of the masking of the calls of all but a male's closest neighbors by the background din of the chorus. Although masking certainly circumscribes the range of male vocal interactions within aggregations of frogs (Narins 1982b), it seems inadequate to explain the difference in responses of males of *H. microcephala* to call intensities that were nearly equal. Another possibility is that high levels of background noise cause a shift in auditory threshold in chorus members (Table 14.2). Thus calls of distant males that would be audible on a quiet night fall below the threshold for hearing after a male has been in a chorus for some period of time. In fact neurophysiological experiments have revealed such threshold shifts in frogs (Narins 1987; Schwartz and Gerhardt 1998). Although this may be part of the answer, this phenomenon probably cannot explain the extremely small differences in intensity between interruptions that were ignored and those that were not by males of *H. microcephala*. Neurophysiological data obtained from crickets (Pollack 1988) and bushcrickets (Römer 1993) indicate that the auditory system may itself augment differences in the intensity of sounds from different individuals in a way that could promote selective attention. Similar data are not yet available for anurans.

The Role of Fine-Scale Call Timing in Attracting Females

How might alternation of calls or notes help a male attract a female? I tested three hypotheses (Table 14.2) with *Hyla microcephala, H. versicolor, Pseudacris crucifer,* and *Physalaemus pustulosus* using interactive playback and female choice tests (Schwartz 1987a; Schwartz and Rand 1991). The first hypothesis posits that a male attempts to avoid interference because this would impair his ability to detect the calls of other males. Hearing competing males is potentially important because it (1) enables a male to adjust his calling effort in a way that maintains his relative attractiveness, and (2) helps maintain intermale spacing in the chorus. The second hypothesis is that acoustic interference can degrade or obliterate temporal features in calls critical for species discrimination by females. The third hypothesis proposes that it is difficult for females to localize the source of calls that are overlapped. I chose *H. microcephala* and *H. versicolor* because the fine pulse structure of the advertisement calls of these

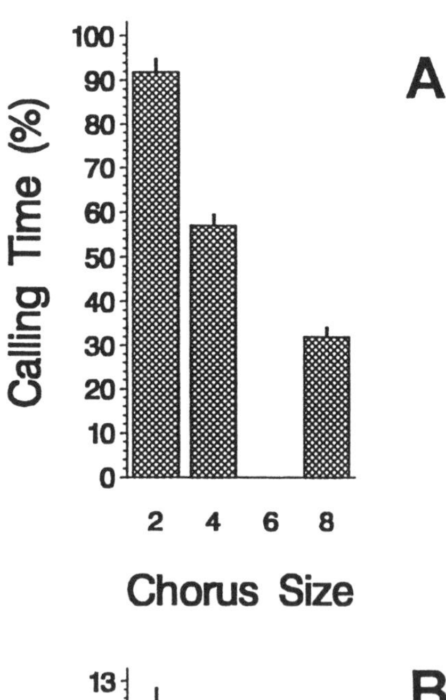

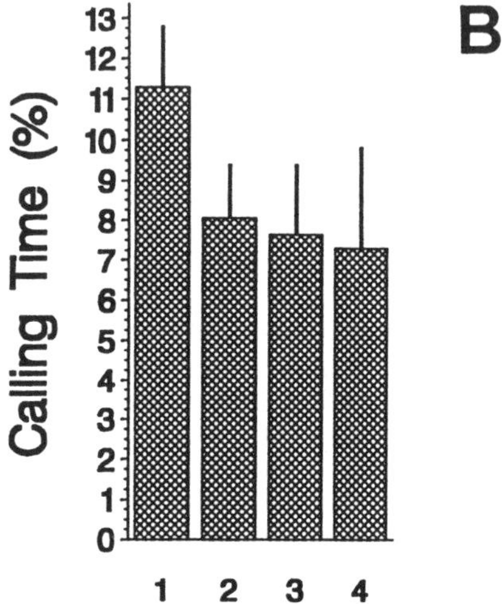

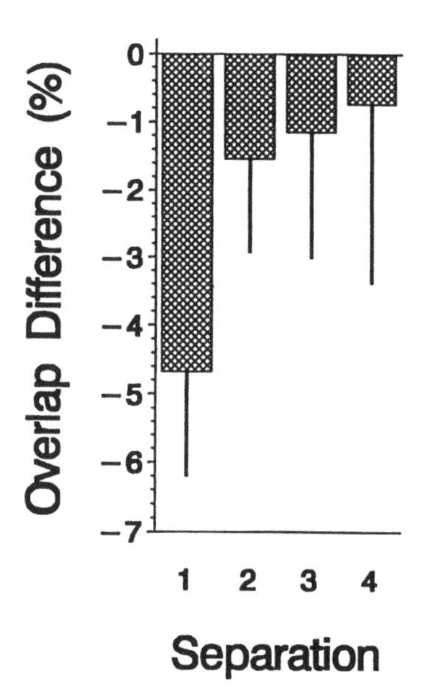

Figure 14.4. (A) Calling time (mean percentage of total time producing sound, n = 10 choruses) of males of *H. versicolor* that was not obstructed by the calls of any other male in the chorus for choruses of different sizes in the Greenhouse pond. (B) Calling time of males that was overlapped by the calls of other males as a function of separation (in terms of calling site positions in the artificial pond; n = 10 choruses of 8 frogs). (C) The difference between expected and observed call overlap among pairs of males as a function of separation. An expected level of acoustic interference was obtained by randomizing listings of the call durations and intercall intervals of each male and then recalculating call overlap. If the observed level of interference was less than that expected for at least 95 of 100 such randomized data sets, I characterized the interaction between a pair of males as showing significant avoidance of interference. The size of the difference between the expected and observed overlap was calculated as: 100 × (Expected overlap[i][j] – Observed overlap[i][j]) ÷ Calling time[i]. Male [i] is the "interfering" male whose calls follow and overlap those of the "leading" male [j].

species is important in species discrimination (Gerhardt 1978; Schwartz 1987b). *Pseudacris crucifer* has a tonal and only weakly frequency-modulated advertisement call that lacks temporal information of this kind, and so it served as an important control. Finally Rand and I tested *Physalaemus pustulosus* because the whine note of the advertisement call is strongly frequency modulated. This frequency modulation can be important in call recognition (Rose et al. 1988; Wilczynski et al. 1995).

I tested the first hypothesis by presenting males with call-triggered synthetic advertisement calls that either alternated with or overlapped their own calls. For all three hylids I found that males were more likely to give aggressive calls to alternating than overlapping stimuli. For *Physalaemus pustulosus,* males were more likely to add chuck notes to their whines if the stimulus was nonoverlapping than overlapping. Both sets of results suggest that males have a more difficult time detecting or gauging the intensity of calls that overlap their own. This may occur because the sound of a male's own call jams the calls of other males coincident with his own or because the increase in buccal pressure accompanying calling stiffens the tympanic membranes (Narins 1992).

I tested the second and third hypotheses with four-speaker choice tests in which females were exposed to simulated pairs of males that either alternated or overlapped one another. When the overlapped calls of *Hyla microcephala* and *H. versicolor* were 180 degrees out of phase so that the pulse rate of the calls seemed to have doubled, females discriminated in favor of the alternating calls. If the overlapped calls were precisely in phase, no discrimination occurred. There was also no discrimination when *Pseudacris crucifer* or *Physalaemus pustulosus* females were tested with either phase relationship. These data support the second hypothesis for species with amplitude-modulated calls while suggesting that frequency-modulated calls are less vulnerable to degradation caused by acoustic interference. The absence of discrimination when overlapping calls were in phase is inconsistent with the prediction of the third (localization) hypothesis. The time of approach to simultaneous versus alternating calls during two-speaker choice tests with *Hyperolius marmoratus* (Passmore and Telford 1981; Grafe 1996) also failed to support this hypothesis.

Perhaps alternation has no function per se but, as argued by Greenfield and his collaborators, is merely an epiphenomenon of signaling interactions that evolved for reasons unrelated to the problem of acoustic interference. In particular it is proposed that a preference by females for leading calls has influenced the evolution of call timing mechanisms that result in the patterns of alternation or even synchrony that are observed among acoustically signaling insects and anurans (Greenfield and Roizen 1993; Greenfield 1997; Greenfield et al. 1997). In this scenario competing males that do not adjust call timing suffer reduced mating success because they have fewer calls in a leading position than males that do adjust call timing. The most recent modeling efforts, incorporating time lags due to the velocity of sound and interactions in simulated choruses of up to 10 males, support this explanation (Greenfield et al. 1997).

The leader preference of females may be a consequence of an innate feature of the auditory system and has been termed the *precedence effect.* In the psychoacoustic literature this terminology has been used when a subject perceives the source of two sounds in a partially overlapping or a close leader-follower timing arrangement to be the location of the leading sound (Zurek 1980; Moore 1982). However in the animal communication literature, the terminology has often been applied when an animal orients towards the leading sound (e.g., Greenfield et al. 1997). The proximate cause of leader preferences in anurans is not known (Grafe 1996). The preference may be mediated in the peripheral or central auditory system and be due to a temporary reduction in sensitivity to following sounds (Yin 1994; Römer et al. 1997). Alternatively both leading and following calls may be equally audible to females, yet processing of the sounds by the brain results in erroneous localization of the calls of followers at the position of the leader.

The call-timing mechanism advocated by Greenfield (1994a, 1994b) relies on an "inhibitory-resetting" of an underlying free-running neuronal oscillator that controls the rhythmic production of calls by a male. Moreover Greenfield's basic model can be elaborated to be consistent with other models based on data from experiments with anurans (e.g., Loftus-Hills 1974; Lemon and Struger 1980; Narins 1982a; Moore et al. 1989). It is likely that an inhibitory-resetting process can explain patterns of fine-scale call timing during interactions between males for species of frogs with fairly simple advertisement calls or patterns of calling (Greenfield 1994a, 1994b; Greenfield et al. 1997). This explanation may also hold for *Physalaemus pustulosus* (Snedden et al., unpublished data). However experiments to elucidate the processes involved in call-timing shifts have been undertaken for only few anuran species, and more are needed. In particular modification of the model may be needed to explain some of the more complex patterns of calling and flexibility found in the vocal behavior of males of certain species.

In *Hyla microcephala,* for example, the inhibition in the production of a secondary note following an interrupting stimulus wanes after about 200 ms of interruption. Perhaps some neural mechanism integrates the output of an oscillator during inhibition and initiates calling following the crossing of a threshold level (Schwartz 1991). Another observation that needs explanation is that the presence of a brief gap in the middle of the interruption results in a strong rebound from vocal inhibition, such that a dilatory effect on vocalization of the second half of a split interrupting note is absent. Furthermore, although the calls of neighbors have an excitatory effect on male calling, chorusing individuals periodically cease vocal activity (see below). Thus there is no stable "free-running" call rhythm. It is likely that in this species different neural circuits may govern elements of the timing of calls and the timing of notes within calls. Finally the production and timing of advertisement and aggressive calls may be under the control of different circuits. In *H. microcephala* aggressive call responses to broadcast calls came significantly later (mode, 1200 ms; median, 2400 ms, n = 812) than did advertisement call responses (mode, 400 ms; median, 960 ms, n = 3511; Schwartz and Wells 1985). This may reflect the need for more sophisticated processing of a signal used in male–male assessment, additional delays in "choosing" a response, and physiological constraints on aggressive call production.

The distribution and causes of leader preferences in anurans needs further investigations using a comparative phylogenetic analysis (Jennions 1994). Studies of such preferences would also benefit from a systematic analysis of the influence of signal form, timing, and amplitude. In some species, such as *Hyla regilla* (Whitney and Krebs 1975), *Hyla cinerea* (Klump and Gerhardt 1992), *Hyperolius marmoratus* (Dyson and Passmore 1988a, 1988b), and *Bufo americanus* (Howard and Palmer 1995), females clearly show leader preferences. In other species, such preferences are absent (e.g., *Hyla versicolor,* Klump and Gerhardt 1987; *Centrolenella granulosa,* Ibáñez 1993) or the data are equivocal (*Pseudacris crucifer,* Forester and Harrison 1987; Schwartz and Gerhardt, unpublished data). In tests with *Hyla ebraccata,* Wells and Schwartz (1984a) found that females approached a speaker broadcasting a two-note call that followed and overlapped the secondary note of a leading two-note call. This pattern of calling was frequently observed in pair-wise vocal interactions among males. In *Physalaemus pustulosus,* Schwartz and Rand (1991) failed to detect a preference for the leading of the partially overlapped calls used in the four-speaker choice tests, although the statistical power was low. However Snedden et al. (unpublished data) observed strong leader preferences in a series of two-choice tests with females of this species. Schwartz and Gerhardt (unpublished data) have found that females of *Pseudacris crucifer* preferred the leading of two partially overlapped calls (by 20 ms of their 150 ms duration; 18:3, $p < 0.002$, two-tailed binomial test) even if the leading call was attenuated by 6 dB (11:1; $p < 0.006$). In *Hyla microcephala,* females did not prefer the leading call when offered two single-note calls separated by 200 ms (Schwartz 1986).

Coarse-Scale Patterns of Call Timing

Shifts in call timing among chorusing males may also occur on a gross scale. For example, Littlejohn and Martin (1969) demonstrated temporal partitioning of calling bouts whereby the Australian frog, *Pseudophryne semimarmorata,* is inhibited from calling in response to the vocal activity of *Geocrinia victoriana.* Schwartz and Wells (1983a, 1983b) also observed such a coarse shift in call timing by males of *Hyla ebraccata* in response to the vocalizations of groups of *H. microcephala* or *H. phlebodes.* The calling of *H. microcephala* was found to reduce both the ability of a male *H. ebraccata* to detect the calls of other males (as evidenced by a reduction in aggressive calling) and his ability to attract a mate. Calling relationships between pairs of species may be asymmetric. For example species that produce the longer calls or longer bouts of calling (and thus the most potent sources of masking interference) may inhibit species producing shorter calls or shorter bouts (Littlejohn and Martin 1969; MacNally 1982; Schwartz and Wells 1983a; Littlejohn et al. 1985).

The choruses of males of some species are active cyclically on a time scale of seconds to minutes such that, on a gross time scale, vocalizations are periodically synchronized before the entire chorus periodically goes silent. The calling part of the cycle is repeated after a male begins to call and so stimulates other members of the chorus to vocalize. In some of the literature on insects, such behavior has been called unison bout singing (e.g., Greenfield and Shaw 1983). Rosen and Lemon (1974) and Whitney and Krebs (1975) have described this behavior in the North American treefrogs, *Pseudacris crucifer* and *Hyla regilla,* respectively. Fellers (1979) indicated that these two species as well as *Hyla cinerea* call in bouts. Similar patterns of simultaneous chorusing have also been reported in the neotropical hylids *Osteocephalus taurinus* (Zimmerman and Bogart 1984), *Phyllomedusa tomoptera* (Zimmerman and Bogart 1984), *Smilisca baudina* (Duellman and Trueb 1966), *Smilisca sordida* (Duellman and Trueb 1966), and *S. sila* (Ibáñez 1991), and the European treefrog *Hyla arborea* (Schneider 1977). I have conducted a series of experiments investigating unison bout singing in *Hyla microcephala* (Schwartz 1991; Schwartz et al. 1995). In this species, bouts of calling typically last less than 30 seconds and intervening quiet periods last approximately the same length of time.

Why do males periodically stop calling (Table 14.2)? Whitney and Krebs (1975) tested the hypothesis that males of *Hyla regilla* stop to repay an oxygen debt incurred during vocalization. However their data were inconsistent with this explanation. Moreover data from subsequent research implicated aerobic pathways in the support of vocal activity in anurans (for review see Pough et al. 1992). Another possibility is that a pattern of group calling followed by cessation of vocal activity makes it more difficult for predators to capture chorus members than would otherwise be the case. Tuttle and Ryan (1982) found that frog-eating bats (*Trachops cirrhosus*) were less likely to approach a speaker broadcasting synchronized than asynchronous calls of *Smilisca sila*. These treefrogs call at low rates ($\bar{x} = 1.7$ calls/minute; Ryan 1986) in rough synchrony (often with partially overlapping calls; also see Ibáñez 1991) for brief periods ($\bar{x} = 488$ ms) followed by longer periods of quiet ($\bar{x} = 15$ seconds; Ibáñez 1991). The predation hypothesis may be less relevant for those species of frogs (e.g., *Hyla microcephala*) or insects that use longer chorusing bouts with less synchronized calls than *Smilisca sila* or are attacked by predators that rely less on auditory cues than frog-eating bats. Nevertheless the hypothesis requires further testing. A third possibility is that the cyclical pattern of chorusing is an emergent property of calling by individuals that can be stimulated to call by other males in the chorus. Such rhythmic patterns of activity appear in groups of ants as density increases and emerges in computer models of their behavior (Cole and Cheshire 1996; Goodwin 1998). This hypothesis does not posit that cyclical calling has a functional basis, but it does not preclude this possibility. Therefore factors such as sensitivity of males to acoustic stimulation might be shaped by selection in part because they could influence features of cyclical calling that affect male mating success.

I have tested three additional hypotheses using *Hyla microcephala*. One hypothesis proposes that females actively discriminate in favor of males that call in discrete bouts relative to males that do not. An arena-based female-choice experiment was conducted using two stimuli with equal long-term note rates. The temporal structure of one stimulus resembled that of a natural chorus with a 15-second bout of high note-rate calling (2 two-note calls per second) alternating with a 15-second block of silence. The second stimulus was a recording of calls with a note rate half that given during bouts in the first stimulus. Females failed to show a preference, and so the hypothesis was rejected.

The "noise hypothesis" proposes that as males join a chorusing bout and add notes to their calls during its progress the benefit of advertisement drops below its cost. This could be because calls are too degraded by interference to attract a female or because males can no longer accurately judge the calling behavior (e.g., the number of notes per call) of their neighbors. Therefore males stop calling briefly. The cycle starts again when a male in the group calls. I tested for an inhibitory role of chorus noise in two ways. First I broadcast a simulated chorus composed of digitized calls at 90, 95, and 100 dB SPL (re 20 μPa) in random order. By using the computer system for playback, it was possible to broadcast simulated bouts of calling and silent periods that were each of random duration within a 5- to 25-second range. If loud chorus noise inhibited calling, I predicted that males would reduce their calling during simulated bouts and increase their calling during quiet intervals as stimulus intensity increased. This did not happen. Second, with the computer in interactive mode, I tested whether the calling of subjects would be inhibited when more males call. The computer was programmed to broadcast calls only when the subject called and simulated the calling of one neighbor, two neighbors, and a large group of up to 12 males. Thus the cyclical calling of the subject controlled the cyclical activity of the simulated chorus, an essential ingredient of the experiment. As chorus size was artificially increased the noise hypothesis predicted that subjects would abbreviate their calling bouts. Again, the results were inconsistent with those predicted.

The work of Wells and Taigen (1989) demonstrated that males of *Hyla microcephala* often have very high rates of energy expenditure during calling. Perhaps fuel reserves were insufficient to meet the metabolic demands of calling at high rates during the entire time period females might arrive to breed during an evening. Therefore by pausing between bouts of vigorous calling a male could extend his period of advertisement over this 4- to 5-hour block of time. To test this "energetic constraint hypothesis," Stephen Ressel, Catherine Robb Bevier, and I used the computer-based call-monitoring device and biochemical analysis of tissue samples (Schwartz et al. 1995). Assays of glycogen levels in the trunk muscles of males gathered at the start of chorusing and of monitored males collected near the end of chorusing enabled us to estimate a cost in muscle carbohydrate of producing a note. With this estimate (0.98 μg glycogen per note) we calculated the length of time each of our monitored males could have called, without pausing, until exhausting his supply of trunk muscle glycogen. We found that with an average starting level of glycogen, over 80% of the males would have exhausted this fuel reserve in less than 3 hours. Since males often call for 4 to 5 hours a night, our results were consistent with the energetic constraint hypothesis.

Because cyclical calling is not especially common among species of frogs, it seems likely that continuous calling was the ancestral pattern of chorusing for *Hyla microcephala*. For cyclical calling to evolve, spread, and be maintained in this species, cyclical callers should have a greater mating success than that of continuous callers. That is, we would expect that a "mutant" male who structured his calling behavior into bouts in order to conserve energy reserves would be more likely to attract a mate than other males who called steadily but were forced by an energetic constraint to stop calling earlier in the evening. In 1996 I initiated a field-based experiment on female choice to test whether this was so.

During July and August ($n = 14$ nights) I assembled an artificial chorus consisting of eight 360-degree speakers (Radio Shack #40–1352) driven by an Amiga 600 computer. Each speaker was suspended inside a 90-cm-tall wire frame tomato stand and a screened enclosure with openings designed to capture approaching females. The chorus simulated four pairs of interacting neighbors giving calls of one to five notes. Each night, one speaker, selected randomly from the array, simulated a male with the "mutant" strategy of cyclical calling. This speaker would broadcast calls for 5 hours; however calls were not broadcast during alternate bouts. The remaining seven speakers mimicked steady callers and would broadcast calls for only 3 hours. The starting and ending times of calling for each of the seven "steady" speakers were staggered over a period of 30 minutes each night. This temporal arrangement approximated the typical pattern of males in a chorus. The "cyclical" speaker began broadcast 15 minutes into the experiment.

I checked the traps at approximately 15-minute intervals and turned off a speaker if I found a female on or inside its screening. Unfortunately the abundance of *Hyla microcephala* in my study site during midsummer 1996 was low: I caught a total of only eight females during the course of the experiment. Six were captured at the speaker broadcasting with the cyclical pattern, whereas just two were captured at speakers broadcasting steadily. All but one of the females at the cyclical speakers "paired" after all the steady speakers went silent. Because the odds favored the "population" strategy by 7:1, this result was statistically significant ($p < 0.0001$, binomial test) and suggests that the mutant strategy could successfully invade. Nevertheless I plan to increase my sample size and also rerun the experiment using the steady calling pattern as the "mutant" strategy.

Elaboration of Calls during Vocal Competition for Mates

Female anurans often discriminate in favor of real or simulated males producing sound at higher amplitudes, greater delivery rates, and for longer time periods than other call sources (Wells 1988; Forester et al. 1989; Ryan and Keddy-Hector 1992; Passmore et al. 1992; Gerhardt 1994; Schwartz et al. 1995; Sullivan et al. 1995; Wagner and Sullivan 1995; Grafe 1997). For example the now classic work of Rand and Ryan (1981) on *Physalaemus pustulosus* demonstrated that females prefer "complex" calls formed from a whine plus chuck notes to those without chucks. Such preferences may be manifest in dense choruses as demonstrated with multichannel call-monitoring systems for call rate in *Hyperolius marmoratus* (Passmore et al. 1992) and for note rate in *Hyla microcephala*. In choruses of *H. microcephala* mean note rates were higher for mated than unmated males when rates were calculated over time intervals ranging from 5 to 60 minutes before pairing (Schwartz 1994). Moreover on a night when all monitored males paired, the order of pairing was inversely related to the males' note rates (Schwartz et al. 1995).

Data from playback experiments indicate that males of chorusing species may increase call rates (e.g., Rosen and Lemon 1974; Ayre et al. 1984; Wells and Schwartz 1984a), notes per call (e.g., Arak 1983; Wells and Schwartz 1984a; Ryan 1985; Jehle and Arak 1998), call note duration (e.g., Wells and Taigen 1986), and even call intensity (Lopez et al. 1988) in response to simulated competitors. Therefore preferences of females may select for males with superior anatomical and physiological attributes related to call production (Table 14.2). Nevertheless because calling may incur significant energetic costs (Pough et al. 1992; Prestwich 1994; Bevier 1997; Wells, this volume), males should modulate calling behavior in ways that reflect the level of competition in the chorus. Here I describe some recent results with *H. versicolor* that demonstrate such behavior.

The grey treefrog, *Hyla versicolor*, is a common species throughout much of the eastern half of North America. The number of pulses in its advertisement call may vary both within and among males in the chorus. In two-stimulus choice tests in the laboratory, females discriminate in favor of higher call rates. They also discriminate among calls that differ by as little as 10% in pulse number (i.e., two pulses for calls about 20 pulses long; Gerhardt et al., unpublished data), preferring those of longer duration. This preference for longer calls is maintained even when longer calls and shorter calls are delivered at equivalent calling efforts (number of pulses per call × call rate; Klump and Gerhardt 1987; Gerhardt et al. 1996). Consistent with data on other species of frogs (Wells 1988; Schwartz 1994), Wells and Taigen (1986) demonstrated that males of *H. versicolor* alter their vocal behavior in response to changes in their competitive environment. In the field and during playback tests males tended to increase call duration with increasing levels of acoustic stimulation while reducing call rate. Accordingly there were usually only small changes in calling effort. Wells and Taigen suggested that the tradeoff between call duration and call rate is a reflection of a physiological limit that males cannot surpass on a long-term basis.

During the night the size and local density of choruses of frogs such as *H. versicolor* change as males pair with females or stop calling for other reasons. How do the remaining callers react to such shifts in the composition of their neighborhood? Bryant Buchanan, Carl Gerhardt, and I recently performed manipulations of chorus size in our greenhouse pond that demonstrated the sensitivity of males of *H. versicolor* to acoustic conditions. In one set of manipulations ($n =$ 12 nights) we modified chorus sizes in steps of 50% (from 8 to 4 to 2, on two of the nights, initial chorus size was 7 males). We acquired data on vocal behavior using the call monitoring system for 10-minute periods at each chorus size before attempting to reverse the pattern of density change. The increasing series of chorus sizes was used to control for any effect of time of night on the calling behavior of our subjects. Some males did not call in the increasing density series, perhaps due to effects of removal and handling. In a second set of density manipulations ($n = 4$ nights) we used step sizes of one male and initial choruses of 6 to 8 males. In these tests we acquired data for 5 rather than 10 minutes and only reduced densities. The order in which we removed males was randomized.

When we changed chorus size in steps of 50%, "focal" males present for all treatments reduced the number of pulses in their calls during the reductions and raised them with increases in chorus size (Kruskal-Wallis test, $\chi^2 =$ 1432.0, $p < 0.0001$, $n = 10$ males from 5 choruses). Males also decreased pulse number in response to removals of one male at a time. For three of the four focal males in the four choruses, a drop in the average pulses per call accompanied each reduction by one male. Although all males altered the duration of their calls during these manipulations, their ranking relative to other chorus males with respect to call duration changed little (50% reductions, mean change in rank between sequential chorus sizes = 0.203; 1 male reductions, $\bar{x} = 0.359$). For example the male that gave the longest calls at the start of the experiment typically continued to give the longest calls at the chorus sizes he experienced.

Our results raise many interesting questions. First, do some males (e.g., those with the longest calls or highest calling effort) exert a disproportionate influence on the vocal behavior of other chorus members? Second, why do lower ranking males maintain their relative rank at lower chorus sizes although they are capable of giving longer calls (as evidenced by their behavior at higher chorus sizes)? Are males reacting primarily to changes in the background noise level or does the spatial distribution of competitors exert a significant effect on vocal performance (Table 14.2)? How does the auditory system integrate and process the complex acoustic environment of the chorus in both spatial and temporal dimensions? Additional analyses and experiments that use interactive playback, multispeaker simulations of a chorus, and neurophysiology should help us find answers.

Aggressive Interactions

Adjacent males may exchange aggressive vocalizations with one another and use these calls almost exclusively if the vocal encounter escalates to a physical conflict. Male frogs often respond to loud broadcasts of conspecific calls by increasing their proportion of aggressive calls and may also respond differentially to advertisement and aggressive calls. For example aggressive calls may elicit a higher level of aggressive calling or even a cessation of calling, and their broadcast may also result in a withdrawal from the experimenter's speaker (see review of Wells 1988). Aggressive calls of many species exhibit a temporal structure that distinguishes them from advertisement calls. For example, in *Hyla ebraccata* (Wells and Schwartz 1984b) and *H. microcephala* (Schwartz 1986), the difference is quantitative: introductory notes of aggressive calls have a higher rate of amplitude modulation than those of advertisement calls. In *Pseudacris crucifer* (Schwartz 1989) and *H. cinerea* (Gerhardt 1978), the difference is qualitative: aggressive calls are amplitude modulated, whereas advertisement calls are not, although occasionally calls intermediate in form occur. In other species, males may simply increase the rate of delivery of advertisement-like calls or notes during agonistic interactions (Lopez et al. 1988; Stewart and Rand 1991). Some species of frogs incorporate an added degree of flexibility in their communication systems by using graded aggressive signals (Arak 1983; Wells and Schwartz 1984b; Schwartz and Wells 1984a; Littlejohn and Harrison 1985; Schwartz 1989; Wagner 1989a; Stewart and Rand 1991; Grafe 1995). In such systems males gradually change the temporal structure of aggressive calls as intermale distance decreases or the sound level of broadcast calls increases. Call duration appears to be a feature commonly altered, with calls or call elements increasing in length in response to the actual or simulated approach of an intruding male. Other call features may change in a correlated fashion with duration (e.g., Wells and Schwartz 1984b). This degree of flexibility may be useful to a male not only in the context of male–male interactions (Wells 1989), but may also allow him to vary those elements of his call that are attractive to females as warranted during a developing agonistic interaction (Wells and Schwartz 1984b; Wells and Bard 1987; Wagner 1989a; Grafe 1995).

A critical question concerns the functional significance of graded aggressive calls in anurans (Schwartz 1989; Wagner 1989a; Wells 1989; Table 14.2) as well as in other taxa that employ similar agonistic displays (Clutton-Brock and Albon 1979; Becker 1982; Bond 1989a, 1989b, 1992; Lambrechts 1992; McGregor and Horn 1992; Bradbury and Vehrencamp 1998; Payne and Pagel 1997). I have followed up earlier investigations of aggressive calling in both *Hyla*

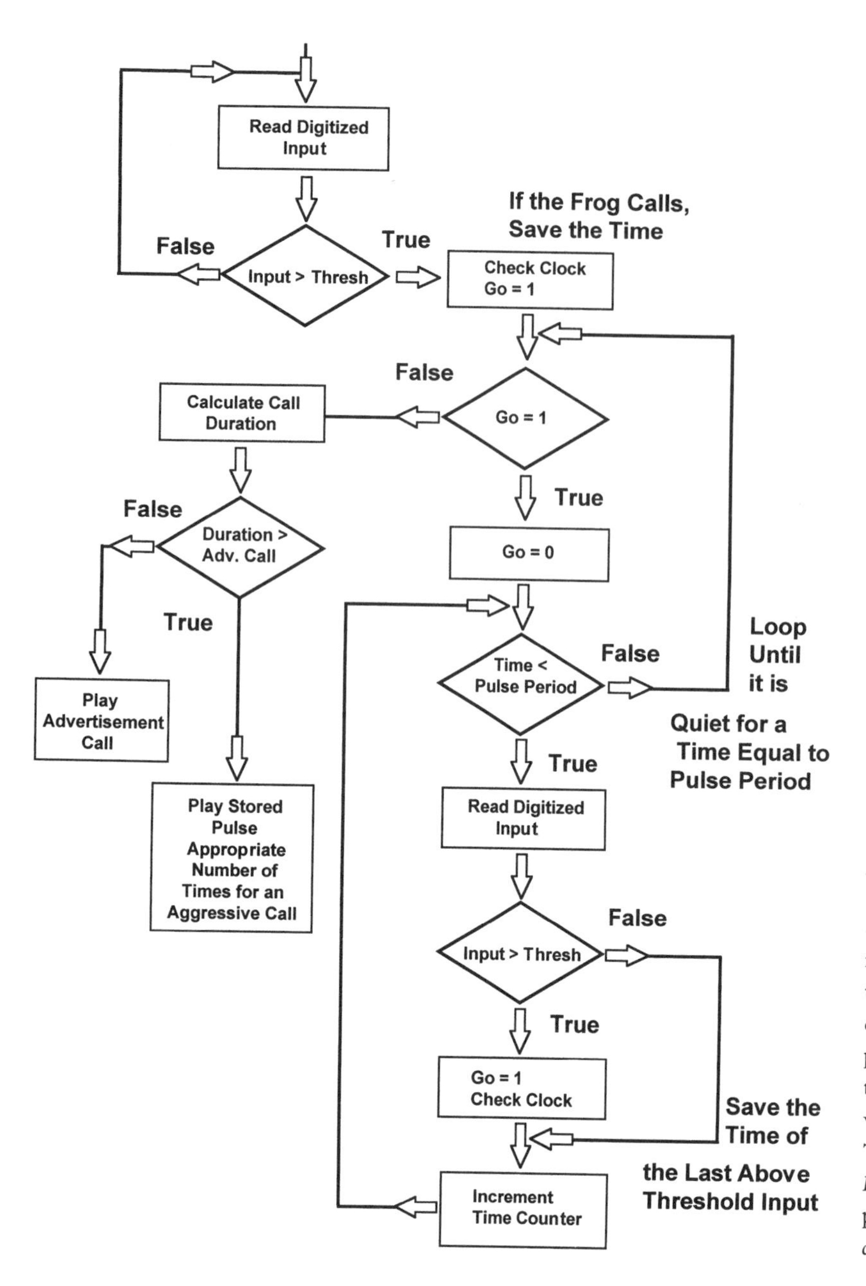

Figure 14.5. Flow chart of an algorithm for interactive playback tests using pulsed calls. In response to above-threshold input, the system time is requested. Data are read until digitized input is below threshold for a period of time equal to the pulse period of the call, at which time the call has ended. The call duration is calculated from the call start time and the time of the last above-threshold input. If the call is longer than some minimum duration, the call is an aggressive call. The call duration is used to calculate the number of pulses in the subject's aggressive call. Response calls of desired duration (e.g., shorter, equal, or longer than the subject's) are created by instructing the computer to play a stored pulse and an inter-pulse interval the appropriate number of times. A second version of this program, which counts pulses directly, is available. The system has been used successfully with *Hyla microcephala* and *Pseudacris crucifer* to play pulsed aggressive calls and with *H. versicolor* to play pulsed advertisement calls.

microcephala (Schwartz and Wells 1985) and *Pseudacris crucifer* (Schwartz 1989) with experiments using computer-based interactive playbacks. The research, not yet completed, is designed to improve our understanding of the functional significance of graded signals.

In my experiments I stage vocal contests between my subject and a simulated competitor (Figure 14.5). In some tests, after a no-stimulus period, males are presented with four stimulus treatments chosen randomly and with intervening recovery periods. In the control treatment the computer is programmed to respond with advertisement calls to all calls given by the subject. In the remaining treatments the computer responds to the subject's advertisement calls as before, but it answers each aggressive call with an aggressive call with an introductory note of duration (1) equal to, (2) twice as long as, or (3) half as long as those of the subject. The program uses note duration to distinguish the subject's advertisement from his aggressive calls. In tests with *Hyla microcephala* the computer distinguishes between introductory notes and secondary notes using note timing information.

If call duration facilitates assessment of an opponent's strength, a male should reduce the intensity of his aggressive response to a stronger individual. Moreover interactions should escalate and assessment of strength should take longer if two opponents have similar vocal behavior. Data currently available on *H. microcephala* are consistent with these predictions. Males gave shorter aggressive calls and had shorter bouts of aggressive calling in response to aggressive

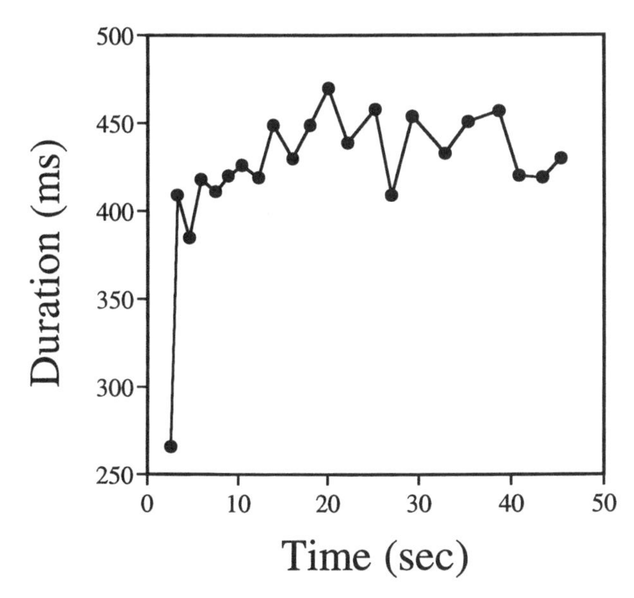

Figure 14.6. Durations of the calls of a male *Pseudacris crucifer* during a bout of aggressive calling. The sequence of calls was recorded during a no-stimulus period in an interactive playback test.

calls that were twice as long as their own aggressive calls as they did to stimulus calls that were equivalent in duration (Friedman test, $p < 0.01$, $n = 15$ males). Stimulus calls that were half as long as the subject's calls also elicited briefer bouts of shorter response calls than equivalent duration stimuli. The changes in call duration during bouts of aggressive calling also support the idea that producing a series of long aggressive calls is physically taxing. Individuals tended to start bouts with a short call, give a series of longer calls, and finally give a short call before ending the bout and returning to advertisement calling (Friedman test, $p < 0.005$).

Although aggressive calls given by males of *Pseudacris crucifer* to the longest stimulus aggressive calls were not significantly shorter than those given to aggressive calls equivalent in duration to their own, males did give longer bouts when presented with stimuli of equivalent duration. This finding and an earlier result suggesting an energetic constraint on aggressive calling (see Figure 14.5 in Schwartz 1989) are also consistent with the strength assessment hypothesis. Intriguingly field data from some individuals were supportive of Bond's (1989a, 1989b; 1992) behavioral efference hypothesis that posits that the primary function of graded aggressive display is modulation of the level of aggressive arousal of the signaler. Not only those males challenged with responses from the computer but also those that initiated aggressive calling during a no-stimulus period tended to increase their call duration (Figure 14.6). Observations that male frogs often continue to give aggressive calls when physically fighting are also consistent with both Bond's behavioral efference model and the strength-assessment hypothesis. Perhaps aggressive calling gradually primes a male for fighting through positive feedback and also advertises to an opponent that he can afford to pay the cost of producing expensive calls even while fighting.

Conclusions

The sound environment in which frogs communicate can often be characterized by considerable heterogeneity along spatial and temporal dimensions. Males of many species respond to acoustic flux in the chorus with rapid adjustments in their vocal behavior. Such responses can be important in maintaining the relative attractiveness of males to gravid females and in allocating energy reserves efficiently. Interactive playback and multichannel call-monitoring systems provide researchers with powerful techniques for generating and testing hypotheses about communication in anurans. Use of a computer obviously provides for greater realism and flexibility in the form of sound delivery than does the use of prerecorded playback stimuli. In particular, interactive playback is a necessity in experiments requiring precise control of stimulus timing relative to the vocalizations of the subject and in tests in which stimuli must change in specific ways in response to changes in the subject's calls. For example I learned from recordings of pair-wise natural interactions and tape-based playbacks that males of *Hyla microcephala* respond to interrupting sounds by increasing the spacing between their notes. However it was only with interactive playback that I was able to efficiently explore details of this behavior. Similarly tests in which I simulated a subject-driven escalated aggressive exchange would not have been possible without computerized playback. This is not to say that playback tests using a pre-recorded sequence of sounds cannot be very informative. Such playback methods have the advantage over call-triggered playback in that all test subjects receive the same number of stimuli (although the relative timing of the subject's calls and stimuli may vary). Unfortunately it is not known how responses of male frogs might differ if presented with stimuli delivered in an interactive (e.g., call-triggered) and noninteractive fashion. Experiments specifically designed to make such a comparison would be very helpful to those planning a study of anuran communication involving sound playbacks to males.

Chorusing is one of the most noteworthy habits of many species of frogs. Indeed it would be incredible if the consequences of calling in aggregations had failed to exert profound effects on the evolution of male vocal behavior. Equipment and software that permit researchers to monitor calling males within a network of interacting individuals are indispensable if we are to elucidate the vocal dynamics within these assemblages. In particular, patterns of male–male vocal competition as well as the details of call-timing adjustments and selective attention are phenomena that require investigation in additional species using approaches outlined in this chapter. Testing the link between male mating success and vocal behavior under chorus conditions is a critically important but difficult exercise that also lends itself well to multichannel monitoring of calling males. Published studies that used this technique indicate that females of two species do well in the field at selecting males producing signals with relatively high energy content (Passmore et al 1992; Schwartz 1994; Schwartz et al. 1995). However work in progress with *H. versicolor* suggests that patterns of selectivity observed in the laboratory sometimes may be much weakened in the chorus (Schwartz, Buchanan and Gerhardt, unpublished data; also see Sullivan and Hinshaw 1992; Bertram et al. 1996).

Interactive playback used with multichannel monitoring could be particularly effective in investigations of communication (McGregor and Dabelsteen 1996). For example the vocal behavior of male frogs in aggressive encounters might be correlated with their relative ability to produce attractive advertisement calls—especially if physical condition or fuel reserves constrain both forms of signaling. To test this idea the attributes of advertisement calls could be ascertained by monitoring chorus members. Then these individuals could be challenged with interactive playbacks. In fact such an experiment is in progress using males of *Hyla versicolor* calling in the artificial pond. It is not certain whether patterns of selective attention as described for call-timing adjustments extend to the elaboration of advertisement calls of males for any species. Do males adjusting the attractiveness of their signals attend to the overall level of background noise in the chorus or fine-tune the energy content of their vocalizations to just their closest neighbors? It is also possible that males might preferentially attend and respond to only the "best" caller they can detect. Thus it is conceivable that there are different spatial arrays of vocal interaction for call timing and call elaboration (Table 14.2)! Call monitoring combined with simultaneous playbacks from one or more speakers should allow us to test these possibilities. Males of some species alter their calling behavior when they detect a female nearby (Wells 1988). This change may be detected by other males, who in turn increase the attractiveness of their calls (e.g., *H. versicolor;* Schwartz and Buchanan, unpublished data). Call monitoring of the chorus could reveal the dynamics and extent of this process, while interactive playback tests during female releases could explore the efficacy of such behavior.

Although only a handful of biologists have used the newer methods in studies of animal communication, I hope that these technologies will be effectively exploited by more scientists in the near future. This should enhance our ability to tackle interesting questions that once may have been considered too difficult to answer.

Acknowledgments

I thank Kent Wells, Carl Gerhardt, Stan Rand, Buck Buchanan, Stephen Ressel, Catherine Robb Bevier, and Michael Jennions for their help with research described in this chapter. I am indebted to Andy Moiseff for introducing me to the C programming language and helping me with various research-related electronics projects. Mike Ryan, Nikki Kime, Mark Bee, and an anonymous reviewer provided valuable comments on the manuscript. A. Stanley Rand has been a good friend and professional colleague over the past 19 years. I am especially grateful to him for his intellectual input and generous allocation of time and resources during my many trips to Panama. I am also grateful to the STRI staff for their valuable assistance. Research described in this chapter was supported by a University of Connecticut Research Foundation Award and Smithsonian Institution Short Term Visitor Awards to JJS, National Science Foundation grants to KDW, HCG, and JJS, and a University of Connecticut Summer Research Fellowship to CRB.

References

Arak, A. 1983. Vocal interactions, call matching and territoriality in a Sri Lankan treefrog, *Philautus leucorhinus* (Rhacophoridae). Animal Behaviour 31:292–302.

Ayre, D. J., P. Coster, W. N. Bailey, and J. D. Roberts. 1984. Calling tactics in *Crinia georgiana* (Anura: Myobatrachidae): Alternation and variation in call duration. Australian Journal of Zoology 32:463–470.

Backwell, P. R. 1988. Functional partitioning in the two-part call of the leaf-folding frog, *Afrixalus brachycnemis*. Herpetologica 44:1–7.

Becker, P. H. 1982. The coding of species-specific characteristics in bird sounds. Pp. 213–252. *In* D. H. Kroodsma, and E. H. Miller (eds.), Acoustic Communication in Birds. Academic Press, New York.

Bee, M. A., and S. A. Perrill 1996. Responses to conspecific advertisement calls in the green frog (*Rana clamitans*) and their role in male–male communication. Behaviour 133:283–301.

Bee, M. A., S. A. Perrill, and P. C. Owen. 1999. Size assessment and simulated territorial encounters between male green frogs (*Rana clamitans*). Behavioral Ecology and Sociobiology 45:177–184.

Bertram, S., M. Berrill, and E. Nol. 1996. Male mating success and variation in chorus attendance within and among breeding seasons in the gray treefrog (*Hyla versicolor*). Copeia 1996:729–734.

Bertram, S., and L. Johnson. 1998. An electronic technique for monitoring the temporal aspects of acoustic signals of captive organisms. Bioacoustics 9:107–118.

Bevier, C. R. 1997. Utilization of energy substrates during calling activity in tropical frogs. Behavioral Ecology and Sociobiology 41:343–352.

Bond, A. B. 1989a. Toward a resolution of the paradox of aggressive displays: I. Optimal deceit in the communication of fighting ability. Ethology 81:29–46.

Bond, A. B. 1989b. Toward a resolution of the paradox of aggressive displays: II. Behavioral efference and the communication of intentions. Ethology 81:235–249.

Bond, A. P. 1992. Aggressive motivation in the midas cichlid: Evidence for behavioral efference. Behaviour 122:135–152.

Bradbury, J. W., and S. L. Vehrencamp. 1994. SingIt! A program for interactive playback on the Macintosh. Bioacoustics 5:308–310.

Bradbury, J. W., and S. L. Vehrencamp. 1998. Principles of Animal Communication. Sinauer Associates, Sunderland, MA.

Brenowitz, E. A., and G. J. Rose. 1994. Behavioural plasticity mediates aggression in choruses of the Pacific treefrog. Animal Behaviour 47:633–641.

Brush, J. S., and P. M. Narins 1989. Chorus dynamics of a Neotropical amphibian assemblage: Comparison of computer simulation and natural behaviour. Animal Behaviour 37:33–44.

Bush, S. L. 1997. Vocal behavior of males and females in the Majorcan midwife toad. Journal of Herpetology 31:251–257.

Clark, C. W. 1989. Call tracks of bowhead whales based on call characteristics as an independent means of determining tracking parameters. Report of the sub-committee on protected species and aboriginal subsistence whaling. Appendix. Report of the International Whaling Commission 39:111–112.

Clutton-Brock, T. H., and S. D. Albon. 1979. The roaring of red deer and the evolution of honest advertisement. Behaviour 69:145–169.

Cole, B. J., and D. Cheshire. 1996. Mobile cellular automata models of ant behavior: Movement activity of *Leptothorax allardycei*. American Naturalist 148:1–15.

Dabelsteen, T., and P. K. McGregor. 1996. Dynamic acoustic communication and interactive playback. Pp. 398–408. *In* D. E. Kroodsma and E. H. Miller (eds.), Ecology and Evolution of Acoustic Communication in Birds. Comstock, Ithaca, NY.

Duellman, W. E., and L. Trueb. 1966. Neotropical hylid frogs, genus *Smilisca*. University of Kansas Publications, Museum of Natural History 17:281–375.

Dyson, M. L., and N. I. Passmore. 1988a. Two-choice phonotaxis in *Hyperolius marmoratus:* The effect of temporal variation in presented stimuli. Animal Behaviour 36:648–652.

Dyson, M. L., and N. I. Passmore. 1988b. The combined effect of intensity and the temporal relationship of stimuli on the phonotactic responses of female painted reed frogs (*Hyperolius marmoratus*). Animal Behaviour 36:1555–1556.

Fellers, G. M. 1979. Aggression, territoriality, and mating behaviour in North American treefrogs. Animal Behaviour 27:107–119.

Forester, D.C., and W. K. Harrison. 1987. The significance of antiphonal vocalisation by the spring peeper, *Pseudacris crucifer.* Behaviour 103:1–15.

Forester, D.C., D. V. Lykens, and W. K. Harrison. 1989. The significance of persistent vocalization by the spring peeper, *Pseudacris crucifer* (Anura: Hylidae). Behaviour 108:197–208.

Gerhardt, H. C. 1978. Discrimination of intermediate sounds in a synthetic call continuum by female green treefrogs. Science 199:1089–1091.

Gerhardt, H. C. 1994. The evolution of vocalization in frogs and toads. Annual Review of Ecology and Systematics 25:293–324.

Gerhardt, H. C., and J. J. Schwartz. 1995. Interspecific interactions and species recognition. Pp. 603–632. *In* H. Heatwole and B. K. Sullivan (eds.), Amphibian Biology. Vol 2. Social Behaviour. Chipping Norton, Surrey Beatty and Sons, New South Wales, Australia.

Gerhardt, H. C., M. L. Dyson, and S. D. Tanner 1996. Dynamic properties of the advertisement calls of gray tree frogs: Patterns of variability and female choice. Behavioral Ecology 7:7–18.

Given, M. F. 1993. Male response to female vocalizations in the carpenter frog, *Rana virgatipes.* Animal Behaviour 46:1139–1149.

Given, M. F. 1999. Frequency alteration of the advertisement call in the carpenter frog, *Rana virgatipes.* Herpetologica 55:304–317.

Goodwin, B. 1998. All for one. New Scientist 158:32–35.

Grafe, T. U. 1995. Graded aggressive calls in the African reed frog, *Hyperolius marmoratus* (Hyperoliidae). Ethology 101:67–81.

Grafe, T. U. 1996. The function of call alternation in the African reed frog *Hyperolius marmoratus:* Precise call timing prevents auditory masking. Behavioral Ecology and Sociobiology 38:149–158.

Grafe, T. U. 1997. Costs and benefits of mate choice in the lek-breeding reed frog, *Hyperolius marmoratus.* Animal Behaviour 53:1103–1117.

Greenfield, M. D. 1994a. Cooperation and conflict in the evolution of signal interactions. Annual Review of Ecology and Systematics 25:97–126.

Greenfield, M. D. 1994b. Synchronous and alternating choruses in insects and anurans: Common mechanisms and diverse functions. American Zoologist 34:605–615.

Greenfield, M. D. 1997. Acoustic communication in Orthoptera. Pp. 197–230. *In* S. K. Gangwere, M. C. Muralirangan, and M. Muralirangan (eds.), The Binomics of Grasshoppers and Their Kin. CAB International.

Greenfield, M. D., and I. Roizen. 1993. Katydid synchronous chorusing is an evolutionarily stable outcome of female choice. Nature 364:618–620.

Greenfield, M. D., and K. C. Shaw. 1983. Adaptive significance of chorusing with special reference to the Orthoptera. Pp. 1–27. *In* G. K. Morris and D. T. Gwynne (eds.), Orthopteran Mating Systems: Sexual Competition in a Diverse Group of Insects. Westview Press, Boulder, CO.

Greenfield, M. D., M. K. Tourtellot, and W. A. Snedden. 1997. Precedence effects and the evolution of chorusing. Proceedings of the Royal Society of London B 264:1355–1361.

Howard, R. D., and J. G. Palmer. 1995. Female choice in *Bufo americanus:* Effects of dominant frequency and call order. Copeia 1995:212–217.

Howard, R. D., and J. R. Young. 1998. Individual variation in male vocal traits and female mating preferences in *Bufo americanus.* Animal Behaviour 55:1165–1179.

Ibáñez R. D. 1991. Synchronized calling in *Centrolenella granulosa* and *Smilisca sila* (Amphibia, Anura). Ph.D. dissertation, University of Connecticut, Storrs, CT.

Ibáñez, R. D. 1993. Female phonotaxis and call overlap in the neotropical glass frog, *Centrolenella granulosa.* Copeia 1993:846–850.

Jehle, R., and A. Arak. 1998. Graded call variation in the Asian cricket frog *Rana nicobariensis.* Bioacoustics 9:35–48.

Jennions, M. D. 1994. Causes of cricket synchrony. Current Biology 4:1047.

Klump, G. M., and H. C. Gerhardt. 1987. Use of non-arbitrary acoustic criteria in mate choice by female gray treefrogs. Nature 326:286–288.

Klump, G. M., and H. C. Gerhardt. 1992. Mechanisms and function of call-timing in male–male interactions in frogs. Pp. 153–174. *In* P. K. McGregor (ed.), Playback and Studies of Animal Communication. Plenum Press, New York.

Lambrechts, M. M. 1992. Male quality and playback in the great tit. Pp. 135–152. *In* P. K. McGregor (ed.), Playback and Studies of Animal Communication. Plenum Press, New York.

Lemon, R. E. 1971. Vocal communication of the frog *Eleutherodactylus martinicensis*. Canadian Journal of Zoology 49:211–217.

Lemon, R. E., and J. Struger. 1980. Acoustic entrainment to randomly generated calls by the frog *Hyla crucifer*. Journal of the Acoustical Society of America 67:2090–2095.

Littlejohn, M. J. 1977. Long-range acoustic communication in anurans: An integrated and evolutionary approach. Pp. 263–294. *In* D. H. Taylor and S. I. Guttman (eds.), The Reproductive Biology of Amphibians. Plenum Press, New York.

Littlejohn, M. J., and P. A. Harrison. 1985. The functional significance of the diphasic advertisement call of *Geocrinia victoriana* (Anura: Leptodactylidae). Behavioral Ecology and Sociobiology 16:363–373.

Littlejohn, M. J., Harrison, P. A., and R. C. MacNally. 1985. Interspecific acoustic interactions in sympatric populations of *Ranidella signifera* and *R. parinsignifera* (Anura: Leptodactylidae). Pp. 287–296. *In* G. Grigg, R. Shine and H. Ehrmann (eds.), The Biology of Australasian Frogs and Reptiles. Surrey Beatty and Sons, Chipping Norton, New South Wales, Australia.

Littlejohn, M. J., and A. A. Martin. 1969. Acoustic interaction between two species of leptodactylid frogs. Animal Behaviour 17:785–791.

Loftus-Hills, J. J. 1974. Analysis of an acoustic pacemaker in Strecker's chorus frog, *Pseudacris streckeri* (Anura: Hylidae). Journal of Comparative Physiology 90:75–87.

Lopez, P. T., P. M. Narins, E. R. Lewis, and S. W. Moore. 1988. Acoustically-induced call modification in the white-lipped frog, *Leptodactylus albilabris*. Animal Behaviour 36:1295–1308.

MacNally, R. C. 1982. Ecological, behavioural, and energy dynamics of two sympatric species of *Ranidella* (Anura). Ph.D. Dissertation, University of Melbourne, Parkville.

Márquez, R., J. M. Pargana, and E. G. Crespo. 1998. Acoustic competition in male *Pelodytes punctatus*. Interactive playback experiments. Abstract 7. *In* Program and Abstracts, Joint Meeting of American Society of Ichthyologists and Herpetologists, Society for Amphibians and Reptiles, Herpetologists League, American Elasmobranch Society, Canadian Association of Herpetologists 1998.

McGregor, P. K., and T. Dabelsteen. 1996. Communication networks. Pp. 409–425. *In* D. E. Kroodsma and E. H. Miller (eds.), Ecology and Evolution of Acoustic Communication in Birds. Comstock, Ithaca, NY.

McGregor, P. K., and A. G. Horn. 1992. Strophe length and response to playback in great tits. Animal Behaviour 43:667–676.

McGregor, P. K., T. Dabelsteen, C. W. Clark, and J. L. Bower. 1997. Accuracy of passive acoustic location system: Empirical studies in terrestrial environments. Ethology Ecology and Evolution 9:269–286.

Minckley, R. L., M. D. Greenfield, and M. K. Tourtellot. 1995. Chorus structure in tarbush grasshoppers: Inhibition, selective phonoresponse and signal competition. Animal Behaviour 50:579–594.

Moore, B. C. J. 1982. An Introduction to the Psychology of Hearing. 2nd ed. Academic Press, New York.

Moore, S. W., E. R. Lewis, P. M. Narins, and P. T. Lopez. 1989. The call-timing algorithm of the white-lipped frog, *Leptodactylus albilabris*. Journal of Comparative Physiology A 164:309–319.

Narins, P. M. 1982a. Behavioral refractory period in Neotropical treefrogs. Journal of Comparative Physiology A 148:337–344.

Narins, P. M. 1982b. Effects of masking noise on evoked calling in the Puerto Rican Coqui (Anura, Leptodactylidae). Journal of Comparative Physiology 147:439–446.

Narins, P. M. 1987. Coding of signals in noise by amphibian auditory nerve fibers. Hearing Research 26:145–154.

Narins, P. M. 1992. Reduction of tympanic membrane displacement during vocalization of the arboreal frog, *Eleutherodactylus coqui*. Journal of the Acoustic Society of America 91:3551–3557.

Narins, P. M., and R. R. Capranica. 1978. Communicative significance of the two-note call of the treefrog *Eleutherodactylus coqui*. Journal of Comparative Physiology 127:1–9.

Narins, P. M., and R. Zelick. 1988. The effects of noise on auditory processing and behavior in amphibians. Pp. 511–536. *In* B. Fritzsch, W. Wilczynski, M. J. Ryan, T. Hetherington, and W. Walkowiak (eds.), The Evolution of the Amphibian Auditory System. John Wiley, New York.

Ovaska, K. E., and J. Caldbeck. 1997. Courtship behavior and vocalizations of the frogs *Eleutherodactylus antillensis* and *E. cochranae* on the British Virgin Islands. Journal of Herpetology 31:149–155.

Passmore, N. I., P. J. Bishop, and N. Caithness. 1992. Calling behavior influences mating success in male painted reed frogs, *Hyperolius marmoratus*. Ethology 92:227–241.

Passmore, N. I., and S. R. Telford. 1981. The effect of chorus organization on mate localization in the painted reed frog (*Hyperolius marmoratus*). Behavioral Ecology and Sociobiology 9:291–293.

Payne, R. J. H., and M. Pagel. 1997. Why do animals repeat displays? Animal Behaviour 54:109–119.

Pina, R. F., and A. Channing. 1981. A portable frog call synthesizer. Monitore Zoologico Italiano. N. S. Supplemento XV 20:387–392.

Pollack, G. S. 1988. Selective attention in an insect auditory neuron. Journal of Neuroscience 8:2635–2639.

Pough, F. W., W. E. Magnusson, M. J. Ryan, K. D. Wells, and T. L. Taigen. 1992. Behavioral energetics. Pp. 395–436. *In* M. E. Feder and W. W. Burggren (eds.), Environmental Physiology of the Amphibians. University of Chicago Press, Chicago.

Popp, J. W. 1989. Methods of measuring avoidance of acoustic interference. Animal Behaviour 38:358–360.

Prestwich, K. N. 1994. The energetics of acoustic signaling in anurans and insects. American Zoologist 34:625–643.

Rand, A. S., and M. J. Ryan. 1981. The adaptive significance of a complex vocal repertoire in a Neotropical frog. Zeitschrift für Tierpsychologie 57:209–214.

Römer, H. 1993. Environmental and biological constraints for the evolution of long-range signalling and hearing in acoustic insects. Philosophical Transactions of the Royal Society of London 340:179–185.

Römer, H., B. Hedwig, and S. R. Ott. 1997. Proximate mechanism of female preference for the leader male in synchronizing bushcrickets (*Megapoda elongata*). P. 322. *In* N. Elsner and H. Wassle (eds.), From Membrane to Mind. Proc. 25th Göttingen Neurobiology Conference Vol. II. Thieme, Stuttgart, New York.

Rose, G. J., R. Zelick, and A. S. Rand. 1988. Auditory processing of temporal information in a neotropical frog is independent of signal intensity. Ethology 77:330–336.

Rosen, M., and R. E. Lemon. 1974. The vocal behavior of spring peepers, *Hyla crucifer*. Copeia 1974:940–950.

Ryan, M. J. 1985. The Túngara Frog: A Study in Sexual Selection and Communication. University of Chicago Press, Chicago.

Ryan, M. J. 1986. Synchronized calling in a treefrog (*Smilisca sila*). Brain Behavior and Evolution 29:196–206.

Ryan, M. J. 1988. Energy, calling, and selection. American Zoologist 28:885–898.

Ryan, M. J., and A. Keddy-Hector. 1992. Directional patterns of female mate choice and the role of sensory biases. American Naturalist 139:S4-S35.

Scharf, B. 1970. Critical bands. Pp. 159–202. *In* J. V. Tbias (ed.), Foundations of Modern Auditory Theory. Academic Press, New York.

Schneider, H. 1977. Acoustic behavior and physiology of vocalization in the European tree frog, *Hyla arborea* (L.). Pp. 295–335. *In* D. H. Taylor and S. I. Guttman (eds.), The Reproductive Biology of Amphibians. Plenum Press, New York.

Schwartz, J. J. 1986. Male calling behavior and female choice in the neotropical treefrog *Hyla microcephala*. Ethology 73:116–127.

Schwartz, J. J. 1987a. The function of call alternation in anuran amphibians: A test of three hypotheses. Evolution 41:461–471.

Schwartz, J. J. 1987b. The importance of spectral and temporal properties in species and call recognition in a neotropical treefrog with a complex vocal repertoire. Animal Behaviour 35:340–347.

Schwartz, J. J. 1989. Graded aggressive calls of the spring peeper, *Pseudacris crucifer*. Herpetologica 45:172–181.

Schwartz, J. J. 1991. Why stop calling? A study of unison bout singing in a neotropical treefrog. Animal Behaviour 42:565–577.

Schwartz, J. J. 1993. Male calling behavior, female discrimination and acoustic interference in the Neotropical treefrog *Hyla microcephala* under realistic acoustic conditions. Behavioral Ecology and Sociobiology 32:401–414.

Schwartz, J. J. 1994. Male advertisement and female choice in frogs: New findings and recent approaches to the study of communication in a dynamic acoustic environment. American Zoologist 34:616–624.

Schwartz, J. J., and H. C. Gerhardt. 1995. Directionality of the auditory system and call pattern recognition during acoustic interference in the gray treefrog, *Hyla versicolor*. Auditory Neuroscience 1:195–206.

Schwartz, J. J., and H. C. Gerhardt. 1998. The neuroethology of frequency preferences in the spring peeper. Animal Behaviour 56:55–69.

Schwartz, J. J. and A. S. Rand. 1991. The consequences for communication of call overlap in the túngara frog, a neotropical anuran with a frequency-modulated call. Ethology 89:73–83.

Schwartz, J. J., S. Ressel, and C. R. Bevier. 1995. Carbohydrate and calling: Depletion of muscle glycogen and the chorusing dynamics of the neotropical treefrog *Hyla microcephala*. Behavioral Ecology and Sociobiology 37:125–135.

Schwartz, J. J., and K. D. Wells. 1983a. An experimental study of acoustic interference between two species of neotropical treefrogs. Animal Behaviour 31:181–190.

Schwartz, J. J., and K. D. Wells. 1983b. The influence of background noise on the behavior of a neotropical treefrog, *Hyla ebraccata*. Herpetologica 39:121–129.

Schwartz, J. J., and K. D. Wells. 1984a. Interspecific acoustic interactions of the neotropical treefrog *Hyla ebraccata*. Behavioral Ecology and Sociobiology 14:211–224.

Schwartz, J. J., and K. D. Wells. 1984b. Vocal behavior of the neotropical treefrog *Hyla phlebodes*. Herpetelogica 40:452–463.

Schwartz, J. J., and K. D. Wells. 1985. Intra- and interspecific vocal behavior of the neotropical treefrog *Hyla microcephala*. Copeia 1985:27–38.

Snedden, W. A., M. D. Greenfield, and Y. Jang. 1998. Mechanisms of selective attention in grasshopper choruses: Who listens to whom? Behavioral Ecology and Sociobiology 43:59–66.

Spiesberger, J. L., and K. M. Fristrup. 1990. Passive localization of calling animals and sensing of their acoustic environment using acoustic tomography. American Naturalist 135:107–153.

Stewart, M. M., and A. S. Rand. 1991. Vocalizations and the defense of retreat sites by male and female frogs, *Eleutherodactylus coqui*. Copeia 1991:1013–1024.

Sullivan, B. K. and S. H. Hinshaw. 1992. Female choice and selection on male calling behaviour in the grey treefrog *Hyla versicolor*. Animal Behaviour 44:733–744.

Sullivan, B. K. and M. R. Leek. 1986. Acoustic communication in Woodhouse's toad (*Bufo woodhousei*). I. Response of calling males to variation in spectral and temporal components of advertisement calls. Behaviour 98:305–319.

Sullivan, B. K., M. J. Ryan, and P. A. Verrell. 1995. Female choice and mating system structure. Pp. 469–517. *In* H. Heatwole and B. K. Sullivan (eds.), Amphibian Biology. Vol 2. Social Behaviour. Chipping Norton, Surrey Beatty and Sons, New South Wales, Australia.

Tuttle, M. D., and M. J. Ryan. 1982. The role of synchronized calling, ambient light, and ambient noise, in anti-bat-predator behavior of a treefrog. Behavioral Ecology and Sociobiology 11:125–131.

Wagner, W. E. 1989a. Graded aggressive signals in Blanchard's cricket frog: Vocal responses to opponent proximity and size. Animal Behaviour 38:1025–1038.

Wagner, W. E. 1989b. Social correlates of variation in male calling behavior in Blanchard's cricket frog, *Acris crepitans blanchardi*. Ethology 82:27–45.

Wagner, W. E., and B. K. Sullivan. 1995. Sexual selection in the Gulf Coast toad, *Bufo valliceps:* Female choice based on variable characters. Animal Behaviour 49:305–319.

Wells, K. D. 1977a. The courtship of frogs. Pp. 233–262. *In* D. H. Taylor and S. I. Guttman (eds.), The Reproductive Biology of Amphibians. Plenum Press, New York.

Wells, K. D. 1977b. The social behavior of anuran amphibians. Animal Behaviour 25:666–693.

Wells, K. D. 1988. The effects of social interactions on anuran vocal behavior. Pp. 433–454. *In* B. Fritszch, W. Wilczynski, M. J. Ryan, T. Hetherington, and W. Walkowiak (eds.), The Evolution of the Amphibian Auditory System. John Wiley, New York.

Wells, K. D. 1989. Vocal communication in a neotropical treefrog, *Hyla ebraccata:* Responses of males to graded aggressive calls. Copeia 1989:461–466.

Wells, K. D., and T. Bard. 1987. Vocal communication in a neotropical treefrog, *Hyla ebraccata:* Responses of females to advertisement and aggressive calls. Behaviour 101:200–210.

Wells, K. D., and J. J. Schwartz. 1984a. Vocal communication in a neotropical treefrog, *Hyla ebraccata:* Advertisement calls. Animal Behaviour 32:405–420.

Wells, K. D., and J. J. Schwartz. 1984b. Vocal communication in a neotropical treefrog, *Hyla ebraccata:* Aggressive calls. Behaviour 91:128–145.

Wells, K. D., and T. L. Taigen. 1986. The effect of social interactions on calling energetics in the gray treefrog (*Hyla versicolor*). Behavioral Ecology and Sociobiology 19:9–18.

Wells, K. D,. and T. L. Taigen. 1989. Calling energetics of the neotropical treefrog, *Hyla microcephala*. Behavioral Ecology and Sociobiology 25:13–22.

Whitney, C. L., and J. R. Krebs. 1975. Mate selection in Pacific treefrogs. Nature 255:325–326.

Wilczynski, W., A. S. Rand, and M. J. Ryan. 1995. The processing of spectral cues by the call analysis system of the túngara frog, *Physalaemus pustulosus*. Animal Behaviour 49:911–929.

Wiley, R. H. 1994. Errors, exaggeration, and deception in animal communication. *In* L. Real (ed.), Behavioral Mechanisms in Evolutionary Ecology. The University of Chicago Press, Chicago.

Yin, T. C. T. 1994. Physiological correlates of the precedence effect and summing localization in the inferior colliculus of the cat. Journal of Neuroscience 14:5170–5186.

Zelick, R., and P. M. Narins 1985. Characterization of the advertisement call oscillator in the frog *Eleutherodactylus coqui*. Journal of Comparative Physiology 156:223–229.

Zimmerman, B. L., and J. P. Bogart. 1984. Vocalizations of primary forest frog species in the Central Amazon. Acta Amazonica 14:473–519.

Zurek, P. M. 1980. The precedence effect and its possible role in the avoidance of interaural ambiguities. Journal of the Acoustic Society of America 67:952–964.

15

CRISTINA GIACOMA AND SERGIO CASTELLANO

Advertisement Call Variation and Speciation in the Bufo viridis *Complex*

Early ethologists focused much of their attention on the origin and evolution of animal communication, arguing that innate (Lorenz 1965) signals were designed to maximize information transfer through ritualization (Huxley 1966; reviewed in Halliday and Slater 1983; Hauser 1996). Within this framework they described signals and applied the comparative method to reconstruct signal evolution (e.g., Lindauer 1961). They also considered that ritualized signals convey more reliable information as they are highly stereotyped, repetitive, exaggerated, and consequently less ambiguous. In sexually reproducing species, communication between partners occurs during preliminary phases of courtship. Thus courtship displays should send unequivocal information about motivation to mate and species identity.

The function of courtship signals as an ethological premating isolating mechanism was also stressed by evolutionary biologists (Dobzhansky 1937; Mayr 1942; Lande 1981; Paterson 1982). Most studies on reproductive behavior have dealt with the species recognition function of courtship and, together with a detailed description of the highly stereotyped sequence of species specific sounds or displays (e.g., Lorenz 1950; Tinbergen 1951), they have often provided evidence that females are able to discriminate between conspecific and heterospecific displays and that they prefer the former to the latter (Littlejohn and Michaud 1959; Gerhardt 1978, 1981, 1988; Ewing 1989; Rand et al. 1992; Ryan and Rand 1993a). Since the pioneering studies by Blair (1964), anuran calling behavior has been a good model by which to study these issues. Furthermore the neural basis of the receiver's ability to decode species-specific information has been well documented in anurans (e.g., Capranica 1976; Wilczynski and Ryan 1988).

Although species recognition is still a crucial issue for many studies on speciation, during the last two decades growing interest in sexual selection has encouraged researchers to focus their attention on within-population variation of courtship displays. Anuran vocalizations proved to be a good model for these studies as well. Calling behavior shows considerable variation among males of the same population. Females can discriminate between different calls and choose males on this basis, thereby generating sexual selection on call traits (reviewed in Arak 1983; Sullivan et al. 1996b; Jennions and Petrie 1997). Both species recognition and mate preference result from an interaction between variation of male signals and female preferences. This interaction can cause nonrandom pairing, both among and within species (Ryan 1990, 1991; Ryan and Rand 1993b). Recent studies also suggest that neural processes resulting in selective phonotaxis toward conspecific calls are not qualitatively different from those that mediate intraspecific mate choice (Gerhardt et al. 1994). Given fundamental similarities in the above-mentioned underlying processes, the notion that individuals recognize features diagnostic of a lineage can be considered problematic. Therefore the term *mate recognition* instead of *species recognition* is now recognized as more appropriate in all settings where it has been clearly established that mechanisms have been selected to avoid mating with heterospecifics (Sullivan et al. 1996a; Gergus et al. 1997).

More recently studies on sexual selection have examined signal variation within individuals. Their main goal has been to outline different patterns of signal property variability to provide insight into the functions of different acoustic properties (Gerhardt 1991) and into possible evolutionary directions of signals and receiver preferences (Boake 1989, 1994; Brenowitz 1994; Wagner and Sullivan 1995). Call properties vary within individuals along a wide continuum (Ryan and Wilczynski 1991; Giacoma et al. 1997; Wollerman 1998) that Gerhardt (1991) categorized into static properties (when their coefficient of variation, CV, is less than 5%) and dynamic properties (when their CV is more than 12%). The within-individual variability of call properties may result from various selective pressures (e.g., directional or stabilizing sexual selection) and constraints (Gerhardt 1991, 1994a; McClelland et al. 1996, 1998; Castellano and Giacoma 1998a; Wollerman 1998).

Studies on patterns of geographic variation represent a more recent approach for evaluating the effects of selection, as well as internal (phylogenetic, morphological, or physiological) and ecological constraints on communication. These studies show substantial differences among populations, both in signals emitted (Ryan and Wilczynski 1991; Ryan et al. 1996; Sullivan et al. 1996a; Gergus et al. 1997; Castellano et al. 2000; Castellano and Giacoma, unpublished data) and in the behavior of receivers (Gerhardt 1994b; Endler and Houde 1995). In some cases the underlying morphological bases of geographic variation in signal and receiver are known (McClelland et al. 1996, 1998). Such studies show that behavioral variation can be partitioned into a hierarchical pattern according to levels of organization and permanency of effects (Boake 1994).

In this chapter we first describe patterns of variation of advertisement call parameters within the *Bufo viridis* complex. We partition this variation into four levels: within a call bout, among bouts of calling of the same individual on different nights, among individuals, and among population levels. Then we discuss the causes of variation at these various levels. Finally we discuss the biological relevance of various types of variation by comparing data obtained at two different stages of the genetic divergence process among natural populations of the *Bufo viridis* complex.

Study Organism

The green toad, *Bufo viridis* complex, ranges from Mediterranean countries of North Africa and Europe (excluding the Iberian Peninsula and southern France) to Central Europe and through Asia Minor as far as eastern Kazakstan, Mongolia, and northwestern China. In Central Asia there are two karyologically different groups of populations: diploid ($2n = 22$) and tetraploid ($2n = 44$) (Pisanetz 1978; Roth 1986). Pisanetz (1978) described the tetraploids as a new species and called them *B. danatensis*. Recent studies have shown that European diploid, Asian diploid, and tetraploid toads differ not only in genetic, cytogenetic, and allozyme characters (Cervella et al. 1997; Lattes 1997; Odierna et al. 1997), but also in morphological features (Stöck 1997; Castellano et al. 1998) and acoustic call properties (Castellano et al. 1998; Stöck 1998). The alpha taxonomy of Central Asiatic toads is not determined. Thus we will refer to all populations as *B. viridis* complex, only distinguishing three lineages: European diploids, Asiatic diploids, and Central Asiatic tetraploids.

Variation in the mating patterns of the green toad depends markedly on the ecology and demography of breeding populations. In Italy the breeding season begins from late February to May along a south-to-north cline, and lasts from three weeks to three months (Laoretti et al. 2000), often with more than one peak of breeding activity. Both males and females modify their behavior in response to male advertisement calls (Giacoma et al. 1993). Females reproduce once during the season and usually initiate sexual contact by moving close to and touching a calling male (Zugolaro et al. 1994; Giacoma et al. 1997). Approaching females can also be clasped on their way to the pond by noncalling males. In high density conditions unpaired males sometimes initiate scramble competition with paired males (unpublished data).

The call is a 3- to 5-second-long train of pulses in a relatively narrow band of frequencies. We have described calls using three temporal properties (call duration, intercall duration, and pulse rate) and one spectral property (fundamental frequency, which is also the dominant frequency) (Castellano and Giacoma 1998a).

From 1993 to 1996 we recorded calls of 325 breeding males from 20 populations, eight in the Italian peninsula, two in Sardinia, and nine in Central Asia. Three of the latter were diploid toads and six were tetraploid toads (for more details on localities see Dujsebayeva et al. 1997; Giacoma et al. 1997; Castellano and Giacoma 1998a; Castellano et al. 1998).

For each male we recorded several calls (3–18) and for each call we measured several acoustic properties. For each of these properties we calculated its mean value, which in turn was employed to characterize that particular male as well as to analyze the linear relationships between call properties and either temperature or body size.

Patterns of Variation at Different Levels

Although the advertisement call is often considered a single unit of information, calls do not necessarily evolve in a unitary fashion. Different acoustic properties of the call are under different constraints (Ryan 1988; Giacoma et al. 1997) and different selective pressures (Gerhardt 1991; Castellano and Giacoma 1998a), and may evolve at different rates (Cocroft and Ryan 1995). These properties may also show different patterns of variability (Castellano and Giacoma 1998a), which can be compared by calculating their coefficient of variation [CV = (SD / Mean) × 100], a dimensionless index for normally distributed parameters (e.g., Gerhardt 1991; Boake 1994). The analysis of these differences may provide a useful insight into the evolution of the advertisement call. Table 15.1 shows the partition of variation for four call characters observed in three groups of populations corresponding to three lineages: Italian diploids, Asiatic diploids, and Asiatic tetraploids. In the computation of CVs at different levels, we adopted an approach similar to that of a nested ANOVA. For each lineage we measured call property variation. At the within-bout level we measured variation of calls emitted by the same male in a single bout and calculated mean CVs for each of the three lineage groups. To analyze the pattern of variation at the within-individual level we considered a sample of 24 males from a population of the Italian peninsula and from a Sardinian population. We recorded these males on more than one night and could therefore calculate CVs for their within-bout mean call properties (corrected for body temperature). At the within-population level we calculated CVs for the mean acoustic properties of different males, corrected for body temperature. Finally at the among-population level we calculated acoustic property means for each population and computed the CVs, adjusting temporal properties for body temperature and fundamental frequency for body size. Since the CVs at one level are calculated from the mean values at the level directly below, they represent the expression of different components of the total variation of these properties (Castellano and Giacoma 1998a).

Coefficients of variation varied among call properties. The call properties with highest to lowest variation were intercall duration, call duration, pulse rate, and fundamental frequency (Table 15.1). Call and intercall duration are always more variable than pulse rate and fundamental frequency regardless of the level of analysis (Table 15.1). Following the criteria proposed by Gerhardt (1991), call and intercall duration would be classified as dynamic properties, whereas the two more-stereotyped parameters, pulse rate and fundamental frequency, would be classified as static properties of the call. Static and dynamic acoustic properties show different patterns: the CVs of both static properties, pulse rate and fundamental frequency, increase with the level of analysis, from individual to population. In contrast CVs of the two dynamic properties, call and intercall duration, exhibit most of their variation at the within-bout level (Castellano and Giacoma 1998a).

Table 15.1 Coefficients of variation of advertisement call acoustic properties at different levels of analysis for Italian diploids (I-2N); Asiatic diploids (A-2N); and Asiatic tetraploids (A-4N)

	Coefficient of Variation							
	Frequency		Pulse Rate		Call Duration		Intercall Duration	
Level of Analysis	*n*	Mean ± SD	*n*	Mean ± SD	*n*	Mean ± SD	*n*	Mean ± SD
Within-Bout								
I-2N	48	2.2 ± 1.6	34	2.8 ± 2.2	34	22.5 ± 7.1	34	52.0 ± 22.7
A-2N	14	1.6 ± 1.1	8	2.9 ± 1.4	8	18.7 ± 5.7	8	54.2 ± 36.2
A-4N	39	1.8 ± 1.2	23	4.3 ± 3.6	23	30.1 ± 9.9	23	76.9 ± 42.9
Within-Individual								
I-2N	24	3.3 ± 2.6	24	5.7 ± 5.2	24	19.3 ±14.5	24	29.9 ± 25.9
Within-Population								
I-2N	8	6.2 ± 1.4	8	6.1 ± 2.8	8	15.3 ± 9.1	8	44.3 ± 15.6
A-2N	3	6.8 ± 1.4	3	7.6 ± 3.9	3	27.6 ± 15.5	3	40.1 ± 8.4
A-4N	9	5.8 ± 2.5	9	5.8 ± 3.5	9	18.4 ± 10.3	9	65.5 ± 34.4
Among-Populations								
I-2N	1	9.0	1	14.9	1	15.2	1	24.1
A-4N	1	8.0	1	9.3	1	8.0	1	48.6

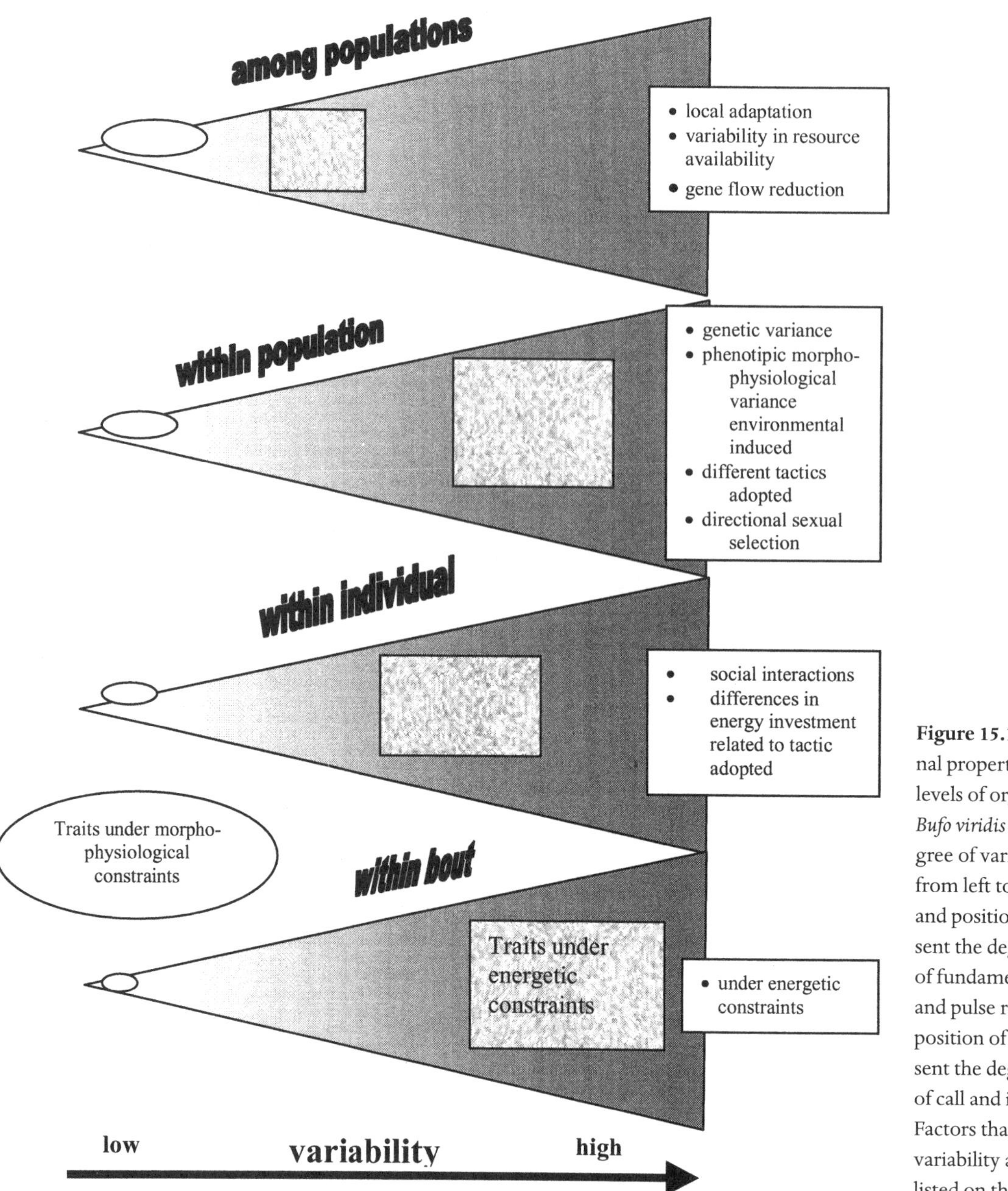

Figure 15.1. Pattern of signal property variation at four levels of organization in the *Bufo viridis* complex. The degree of variability increases from left to right. The size and position of ellipses represent the degree of variability of fundamental frequency and pulse rate. The size and position of rectangles represent the degree of variability of call and intercall duration. Factors that cause increased variability at each level are listed on the right.

To highlight the biological relevance of different types of variance we will discuss in more detail the environmental causes of variation acting at different levels on the expression of fundamental frequency, pulse rate, and call and intercall duration.

Variance Pattern of Fundamental Frequency

Variability increases from the within-bout to the among-population level (Figure 15.1; Table 15.1). The low level of within-bout variance in fundamental frequency may be interpreted as the result of morphological constraints. In many anurans (reviewed in Ryan 1986, 1988) as well as in green toads, body size has a strong negative correlation with frequency (Nevo and Schneider 1976; Giacoma et al. 1997; Castellano and Giacoma 1998a).

Variation is higher at the within-individual level than at the within-bout level (Table 15.1). The fundamental frequency significantly differs among bouts emitted by the same individual, at least in the Sardinian population ($F = 2.426$; $df = 8, 124$; $p = 0.018$). Since in green toads the fundamental frequency is not significantly influenced by the body temperature of calling males (Giacoma et al. 1997; Castellano and Giacoma 1998a) this variability probably results from vocal interactions with neighboring males (Wagner 1989; Howard and Young 1998). In fact in a south-Italian

Table 15.2 Fundamental frequency (Hz) and intercall duration (seconds) of advertisement calls recorded in southern Italy

	Mean ± SD	Min.	Max.	Range	CV (%)
Fundamental Frequency					
Non-overlapping calls	1420 ± 99.9	1280	1600	320	7
Temporally overlapping calls	1377 ± 149	1100	1600	500	11
Intercall Duration					
Non-overlapping calls	10.1 ± 1.42	1.0	32.0	31.0	14
Temporally overlapping calls	13.7 ± 5.46	1.4	29.9	28.5	40

population (Amendolea) we found that males call at lower frequencies (minimum value 1100 Hz vs. 1280 Hz) when their calls temporally overlap with those of other nearby calling males (Table 15.2). Martin (1972) suggested that, in *Bufo,* changes in the air pressure within the vocal tract may be responsible for within-individual changes in dominant frequency.

When we compare calls emitted by the same male recorded on different nights with those of other males from the same population recorded more than once during the breeding season we find significant differences among males ($n = 24$; $F = 3.607$; $df = 13, 27$; $p < 0.01$). Within-population variation is higher than that at lower levels. Since body size has a strong influence on call frequency, fundamental frequency variation among individuals of the same population may be mainly a consequence of variation in male body size. Fundamental frequencies regressed against body length generate high values of squared multiple *r*'s, such as $r^2 = 0.246$ for the Italian island diploid population of Portoscuso ($n = 71$); $r^2 = 0.368$ for the Italian mainland diploid population of Zucchea ($n = 75$); $r^2 = 0.474$ for the Asiatic diploid population of Kopa ($n = 15$); and $r^2 = 0.463$ for the Asiatic tetraploid population of Almaty ($n = 15$).

Phenotypic variation is a prerequisite for selection (Darwin 1859). Fundamental frequency can influence male reproductive success by influencing female phonotactic behavior. In playback experiments *Bufo viridis* females showed a slight preference for frequencies 150 Hz lower than the population mean over frequencies 150 Hz higher than the mean (only the sample of females that were larger than the population average body size showed a statistically significant preference for lower frequency stimulus), although they significantly preferred the mean frequencies of the population over frequencies below the population range (Castellano and Giacoma 1998a). Call variation among males may also influence success in male–male competition for mates. To test the role of call frequency in male–male competition we played back artificial calls with a low (1300 Hz) or a high (1600 Hz) fundamental frequency while slowly approaching a calling male with the loudspeaker. Males that responded by attacking the source of the lower-frequency stimulus were significantly larger than males that moved away or did not react (Doglio et al. 2000). Call dominant frequency may be involved in the assessment of competitive ability between rival males, thereby favoring bigger males. Therefore lower frequency calls should favor larger-than-average males both in male–male competition and in female choice contexts. In *Bufo viridis* larger males reach sexual maturity earlier (Giacoma et al. 1997). The onset of sexual maturity is an adaptive life-history trait that can vary among individuals of the same population. Therefore it is possible that female preference for lower frequencies and larger body size favors early-maturing males and results in an increase of female fitness: the younger the males start reproduction, the faster will be the recruitment rate of breeding males in the population (Giacoma et al. 1997).

Females show size-based variation in mating preferences (Giacoma et al. 1997; Castellano and Giacoma 1998a). Increased variation in a preference will reduce its strength of selection (Jennions et al. 1995; Jennions and Petrie 1997; Howard and Young 1998). This preference exhibited by larger females might result from larger females being tuned to lower frequencies, as has been shown in cricket frogs (Ryan et al. 1992). In nature female preferences for low frequencies may have weak selective effects, since only one of six populations studied shows a significant large-male advantage in reproductive success (Giacoma, unpublished data). These data on the reproductive success of natural populations indicate that, even when females show a consistent pattern, the effects of their preferences may be masked by the effects of ecological or social factors on the mating system.

At the among-population level, call frequency variation may be the allometric effect of body size variation. A westward increase of body size in *Acris crepitans* (Nevo and Capranica 1985) and a southward increase in Israeli green toads (Nevo 1972) (both explained as an adaptation to arid climates) result in a consistent reduction in fundamental frequency. Climate is also associated with body size in Asian tetraploid green toad populations. In this case, however, toads from the arid lowlands are smaller than those from the more humid highlands, suggesting that body size variation, rather than being adaptive, can be the result of climatic constraints on growth (Castellano 1996; Castellano and Giacoma 2000).

Not all geographic variation is the effect of differences in body size. Italian diploid and Asian tetraploid populations still differ in call frequency even after correction for body size (Castellano 1996; Castellano and Giacoma 2000), and the among-population variation for call frequencies is higher than that observed at lower levels (Table 15.1). Significant size-adjusted call frequency differences have also been observed in *Acris crepitans* (Ryan and Wilczynski 1991). In Asian tetraploid toads the pattern of call variation, described by considering together the temperature-adjusted pulse rate (see below) and the size-adjusted fundamental frequency, is not congruent with climate, but is congruent with geographic distances. Partial Mantel tests of matrix association show a significantly positive correlation between geographic and call distances (partial regression coefficient = 0.472; $p = 0.012$, after 10 000 randomizations; Castellano and Giacoma 2000). Unlike *A. crepitans,* the pattern of call frequency variation in tetraploid toads can be explained by restriction of gene flow rather than by adaptation to different environments.

Variance Pattern of Pulse Rate

The hierarchical pattern of pulse rate variability shows a trend similar to that of the fundamental frequency, with the lowest values at the within-bout level and the highest values at the among-population level (Figure 15.1; Table 15.1).

In green toads pulse rate is weakly influenced by body size: a positive correlation between pulse rate and body size was found in 17 of 20 populations, but was statistically significant in 3 samples only (Giacoma et al. 1997; Castellano and Giacoma 1998a). A similar trend occurs in other species of *Bufo* (Blair 1964; Howard and Young 1998). Pulse rate production in *Bufo viridis* involves a series of active contractions of the dilator laryngeal muscles, which are well adapted to generate rapid movements with the periodicity of the pulse rate (Martin 1971, 1972; Martin and Gans 1972). In the genus *Bufo,* the smaller the animal, the smaller the larynx and the smaller the muscles. In green toads larynx size averages 5.5% of body size (Martin 1972). Laryngeal muscle size influences temporal parameters by affecting the strength of the contractions; that is, bigger individuals, such as Asiatic diploids, have bigger muscles and consequently can maintain a higher contraction rate. In *Acris crepitans* the pulse rate is biomechanically related to the volume of the constrictor muscles, which is largely mediated by body size (McClelland et al. 1998). In the case of the fundamental frequency, the strong morphological constraint of body size can plausibly explain the low values of within-bout variability, ranging from 1.6 to 2.2%, depending on the group of populations. The weaker effect of morphology on pulse rate can explain the relatively higher variation with respect to fundamental frequency (2.8 to 4.3%).

Pulse rate variation increases from the within-bout to the within-individual level. Pulse rates of different bouts differ within individuals, mostly because of their different temperatures. In each of the 20 populations we studied, a positive correlation was found between pulse rate and body temperature, and in 16 of them the relationship was statistically significant (Giacoma et al. 1997; Castellano and Giacoma 1998a). These results are consistent with those observed in many different anuran species that exhibit a pulsed call (reviewed in Ryan 1986, 1988) and particularly in bufonids (Zweifel 1968; Martin 1972; Nevo and Schneider 1976; Sullivan and Wagner 1988; Sullivan 1992; Howard and Young 1998). Temperature constrains the contractile properties of the trunk and laryngeal muscles involved in the production of the call (Martin 1972; Wells and Taigen 1992) and can explain the positive relationship between temperature and pulse rate. When we control for temperature effects we still observe differences among bouts in the same individual ($F = 11.99$; $df = 10, 207$; $p < 0.001$); this may be due to differences in physiological conditions or to differences in the social contexts.

Within-population variation is higher than at the lower levels. Differences in pulse rate among males are not significant, although there is a strong trend in that direction ($F = 1.925$; $df = 13, 27$; $p = 0.073$). Since CVs were calculated on temperature-adjusted values, they do not reflect different thermal conditions, but may be due to morphophysiological differences among males.

Gerhardt (1991) suggested that the typically narrow range of variation of pulse rate within a single population is maintained by female choice. Our playback experiments, carried out within the range of variation of each population, clearly show that mean values in green toads are weakly preferred to extreme values. As no directional selection is acting, present mechanisms of female choice do not result in a reduction of the within-population variability of pulse rate (Castellano and Giacoma 1998a).

At the among-population level, CVs are higher than those at the lower levels. Since all values have been temperature-adjusted, the variation they express is independent of temperature. Both Italian diploid (Castellano 1996) and Asiatic tetraploid populations (Castellano et al. 2000) show significant differences in pulse rate mean values. As already discussed, Mantel tests show a significantly positive association between the geographic patterns of pulse rate and fundamental frequency on the one hand and geographic distances among populations on the other, thereby suggesting that gene flow explains the pattern of variation.

Variance Pattern of Call and Intercall Duration

Since call and intercall duration are two complementary components of call rate, they will be discussed together. The hierarchical pattern of variation of call and intercall duration is quite distinct from those of pulse rate and fundamental frequency (Figure 15.1). The highest variation occurs at the within-bout level and the lowest at the among-population level (Table 15.1). Call and intercall duration can be influenced by environmental temperature as well as energy expenditures. Both will tend to increase variation in these call parameters.

At the within-bout level both properties show high variability, 10 and 20 times higher than those of fundamental frequency and pulse rate, respectively. Since calling in anurans has a high energetic cost (reviewed in Wells and Taigen 1992; Wells, this volume), call and intercall durations are assumed to be under strong energetic constraints. Energetic costs limit the maintenance of the maximum performance level over time. Our results show a positive correlation between body size and call duration in 9 of 20 populations (only 2 associations are significant) and between body size and intercall duration in 10 of 20 populations (1 association is significant) (Giacoma et al. 1997; Castellano and Giacoma 1998a). We found a negative correlation between body temperature and call duration in 14 (9 are significant) of 20 populations and between body temperature and intercall duration in 13 (only 3 are significant) of 20 populations. A similar trend was found in Israeli populations of green toads (Nevo and Schneider 1976), as well as in other bufonids (Zweifel 1968; Fairchild 1981; Sullivan and Wagner 1988; Howard and Young 1998). The high levels of variation in call and intercall duration seem to result from a strong influence of energetic constraints and a weak influence of morphological constraints and temperature.

At the within-individual level, variation is comparable to within-bout variation. Measures were calculated on data corrected for air temperature using the pooled within-group regression coefficients. Bouts of calling recorded from the same male on different nights differed significantly with respect to call length ($F = 2.981$; $df = 10, 207$; $p = 0.002$) and intercall duration ($F = 2.484$; $df = 10, 191$; $p = 0.008$). In Italian populations of *Bufo viridis* (Giacoma, unpublished data), as well as in a number of other species, the amount of time and energy spent in reproductive activity are directly correlated with reproductive success (Parker 1983; Halliday and Tejedo 1995). Given the prolonged mating period and the low probability of encountering a receptive female, selection may have favored males that did not incur the high energetic expenses of maintaining continually elevated calling activity. It is not surprising therefore that individuals reach the upper level of performance and endure the consequent maximum energetic investment only in response to a challenge (e.g., the presence of a female or a rival male or both), thereby increasing the benefit/cost ratio of calling activity. The same individual can change his energetic investment by changing the tactics adopted during the same night or during the breeding season. In *Bufo viridis* the proximity of another calling male does not significantly change the mean call duration, whereas it causes a significant increase in intercall duration, thereby avoiding overlap between calls of neighboring males (from 10.1 seconds $\pm$ 1.42 SD to 13.7 seconds $\pm$ 5.46 SD, $p < 0.006$; Table 15.2). Without strong morphological and physiological constraints, energetic requirements put an upper limit but not a lower limit on the degree of expression of call duration, yielding a substantial random variance at both within-bout and within-individual levels. Wagner and Sullivan (1995) simulated the effect of increasing maximum possible within-male deviation of a trait and consequently found a decrease in the strength of sexual selection on the trait.

Variation at the within-population level is lower than at the within-male level. CVs were calculated on data corrected for air temperature using the pooled within-group regression coefficients. Within the same population we observed significant differences among males both in call length ($F = 5.404$; $df = 14, 207$; $p < 0.001$) and in intercall duration ($F = 2.167$; $df = 14, 191$; $p = 0.01$).

Since directional selection causes a loss in variability (Somero and Soulé 1974), mate choice can impose a selective pressure that influences variability. In playback experiments based on a choice between artificial stimuli having the same call rate but with a longer or shorter call duration, females significantly preferred signals with the longest call durations (Castellano and Giacoma 1998a). These results are consistent with those found in other species with pulsed calls (reviewed in Gerhardt 1988; see also Ryan and Keddy-Hector 1992; Sullivan 1992; Cherry 1993). Recent data show that call duration reliably reflects genetic quality in *Hyla versicolor,* which implies that preference for long calls should enable females to increase their inclusive fitness (Welch et al. 1998).

Among-population variation of call and intercall duration, calculated on temperature-adjusted data, tends to be lower than those found at lower levels. Unlike pulse rate and fundamental frequency, neither call nor intercall durations differ among populations. At the among-population level short-term ecological factors, such as local differences in food availability, can cause random differences in possible energetic investment, which in turn may affect call properties more than genetic differences (Figure 15.1).

Trait Variation and Patterns of Character Evolution

The evolution of advertisement calls, like the evolution of any other character, is the result of both selection on phenotypic variation and the inheritance of its variants (Fisher 1930). We calculated the repeatability of each call property, a normalized index of among-individual in relation to within-individual variability (Lessels and Boag 1987; Boake 1989). This index is computed on phenotypic characters; therefore it does not measure heritability per se but puts an upper limit on it (Falconer 1981). Generally no direct correlation has been found between the values of repeatability and heritability, so that a high repeatability means that it is possible but not necessarily true that heritability is high (Shaffer and Lauder 1985; Boake 1989). However a low value of repeatability means that heritability is necessarily low. A second method to compare among-individual variation to within-individual variation consists of calculating the ratio between the estimates of among-individual and within-bout CVs (Table 15.3). All these indexes are good measures of variation and in *Bufo viridis* gave the same general pattern as the within-bout CVs, providing relatively high stereotypy values for fundamental frequency and pulse rate (Table 15.3). These estimates show that the potential heritabilities of call and intercall length are low and that it is possible, but not necessarily true, that heritabilities of fundamental frequency and pulse rate are higher (the exception is the relatively high value of call duration repeatability). Heritabilities of morphological traits are generally higher than those of behavioral traits (Mousseau and Roff 1987; Roff and Mousseau 1987). Provided that no variable

Table 15.3 Factors influencing the rate of evolution of properties of the advertisement call

	Fundamental Frequency	Pulse Rate	Call Duration	Intercall Duration
Among-Individuals to Within-Bout CV Ratio[a]				
I-2N	2.82	2.18	0.68	0.85
A-2N	4.25	2.62	1.48	0.74
A-4N	3.23	1.35	0.61	0.85
Repeatability	0.45	0.24	0.31	0.15
Among-Populations to Within-Population CV Ratio[a]				
I-2N	1.45	2.44	0.99	0.55
A-4N	1.38	1.60	0.43	0.74
CCI[b]	0.19	0.03	0.36	0.78[c]

Constraints	Morphological	Morpho-Physiological	Energetic and Indirect Morpho-Physiological	Energetic
Under long-term environmental factors	Strong side effects, mediated by body size	Weak side effects, mediated by body size	—	—
Under short-term environmental factors	Social interactions	Strong effect of temperature	Weak effect of temperature; food availability	Food availability
Female selective pressure	Stabilizing or weakly directional	Stabilizing	Directional	Directional

[a] CV = coefficient of variation; I-2N = Italian diploids; A-2N = Asiatic diploids; A-4N = Asiatic tetraploids.

[b] CCI = character change index, from Cocroft and Ryan (1995).

[c] Refers to call rate.

pleiotropy is acting (Maynard Smith 1959), if the expression of a behavioral trait is strictly dependent on a morphological one, then as a general trend, heritability of the behavioral trait under morphological constraint should be higher than that of traits depending on context. This last point supports the hypothesis that traits under morphological and morpho-physiological constraints, such as fundamental frequency and pulse rate, have higher heritabilities than parameters under strong energetic constraints, such as call and intercall duration.

There is no doubt that genetically based variation among individuals is a prerequisite for evolutionary change, but discussing signal evolution also requires a view of organisms as variable within the population or species. Because of the species-specific nature of mate recognition signals, some authors have suggested that these characters should show little within-species variation (Paterson 1985, among others). However the generalization that species must differ in mate recognition systems is not necessarily true, as exemplified by the *Bufo microscaphus* complex and by *Bufo americanus* group taxa (Sullivan et al. 1996a; Gergus et al. 1997). Moreover anurans show significant among-population call differences (Nevo and Schneider 1976; Sullivan 1989; Ryan and Wilczynski 1991; Ryan et al. 1996; Castellano et al. 2000). The study of the geographic variability of different components of mate choice signals allows us to verify whether evolutionary forces responsible for population divergence produce the same patterns of different mate recognition characters (Ryan et al. 1996). In the green toad there are significant among-population differences in size-adjusted fundamental frequency and temperature-adjusted pulse rate, but not in temperature-adjusted call and intercall durations (Castellano et al. 2000). Apart from analysis of the variance among populations, an index of the degree of differentiation at the among-population level can be the ratio of among-population CV to within population CV (Table 15.3). The trend of this index is consistent with patterns of variation among individuals; that is, populations differ significantly with respect to those properties that show the highest morphological and physiological constraints, and that those properties show the lowest variability within individuals (Table 15.3). In central Asiatic tetraploids, the pattern of variation of call and intercall durations did not show a significant correlation with geographic and climatic distances; fundamental frequency and pulse rate showed a positive association with geographic distances but not with climatic distances, suggesting that the main causal agent responsible for call variation might have been isolation by distance (Castellano et al. 2000). These results are not inconsistent with the hypothesis that fundamental frequency and pulse rate have a higher additive genetic component than call and intercall duration.

Trait Variation, Rate of Evolution, and Function

We will consider the hypothesis that traits with high variation and potentially lower heritabilities, such as call and intercall duration, also have a potentially lower rate of evolution than traits with a lower variability, such as fundamental frequency and pulse rate. Furthermore we will discuss the hypothesis that in recognition females rely on call variables that show high levels of evolutionary divergence.

In *Bufo viridis* we found that fundamental frequency, pulse rate, and call and intercall duration show an increasing gradient of variability: traits characterized by low variation at the individual level are under morpho-physiological constraints, and traits characterized by high variation at the individual level are strongly energy dependent. The ellipses in Figure 15.1 represent fundamental frequency and pulse rate, the rectangles represent call and intercall duration. Fundamental frequency and pulse rate are characterized by increasing levels of variability according to the levels of organization. Their phenotype depends on environmental, long-lasting effects such as morpho-physiological adaptation to environmental factors occurring during ontogeny. These traits also have high repeatability and show geographic variation. Call and intercall duration show a higher level of variation at any level, are under the influence of random and/or short-term environmental factors (e.g., density, physiological condition, tactics adopted by the competing male nearby), do not show an increase in variability according to levels of organization, have low heritability, and do not show significant geographic variation.

Gerhardt (1991, 1995) suggested that "static" and "dynamic" properties convey different kinds of biologically significant information. The "good genes" model of sexual selection predicts that some attributes of male displays advertise genetic quality. Traits under energetic constraints can be honest indicators of genetic quality, since the level of their performance depends on the amount of energy that males can afford to invest. Parameters under energetic constraints show a high within-individual variability in comparison to among-individual variability. If females sample males over a limited period, an increase in within-male variation in the traits increases the probability that males with low average call duration and/or high average intercall duration will obtain matings because these males will occasionally call with longer signals and/or shorter intervals than males with higher average values of these traits. Therefore behavioral variation can reduce evolutionary rates of the variable characters and make selection effects more difficult to detect (Wagner and Sullivan 1995). Within the "good genes" framework females should select the extreme trait in order to select the best mate available among many conspecific males. Female preferences for extreme values of dynamic parameters are known to exist in the green toad (Castellano and Giacoma 1998a) as well as in other species (reviewed in Ryan and Keddy-Hector 1992). Strong directional selection reduces character variability and consequently further slows down the rate of evolution of "dynamic" traits.

From information theory we know that the stronger the constraint on a property of a signal the higher the redundancy of the information that can be transmitted and the lower the risk of transmission errors (Nauta 1972). If this general rule holds true also for biological systems, then the highly constrained properties of the calls should be the most suitable to transmit those messages whose erroneous interpretation bears a high fitness cost for both senders and receivers. Our analysis at both individual and population levels of variability suggest that static parameters such as fundamental frequency and pulse rate have a lower variability at any level. As the strength of phenotypic selection on a trait appears to depend partly on the pattern of variability of that trait, above all on the within-individual variability in comparison with the among-individual variability, pulse rate and fundamental frequency should have a potentially high evolutionary rate. Thus rapid divergence in this character could result in the rapid evolution of prezygotic reproductive isolation (Coyne and Orr 1989). Female stabilizing or weakly directional preferences for mean values of "static" parameters are known to exist in the green toad (Castellano and Giacoma 1998a) as well as in other species (reviewed in Ryan and Keddy-Hector 1992). As a general pattern those properties whose mean values are preferred to extreme values may encode important information for species recognition.

It is also true that evolutionary hypotheses can be successfully tested by applying the comparative method (Giacoma and Balletto 1988; Ryan, Fox et al. 1990; Brooks and McLennan 1991; Harvey and Pagel 1991; Cocroft and Ryan 1995; Ryan and Rand 1995; Ryan 1996). Cocroft and Ryan (1995) used an index of character conservation to measure the likelihood that close relatives will share the same character state and showed that in clades of North American *Bufo* and hylids the various characters that make up a call evolve at different rates (Table 15.3). The lowest values of character conservation that characterize the highest evolutionary rate in *Bufo* are found for pulse rate (0.03) and dominant frequency/size ratio (0.19), whereas call duration (0.36) and call rate are more conservative (0.78) (Table 15.3).

To verify the hypothesis that less variable properties should be more strictly associated with a species recognition function than properties with higher variability, we compared the degree of call trait divergence between Italian island and mainland populations, and between Asiatic diploids and tetraploids. If this hypothesis is correct, low variability properties (pulse rate and fundamental frequency) should differ more than "dynamic" properties, and the degree of call divergence in these properties should be positively correlated with the degree of genetic divergence and should therefore differ more between Asiatic diploids and tetraploids than between mainland and island Italian toads.

The comparison between Italian island and mainland toads shows that Italian island toads are significantly larger (Castellano and Giacoma 1998b), call at lower fundamental frequencies, with longer call and intercall durations and higher pulse rates adjusted for temperature (Castellano et al. 1998); fundamental frequency regressed against body length and call duration regressed against body temperature do not differ significantly either in slopes or intercepts. In contrast pulse rate regressed against body temperature differs significantly in slope between mainland and island toads (Castellano et al. 1999). Intercall duration is not significantly correlated with temperature and does not differ significantly between Italian mainland (n = 129, mean = 14.5 seconds, 9.6 SD) and Italian islands toads (n = 67, mean = 15.7 seconds, 6.7 SD) ($F = 0.841$; $df = 1, 194$; $p = 0.360$).

Asiatic diploids are bigger than tetraploids and call at lower fundamental frequencies; at the same temperature Asiatic diploids call with longer call and intercall durations and higher pulse rates (Castellano et al. 1999). Covariance analyses show that diploids and tetraploids differ in the slopes at regressions of dominant frequency against body size, pulse rate against body temperature significantly differ both in slopes and in intercepts, and call duration regressed against body temperature significantly differ in intercepts. Since intercall duration is not significantly correlated with body temperature, we compared intercall duration of diploids (n = 47, mean = 9.9 seconds, 4.0 SD) with tetraploids mean values (n = 86, mean = 7.9 seconds, 6.5 SD). Intercall durations do not differ significantly between the two groups ($F = 3.204$; $df = 1, 131$; $p = 0.072$), although there is a strong trend in that direction.

These data partly support the hypothesis that low variability traits differ more than high variability traits. Pulse rates differ significantly between the two groups of populations in the two case studies, fundamental frequency and call duration are significantly different only between Asiatic diploids and tetraploids (i.e., in the case study characterized by higher genetic divergence; Cervella et al. 1997; Lattes 1997), whereas intercall duration does not differ.

The prediction that the degree of divergence of static properties between independently evolving lineages should be positively correlated with the degree of genetic divergence is supported by the comparison between the regres-

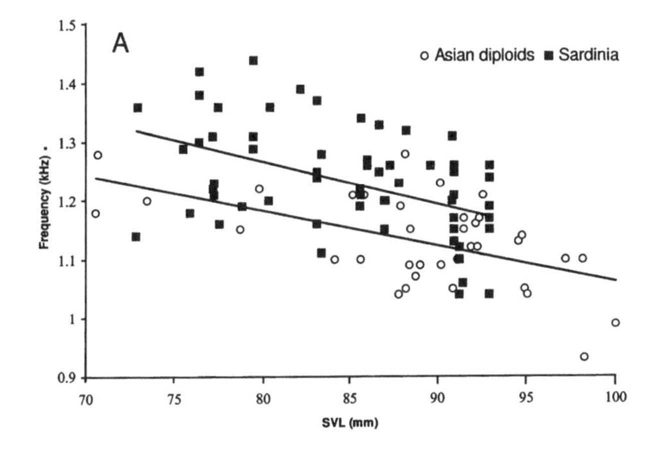

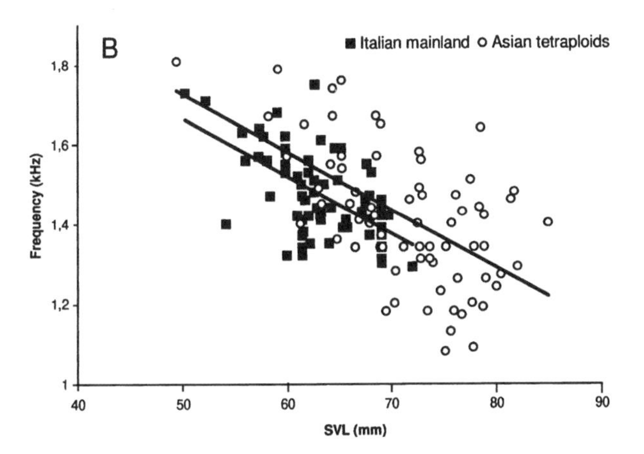

Figure 15.2. Fundamental frequency of the advertisement call regressed against body length in the four groups of populations analyzed: (A) Asian diploids are compared with Italian island diploids (Sardinia). (B) Italian mainland diploids are compared with Asian tetraploids.

sion lines of Asiatic diploids and Italian island diploids: within the same body size range Asiatic diploids show a lower call frequency than Italian island toads (Figure 15.2A). If we compare directly the regression lines of Asiatic tetraploids and Italian mainland diploids, we find that within the same body size range Asiatic tetraploids show a higher call frequency than Italian mainland toads (Figure 15.2B). Therefore the higher degree of genetic divergence between Asiatic diploids and tetraploids in comparison with Italian mainland and island toads results in a similar divergence in body size in both cases, but in a more pronounced divergence in call parameters. Independent of body size modifications, the evolutionary divergence of the fundamental frequency of the advertisement call of *Bufo* can be the result of changes in the weight of the fibrous masses, as well as of allometric evolution of the larynx size relative to body size, mediated by changes in vocal cord tension (Martin 1972). Pulse rate regressed against body temperature gave a similar scenario: the higher degree of genetic divergence between Asiatic diploids and tetraploids in comparison with Italian mainland and island toads results in a more pronounced divergence in pulse rate (Table 15.3).

To verify whether the significant differences found between advertisement calls have a functional role in mate recognition, we played back an average diploid advertisement call for pulse rate and fundamental frequency against a typical tetraploid call for the same properties. Our experiments clearly show that diploid females significantly discriminate against the tetraploid calls (13 of 16 females chose their own species average call), thereby suggesting that some combination of fundamental frequency and pulse rate are important in influencing female responses in the green toad. Ryan and Rand (this volume) show that female túngara frogs rely on call variables that show high levels of evolutionary divergence for recognition.

Conclusions

Studies of the evolution of mate recognition initially addressed the process of speciation and the importance of signal variation among species. More recently however the emphasis has shifted to the process of sexual selection and signal variation within species. Without denying the importance of each type of variation in isolation, we suggest that the simultaneous comparison of phenotypic variation at multiple levels of organization can be especially rewarding.

We used coefficients of variation to compare the variability of different signal properties of green toad advertisement calls at different levels of organization. Within-individual variability of fundamental frequency was low and increased for pulse rate, call duration, and intercall duration, in that order. Fundamental frequency might have low within-individual variability because it is under strong morphological constraints and is little affected by environmental variables such as temperature. The other three call properties however are all influenced by temperature and are under different morphological and physiological constraint; perhaps consequently pulse rate is twice, call duration is about 10 times, and intercall duration is nearly 20 times more variable within individuals than is fundamental frequency.

In Figure 15.1 we illustrate two general patterns of signal property variation at four levels of organization within the *Bufo viridis* complex. Call properties that have low variability within calling bouts of the same individual exhibited progressively greater variation at higher levels of organization (i.e., from within-bout to among-populations). In contrast call properties that were more variable within calling bouts tended to have similar or lower levels of variation at higher levels of organization.

Although the advertisement call is often considered a single unit of information, calls do not necessarily evolve in a unitary fashion. Instead call traits differ in how their variation is partitioned among levels of organization, which might have significant consequences for signal evolution. When variation among individuals is greater than that within individuals, traits should be more likely to exhibit an evolutionary response to selection. We tested this hypothesis by comparing data obtained from three lineages of the *Bufo viridis* complex. Specifically we predicted that call properties with higher variation among individuals than within calling bouts will have a higher rate of evolution and be important in species recognition, whereas call properties that vary more within calling bouts have a lower rate of evolution and can convey information relevant to intraspecific mate choice.

The comparison between island and mainland toads of the Italian lineages (a case of relatively little genetic divergence) shows that only pulse rate regressed against body temperature differs significantly in slope among populations (Castellano et al. 1999). Comparing the Asiatic diploid *Bufo viridis turanensis* with the Asiatic tetraploid *Bufo danatensis* (a case of greater genetic divergence than the previous one), we found that populations differ in the slopes of regressions of dominant frequency regressed against body size, in slopes and intercepts of pulse rate regressed against body temperature, and in the intercept of call duration regressed against body temperature. Four of five significant differences involve low-variability traits. These data support the hypothesis that low-variability traits should have a higher probability of divergence between lineages.

The second prediction is that the degree of divergence of low-variability traits, characterized by a high ratio of among- to within-individual and among- to within-population variation, should be positively correlated with the degree of genetic divergence. This prediction is supported by the comparison of fundamental frequency regressed against body size, and pulse rate regressed against body temperature; in both cases the higher degree of genetic divergence between Asiatic diploids and tetraploids in comparison with Italian mainland and island toads results in a more pronounced divergence of these signal properties (Figure 15.2; Table 15.3).

Since rapid divergence in low variability characters could result in the rapid evolution of prezygotic reproductive isolation (Coyne and Orr 1989), we tested whether the significant differences found between advertisement calls of Asiatic diploids and tetraploids have a functional role in mate recognition. We played back an average Central Asiatic diploid advertisement call for pulse rate and fundamental frequency against a typical tetraploid call for the same properties. Diploid females from Central Asia significantly discriminate against the tetraploid calls, suggesting that some combinations of fundamental frequency and pulse rate are important in the mate recognition processes of the green toads.

Within the "good genes" framework, females should select the extreme values of traits under energetic constraints, such as call duration, to select the best mate available among many conspecific males. Evidence of directional selection was obtained in playback experiments based on a choice between artificial stimuli having the same call rate but with a longer or shorter call duration. Italian diploid females significantly preferred signals having the longest call durations (Castellano and Giacoma 1998a). Female preferences for extreme values of highly variable parameters are known to exist in other species as well (reviewed in Ryan and Keddy-Hector 1992).

Some of the issues raised here are speculative at the moment but lead to testable predictions regarding the influence of patterns of variability on trait evolution as well as on the predictability of associations between signal properties and functions. Further data on the relative heritabilities of different call properties, together with estimates of patterns of signal evolution in clades with different sound-producing systems, will be necessary to estimate the general significance of the qualitative patterns identified in the *Bufo viridis* complex.

Acknowledgments

We thank M. Ryan for giving us the stimulus to review all the results of our ongoing research on the evolution of advertisement call within the *Bufo viridis* complex and for supporting our attendance of the symposium held in Guelph. We are grateful to E. Balletto, V. Cameron Curry, N. Kime, M. Ryan, G. Sella, and B. Sullivan for reviews of early drafts of this paper, and to M. Gamba for graphical help. This research was supported by grants of MURST (Ministero Ricerca Scientifica e Tecnologica) and of INTAS (International Association for the Promotion of Cooperation with Scientists from the New Independent States of the former Soviet Union).

References

Arak, A. 1983. Male–male competition and mate choice in anuran amphibians. Pp. 181–240. *In* P. G. Bateson (ed.), Mate choice. Cambridge University Press, Cambridge.

Blair, W. F. 1964. Isolating mechanisms and interspecific interactions in anuran amphibians. Quarterly Review of Biology 39:334–344.

Boake, C. R. B. 1989. Repeatability: Its role in evolutionary studies of mating behavior. Evolutionary Ecology 3:173–182.

Boake, C. R. B. 1994. Behavioral variation and speciation: Flexibility may be a constraint. Pp. 259–267. *In* R. J. Greenspan, and C. P. Kyriacou (eds.), Flexibility and constraint in behavioral systems. John Wiley and Sons, New York.

Brenowitz, E. A. 1994. Flexibility and constraint in the evolution of animal communication. Pp. 247–258. *In* R. J. Greenspan, and C. P. Kyriacou (eds.), Flexibility and constraint in behavioral systems. John Wiley and Sons, New York.

Brooks, D. R., and D. A. McLennan. 1991. Phylogeny, Ecology and Behavior: A Research Program in Comparative Biology. University of Chicago Press, Chicago.

Capranica R. R. 1976. The auditory system. Pp. 552–575. *In* B. Lofts (ed.), Physiology of the Amphibia. Academic Press, New York.

Castellano, S. 1996. Biologia evolutiva e strategia riproduttiva di *Bufo viridis* (Anura: Bufonidae). Ph.D. thesis, University of Pavia, Italy.

Castellano, S., and C. Giacoma. 1998a. Stabilizing and directional female choice for male calls in the European green toad. Animal Behaviour 56:275–287.

Castellano, S., and C. Giacoma. 1998b. Morphological variation of the green toad, *Bufo viridis,* in Italy: A test of causation. Journal of Herpetology 32:540–550.

Castellano, S., and C. Giacoma. 2000. Morphometric and advertisement call geographic variation in polyploid green toads. Biological Journal of the Linnean Society. In press.

Castellano, S., C. Giacoma, T. Dujsebayeva, G. Odierna, and E. Balletto. 1998. Morphological and acoustical comparison between diploid and tetraploid green toads. Biological Journal of the Linnean Society 63:257–281.

Castellano, S., A. Rosso, S. Doglio, and C. Giacoma. 1999. Body size and calling variation in the green toad, *Bufo viridis.* Journal of Zoology London 248:83–90.

Cervella, P., M. Delpero, and E. Balletto. 1997. Characterisation of Asiatic populations of the *Bufo viridis* complex by random amplified polymorphic DNA (RAPD) analysis. Pp. 38–39. *In* Rocek, Z., and S. Hart (eds.), Abstracts of the 3rd World Congress of Herpetology (Prague: 2–10 August, 1997).

Cherry, M. I., 1993. Sexual selection in the raucous toad, *Bufo rangeri.* Animal Behaviour 45:359–374.

Cocroft, R. B., and M. J. Ryan. 1995. Patterns of advertisement call evolution in toads and chorus frogs. Animal Behaviour 49:283–303.

Coyne, J. A., and H. A. Orr. 1989. Patterns of speciation in *Drosophila.* Evolution 43:362–381.

Darwin, C. 1859. On the Origin of Species. John Murray, London.

Dobzhansky, T. 1937. Genetics and the Origin of Species. Columbia University Press, New York.

Doglio, S., S. Castellano, and C. Giacoma. 2000. The role of the advertisement call in the competitive interactions among males of the European green toad, *Bufo viridis.* Pp. 117–121. *In* C. Giacoma (ed.), Atti I° Convegno Societas Herpetologica Italica.

Dujsebayeva, T., S. Castellano, C. Giacoma, E. Balletto, and G. Odierna. 1997. On the distribution of diploid and tetraploid green toads of the *Bufo viridis* complex (Anura: Bufonidae) in Southern Kazakhstan. Asiatic Herpetological Research 7:27–31.

Endler, J. A., and E. Houde. 1995. Geographic variation in female preferences for male traits in *Poecilia reticulata.* Evolution 49:456–468.

Ewing, A. W. 1989. Arthropod Bioacoustics: Neurobiology and Behavior. Comstock/Cornell, Ithaca, NY.

Fairchild, L. 1981. Mate selection and behavioural thermoregulation in Fowler's toads. Science 212:950–951.

Falconer, D. S. 1981. Introduction to Quantitative Genetics. Longman, New York.

Fisher, R. A. 1930. The Genetical Theory of Natural Selection. Clarendon Press, Oxford.

Gergus, E. W. A., B. K. Sullivan, and K. B. Malmos 1997. Call variation in the *Bufo microscaphus* complex: Implications for species boundaries and the evolution of mate recognition. Ethology 103:979–989.

Gerhardt, H. C. 1978. Mating call recognition in the green treefrog (*Hyla cinerea*): The significance of some fine temporal properties. Journal of Experimental Biology 74:59–73.

Gerhardt, H. C. 1981. Mating call recognition in the green treefrog (*Hyla cinerea*): Importance of two frequency bands as a function of sound pressure level. Journal of Comparative Physiology 144:9–16.

Gerhardt, H. C. 1988. Acoustic properties used in call recognition by frogs and toads. Pp. 455–483. *In* B. Fritzsch, M. J. Ryan, W. Wilczynski, T. Hetherington, and W. Walkowiak (eds.), The evolution of the amphibian auditory system. John Wiley and Sons, New York.

Gerhardt, H. C. 1991. Female mate choice in treefrogs: Static and dynamic acoustic criteria. Animal Behaviour 42:615–635.

Gerhardt, H. C. 1994a. The evolution of vocalization in frogs and toads. Annual Review of Ecology and Systematics 25:293–324.

Gerhardt, H. C. 1994b. Reproductive character displacement of female mate choice in the grey treefrog, *Hyla chrysoscoelis.* Animal Behaviour 42:615–635.

Gerhardt, H. C. 1995. Phonotaxis in female frogs and toads: Execution and design of experiments. Pp. 209–220. *In* G. M. Klump, R. J. Dooling, R. R. Fay, and W. C. Stebbins (eds.), Methods in Comparative Psychoacoustics. Birkhauser-Verlag, Basle.

Gerhardt, H. C., M. L. Dyson, S. D. Tanner, and C. G. Murphy. 1994. Female treefrogs do not avoid heterospecific calls as they approach conspecific calls: Implications for mechanisms of mate choice. Animal Behaviour 47:1323–1332.

Giacoma, C., and E. Balletto. 1988. Phylogeny of the salamandrid genus *Triturus.* Bollettino Zoologico 55:337–360.

Giacoma, C., F. Kozar, C. Zugolaro, and I. Pavignano. 1993. Reproductive behaviour of *Bufo viridis.* Ethology, Ecology & Evolution 5:396–397.

Giacoma, C., C. Zugolaro, and L. Beani. 1997. The advertisement calls of the green toad (*Bufo viridis*): Variability and role in mate choice. Herpetologica 53:454–464.

Halliday, T. R., and P. J. Slater. 1983. Animal Behaviour: Communication. W. H. Freeman, New York.

Halliday, T. R., and M. Tejedo. 1995. Intrasexual Selection and Alternative Mating Behaviour. Pp. 420–468. *In* H. Heatwole and B. K. Sullivan (eds.), Amphibian Biology: Social Behaviour. Surrey Beatty and Sons, Chipping Norton, Australia.

Harvey, P. H., and M. D. Pagel. 1991. The Comparative Method in Evolutionary Biology. Oxford University Press, Oxford.

Hauser, M. D. 1996. The Evolution of Communication. MIT Press, Cambridge, MA.

Howard, R. D., and J. R. Young. 1998. Individual variation in male traits and female mating preferences in *Bufo americanus.* Animal Behaviour 55:1165–1179.

Huxley, J. 1966. A discussion on ritualization of behaviour in animals and man: Introduction. Philosophical Transactions of the Royal Society Series B 251:249–271.

Jennions, M. D., P. R. Y. Backwell, and N. I. Passmore. 1995. Repeatability of mate choice: The effect of size in the African painted reed frog, *Hyperolius marmoratus*. Animal Behaviour 49:181–186.

Jennions, M. D., and M. Petrie. 1997. Variation in mate choice and mating preferences: A review of causes and consequences. Biological Review 72:283–327.

Lande, R. 1981. Modes of speciation by sexual selection on polygenic characters. Proceedings of the National Academy of Sciences 78:3721–3725.

Laoretti, F., S. Castellano, and C. Giacoma. 2000. Struttura e dinamica stagionale di alcune popolazioni di rospo smeraldino (*Bufo viridis*) del Savonese. Pp. 109–116. *In* C. Giacoma (ed.), Atti I° Convegno Societas Herpetologica Italica.

Lattes, A. 1997. The central Asiatic populations of the *Bufo viridis* complex are genetically different from the European taxon. Pp. 152–153. *In* Z. Rocek, and S. Hart (eds.), Abstracts of the 3rd World Congress of Herpetology (Prague: 2–10 August, 1997).

Lessels, C. M., and P. T. Boag. 1987. Unrepeatable repeatabilities: A common mistake. Auk 104:116–121.

Lindauer, M. 1961. Communication among Social Bees. Harvard University Press, Cambridge, MA.

Littlejohn, M. J., and T. C. Michaud. 1959. Call discrimination by females of Strecker's chorus frog. Texas Journal of Science 11:86–92.

Lorenz, K. 1950. The comparative method in studying innate behavior patterns. Symp. Soc. Exp. Biol. IV, Cambridge, MA.

Lorenz, K. 1965. Evolution and Modification of Behavior. University of Chicago Press, Chicago.

Martin, W. F. 1971. Mechanics of sound production in toads of the genus *Bufo:* Passive elements. Journal of Experimenta Zoology 176:273–294.

Martin, W. F. 1972. Evolution of vocalization in the genus *Bufo*. Pp. 279–309. *In* W. F. Blair (ed.), Evolution in the Genus *Bufo*. University of Texas Press, Austin.

Martin, W. F., and C. Gans. 1972. Muscular control of the vocal tract during release signalling in the toad *Bufo valliceps*. Journal of Morphology 137:1–27.

McClelland, B. E., W. Wilczynski, and M. J. Ryan. 1996. Correlations between call characteristics and morphology in male cricket frogs (*Acris crepitans*). Journal of Experimental Biology 199:1907–1919.

McClelland, B. E., W. Wilczynski, and M. J. Ryan. 1998. Intraspecific variation in laryngeal and ear morphology in male cricket frogs (*Acris crepitans*). Biological Journal of the Linnean Society 63:51–67.

Maynard Smith, J. 1959. Sex-limited inheritance of longevity in *Drosophila subobscura*. Journal of Genetics 56:227–235.

Mayr, E. 1942. Systematics and the Origin of Species from the Viewpoint of a Zoologist. Columbia University Press, New York.

Mousseau, T. A., and D. A. Roff. 1987. Natural selection and the heritability of fitness components. Heredity 59:181–197.

Nauta, D. 1972. The Meaning of Information. Mouton, The Hague.

Nevo, E. 1972. Climatic adaptation in size of the green toad. Israel Journal of Medical Sciences, 1010.

Nevo, E., and R. R. Capranica. 1985. Evolutionary origin of ethological reproductive isolation in cricket frogs, *Acris*. Evol. Biol. 19:147–214.

Nevo, E., and H. Schneider. 1976. Mating call pattern of green toads in Israel and its ecological correlates. Journal of Zoology London 178:133–145.

Odierna, G., G. Aprea, E. Balletto, T. Capriglione, and A. Morescalchi. 1997. Polyploidy in the *Bufo viridis* complex: A cytogenetic approach. Pp. 152–153. *In* Rocek, Z., and S. Hart (eds.), Abstracts of the 3rd World Congress of Herpetology (Prague: 2–10 August, 1997).

Parker, G. A. 1983. Mate quality and mating decision. Pp. 141–166. *In* P. Bateson (ed.), Mate Choice. Cambridge University Press, Cambridge.

Paterson, H. E. H. 1982. Perspectives on speciation by reinforcement. South African Sci. 78:53–57.

Paterson, H. E. H. 1985. The recognition concept of species. Pp. 21–29. *In* E. S. Vrba (ed.), Species and Speciation. Transvaal Museum Press, Pretoria.

Pisanetz, E. M. 1978. On new poliploid species of *Bufo danatensis* Pisanetz sp.n. from Turkmenia. Dokl. Ukrain. Acad. Sci. 3B:280–284.

Rand, A. S., M. J. Ryan, and W. Wilczynski. 1992. Signal redundancy and receiver permissiveness in acoustic mate recognition by the túngara frog, *Physalaemus pustulosus*. American Zoologist 32:81–90.

Roff, D. A., and T. A. Mousseau. 1987. Quantitative genetics and fitness: Lessons from *Drosophila*. Heredity 58:103–118.

Roth, P. 1986. An overview of the systematics of the *Bufo viridis* group in Middle and Central Asia. Pp. 127–130. *In* Z. Rocek (ed.), Studies in Herpetology. Charles University Press, Prague.

Ryan, M. J. 1986. Factors influencing the evolution of acoustic communication: Biological constraints. Brain, Behavior and Evolution 28:70–82.

Ryan, M. J. 1988. Constraints and patterns in the evolution of anuran acoustic communication. Pp. 455–483. *In* B. Fritzsch, M. J. Ryan, W. Wilczynski, T. Hetherington, and W. Walkowiak (eds.), The Evolution of the Amphibian Auditory System. John Wiley and Sons, New York.

Ryan, M. J. 1990. Sexual selection, sensory systems and sensory exploitation. Oxford Surveys in Evolutionary Biology 7:157–195.

Ryan, M. J. 1991. Sexual selection and communication in frogs. Trends in Ecology and Evolution 6:351–355.

Ryan, M. J. 1996. Phylogenetics in behaviour: Some cautions and expectations. Pp. 1–21. *In* E. P. Martin (ed.), Phylogenies and the Comparative Method in Animal Behaviour. Oxford University Press, Oxford.

Ryan, M. J., J. H. Fox, W. Wilczynski, and A. S. Rand. 1990. Sexual selection for sensory exploitation in the frog *Physalaemus pustulosus*. Nature 343:66–67.

Ryan, M. J., and A. Keddy-Hector. 1992. Directional pattern of female mate choice and the role of sensory biases. American Naturalist 139:S4–S35.

Ryan, M. J., S. A. Perrill, and W. Wilczynski 1992. Auditory tuning and call frequency predict population-based mating preferences in the cricket frog, *Acris crepitans*. The American Naturalist 139:1370–1383.

Ryan, M. J., and A. S. Rand. 1993a. Phylogenetic patterns of behavioural mate recognition systems in the *Physalaemus pustulosus* species group (Anura: Leptodactylidae): The role of ancestral and derived characters and sensory exploitation. Pp. 251–267. *In* D. Lees and D. Edwards (eds.), Evolutionary Patterns and Processes. Linnean Society Symposium Series No 14. Academic Press, New York.

Ryan, M. J., and A. S. Rand. 1993b. Species recognition and sexual selection as a unitary problem in animal communication. Evolution 47:647–657.

Ryan, M. J., and A. S. Rand. 1995. Female responses to ancestral advertisement calls in túngara frog. Science 269:390–392.

Ryan, M. J., A. S. Rand, and L. A. Weigt. 1996. Allozyme and advertisement call variation in the túngara frog, *Physalaemus pustulosus.* Evolution 50:2435–2453.

Ryan, M. J., and W. Wilczynski. 1991. Evolution of intraspecific variation in the advertisement call of a cricket frog (*Acris crepitans,* Hylidae). Biological Journal of the Linnean Society 44:249–271.

Shaffer, H. B., and G. V. Lauder. 1985. Patterns of variation in aquatic ambystomatid salamanders: Kinematics of the feeding mechanism. Evolution 39:83–92.

Somero, G., and M. Soulé. 1974. Genetic variation in marine fishes as a test of the niche-variation hypothesis. Nature 249:670–672.

Stöck, M. 1997. Unterschungen zur Morphologie und Morphometrie di- und tetraploider Grünkröten (*Bufo viridis* komplex) in Mittelasien (Amphibia: Anura: Bufonidae). Zool. Abh. Mus. Tierkd. Dresden 49:193–222.

Stöck, M. 1998. Mating call differences between diploid and tetraploid green toads (*Bufo viridis* complex) in Middle Asia. Amphibia-Reptilia 19:29–42.

Sullivan, B. K. 1989. Interpopulational variation in vocalizations of *Bufo woodhousei.* Journal of Herpetology 23:368–373.

Sullivan, B. K. 1992. Sexual selection and calling behavior in the American toad (*Bufo americanus*). Copeia 1992:1–8.

Sullivan, B. K., K. B. Malmos, and M. F. Given. 1996a. Systematics of the *Bufo woodhousii* Complex (Anura: Bufonidae): Advertisement call variation. Copeia 1996:274–280.

Sullivan, B. K., M. J. Ryan, and P. A. Verrell. 1996b. Female choice and mating system structure. Pp. 470–517. *In* H. Heatwole and B. K. Sullivan (eds.). Amphibian Biology, Vol. 2. Social Behavior. Surrey Beatty and Sons, Chipping Norton, Australia.

Sullivan, B. K., and W. E. Wagner Jr. 1988. Variation in advertisement and release calls, and social influences on calling behavior in the gulf coast toad (*Bufo valliceps*). Copeia, 4:1014–1020.

Tinbergen, N. 1951. The Study of Instinct. Oxford University Press, Oxford.

Wagner, W. E., Jr. 1989. Graded aggressive signals in Blanchard's cricket frog: Vocal responses to opponent proximity and size. Animal Behaviour 38:1025–1038.

Wagner, W. E., and B. K. Sullivan. 1995. Sexual selection in the gulf coast toad, *Bufo valliceps:* Female choice based on variable characters. Animal Behaviour 49:305–319.

Welch, A. M., R. D. Semlitsch, and H. C. Gerhardt. 1998. Call duration as an indicator of genetic quality in male gray tree frogs. Science 280:1928–1930.

Wells, K. D., and T. L. Taigen. 1992. The energetics of reproductive behavior. Pp. 410–426. *In* M. E. Feder and W. W. Burggren (eds.), Environmental Physiology of the Amphibians. Chicago University Press, Chicago, Illinois.

Wilczynski, W., and M. J. Ryan. 1988. The amphibian auditory system as a model for neurology, behavior, and evolution. Pp. 455–483. *In* B. Fritzsch, M. J. Ryan, W. Wilczynski, T. Hetherington, and W. Walkowiak (eds.), The Evolution of the Amphibian Auditory System. John Wiley and Sons, New York.

Wollerman, L. 1998. Stabilizing and directional preferences of female *Hyla ebraccata* for calls differing in static properties. Animal Behaviour 55:1619–1630.

Zugolaro, C., L. Beani, and C. Giacoma. 1994. Vocal communication in green toads, *Bufo viridis,* Boll. Zool., Suppl:81.

Zweifel, R. G. 1968. Effect of temperature, body size, and hybridization on mating calls of toads, *B. a. americanus* and *B. woodhousei fowleri.* Copeia 2:269–285.

16

RAFAEL MÁRQUEZ AND JAIME BOSCH

Communication and Mating in the Midwife Toads (Alytes obstetricans *and* Alytes cisternasii)

Introduction

This chapter provides a synoptic review of 10 years' study on the behavioral ecology of continental midwife toads in the Iberian Peninsula. The information presented is derived from two doctoral dissertations (Márquez 1990; Bosch 1997), a number of published papers, some unpublished experiments, and many field observations of animals.

Midwife toads belong to the family Discoglossidae, one of the most primitive families of anurans in the suborder Archaeobatrachia. The four extant species of midwife toad form the genus *Alytes,* which is distributed in southwestern Europe; three species are present in the Iberian Peninsula and an insular species is found on the island of Mallorca. Unlike most temperate anurans, these exhibit paternal care of their eggs on land. Their reproductive behavior is also unique in that females repeatedly emit a low-intensity, short, tonal note in response to the louder mating call of the male. This exchange of calls occurs before amplexus.

In 1987 the first author decided to undertake a field study of midwife toads in Spain. At that time two main sources of information were available. The first source consisted of a number of descriptive publications of the mating behavior dating back to the turn of the century (De l'Isle 1873, 1876; Héron-Royer 1878, 1883, 1886; Boscá 1879, 1880; Boulenger 1912). The second source was Dr. E. G. Crespo's doctoral dissertation (Crespo 1979)—a thorough study of the morphology and physiology of the Iberian species of midwife toads that included information on their genetics, phenology, and behavior. It was published in a number of scientific papers (Crespo 1976, 1981a, 1981b, 1981c, 1982a, 1982b, 1982c, 1982d). In addition a few more-recent studies examined aspects of vocal communication (Heinzmann 1970) and reproduction in the field (López-Jurado et al. 1979; Rodríguez-Jiménez 1984; Reading and Clarke 1988). The recently discovered midwife toad from the island of Mallorca (*Alytes muletensis*) had also been the subject of considerable research at the time (see Hemmer and Alcover 1984), but for logistical reasons it was omitted from our studies. Its prolonged breeding season overlapped substantially with that of most of the Iberian populations. The behavior and ecology of this species was recently studied by Bush (1993).

The peculiar reproductive mode of *Alytes* was well known before our studies, even beyond the scientific literature (Kammerer 1924; Koestler 1971). Male *Alytes* call away from water with a very simple advertisement call. Advertisement calls of male midwife toads are short tonal notes (like flute notes, according to some early literature) (Figure 16.1). The female approaches the male and emits a series of short calls, engaging in a duet with the male, who increases his calling

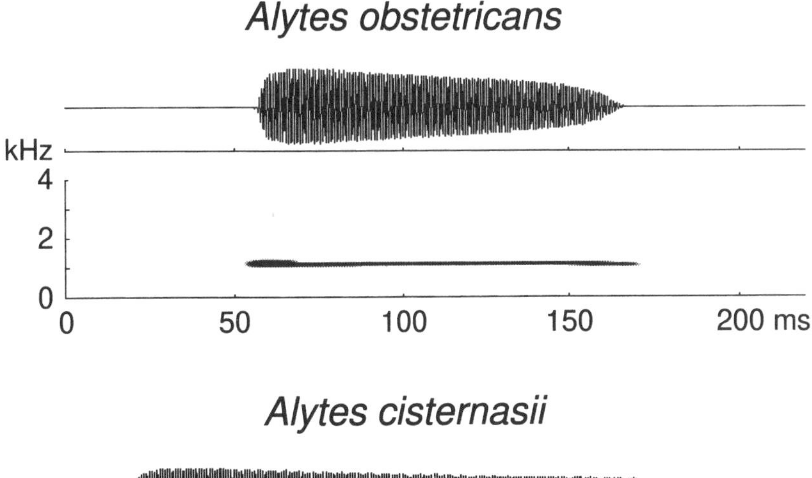

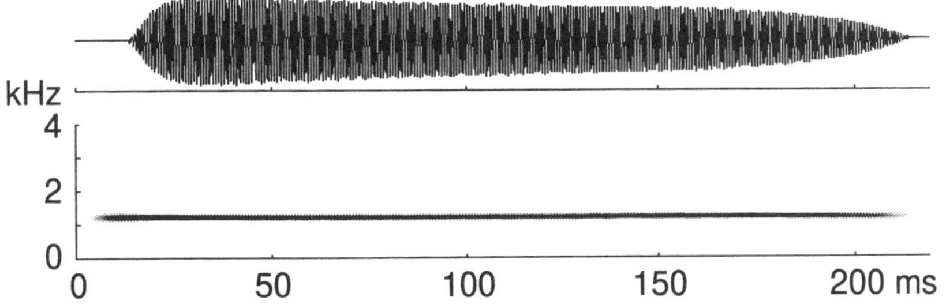

Figure 16.1. Characteristic oscillogram (above) and audiospectrogram (below) of a male advertisement call of midwife toads (*Alytes obstetricans* from Madrid, central Spain, 10.3°C; *A. cisternasii* from Mérida, central western Spain, 15°C).

rate until they engage in amplexus. The complex amplexus occurs on land (Figure 16.2). The male clasps the female at the inguinal region and, with the feet positioned below the lower abdomen of the female, he extends his hind legs alternately causing a lateral rocking motion. After more than half an hour of this activity, the female releases the eggs connected by a flexible thread. The male then moves to an axillary and later to a cervical amplexus and proceeds to fertilize the eggs. He then engages in complex movements of the hind legs and twines the egg string around his ankles. Males can obtain more than one clutch of eggs and carry the eggs for an undetermined time until fully formed tadpoles are ready to hatch. At this time males walk to a nearby body of water and release the egg batch. Additionally males were reported to walk to the edge of the water during the tending period to moisten their progeny (this behavior was never seen in the many years of study of midwife toads in the Iberian Peninsula). Despite the bounty of descriptive information, the evolutionary relevance of the behavior and ecology of midwife toads has not been sufficiently addressed.

Reproductive Behavior

The initial motivation for the study of the midwife toads stemmed from the interest in species with male parental care as tests for the hypothesis proposed by Williams (1966, 1975) and Trivers (1972) that parental effort or investment was the key variable controlling the operation of sexual selection. The fact that in these species males have obligate uniparental care and that the care is performed with the offspring physically attached to the father (as opposed to in a tending site) appeared to be a clear and unambiguous example of paternal care. Thus sex role reversal was expected in *Alytes* if male paternal investment exceeded female investment (see below). Furthermore, if it occurred this reversal would be more marked in species where paternal investment was larger (relative to female investment). Evidence for reversal of sexual selection was to be quantified as the difference in variance of mating success between the sexes (Bateman 1948).

After completing a detailed account of the mating behavior of the lesser-studied Iberian midwife toad (Márquez and Verrell 1991), the next step was to select two focal populations for the initial study—one of *A. obstetricans* and one of *A. cisternasii*. We expected these two species to have differences in relative parental investment and that these differences would be reflected in differences in other parameters related to sexual selection. The choice was based partly on the fact that their breeding seasons should not overlap. Thus an inland population of *A. cisternasii,* in Mérida (Badajoz) with its short, fall breeding season, and a montane population of *A. obstetricans* in Formigal (Huesca), with its summer breeding season, were selected. Basic phenological information was obtained from these populations. The emerging picture was that the studied population of *A. cisternasii* was an explosive breeder, whereas the population of *A. ob-*

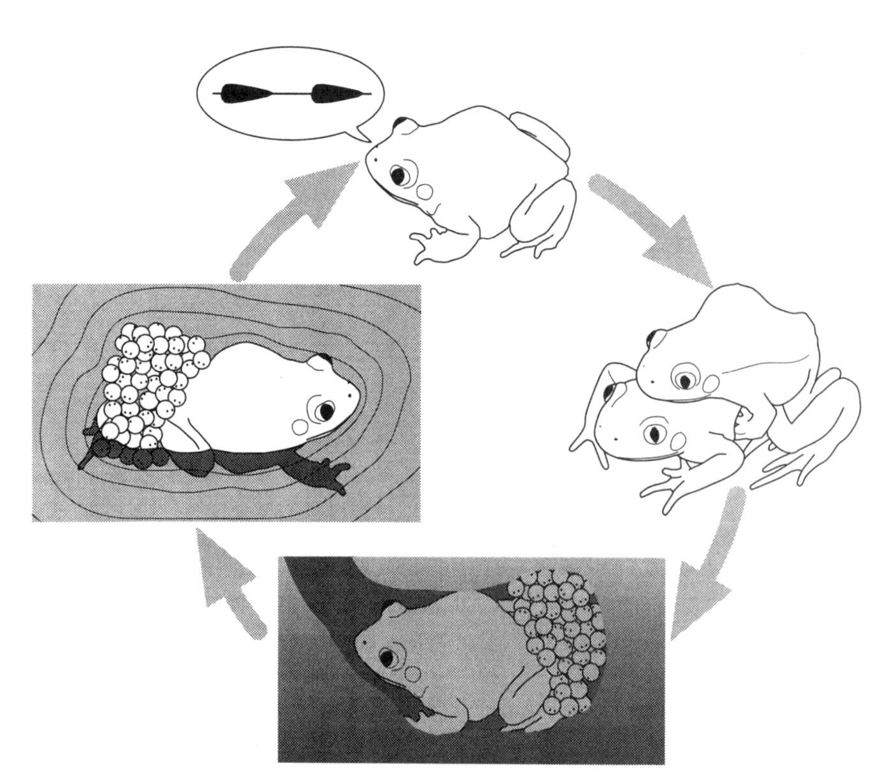

Figure 16.2. Males produce advertisement calls that attract females, and close-range exchange of vocalizations between females and males has been reported. Matings occur on land. A rosary of eggs is produced by the female in amplexus. Males carry the eggs attached to their legs; tending occurs mainly underground and can last about a month. The mature egg masses are released at the shore of a body of water. (Drawing by J. Bosch)

stetricans, despite its high elevation, appeared to have a more prolonged breeding season (Márquez 1992). Caution should be used in extrapolating the phenological information to other populations. It is well known that in other sites in the Iberian Peninsula the breeding season of both species of *Alytes* is markedly longer (Crespo 1982d).

Considering time and energy investment, and variance in mating success together with the first phenological information, we obtained estimates of the length of the paternal tending period. It was similar in both species (23–30 days for *A. cisternasii* and 26–32 days for *A. obstetricans;* Márquez 1992). From an energetic standpoint, a comparison of caloric content of abdominal fat bodies of males before and after the tending period was not significantly different in either species. This was probably due to the extremely high variability of caloric content between males both at the beginning and at the end of the tending season in both species. Also relevant in relation to the male tending period was that males of both species could carry eggs from one to four different females in the same egg batch, but that they did not carry more than one batch per reproductive season (Márquez 1993).

To address the investment of the females we compared the energetic content of the oviducal clutch of eggs (mature ova) with that of the ovaries and of the rest of the body. Given that the caloric value of the oviducal clutch was superior in both species to the caloric contents of the sum of the remaining two ovaries (including abdominal fat bodies), it was inferred that females could only produce a single clutch of eggs per season (Márquez 1990). Furthermore field observations of females after amplexus did not reveal any visible oviducal eggs (which can be observed through the transparent lower skin of the abdomen before mating). This indicates that they released all the mature ova at once, rather than partitioning it between different males as was suggested by Reading and Clarke (1988). Altogether the results suggest that females in the populations studied did not actually exploit their "iteroparous potential" (Jørgensen et al. 1986).

Considering that reproductive males fathered on average more eggs than females laid, even if all males decreased from the best possible state of energetic reserves to the worst possible state, the energetic investment per egg would still be larger in females than in males (Márquez 1990).

Thus sex role reversal in either species was not predicted since female energetic investment per egg greatly exceeded the potential investment of males. Also the variance in mating success of reproductive males (1–4 mates per season) greatly exceeded that of females (1 clutch per season). Ultimately male parental care does not appear to exert an important limitation on the number of mates per male. When compared with other species of anurans without parental care, the average number of mates per season of *Alytes* males that obtained at least one mate (2.23 in 1988 and 1.39 in 1989 for *A. cisternasii,* and 1.75 in 1988 and 1.62 in 1989 for *A. obstetricans*) was not significantly lower than the same value for 22 different studies of 9 different species of anurans without male parental care (Márquez 1993).

Acoustic Communication and Mate Choice

Description of the Advertisement Calls

The interest in the focal species thus shifted from their parental behavior to the characteristics of their communication. The first goal of the acoustic analyses of the mating calls was to determine whether a large-male mating advantage observed in both focal populations (Márquez 1993) could be due to female preference for call characteristics correlated with male size. Although the mating calls of both species were adequately described (Heinzmann 1970; Crespo et al. 1989; Márquez and Verrell 1991), it was necessary to obtain population data to determine the male call range from which females could express a choice (Gerhardt 1988, 1994a, 1994b). Additionally a study of female preferences relative to call variation was undertaken. The quantitative description of the calls was accomplished for the two focal populations initially (Márquez 1990), and later for several additional populations (Márquez and Bosch 1995). In all cases the high average within-individual coefficients of variation obtained for intercall interval (> 0.25) contrast with the relatively low within-individual coefficients of variation obtained for duration and dominant frequency (< 0.06). Cloacal temperature was correlated with call duration, and call dominant frequency was correlated with male size (Márquez and Bosch 1995). Some preliminary playback tests in the field showed that more females of *A. obstetricans* approached a speaker emitting a low-frequency synthetic call than one emitting a similar call with higher frequency (Márquez 1995a). These results were later confirmed by two-speaker playback tests that showed that females of both species preferentially approached a low-frequency call rather than a high- or average-frequency call (Márquez 1995b). Therefore it was shown that (1) male calls were correlated with male size and (2) females exerted a choice based on this information. This suggests that the observed trend of mating advantage of large males could be at least partially accounted for by female preferences for lower frequency calls. This is particularly interesting because it is a rare example of female choice exerting directional selection on a call characteristic with a low degree of intra-individual variation.

Is Female Choice Adaptive?

The study of the reproductive behavior of these species of *Alytes* allowed us to address other hypotheses of more general interest. When female preferences influence immediate fecundity, they should be under direct selection rather than indirect selection, as occurs in runaway selection or selection for good genes (Kirkpatrick and Ryan 1991; Ryan 1997). Given paternal care in this species there is greater potential for female preference to be under direct selection (see also Bateson 1983; Halliday 1983; Bradbury and Andersson 1987). For this to occur two conditions are necessary: (1) that enough variability in paternal care exists and (2) that females are able to assess paternal quality through some correlated characteristic before the time of mating (Verrell 1985, 1986). In both species of *Alytes* the first condition was met. The hatching proportion of eggs was lower for *A. obstetricans* and its variance was higher ($n = 591$, mean $= 0.724$, var. $= 0.057$) than in *A. cisternasii* ($n = 261$, mean $= 0.885$, var. $= 0.013$; Márquez 1990, 1996). The second criterion was not met because none of the measured acoustic parameters of the advertisement call was significantly correlated with male hatching success (unpublished data). Furthermore the advertisement calls of males with eggs (seeking to attract additional mates) did not differ from those of males without eggs (seeking to attract their first mate; Márquez 1990).

Female Preference Function

The study of *Alytes* also contributed to the general theory of mate choice by describing the female preference function's shape and its relationship to the population distribution of male call characteristics. Neurological studies have provided frequency tuning curves for several species of anurans (Wilczynski and Ryan 1988) and there is evidence of mate choice correlation with some tuning parameters (Ryan et al. 1992; Gerhardt and Schwartz, this volume). Despite the observed concordance between optimum frequencies and female preferences, the shapes of tuning curves are not necessarily similar to the shapes of the female preference function because there are complex interactions with background noise and conflicting effects of other frequencies in the call (Ehret and Gerhardt 1980; Gerhardt and Klump 1988; Penna and Narins 1989). Also, females generally express their choices at intensities well above their threshold of auditory hearing (Gerhardt 1994a). Thus a study of female preference when exposed to a variety of stimuli spanning the natural variation of the call and beyond is likely to generate a more reliable estimate of the behavioral preference function. To examine this we performed multispeaker playback tests. These tests also sought to determine if a more complex acoustic environment affected the female preference for lower frequencies than was previously reported with two-speaker tests in the field for *A. obstetricans* (Márquez 1995a) and in seminatural conditions for both species (Márquez 1995b). In these tests a female was exposed to seven

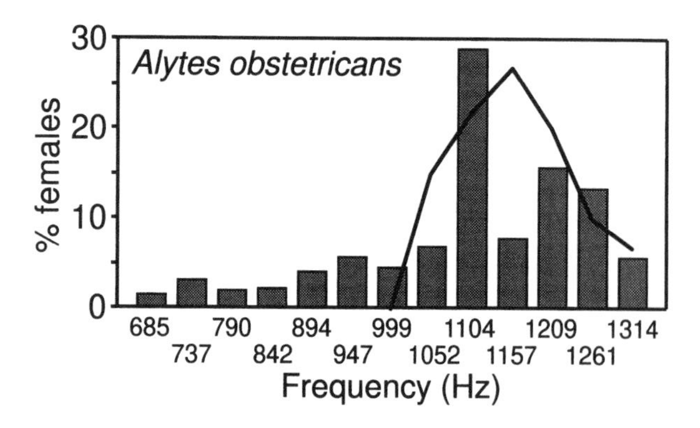

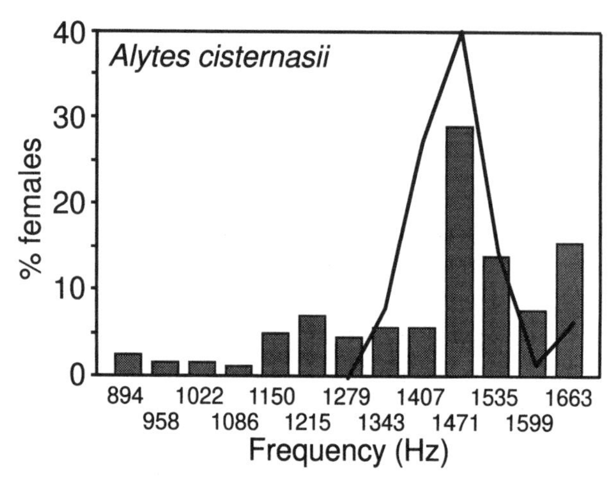

Figure 16.3. Results for the multispeaker playback tests. Numbers indicate the frequency of calls broadcast for each speaker. Shaded bars represent the distribution of females attracted to each speaker. Line shows the distribution of available frequencies of males in the natural population (from Márquez and Bosch 1997a).

synthetic calls, differing in frequency, broadcast from seven different speakers, all with the same intensity, duration, and repetition rate. The results of the tests showed, in both species, that the frequencies that were most attractive to females were within the limits of the distribution of male calls in their respective populations (Figure 16.3). Thus preference for the character "frequency" is not open-ended. In *A. cisternasii* the most attractive frequency was significantly lower than the population average, whereas in *A. obstetricans* it was centered on the average male frequency (Márquez and Bosch 1997a). Therefore in *A. cisternasii* the directionality of the preference remained under more complex acoustic conditions, whereas in *A. obstetricans* the directionality observed in two-speaker tests was not measurable in more complex environments. However the two-speaker test performed in nature (Márquez 1995b) and the resulting large-male mating advantage (Márquez 1993) suggests that perhaps in nature the choice is made between less than seven males, given that the choruses of this population were usually less dense and males were dispersed over a wide area. This result is similar to that observed by Telford et al. (1989) who found that female *Hyperolius marmoratus* only expressed their preferences in small choruses.

Overall the existence of female preference for larger males may be explained by one of two different male characteristics associated with size. On one hand a large male may be an old male, proving a greater survival ability. *Alytes* are known to live to 7 or 8 years (Márquez et al. 1997) and superior survival ability may result in additional reproductive seasons. On the other hand larger males may just be males with a higher growth rate. Extremely small males of both species are known to carry fertile eggs, resulting in an extremely wide variability of male sizes (Márquez 1993 and unpublished data). If growth rate were the favored character, females would gain only the advantage derived from any potential heritability of this character that may provide an advantage for their progeny of reaching reproductive size earlier. Multiple regressions show that in both species the dominant frequency of the male call is correlated more with male size than with age. Similarly male reproductive and mating success is correlated more with size than with age (unpublished data).

Reproductive Character Displacement in Male Calls and Female Preference

The two focal species have similar advertisement calls, are congeneric, and share areas of strict syntopy. This provided an appropriate setting for a study of the variation of male call characteristics and female preferences between allopatry and sympatry as a test of the evolutionary concept of "reproductive character displacement." This concept predicts an increase in the differences in premating recognition behavior in the area of contact of two closely related taxa that may potentially mate with the wrong species (Dobzhansky 1951; Brown and Wilson 1956). The distribution of two important call parameters of the advertisement calls of the two species (frequency and duration) overlapped substantially in allopatry (Márquez and Bosch 1995). The comparison of male advertisement calls between sympatry and allopatry did not yield the expected increment in differences between species in sympatry. However female preference for longer calls in *A. obstetricans,* which was common in allopatry, did not occur in sympatry, in which the longest calls of the mixed chorus would be emitted by *A. cisternasii* males. Similarly female preference for lower-frequency calls, which was found in allopatric populations of *A. cisternasii* both in two-speaker tests (Márquez 1995b) and in multispeaker tests (Márquez and Bosch 1997a), did not occur in the sympatric population where the lower-frequency calls of the mixed chorus could be emitted by heterospecifics (Márquez

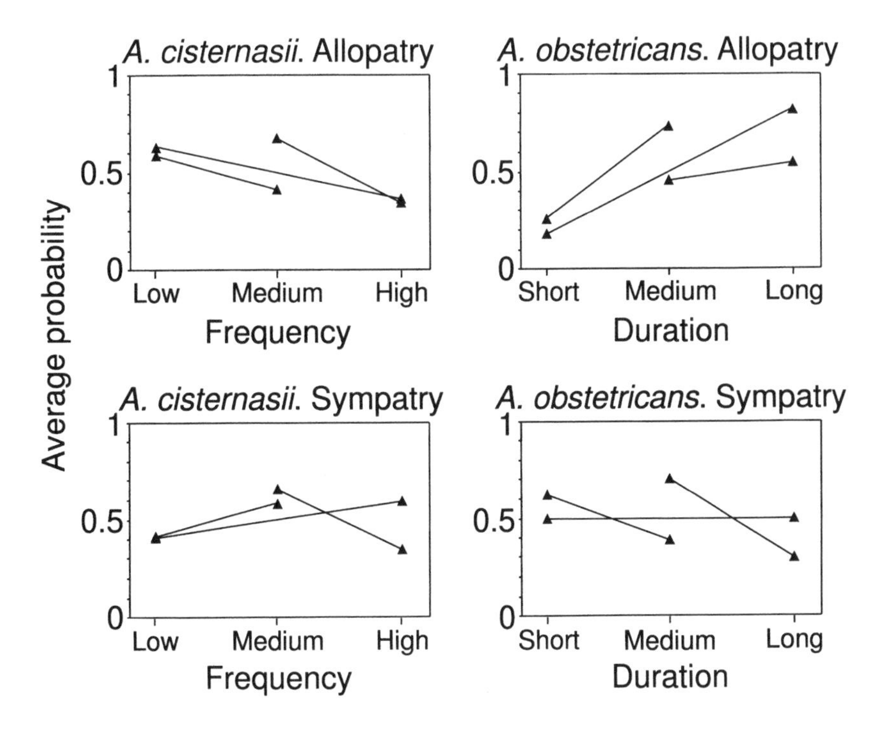

Figure 16.4. Female preferences for call duration and call frequency in sympatry and allopatry. The vertical axis represents the average probability for individual females. Each connected pair of triangles represents a set of playback tests between the options shown on the *x*-axis (from Márquez and Bosch 1997b).

and Bosch 1997b) (Figure 16.4). This result was in agreement with the only previous study of female preferences and call characteristics in a hylid (Gerhardt 1994c). Both studies suggest character displacement in female preferences rather than in male call characteristics.

In general, studies of female choice in *Alytes* yielded answers that downplayed the expected effect of male parental care on the mating system. Males may obtain several matings per season, females only obtain one. The energetic investment of females in their progeny is far greater than that of the males, and at least in the populations studied, the temporal investment in egg tending by males (one month) is less than that of females, which are only able to produce a second clutch of eggs the following season. There is little evidence of reversal of sex roles in nature. The two examples of females fighting for males (Verrell and Brown 1993; Bush and Bell 1997) were observed under artificial captive conditions and female fighting has never been observed in the wild. Female preference need not result in the most adaptive mate choice (females are not selecting superior caretakers), possibly because they cannot evaluate this quality of the male a priori. Although one would expect males that are providing costly parental care to experience strong selection against attracting heterospecific females in sympatry, no evidence suggests divergence between male calls in syntopic populations of continental *Alytes*. In general we consider male parental care in *Alytes* an adaptation to uncertainty in the permanence of the bodies of water (Márquez 1990; Márquez and Verrell 1991) rather than a behavioral trait shaped by sexual selection.

Male–Male Competition

We also studied the importance of male–male competition in the mating system of the continental *Alytes*. Male–male competition through acoustic communication has been considered as agonistic behavior (preceding possible physical combat) or as investment competition when calling activity was considered to entail a significant energetic expenditure or a greater risk to exposure to danger. In the case of *A. cisternasii* and *A. obstetricans* however there is an absence of any sort of physical combat between males (or females); male competition is strictly of an acoustic nature. Males thus could only compete to make their calls more attractive. To study acoustic competition, three questions are important: (1) Does the presence of a nearby calling competitor alter the call characteristics and calling behavior of a calling male? (2) Do alterations depend on the acoustic characteristics of the intruder? and (3) Are alterations observed likely to affect the attractiveness of males to females?

Male Response to Synthetic Calls

Breeding aggregations of *Alytes* are composed of choruses of variable densities. Population studies of three species of midwife toads (Márquez and Bosch 1995, 1996) have shown that although dominant frequency and call duration appeared to be constant within individuals (although they may vary with temperature), call rate was much more variable (even at the same temperature). The mating system of *Alytes* has been described in considerable detail; however male combat or displacement has never been reported and there is some evidence suggesting that acoustically oriented predators such as owls do not prey significantly on either species (Martín and López 1990), thus diminishing the cost of calling that may be derived from increased risk of predation.

Anuran mating calls may provide information about individual characteristics. In many anuran species, including *Alytes*, male call frequency is correlated with body size. Only in a few species of frogs has a change in dominant frequency been reported in the presence of competing males (Lopez et al. 1988; Wagner 1989a, 1989b; Grafe 1995). Our playback studies in the field (Bosch and Márquez 1996) show that this phenomenon does not occur in *Alytes*. Thus in *Alytes* dominant frequency could be considered an honest indicator of male size. This indicator may be honest but is not highly reliable since the regression coefficient between male size and frequency is not very high (r^2 range: 0.16–0.49; Márquez and Bosch 1995). Calling males exposed to a competing synthetic call responded by increasing their calling rate (Bosch and Márquez 1996). However the call characters with a low degree of intra-individual variation—dominant frequency and duration—do not change in the presence of a competitor and are hence not dynamically involved in male–male competition in *Alytes* (Figure 16.5). This character is clearly related to the energetic cost of the production of the call and could therefore be considered an indicator of (energetic) reproductive investment on the part of the male.

Males responded with a more marked increase in call rate to playbacks of synthetic calls with low dominant frequency than to playbacks of calls with a high dominant frequency (Bosch and Márquez 1996). This shows that males as well as receptive females respond differently to calls with different frequencies (Márquez 1995a, 1995b). This could suggest that males perceive the intruder's call frequency as an indicator of male competitive ability or attractiveness to females and respond with a graded increase in call rate. Alternatively this could be explained mechanistically by the possibility that lower frequency calls would be perceived as more intense sounds by the hearing apparatus of males and that the differences in response to the stimulus with two frequencies would be the same as to stimuli of different intensities (of similar males at different distances). In addition males responded with a more marked increase in call rate to playbacks of synthetic calls with high repetition rate than to playbacks of calls with a low repetition rate (Bosch and Márquez 1996). This graded increase in emission has been reported in several species of anurans (e.g., Wagner 1989a). Overall the maximum increases in calling rate observed (43% in *A. cisternasii* and 28% in *A. obstetricans*) are within the ranges observed in other species of anurans. The increase in calling rate in response to an acoustic competitor

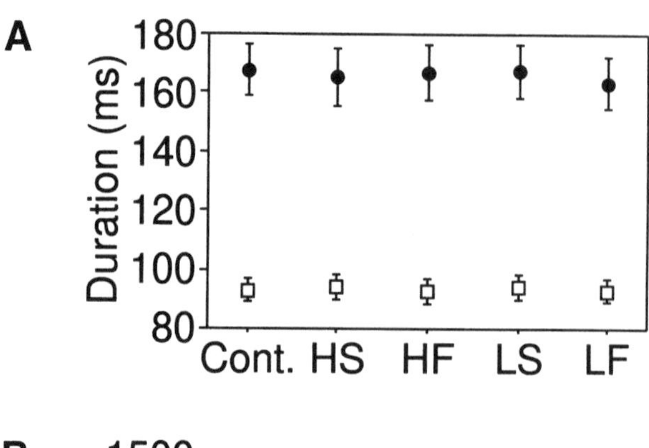

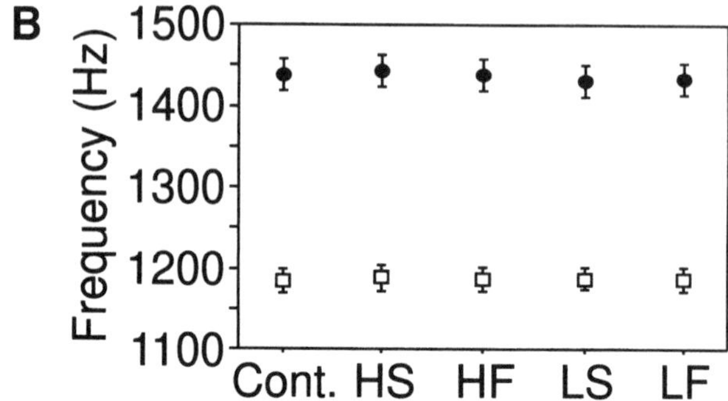

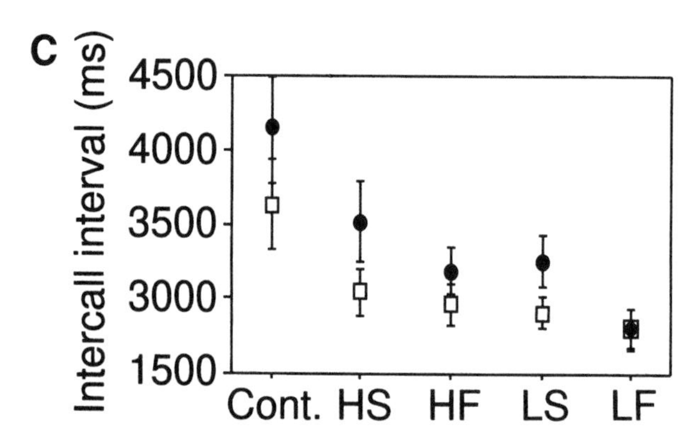

Figure 16.5. Results for the playback tests with males. Mean and standard error for (A) call duration, (B) dominant frequency, and (C) interval between calls, before (*Cont.*) and during the four playback tests (*HS* = high frequency-slow rate stimulus, *HF* = high-fast, *LS* = low-slow, and *LF* = low-fast). From Bosch and Márquez (1996).

is graded; a male increases his calling rate more as the stimulus is emitted at a higher relative rate. On the contrary the response to a stimulus of low frequency is not graded; it is not correlated with the difference in call frequencies between the resident male and the stimulus. This finding suggests that the resident male is shifting his calling rate to a maximum when in the presence of a truly superior competitor (having lower fundamental frequency). Conversely the response could be graded in the presence of an intruder calling at a different calling rate, the resident male increasing his calling rate just enough to counteract the difference in attractiveness with the intruder. The relevance of the changes in male call repetition rate for female choice is obvious when we consider that females are significantly more attracted to the speaker emitting synthetic calls at a faster rate (Bosch and Márquez 1996). However this type of competition may be different than that described for other anurans where acoustic interactions between males have been related to territoriality or to a potential escalation to physically aggressive interactions (e.g., Fellers 1979; Arak 1983). In all known species of *Alytes,* males are often found calling in close vicinity of each other (even under the same rock or in the same refugium) and physical combat between males has never been reported.

Natural Interactions between Two Males

Many anuran males increase their calling activity in dense choruses (Wells 1977, 1988). In such cases it is crucial that males are able to ensure that their advertisement calls are adequately transmitted through the acoustic space. The temporal adjustment of a calling male in relation to the calling pattern of neighboring calling males (e.g., Walkowiak 1992; Given 1993; Grafe 1996) is the most common type of acoustic interaction. Male–male interactions in call timing in *Alytes* are important in competition in the context of sexual selection because the presence of a competing calling male affects the timing of emission of the calls of males (Bosch and Márquez 2000a). The emission of the advertisement call must involve an internal timing mechanism since the perception of calls from neighboring males appears to have little effect on the emission periodicity of the advertisement call (Bosch and Márquez 2000a). The emission of the call of a male is not simply triggered by the perception of the neighboring male, because the lowest adjustment levels between calls were reached with playback calls emitted at irregular intervals (Bosch and Márquez 2000a).

Recording the timing of call exchange between a pair of relatively isolated males in the field is rather simple with a stereo recorder. However the analysis and quantification of acoustic interactions between males is not simple (Klump and Gerhardt 1992). Call adjustment in *Alytes* is not easy to measure compared with other species that have been studied previously. This may result from the fact that in nature, male *Alytes* duets often interrupt the regular alternation of calls, with one male emitting two or more calls sequentially, thus inverting the sequence between males; when the alternation is re-established the leader has become the follower. Call adjustment may be less crucial in this species than in previously studied anurans (mainly hylids) because call duration in relation to call interval (or duty cycle) is extremely low in *Alytes* and the opportunity for overlap or acoustic interference is much lower than in species with more prolonged calls, such as many hylids (e.g., Klump and Gerhardt 1992). In addition many hylids have louder calls than *Alytes* and may thereby interfere with a larger number of neighboring calling males. The phase angle is the latency of the response of a male to another male in relation to the periodicity of the first male, and is 78 ± 13.6 degrees ($n = 9$) for *A. cisternasii* and 83 ± 16 degrees ($n = 7$) for *A. obstetricans.* The angles were not significantly different between species (Bosch 1997).

Our results support the observation of Narins (1982) who found that call adjustment was dependent on call rate. Male *Alytes* are thus more successful at adjusting their calling rate to the stimulus when their body temperature is higher and when their calling rate is lower. Our results also indicate that call adjustment can be maintained over a prolonged sequence of time. The playback experiments suggest that there is some progression in the adjustment to the frequency of the stimulus. We found that call adjustment is more precise with a stimulus of high dominant frequency (Bosch and Márquez 2000a). Thus when the individual that calls first is smaller (and therefore has a higher call dominant frequency) the situation is more favorable for another male that calls second. However several studies of insects and anurans have identified "precedence effects" (sensu Wallach et al. 1949; Zurek 1980; Wyttenbach and Hoy 1993; Howard and Palmer 1995) wherein females orient preferentially toward the leading of two identical calls presented in close succession. The results obtained in our phonotaxis experiments suggest that the temporal sequence of the calls in male duets have an effect on female choice, although our preliminary evidence shows that temperature is likely to play an important role in this process to the point of reversing preferences.

Acoustic Interference

Efficient transmission of acoustic signals is crucial for the reproduction of animals that rely on advertisement calls for mate recognition. It is generally accepted that call alternation between males is a mechanism that diminishes call overlap between calls of adjacent males (e.g., Littlejohn and Martin 1969; Whitney and Krebs 1975; Narins and Capranica 1978; Narins 1982; Zelick and Narins 1982). Few empirical tests demonstrate that call overlap between adjacent conspecific males occurs less often than if the males were emitting their calls at intervals that did not depend on their neighbor's calling rate (Klump and Gerhardt 1992). Schwartz (1987) found that the alternation of calls could play a role in the maintenance of male–male distances and could also be a useful mechanism to preserve the information encoded in the temporal structure of the mating calls. On the other hand he rejected the hypothesis that call alternation improved the detection of calling males. Grafe (1996) found that call overlap did not influence localization by female *Hyperolius marmoratus*. He also found that females preferentially approached calls with low-amplitude modulation. He explained his observations by stating that alternating calls had less amplitude modulation than overlapping calls, and therefore the former are more attractive (the sum of two pulsed calls increased the degree of amplitude modulation).

The study of call overlap has usually been addressed under ideal experimental conditions, with only two individuals interacting naturally or with an individual responding to playback tapes. Although these studies may seem rather reductionist, it has been shown that the calling behavior of *Eleutherodactylus coqui* males in dense choruses can be explained by a model in which a male avoids call overlap with only the two nearest calling males (Brush and Narins 1989). Moreover Gerhardt and Klump (1988) found that *Hyla cinerea* females were not able to locate the exact position of a calling male when the sound pressure level of the background chorus was similar to that of the calls of the calling male. This suggested that in dense choruses females must be choosing among a limited number of males. The experimental study of the effect of call overlap in *Alytes* used interactive playback with a fast enough response time so that the response emitted by the experimental apparatus overlapped with the call emitted by the experimental animal. Our tests showed that call overlap had a clear effect on male *Alytes*, since males decreased their call rate when a nearby calling male consistently overlapped their call (Bosch and Márquez 2000b). This effect was not observed however when males were subjected to the sounds of a loud chorus (Bosch 1997).

Call timing between two male midwife toads in nature did not result in less overlap between calls than in a hypothetical situation where the same males called randomly to each other (Bosch and Márquez 2000b). This result is surprising as it rejects the hypothesis that call timing between males reduces call overlap (Schneider 1967, 1968). Actually the short duration of *Alytes* calls, and their large intercall interval, render the likelihood of random call overlap between two males extremely low. Call overlap however may cause a decrease in call attractiveness. In playback tests females preferentially approach male duets without call overlap rather than male duets with overlap (Bosch and Márquez 2000b). This preference may be interpreted as resulting from differences in duty cycle between the stimuli (overlapping calls had a duty cycle 25% lower than nonoverlapping calls). However other two-speaker tests showed that females did not discriminate significantly between speakers emitting two overlapped calls and speakers emitting a single call (Bosch and Márquez 2000b). This suggests that females do not select calls solely by call duration or duty cycle. Grafe (1996) found that females preferred males calling first in call duets. However our female playback tests suggest that females are not able to distinguish between males that call first and males that call second in duets when there is call overlap (Bosch and Márquez 2000b).

It is difficult to apply to *Alytes* the hypotheses suggested by Schwartz (1987) to explain the advantages of non-overlapping calls over calls that overlap. The first hypothesis is that this mechanism enables adequate spacing between calling males. This is not likely to be applicable to midwife toads where several males can call from the same location, underground or in a same crack of a rock, and amplexus can even occur in such settings. The second hypothesis suggested by Schwartz (1987) is that call alternation may serve to preserve the temporal characteristics of the calls. This hypothesis too may be of limited relevance in the case of the midwife toad and may be more adequate for species of anurans that have an important part of the message codified in the amplitude modulations of the calls. Although overlapping calls are clearly less attractive to female *Alytes* than non-overlapping calls, this is not likely to be due to the fact that differences in call duration caused by the overlapping call render male calls unrecognizable to females. The extreme simplicity of the temporal characteristics of the calls of *Alytes*, the extreme variability of call duration (related in part to temperature, Márquez and Bosch 1995), and female preference for longer calls in most populations (Márquez and Bosch 1997a, 1997b) suggest that an alteration of the duration of the sound may not render it unrecognizable. On the other hand the spectral component of the calls may suffer substantial alterations as a consequence of the overlap of the calls. Depending on the differences in phases of the overlapping calls, the sum of the waves may result in the formation of harmonics that are well beyond the spectral range of the calls, and this would dramatically limit the available energy within the preference range and therefore may even render the calls unrecognizable to females. Finally the third hypothesis tested by Schwartz (1987), for which neither he nor Grafe (1996) found any support, was that call alternation was useful for better localization of calling males. This hypothesis appears to be more relevant for the specific case of *Alytes*. The fact that females call during the precopulatory period in *Alytes* has been interpreted by Márquez and Verrell (1991) as a mechanism to improve the chances of an encounter between an incoming female and a calling male that may be calling buried or in a crack of a rock. If this example of female behavior, extremely rare among anurans, developed as a consequence of the difficulties involved in call localization, it is reasonable to believe that other mechanisms such as the non-overlapping of the calls may have been selected as a means to improve the locatability of calling males or to facilitate female choice (Bush et al. 1996). However phonotaxis tests where a female was offered a choice of a high frequency (thus less attractive) call and two overlapped calls (one of which was low in frequency, thus highly attractive), showed that half of the females chose the two overlapped calls option and did not appear to have problems localizing these calls.

Conclusions

Study of the behavioral ecology of the continental *Alytes* has evolved from an initial stage in which the questions addressed related to its peculiar reproductive mode, male parental care, to issues related to communication and evolution in general. Although these animals present logistical challenges, the investment in the description of the mating system, acoustic communication, and general ecology should still produce some dividends in the near future. Currently additional studies on the timing of the interactions between competing males are underway, focusing particularly on the precedence effect and the importance of variation in phase angle or latency in male duets. In addition we have recently completed a study where we found preliminary evidence of a correlation between tympanum fluctuating asymmetry and precision of sound localization in females. In addition to these studies, measurements on auditory tuning of *Alytes* are being pursued. The extreme simplicity of the call of these species, together with their very primitive evolutionary characteristics, suggests that it may be a valuable model for the study of acoustic communication in the frequency domain. Eventually the discovery of new taxa related to *Alytes* (Arntzen and García-París 1995) and further studies on the behavioral ecology of species in the genus *Bombina* may provide sufficient information on the phylogeny of the group to foster interesting studies on the evolution of the calls of the most primitive anurans.

Acknowledgments

The authors are grateful to the many people that helped with the recordings and playback experiments in the field and laboratory. The authors are also grateful to M. J. Ryan and an anonymous referee for comments on the manuscript, and the Consejería de Medio Ambiente de la Comunidad de Madrid, de la Comunidad Autónoma de Andalucía, de Castilla-León, del Gobierno de Aragón, de la Junta de Extremadura, and de la Junta de Castilla-la Mancha for extending the permits to work in the field. Field work was funded by projects N.S.F. Doctoral Dissertation Improvement Grant BSR-8714956 (S. J. Arnold, P. I.), Hinds Fund Grants (University of Chicago) and CYCIT PB 92–0091 (P.I.: P. Alberch) Ministerio de Educación y Cultura (Spain). JB was the recipient of a predoctoral fellowship from the field station Estación Bio-Geológica El Ventorrillo (C.S.I.C.) and of a postdoctoral fellowship from the Ministerio de Cultura of Spain. Partial funding for JB was provided by program Praxis XXI/BCC11965/97 (Portugal).

References

Arak, A. 1983. Vocal interactions, call matching and territoriality in a Sri Lankan treefrog, *Philautus leucorhinus* (Rhacophoridae). Animal Behaviour 31:292–302.

Arntzen, J. W., and M. García-París. 1995. Morphological and allozyme studies of midwife toads (Genus *Alytes*), including the description of two new taxa from Spain. Contributions to Zoology 65:5–34.

Bateman, A. J. 1948. Intra-sexual selection in *Drosophila*. Heredity 2:349–368.

Bateson, P. 1983. Mate Choice. Cambridge University Press, Cambridge.

Boscá, E. 1879. *Alytes cisternasii*, descripción de un nuevo batracio de la fauna Española. Anales de Historia Natural (4 Junio):216–227.

Boscá, E. 1880. *Alytes obstetricans* Laur. var. *Boscai* Lataste. Actas de la Sociedad Española de Historia Natural 8:4–8.

Bosch, J. 1997. Competencia e interacciones acústicas en *Alytes obstetricans* y *Alytes cisternasii*. Implicaciones en la selección de pareja. PhD. dissertation, Universidad Complutense de Madrid, Spain.

Bosch, J., and R. Márquez. 1996. Acoustic competition in male midwife toads *Alytes obstetricans* and *A. cisternasii* (Amphibia, Anura, Discoglossidae): Response to neighbour size and calling rate. Implications for female choice. Ethology 102:841–855.

Bosch, J., and R. Márquez. 2000a. Acoustical interference in the advertisement calls of the midwife toads (*Alytes obstetricans* and *A. cisternasii*). Behaviour 137:249–263.

Bosch, J., and R. Márquez. 2000b. Call timing in male–male acoustical interactions and female choice in the midwife toad *Alytes obstetricans*. Copeia 2000:00–00.

Boulenger, G. A. 1912. Observations sur l'accouchement et la ponte de l'alyte accoucheur *Alytes obstetricans*. Académie Royale des Sciences, Lettres, et Beaux Arts de Belgique. Bulletin. 570–579.

Bradbury, J. W., and M. B. Andersson. 1987. Sexual Selection: Testing the Alternatives. John Wiley and Sons, New York.

Brush, J. S., and P. M. Narins. 1989. Chorus dynamics of a neotropical amphibian assemblage: Comparison of computer simulation and natural behaviour. Animal Behaviour 37:33–44.

Brown, W. L. J., and E. O. Wilson. 1956. Character displacement. Systematic Zoology 5:49–64.

Bush, S. L. 1993. Courtship and male parental care in the Mallorcan midwife toad *Alytes muletensis*. Ph.D. dissertation, University of East Anglia, Norwich, UK.

Bush, S. L., and D. J. Bell. 1997. Courtship and female competition in the Majorcan midwife toad *Alytes muletensis*. Ethology 103:292–303.

Bush, S. L., M. L. Dyson, and T. R. Halliday. 1996. Selective phonotaxis by males in the Majorcan midwife toad. Proceedings of The Royal Society of London B 263:913–917.

Crespo, E. G. 1976. Contribuição para o conhecimento da biología das espécies Ibéricas de *Alytes*, *Alytes obstetricans boscai* (Lataste 1879) e *Alytes cisternasii* (Boscá 1879) (Amphibia Discoglossidae). Testes de preçipitaçao e electroforéticos. Boletim da Sociedade Portuguesa de Ciencias Naturais 17:39–54.

Crespo, E. G. 1979. Contribuição para o conhecimento da biología das espécies s Ibéricas de *Alytes*, *Alytes obstetricans boscai* (Lataste 1879) e *Alytes cisternasii* (Boscá 1879) (Amphibia Discoglossidae). A problematica da especiação de *Alytes cisternasii*. PhD. dissertation, Universidade de Lisboa, Portugal.

Crespo, E. G. 1981a. Contribuição para o conhecimento da biología das espécies Ibéricas de *Alytes, Alytes obstetricans boscai* (Lataste 1879) e *Alytes cisternasii* (Boscá 1879) (Amphibia Discoglossidae). Regulaçao hídrica-Balanço osmótico. Arquivos do Museu Bocage (ser. c) 1 (4):77–132.

Crespo, E. G. 1981b. Contribuição para o conhecimento da biología das espécies Ibéricas de *Alytes, Alytes obstetricans boscai* (Lataste 1879) e *Alytes cisternasii* (Boscá 1879) (Amphibia Discoglossidae). Emissoes sonoras. Arquivos do Museu Bocage (ser. c) 1:57–75.

Crespo, E. G. 1981c. Contribuição para o conhecimento da biología das espécies Ibéricas de *Alytes, Alytes obstetricans boscai* (Lataste 1879) e *Alytes cisternasii* (Boscá 1879) (Amphibia Discoglossidae). Tegumento (histologia e polipeptidos activos). Arquivos do Museu Bocage (ser. c) 1 (2):33–56.

Crespo, E. G. 1982a. Contribuição para o conhecimento da biología das espécies Ibéricas de *Alytes, Alytes obstetricans boscai* (Lataste 1879) e *Alytes cisternasii* (Boscá 1879) (Amphibia Discoglossidae). Desenvolvimento embrionario e larvar. Arquivos do Museu Bocage (ser. C) 1:313–352.

Crespo, E. G. 1982b. Contribuição para o conhecimento da biología das espécies Ibéricas de *Alytes, Alytes obstetricans boscai* (Lataste 1879) e *Alytes cisternasii* (Boscá 1879) (Amphibia Discoglossidae). Ciclos espermatogénicos e ováricos. Arquivos do Museu Bocage (ser. C) 1:353–379.

Crespo, E. G. 1982c. Contribuição para o conhecimento da biología das espécies Ibéricas de *Alytes, Alytes obstetricans boscai* (Lataste 1879) e *Alytes cisternasii* (Boscá 1879) (Amphibia Discoglossidae). Morfología dos adultos e dos girinos. Arquivos do Museo Bocage (ser. c) 1:255–312.

Crespo, E. G. 1982d. Contribuição para o conhecimento da biología das espécies Ibéricas de *Alytes, Alytes obstetricans boscai* (Lataste 1879) e *Alytes cisternasii* (Boscá 1879) (Amphibia Discoglossidae). Ovos, posturas (Epocas de reprodução). Arquivos do Museo Bocage (ser. a) 1:453–466.

Crespo, E. G., M. E. Oliveira, H. C. Rosa, and M. Paillette. 1989. Mating calls of the Iberian midwife toads *Alytes obstetricans boscai* and *Alytes cisternasii*. Bioacoustics 2:1–9.

De l'Isle, A. 1873. Mémoire sur l'alyte accoucheur et son mode d'accouplement. Annales des Sciences Naturelles (5ème Série, Zoologie et Paléontologie) XVII:1–12.

De l'Isle, A. 1876. Mémoire sur les moeurs et l'accouchement de l'*Alytes obstetricans*. Annales des Sciences Naturelles (6ème Série, Zoologie et Paléontologie) 3:1–51.

Dobzhansky, T. 1951. Genetics and the Origin of Species. Columbia University Press, New York.

Ehret, G., and H. C. Gerhardt. 1980. Auditory masking and effects of noise on responses of the green treefrog (*Hyla cinerea*) to synthetic mating calls. Journal of Comparative Physiology A 141:13–18.

Fellers, G. M. 1979. Aggression, territoriality, and mating behaviour in North American treefrogs. Animal Behaviour 27:107–119.

Gerhardt, H. C. 1988. Acoustic properties used in call recognition by frogs and toads. Pp. 455–484. *In* B. Fritzsch, M. J. Ryan, W. Wilczynski, T. E. Hetherington and W. Walkowiak (eds.), The Evolution of the Amphibian Auditory System. John Wiley and Sons, New York.

Gerhardt, H. C. 1994a. The evolution of vocalizations in frogs and toads. Annual Review of Ecology and Systematics 25:293–324.

Gerhardt, H. C. 1994b. Selective responsiveness to long-range acoustic signals in insects and anurans. American Zoologist 34:706–714.

Gerhardt, H. C. 1994c. Reproductive character displacement on female mate choice in the grey treefrog *Hyla chrysoscelis*. Animal Behaviour 47:959–969.

Gerhardt, H. C., and G. M. Klump. 1988. Masking of acoustic signals by the chorus background noise in the green treefrog: A limitation on mate choice. Animal Behaviour 36:1247–1249.

Given, M. F. 1993. Vocal interactions in *Bufo woodhousii fowleri*. Journal of Herpetology 27:447–452.

Grafe, T. U. 1995. Graded aggressive calls in the African painted reed frog *Hyperolius marmoratus* (Hyperoliidae). Ethology 101:67–81.

Grafe, T. U. 1996. The function of call alternation in the African reed frog (*Hyperolius marmoratus*): Precise call timing prevents auditory masking. Behavioral Ecology and Sociobiology 38:148–158.

Halliday, T. 1983. The study of mate choice, Pp. 3–32. *In* P. Bateson (ed.), Mate Choice. Cambridge University Press, Cambridge.

Heinzmann, U. 1970. Untersuchungen zur bio-akustik und ökologie der geburtshelferkröte. Oecologia 5:19–55.

Hemmer, H., and J. A. Alcover. 1984. Història biològica del ferreret (Life history of the Mallorcan midwife toad). Editorial Moll, Palma de Mallorca, Spain.

Héron-Royer. 1878. Recherches sur la fécondité des batraciens anoures *Alytes obstetricans, Hyla viridis* et sur la fécondation des oeufs du *Bufo vulgaris* dans l'obscurité. Bulletin de la Société Zoologique de France 3:278–285.

Héron-Royer. 1883. Recherches sur les caractéres embryonnaires externes de l'alyte accoucheur (*Alytes obstetricans*) à partir de la ponte jusqu'à l'éclosion de la larve. Bulletin de la Société Zoologique de France 8:417–436.

Héron-Royer. 1886. Sur la reproduction de l'albinisme par voie héréditaire chez l'alyte accoucheur et sur l'accouplement de ce batracien. Bulletin de la Société Zoologique de France 11:671–679.

Howard, R. D., and J. G. Palmer 1995. Female choice in *Bufo americanus:* Effects of dominant frequency and call order. Copeia 1995:212–217.

Jørgensen, C. B., K. Shakuntala, and S. Vijayakumar. 1986. Body size, reproduction and growth in a tropical toad *Bufo melanostictus,* with a comparison of ovarian cycles in tropical and temperate zone anurans. Oikos 46:379–389.

Kammerer, P. 1924. The Inheritance of Acquired Characteristics. Boni and Liveright, New York.

Kirkpatrick, M., and M. J. Ryan. 1991. The evolution of mating preferences and the paradox of the lek. Nature 350:33–38.

Koestler, A. 1971. The Case of the Midwife Toad. Random House, New York.

Klump, G. M., and H. C. Gerhardt. 1992. Mechanisms and function of call-timing in male–male interactions in frogs. Pp. 153–174. *In* P. K. McGregor (ed.), Playback and Studies of Animal Communication, Plenum, New York.

Littlejohn, M. J., and A. A. Martin, 1969. Acoustic interaction between two species of leptodactylid frogs. Animal Behaviour 17:785–791.

Lopez, P. T., P. M. Narins, E. R. Lewis, and S. W. Moore. 1988. Acoustically-induced call modification in the white-lipped frog *Leptodactylus albilabris*. Animal Behaviour 36:1295–1308.

López-Jurado, L. F., M. R. Caballero, and L. D.-S. Freitas. 1979. Biología de la reproducción de *Alytes cisternasii Boscá* 1879. Doñana, Acta Vertebrata 6 (1):6–17.

Márquez, R. 1990. Male parental care, sexual selection, and the mating system of the midwife toads (*Alytes cisternasii* and *Alytes obstetricans*). PhD. dissertation, University of Chicago, Chicago.

Márquez, R. 1992. Terrestrial paternal care and short breeding seasons: Reproductive phenology of the midwife toads *Alytes obstetricans* and *A. cisternasii.* (Ecography) Holartic Ecology 15:279–288.

Márquez, R. 1993. Male reproductive success in two midwife toads (*Alytes obstetricans* and *A. cisternasii*). Behavioral Ecology and Sociobiology 32:283–291.

Márquez, R. 1995a. Preferencia de las hembras por cantos de frecuencia dominante baja en el sapo partero común *Alytes obstetricans* (Anura, Discoglossidae). Experimentos in situ. Revista Española de Herpetología 9:77–83.

Márquez, R. 1995b. Female choice in the midwife toads (*Alytes obstetricans* and *A. cisternasii*). Behaviour 132:151–161.

Márquez, R. 1996. Egg mass and size of tadpoles at hatching in the midwife toads *A. obstetricans* and *A. cisternasii.* Implications for female choice. Copeia 1996:824–831.

Márquez, R., and J. Bosch. 1995. Advertisement calls of the midwife toads *Alytes* (Amphibia, Anura, Discoglossidae) in continental Spain. Journal of Zoological Systematics and Evolutionary Research 33:185–192.

Márquez, R., and J. Bosch. 1996. Advertisement call of the midwife toad from the sierras béticas *Alytes dickhilleni* Arntzen & García-París, 1995 (Amphibia, Anura, Discoglossidae). Herpetological Journal 6:9–14.

Márquez, R., and J. Bosch. 1997a. Female preference in complex acoustical environments in the midwife toads *Alytes obstetricans* and *Alytes cisternasii.* Behavioral Ecology 8:588–594.

Márquez, R., and J. Bosch. 1997b. Male advertisement call and female preference in sympatric and allopatric midwife toads (*Alytes obstetricans* and *Alytes cisternasii*). Animal Behaviour 54:1333–1345.

Márquez, R., M. Esteban, and J. Castanet. 1997. Sexual size dimorphism and age in midwife toads *A. obstetricans* and *A. cisternasii.* Journal of Herpetology 31:52–59.

Márquez, R., and P. Verrell. 1991. The courtship and mating of the Iberian midwife toad, *Alytes cisternasii* (Amphibia, Anura, Discoglossidae). Journal of Zoology of London 225:125–139.

Martín, J., and P. López. 1990. Amphibians and reptiles as prey of birds in Southwestern Europe. Smithsonian Herpetological Information Service 82:1–43.

Narins, P. M. 1982. Behavioral refractory period in neotropical treefrogs. Journal of Comparative Physiology A 148:337–344.

Narins, P. M., and R. R. Capranica. 1978. Communicative significance of the two-note call of the treefrog *Eleutherodactylus coqui.* Journal of Comparative Physiology A 127:1–9.

Penna, M., and P. M. Narins. 1989. Effect of acoustic overstimulation on spectral and temporal processing in the amphibian auditory nerve. Journal of the Acoustical Society of America 85:1617–1629.

Reading, C. J., and R. T. Clarke. 1988. Multiple clutches, egg mortality and mate choice in the mid-wife toad, *Alytes obstetricans.* Amphibia-Reptilia 9:357–364.

Rodríguez-Jiménez, A. J. 1984. Fenología del sapo partero ibérico (*Alytes cisternasii* Boscá. 1879). Alytes (Spain) 2:9–23.

Ryan, M. J. 1997. Sexual selection and mate choice. Pp. 179–202. *In* J. R. Krebs, and N. B. Davies (eds.), Behavioural Ecology, An Evolutionary Approach. Blackwell, Oxford.

Ryan, M. J., S. A. Perrill, and W. Wilcynski. 1992. Auditory tuning and call frequency predict population-based mating preferences in the cricket frog *Acris crepitans.* American Naturalist 139:1370–1383.

Schneider, H. 1967. Rufe und Rufverhalten des Laubfrosches *Hyla arborea arborea* (L.). Zeitschrift für Vergleichende Physiologie 57:174–189.

Schneider, H. 1968. Bio-akustische Untersuchungen a Mittelmeerlaubfrosch Zeitschrift für Vergleichende Physiologie 61:369–385.

Schwartz, J. J. 1987. The function of call alternation in anuran amphibians: A test of three hypotheses. Evolution 41:461–471.

Telford, S. D., M. L. Dyson, and N. I. Passmore. 1989. Mate choice occurs only in small choruses of painted reed frogs *Hyperolius marmoratus.* Bioacoustics 2:47–53.

Trivers, R. L. 1972. Parental investment and sexual selection. Pp. 136–179. *In* B. Campbell (ed.), Sexual Selection and the Descent of Man. Aldine, Chicago.

Verrell, P. A. 1985. Male mate choice for large, fecund females in the red-spotted newt, *Notophthalmus viridescens:* How is size assessed? Herpetologica 41:382–386.

Verrell, P. A. 1986. Male discrimination of larger, more fecund females in the smooth newt, *Triturus vulgaris.* Journal of Herpetology 20:416–422.

Verrell, P. A., and L. E. Brown. 1993. Competition among females for mates in a species with male parental care in the midwife toad *Alytes obstetricans.* Ethology 93:247–257.

Wagner, W. E., Jr. 1989a. Graded aggressive signals in Blanchard's cricket frog: Vocal responses to opponent proximity and size. Animal Behaviour 38:1025–1038.

Wagner, W. E., Jr. 1989b. Fighting, assessment, and frequency alteration in Blanchard's cricket frog. Behavioral Ecology and Sociobiology 25:429–436.

Walkowiak, W. 1992. Acoustic communication in the fire-bellied toad: An integrative neurobiological approach. Ethology Ecology and Evolution 4:63–74.

Wallach, H., E. B. Newman, and M. R. Rosenzweig. 1949. The precedence effect in sound localization. American Journal of Psychology 62:315–336.

Wells, K. D. 1977. The courtship of frogs. Pp. 233–262. *In* D. H. Taylor, and S. I. Guttman (eds.), The Reproductive Biology of Amphibians. Plenum, New York/London.

Wells, K. D. 1988. The effect of social interactions on anuran vocal behavior. Pp. 433–454. *In* B. Fritzsch, M. J. Ryan, W. Wilczynski, T. E. Hetherington, and W. Walkowiak (eds.), The Evolution of the Amphibian Auditory System. John Wiley and Sons, New York.

Whitney, C. L., and J. R. Krebs. 1975. Mate selection in Pacific tree frogs. Nature 255:325–326.

Wilczynski, W., and M. J. Ryan. 1988. The amphibian auditory system as a model for neurobiology, behavior and evolution. Pp. 3–12. *In* B. Fritzsch, M. J. Ryan, W. Wilczynski, T. E. Hetherington and W. Walkowiak (eds.), The Evolution of the Amphibian Auditory System. John Wiley and Sons, New York.

Williams, G. C. 1966. Adaptation and Natural Selection. Princeton University Press, Princeton, NJ.

Williams, G. C. 1975. Sex and Evolution. Princeton University Press, Princeton, NJ.

Wyttenbach, R. A., and R. R. Hoy. 1993. Demonstration of the precedence effect in an insect. Journal of the Acoustical Society of America 94:777–784.

Zelick, R. D., and P. N. Narins. 1982. Analysis of acoustically evoked call suppression behaviour in a neotropical treefrog. Animal Behaviour 30:728–733.

Zurek, P. M. 1980. The precedence effect and its possible role in the avoidance of interaural ambiguities. Journal of the Acoustical Society of America 67:952–964.

17

BRUCE WALDMAN

Kin Recognition, Sexual Selection, and Mate Choice in Toads

Amphibians, unlike other vertebrates, have complex life histories. Larvae and adults have evolved adaptations to very different environments, and metamorphosis demands a radical reorganization of anatomy, physiology, ecology, and behavior. Mechanisms of kin recognition are arguably more thoroughly understood in amphibians than in other vertebrates because amphibian larvae can be manipulated with relative ease for studies of the development and sensory bases of discrimination. Recognition abilities are widely distributed among diverse taxa, as demonstrated in behavioral studies of larvae of at least 18 anurans and four urodeles. Kin discrimination may confer benefits on larvae (Waldman 1991; Pfennig 1997, 1999) but to be favored by natural selection, larval adaptations ultimately must increase the chance that individuals or their close kin reproduce.

Might adults also recognize their kin? Amphibians often show high levels of natal philopatry, and individuals sometimes return to breed just meters from where they originally metamorphosed (Sinsch 1990). Under these conditions close kin are likely to encounter one another as potential mates. Any ability to recognize and avoid breeding with these kin might be strongly selected to avoid deleterious consequences associated with close inbreeding (Waldman and McKinnon 1993). In this chapter I review our current understanding of the mechanisms by which amphibians recognize their kin, the contexts in which kin discrimination occurs, and the functional significance of kin discrimination. I discuss recent work that suggests that females do not necessarily choose as mates the "best" males, but rather those with which they are most genetically compatible.

What Is Kin Recognition?

Animals use communication systems to recognize members of their own species, members of their own population, and specific individuals or classes of individuals that may comprise potential mates or members of their social group (Waldman 1987). Within this framework kin recognition refers specifically to the classification of individuals based on their genetic relatedness, that is, the proportion of genes they share in common. Often kin recognition denotes the process by which individuals assess their own relatedness to target conspecifics, but more generally any evaluation of relatedness among conspecifics constitutes kin recognition (e.g., among members of a social group, see Cheney and Seyfarth 1990).

Because of their genetic similarity relatives should phenotypically resemble one another more than do nonrelatives (Getz 1981; Lacy and Sherman 1983). In addition to the expression of similar genetically determined traits (e.g., Greenberg 1979, 1988) relatives often share traits that they have acquired from a common environment (Gamboa et al. 1986; Porter et al. 1989) and behaviors that they have learned during ontogenetic stages in which kin groups exist (Waldman 1981). Morphological characteristics, behavioral tendencies, and the generation of specific signals (e.g., vocalizations, chemical secretions, or visual displays) all serve as possible means for communicating kinship identity (Waldman 1987).

All these traits or signals, together termed *labels*, propagate through the environment, are perceived by conspecifics, and then are evaluated by comparison with some representation of kin stored in a model *template*. In principle this template might be genetically determined, learned from oneself, or learned from others. Social learning may be based on traits of neighbors encountered during life stages in which kin groups normally occur, or may be based on those characteristics that appear most salient because of genetically constrained stimulus filtering. The match of labels and template may be accomplished by a variety of algorithms. Common characters may be accepted, foreign characters may be rejected, or some combination of both may occur. Many animals are able to recognize relatives with which they have had no previous contact. Phenotype matching, in which individuals assess kinship based on the proportion of matching labels, is the most parsimonious explanation (see Waldman 1987 for further discussion).

Even if an individual recognizes its kin, it might not treat them any differently than nonkin (Waldman 1988). Natural selection should only favor kin discrimination when its benefits exceed its costs, measured in terms of inclusive fitness. The expression of kin-recognition abilities is thus expected to be context-dependent (Waldman 1988; Reeve 1989).

Kin Recognition in Anuran Amphibians

Most research on kin recognition in amphibians has focused on larval anurans and I summarize the major findings here. Recent work has revealed kin-recognition abilities in some urodeles (e.g., Pfennig et al. 1993; Masters and Forester 1995; Carreno et al. 1996; Gabor 1996; Walls et al. 1996). For more complete reviews see Waldman (1991), Blaustein and Waldman (1992), and Blaustein and Walls (1995).

Ontogeny

How kin-recognition abilities develop has been studied by rearing larvae under varied social regimens and assaying the subsequent responses of tadpoles to kin and nonkin in laboratory pools (e.g., Waldman and Adler 1979; Waldman 1981), choice tanks (e.g., Blaustein and O'Hara 1981; Cornell et al. 1989), or natural ponds (Waldman 1982; O'Hara and Blaustein 1985). Results vary among species and sometimes among studies. Tadpoles reared in groups with their siblings associate preferentially with these siblings (e.g., *Bufo americanus*, Waldman 1981; *Bufo boreas*, O'Hara and Blaustein 1982; *Rana cascadae*, O'Hara and Blaustein 1981; *Rana sylvatica*, Waldman 1984; *Rana aurora*, Blaustein and O'Hara 1986), but some species fail to show evidence of kin discrimination (*Hyla regilla*, *Rana pretiosa*, O'Hara and Blaustein 1988; *Pseudacris crucifer*, *Rana pipiens*, Fishwild et al. 1990; *Bufo bufo*, *Hyla arborea*, *Rana dalmatina*, *Rana italica*, Bonini et al. 1994; *Scaphiopus intermontanus*, Hall et al. 1995). Reared in social isolation, *Bufo americanus* and *Rana cascadae* tadpoles also preferentially associate with their siblings (Blaustein and O'Hara 1981; Waldman 1981), but *Rana aurora* and *Rana pretiosa* tadpoles do not (Blaustein and O'Hara 1986; O'Hara and Blaustein 1988). In laboratory tests *Rana sylvatica* tadpoles reared in mixed genetic groups discriminate between familiar siblings and familiar nonsiblings (Waldman 1984) but *Rana cascadae* tadpoles do not (Blaustein and O'Hara 1981). *Bufo americanus* tadpoles discriminate between familiar siblings and familiar nonsiblings, but only if reared in sibling groups during a sensitive period following hatching (Waldman 1981).

Kin-recognition abilities in some cases may be retained through metamorphosis. *Rana cascadae* froglets, tested up to 47 days after metamorphosis, recognize and preferentially associate with siblings (Blaustein et al. 1984). In contrast *Rana sylvatica* froglets show kin preference within 24 hours of metamorphosis (Cornell et al. 1989), but not at later developmental stages (Waldman 1989; see Walls 1991 for a similar study in urodeles). The expression of kin-recognition abilities in larvae also may be limited to specific developmental periods (Blaustein et al. 1993).

What Are the Cues?

Larval kin recognition occurs principally through chemosensory communication. Kin association is facilitated by visual attraction to conspecifics and perhaps through additional sensory modalities, but to date only chemical cues have been shown to encode information concerning kinship. When tested in a Y-maze apparatus in which they are exposed only to waterborne cues, *Bufo americanus* tadpoles orient toward their siblings rather than nonsiblings, and they demonstrate some aversion to odors emanating from nonsiblings (Waldman 1985). *Rana cascadae* tadpoles do not discriminate between siblings and nonsiblings when they are in visual contact only; however, when communication is possible through chemical, auditory, and lateral line senses, tadpoles associate with siblings (Blaustein and O'Hara 1982a).

Kin-recognition labels may be derived from a variety of sources, including gene products, environmental cues, and maternal factors. Maternal contributions to the egg appear to influence recognition abilities both in *Bufo americanus* and *Rana cascadae* tadpoles (Waldman 1981; Blaustein and O'Hara 1982b) but recognition of paternal half-siblings (*Rana cascadae, Rana sylvatica*) provides compelling evidence that recognition cues are in part genetically determined (Blaustein and O'Hara 1982b; Cornell et al. 1989). Kin recognition based on genetically determined cues may be important in situations in which multiple paternity occurs (D'Orgeix and Turner 1995; Laurilia and Seppa 1998), although this has yet to be studied. Dietary and other environmental odors also may be incorporated into labels (Waldman 1984; Pfennig 1990; Gamboa, Berven et al. 1991; Hepper and Waldman 1992; Hall et al. 1995), and natural differences in these cues among sibling groups may be sufficient to serve as a basis for kin recognition.

Phenotype Matching

The ability of larvae reared in isolation to recognize their siblings might be interpreted as evidence of a genetic recognition mechanism (Blaustein and O'Hara 1981; Blaustein 1983). Yet ontogenetic studies demonstrate multiple factors in addition to genetic ones that affect kinship preferences. Individuals might learn traits they encounter during their development and recognize as close kin those individuals that share the same traits, even if they have not previously encountered those particular individuals. In social isolation individuals can learn their own traits, which incorporate genetic, maternal, and environmental effects. Waldman (1981) described this process of "phenotype matching" on which others later elaborated (see, e.g., Blaustein 1983; Sherman and Holmes 1985; Waldman 1987; Gamboa, Reeve et al. 1991).

Functions

Originally kin recognition was studied in larval *Bufo* (Waldman and Adler 1979) because toad tadpoles are distasteful, appear aposematically colored, and form conspicuous schools—traits most likely to evolve under kin selection (Fisher 1958; Wassersug 1973). However kin-recognition abilities were quickly uncovered in species that lacked these traits, and numerous ideas were proposed to explain this (Waldman 1982; Blaustein et al. 1991). Larval kin recognition may facilitate the regulation of growth and development in ephemeral habitat (Waldman 1982, 1986, 1991; Jasienski 1988; Smith 1990; Hokit and Blaustein 1994, 1997), the reduction of competitive interference and cannibalism among close relatives (Walls and Roudebush 1991; Pfennig et al. 1993, 1994; Walls and Blaustein 1995; Pfennig 1999), or cooperation among close relatives in accruing resources (Waldman 1991) and in avoiding predation (Hews and Blaustein 1985; Hews 1988). Alternatively apparent larval kin recognition may be epiphenomenal, reflecting species recognition (O'Hara and Blaustein 1982; Grafen 1990), philopatry (Pfennig 1990), or an epigenetic process with fitness consequences only later in development (Waldman 1984).

The larval stage is the first in the complex life cycle of amphibians, and the contexts in which kin recognition might confer benefit after metamorphosis need to be considered. Parental care is widespread in urodeles, and less common but present in several groups of anurans (reviewed in Waldman 1991). Parents may avoid cannibalizing their offspring (Gabor 1996) or preferentially brood them (Masters and Forester 1995). Some amphibians are territorial and their tendencies to act more agonistically toward nonneighbors than neighbors may be functionally equivalent to kin recognition (Madison 1975; Jaeger 1986) as demonstrated in experimental tests (Walls 1991; Walls and Roudebush 1991).

Kin recognition can be an important determinant of mate choice (Bateson 1983), especially when close kin are readily accessible as mates and inbreeding (or outbreeding) incurs some cost. Natal philopatry is common among amphibians (reviewed in Waldman and McKinnon 1993) and can give rise to highly structured demes. Unless effective population sizes are huge, individuals that return to their natal pond to breed should encounter close relatives as potential mates. Recent molecular work on natural populations suggests that some amphibians avoid mating with their close relatives and that females may choose mates based on genetically specified components of males' calls (Waldman et al. 1992; Waldman and Tocher 1998). I provide an overview of this work in the following sections.

Genetic Population Structure and Behavior

Behaviors such as homing determine genetic population structure. In turn behavioral interactions among individuals may be affected by the level of inbreeding within populations, the degree of genetic differentiation among populations, and their relatedness to individuals with which they interact.

Philopatry

The abilities of frogs, toads, and salamanders to home over long distances after displacements are legendary (e.g., Twitty

1966). Some frogs migrate year after year to particular sites to breed (Bogert 1947; Jameson 1957; reviewed in Sinsch 1990; Waldman and McKinnon 1993), even if the ponds are paved over as parking lots (Heusser 1969). Fowler's toads, *Bufo woodhousei fowleri*, marked at metamorphosis usually return as adults to their natal pond to breed (Breden 1987). Similar findings were reported by Berven and Grudzien (1990) on *Rana sylvatica* and Reading et al. (1991) on *Bufo bufo*. Not all frogs, or even toads (Sinsch 1992; Schwarzkopf and Alford 2001), necessarily show uniformly high levels of philopatry. Yet when philopatry is strong, genetic analyses often reveal highly structured local populations.

Genetic Differentiation

Local populations of the common frog, *Rana temporaria*, show pronounced genetic differences, not only when separated by formidable barriers (Reh and Seitz 1990), but even when dispersal appears readily possible between ponds just 2.3 km apart (Hitchings and Beebee 1997). Genetic distances correlate with geographical distances, but ponds near towns show more genetic differentiation than do those in rural districts (Hitchings and Beebee 1997). Strong philopatry in the Australian frogs *Geocrinia alba* and *Geocrinia vitellina* is accompanied by significant differences in allele frequencies among populations (Driscoll 1997, 1998a). Even more strikingly, Driscoll (1998b) found fixed allelic differences between populations of *Geocrinia rosea* just 4 km apart and of *Geocrinia lutea* 1.25 km apart. *Rana pipiens* populations show genetic differentiation over 3.5 km (Kimberling et al. 1996). Work currently underway on populations of the New Zealand frog *Leiopelma archeyi* suggests that populations are genetically subdivided over distances as small as 20 m (B. Waldman, unpublished data). Even when genetic measures show substructuring within ponds attributable to philopatry, genetic differentiation among populations is not an inevitable consequence (Buckley et al. 1996). Moreover genetic differentiation that may occur temporally among cohorts within localities can be greater than that observed among localities (e.g., *Bufo calamita*, Sinsch 1997).

Inbreeding

The typical amphibian population structure is conducive to high levels of inbreeding. Some individuals disperse and are unlikely to inbreed. But in many species, most individuals—of both sexes—probably return to natal localities to breed (e.g., Berven and Grudzien 1990). Once there they encounter siblings and other relatives as potential mates. Close relatives that mate are likely to produce offspring that suffer from inbreeding depression and possibly increased vulnerability to environmental changes and disease (Waldman and Tocher 1998).

Although good data are still lacking for amphibians, in many ectothermic vertebrates, traits that are correlated with fitness (e.g., hatching success, growth rates, developmental stability, size at maturity, and courtship behaviors) show evidence of inbreeding depression (Waldman and McKinnon 1993; Vrijenhoek 1994; Madsen et al. 1996). Conversely outbreeding can benefit progeny by conferring on them fitness benefits associated with increased heterozygosity (Naylor 1962; Falconer 1989). Heterozygous tiger salamanders, *Ambystoma tigrinum*, grow faster and have greater metabolic efficiency than homozygous individuals (Pierce and Mitton 1982; Mitton et al. 1986; cf. McAlpine and Smith 1995; Wright and Guttman 1995). Survival correlates positively with the level of multilocus heterozygosity in overwintering juvenile western toads, *Bufo boreas* (Samallow and Soulé 1983; cf. McAlpine and Smith 1995), and reproductive success of green treefrogs, *Hyla cinerea*, correlates positively with their heterozygosity (McAlpine 1993).

When habitat becomes fragmented, as increasingly is occurring because of urban expansion, agricultural development, logging, and the introduction of exotic predators, metapopulation structure becomes disrupted and levels of inbreeding within frog populations increase (Waldman and Tocher 1998). Hitchings and Beebee (1998) found that levels of genetic diversity and heterozygosity decrease within populations of larval common toads, *Bufo bufo*, as a function of pond isolation. In turn as genetic diversity decreases, populations show increased numbers of physical abnormalities and reduced survivorship. Similarly inbreeding depression is increasingly apparent in isolated populations of *Rana temporaria* (Hitchings and Beebee 1997). In part attributable to these effects, the probability that local populations go extinct increases in proportion to the distance between them (Sjögren 1991a, 1991b; Sjögren Gulve 1994; Edenhamn 1996).

The American Toad: A Case Study

If amphibians can recognize their kin, they might avoid mating with them. Over many years we have studied natural breeding populations of *Bufo americanus* (Waldman et al. 1992; Waldman and McKinnon 1993; Waldman 1997) to examine genetic population structure, frequencies of matings between close relatives, and means by which individuals might selectively outbreed. The mating behavior of this species has been reasonably well studied across its range, but with variable results.

Sexual Selection and Mate Choice

Bufo americanus often breeds explosively, in periods as short as 48 hours (Wells and Taigen 1984; Howard 1988), but sometimes over 5 weeks or longer (Forester and Thompson 1998). Consistent with expectations that short seasons promote scramble competition (Wells 1977), mate choice at first appears improbable. Most males either call for females or clasp indiscriminately on the water surface (Licht 1976; Christein and Taylor 1978; Gatz 1981; Fairchild 1984; Howard 1988; Waldman et al. 1992). Yet careful observation suggests that females behave more selectively. Licht (1976) was first to propose that males' advertisement calls serve not simply as a mechanism of species recognition but also as a criterion for female mate choice. He observed that females swam slowly underwater, then surfaced near calling males apparently to listen to their calls, and subsequently initiated amplexus by approaching particular individuals. These observations have been confirmed in other populations (Howard 1988; Sullivan 1992; Waldman et al. 1992). Alternative mating tactics, such as terrestrial interception of females by males (Forester and Thompson 1998), may be more common in certain localities or as the breeding season ensues. Other males frequently attempt to dislodge males amplexed to females but rarely succeed (Gatz 1981; Howard 1988; Sullivan 1992).

If mate choice is possible, sexual selection should favor females that choose larger males whose "good genes" will confer on their offspring faster growth, larger body size, longer life, higher fecundity, and the ability to attract more mates (Woodward 1986, 1987; Mitchell 1990; cf. Howard et al. 1994). Alternatively females might be selected to choose males of some optimal size to achieve higher fertilization rates (Davies and Halliday 1977; but see Kruse 1981). Despite ample studies, field evidence for differential mating success by size is weak. Mating success of male *Bufo americanus* was found to be random with respect to body size in three studies (Wilbur et al. 1978; Kruse 1981; Sullivan 1992). Gatz (1981) found that larger males accrued some reproductive advantage, and Howard (1988) also found marginal evidence of a large-male advantage but not in every year of his study. Although Licht's (1976) data suggest some assortative mating by size, other studies fail to corroborate this (Wilbur et al. 1978; Gatz 1981; Kruse 1981).

Both calling males, which produce advertisement calls from mostly stationary positions, and noncallers, which swim on the pond surface and attempt to clasp females that they encounter, are usually present in breeding groups. The proportion of callers decreases as the number of males in the breeding assemblage increases (Waldman et al. 1992). Although noncallers are occasionally successful, callers have much higher mating success (Gatz 1981; Fairchild 1984; Howard 1988). In some populations call dominant frequency is inversely correlated with male body size (Sullivan 1992; Howard and Young 1998) and thus might serve as a cue for female choice. Indeed in laboratory playback experiments, females prefer males with low frequency calls (Howard and Palmer 1995; Howard and Young 1998). In experiments under field conditions however females fail to discriminate between males based on dominant call frequency but do prefer males with high call efforts, achieved by calling at high rates or over longer durations (Sullivan 1992).

Mate Choice and Genetic Compatibility

Aside from choosing a mate with good genes, selection should favor individuals that choose to mate with conspecifics whose genes are compatible with their own. Genomic compatibility long has been considered an important selective force in mate choice between species (e.g., Littlejohn 1981) but it may also be important in mate choice among conspecifics (Bateson 1983; Penn and Potts 1999). Close inbreeding entails genetic costs, but so too might outbreeding: the breakup of coadapted gene complexes might lead to reduced viability and to the loss of local adaptations (Partridge 1983). A preference for mates of intermediate relatedness then would be selected (Shields 1982; Bateson 1983). Philopatry, coupled with behaviorally mediated inbreeding avoidance, could effectively result in "optimal outbreeding."

Genetic Population Structure

During five breeding seasons, from 1986 to 1990, my collaborators and I surveyed the genetic structure of *Bufo americanus* breeding populations at several localities bordering the Estabrook Woods, a 650-acre mixed oak-birch forest in Concord, Massachusetts (Waldman et al. 1992). We examined molecular markers to attempt to discern patterns of dispersal consistent with natal philopatry.

Although suitable breeding habitats are absent within the reserve, toads are found in large numbers there during the spring, summer, and autumn. During late April and May, depending on weather conditions, individuals migrate to breeding pools at the edge of the woods. Population and breeding structure were intensively studied at three primary localities: Beecher Pond, Concord Center, and Mink Pond (Figure 17.1). Distances between these ponds range from 0.8 to 2.2 km.

Breeding ponds were monitored nightly during April and May each year. During the study we collected a total of 200 amplectant pairs from the three primary localities, most after pairs had begun laying eggs. To compare levels of genetic variation within and among ponds we analyzed mitochondrial DNA (mtDNA) extracted from the pairs. Pure mtDNA was isolated, digested with various enzymes, and then run

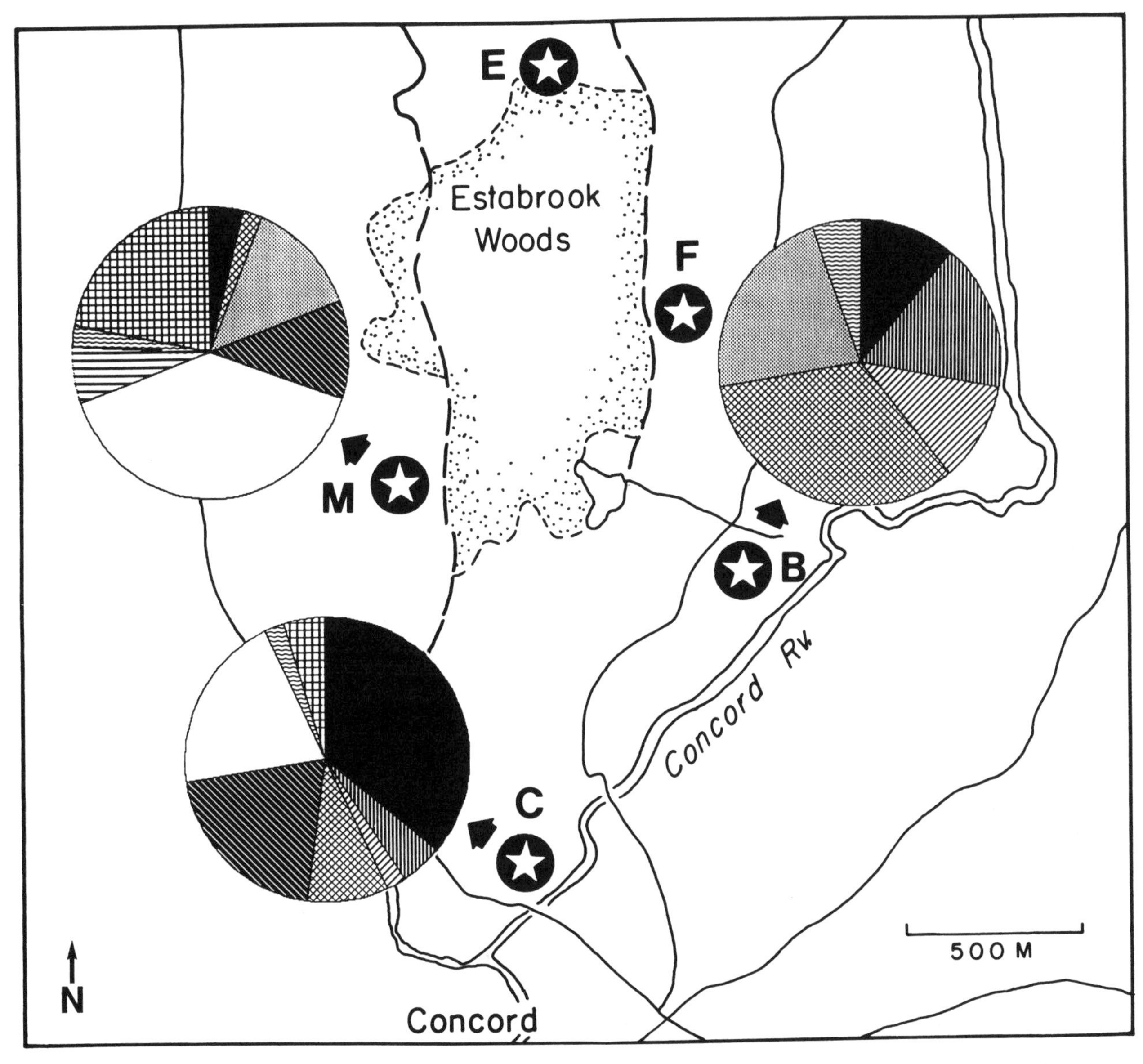

Figure 17.1. Distribution of common mtDNA haplotypes among toads mating at three primary breeding sites surrounding the Estabrook Woods (*B* = Beecher Pond; *C* = Concord Center; *M* = Mink Pond). Advertisement vocalizations of males were recorded at Beecher and Mink ponds, and two additional localities (*E* = Evans Pond; *F* = Freeman Pond). Frequencies of haplotypes (each denoted by a unique shading pattern) are shown in pie charts. Haplotypes were determined as composite restriction fragment length polymorphisms, based on digests with four restriction enzymes (from Waldman et al. 1992).

on electrophoretic gels to separate fragments of different size that result from the digestion (Densmore et al. 1985). The banding patterns generated reveal mitochondrial restriction fragment length polymorphisms, or fingerprints, from which we can infer matrilineal relationships within and among populations (Harrison 1989).

Mitochondrial DNA is maternally inherited, so siblings, as well as more distant matrilineal relatives, share identical fragment patterns. If individuals differ in mtDNA haplotypes, barring mutations, they cannot be siblings. The application of mtDNA to studies of population structure would be limited if particular haplotypes predominated, but this is often not so. In many vertebrates, mtDNA undergoes more rapid evolution than nuclear DNA. Consequently mitochondrial markers can vary extensively among individuals, both within and between populations. Ectothermic vertebrates have particularly variable mitochondrial genomes (Bermingham et al. 1986; Avise et al. 1989; Waldman et al. 1992), sometimes approaching hypervariable minisatellite regions of the nuclear genome (Jeffreys 1987) in their diversity.

Comparisons of the mtDNA haplotypes represented among breeding individuals in the three ponds indicate significant differences (Figure 17.1). Over three-quarters of the 44 individuals present in temporary ponds at Concord

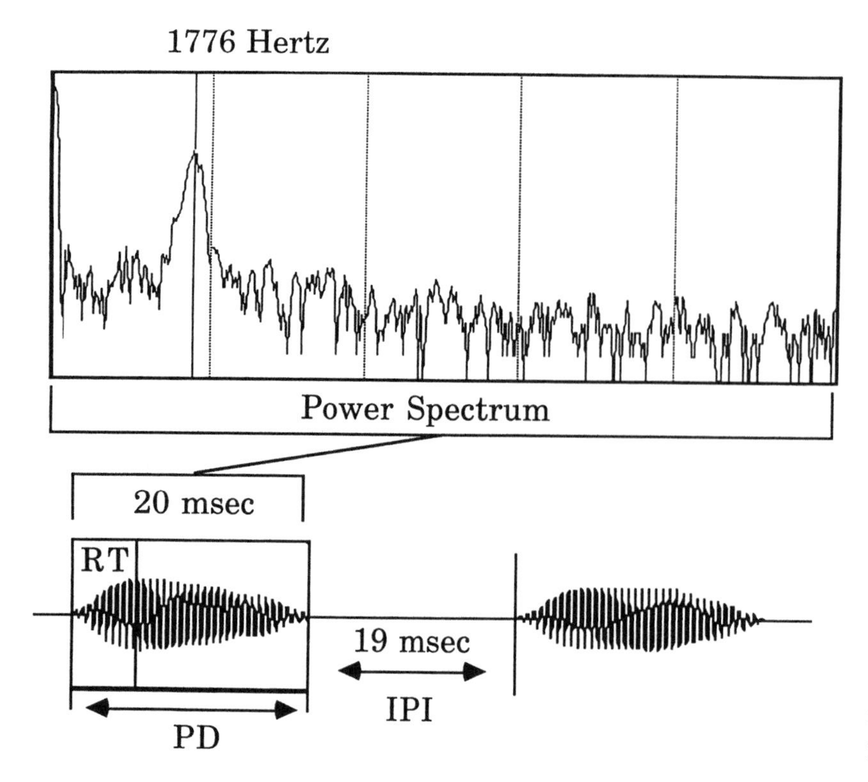

Figure 17.2. Components of typical *Bufo americanus* advertisement call. Top: Power spectrum taken from 20-ms section, showing narrow frequency bands; 1776 Hz dominant frequency. Bottom: Oscillogram (energy vs. time) showing two pulses; *RT* = rise time; *PD* = pulse duration; *IPI* = interpulse interval (from Waldman et al. 1992).

Center, south of the woods, belonged to one of three haplotypes (*A1, B2, B3*), and the remaining individuals comprised five other haplotypes. Of 18 toads spawning in similarly ephemeral habitat due east of the woods (Beecher Pond, less than 1 km northeast of Concord Center), six haplotypes were represented. Neither haplotype *B2* nor *B3*, two of the most common at Concord Center, were present in Beecher Pond. West of the woods, in Mink Pond (less than 1 km from Beecher Pond and Concord Center), eight haplotypes were represented among 46 individuals, but again in clearly different proportions from the other ponds. Haplotype *C1*, for example, was found in Mink Pond but nowhere else. Haplotype frequencies differed significantly among the three localities, but variation among years within localities was not significant (Waldman et al. 1992). Natal philopatry, facilitated by well-developed homing abilities (Dole 1972), probably generates these patterns of genetic variation.

Do Toads Mate with Close Relatives?

Given their propensity to return to their natal pond to breed, toads are likely to encounter their brothers and sisters as potential mates. By comparing mtDNA haplotypes of males and females captured in amplexus we constructed minimal estimates of the frequency of matings between close relatives (Waldman et al. 1992). Should mates share haplotypes they may be siblings, but they also could be related through a more distant female ancestor. Despite the imprecision with which we can determine ancestry we conclude that close inbreeding is exceedingly rare. Of 86 mated pairs collected near Concord, Massachusetts (from the three focal breeding populations and additional breeding sites), members of only two pairs had identical haplotypes. Randomly generated pairings of males and females present at each pond during each season however lead to a null expectation of 12 matings between individuals bearing identical haplotypes. Thus, fewer individuals mated with close relatives than would be expected if pairing were random, but the expected frequencies are low (Waldman et al. 1992). As the null expected probabilities are based on those individuals that return to each pond, the results imply that toads use behavioral mechanisms to avoid close inbreeding (see Pärt 1996).

Vocalizations as Potential Kin-Recognition Cues

Recent work suggests that anuran advertisement calls sometimes provide an indicator of heritable genetic quality (Welch et al. 1998). Yet evidence of mate choice in natural populations of *Bufo americanus* is weak, and studies of the responses of female toads to broadcast calls have yielded inconsistent data. Females seem to prefer calling males but not necessarily larger males or those with "good genes." Perhaps females evaluate male calls in part to avoid close inbreeding or to "optimally outbreed." To the extent that the advertisement calls of males are heritable, and thus encode information about the caller's genotype, they might serve as potential kin-recognition cues that would enable females to evaluate their genetic relationship to callers.

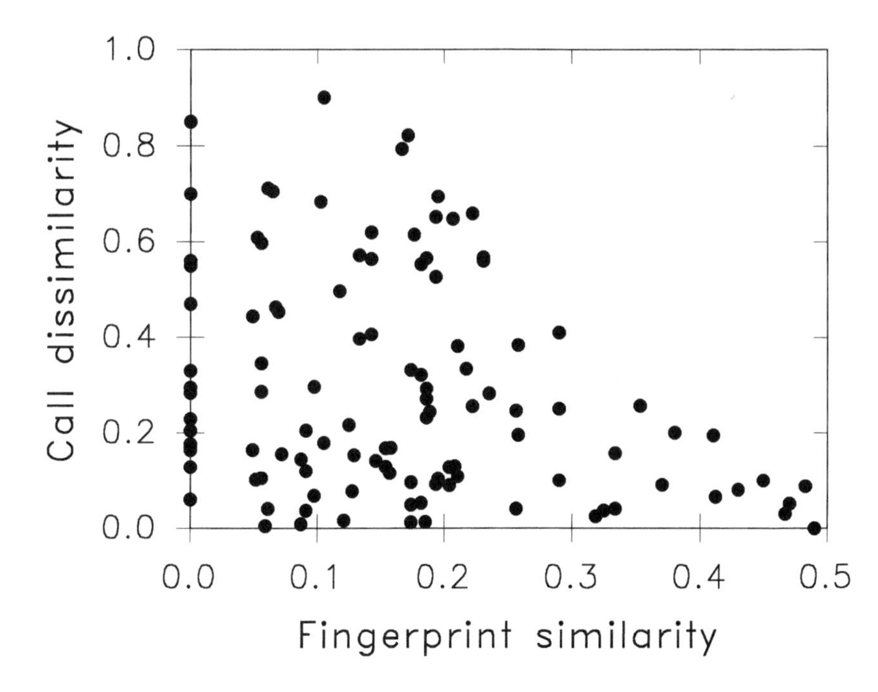

Figure 17.3. Call dissimilarity as a function of genetic similarity among calling males. Shown here are analyses of one component of the advertisement call, the interpulse interval, from males collected at Freeman Pond. Similar results are obtained with other call parameters and at each pond. Identical calls have a dissimilarity value of 0, and increasingly dissimilar calls have higher values. Fingerprint similarity values increase with relatedness (r) and inbreeding (F) coefficients (from Waldman et al. 1992).

Table 17.1 Genetic determinants of call differences (Mantel analyses)

Pond	Pulse Duration	Interpulse Interval	Rise Time	Call Duration	Dominant Frequency
Beecher	$p < 0.005$	$p < 0.03$	$p < 0.01$	$p < 0.05$	ns[a]
Evans	$p < 0.03$	$p < 0.05$	$p < 0.05$	$p < 0.01$	ns
Freeman	$p < 0.03$	$p < 0.03$	ns	ns	ns

[a]ns = not significant.

To investigate this possibility we recorded calls of 15 males in each of four breeding populations around the Estabrook Woods (Waldman et al. 1992; see Figure 17.1). After recording each male's calls we measured his snout-vent length and cloacal temperature, as well as water and air temperatures at the calling sites. When regressions were significant we adjusted call parameters to control for these effects. We recorded most males on multiple occasions; this permitted us to compare variation in call parameters within and among individuals as well as within and among ponds. Later we genetically typed each male by obtaining nuclear DNA fingerprints of each using multilocus minisatellite probes (33.6 and 33.15; Jeffreys 1987). We estimated the genetic similarity of calling males within ponds by the proportion of bands they shared (Wetton et al. 1987), and we determined proportions of bands shared by siblings and half-siblings by examining progeny of known parents in the laboratory. Next we examined similarities in both temporal and frequency components of males' calls (Figure 17.2) as a function of the callers' genetic similarity by means of Mantel analyses (Mantel 1967).

Genetically similar males, including brothers and more distant relatives, do indeed produce similar advertisement vocalizations (Table 17.1). Only the temporal components of the call appear genetically constrained, however. Although pulse duration, interpulse interval, rise time, and call duration all show significant correlations with the genetic relatedness of callers, the dominant frequency of the call appears to provide no information at all about kinship. Genetically similar individuals had very similar calls, whereas calls of genetically dissimilar individuals were much more variable, even when temperature and size were held constant (Figure 17.3). Consistent with the genetic differences observed among localities, temporal properties of advertisement calls differed significantly among ponds, and among individuals within ponds (see Waldman et al. 1992).

Kin Recognition and Mate Choice

Kinship information encoded in males' calls might enable females to recognize their close relatives and to avoid mating with them or even to choose an "optimally" related mate (Bateson 1983). But can females decode this information and use it? To answer this question we individually tested 29 females in a laboratory arena, observing their re-

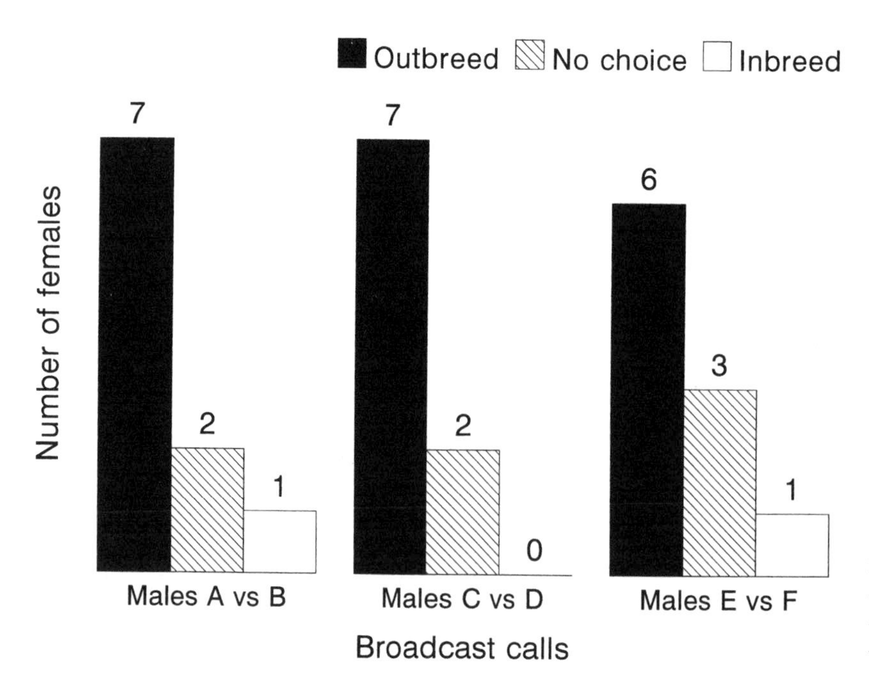

Figure 17.4. Responses of females to calls of males broadcast alternately from speakers on either side of an indoor test arena. Calls of three pairs of males were broadcast and different females were tested with each pair. Females were judged to have made a choice if they moved within 20 cm of a speaker. DNA fingerprints of males and females were compared only after behavioral tests had been completed. Twenty females approached the speaker broadcasting the call of the male to which they were less closely related (i.e., they "outbred"). Only two females approached the speaker broadcasting the call of the male to which they were more closely related (i.e., they "inbred"). Seven females failed to make a choice, but they were equally closely related to the two males whose calls were being broadcast (from Waldman and Tocher 1998).

sponses to recorded calls of two males, alternately broadcast from speakers on either side of the arena (Waldman 1997; Waldman and Tocher 1998). In each test both males were from the same pond as the female subject, but they were probably unfamiliar to her because we had recorded the males in the field during a previous year. Additionally we matched stimulus calls for call duration and sound intensity. We genetically typed the females, comparing their multilocus nuclear DNA fingerprints (Jeffreys 1987) with those of the males only after the behavioral tests had been completed. Thus the behavioral tests were conducted blindly.

Females can discriminate between relatives and nonrelatives by their calls. Subjects showed a strong preference to approach the speaker broadcasting the call of the male genetically less similar to themselves (Figure 17.4). When calls were switched between speakers, subjects approached the new speaker, clearly indicating that they were responding to the call and no other aspect of the arena. Only two of 29 females preferred the call of the genetically more similar male ($p < 0.0001$, binomial test). Seven females showed no consistent preference for either call, but their DNA fingerprints reveal that they were no more closely related to one male than to the other.

Although playback experiments using calls recorded in natural conditions provide a realistic choice situation for the test subjects, confounding variables may have unknowingly been introduced into the experiment because of the normal variation inherent in the calls. To control for this possibility we next generated synthesized calls from two pairs of males. The synthesized calls held total call effort (Sullivan 1992) and dominant frequency (Howard and Palmer 1995) constant but preserved all temporal call characteristics including rise time, pulse duration, and interpulse interval. Females tested with synthesized calls demonstrated preferences for calls derived from the male to which they were less closely related ($p = 0.0017$, binomial test; Figure 17.5), much as in the experiments with natural recordings.

We still do not know how females assess their relatedness to calling males. Unlike tadpoles, which can learn their own odors and recognize similarly smelling conspecifics as their siblings, female toads appear to lack a contextually reliable opportunity to learn characteristics of their brothers' calls. Perhaps siblings initially recognize one another through other cues, such as odors, that they had previously learned (e.g., at metamorphosis), and then selectively learn calls of their brothers (Waldman et al. 1992). Genetic partitioning within ponds (Christein and Taylor 1978) might be perpetuated through metamorphosis so that kin groups disperse together away from ponds. Some frogs call outside their breeding seasons, and if this occurs females might have an opportunity to hear their brothers' calls (Waldman et al. 1992). The most likely explanation however is that the temporal components of males' calls are generated by a neural pattern generator (template) that is in part genetically encoded. Even though females do not call they might use the same template to evaluate males' calls. In this manner they might recognize calls of their brothers and other close relatives.

Conclusions

Amphibians have proven a useful model system for studying kin recognition. Larvae of many species show remarkably precise abilities to recognize their close relatives. Kin recog-

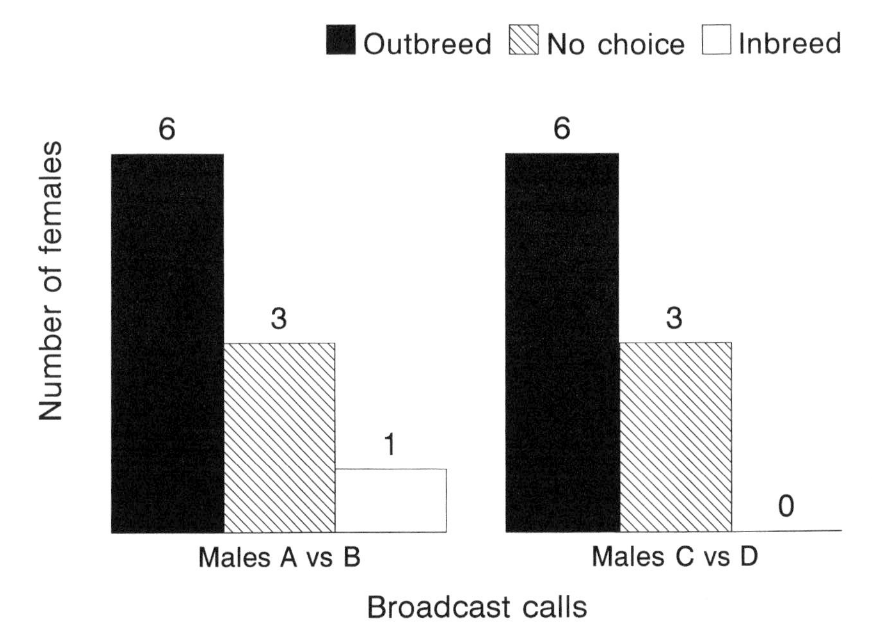

Figure 17.5. Responses of females to synthesized calls, tested as in Figure 17.4. Pulses were synthetically generated that matched the shape (rise time) and temporal features of calling males, but total call effort and dominant frequency were matched to prevent possible discrimination based on these parameters. DNA fingerprints of males and females were compared only after behavioral tests had been completed. Twelve females approached the speaker broadcasting the synthesized call of the male to which they were less closely related (i.e., they "outbred"). Only one female approached the speaker broadcasting the synthesized call of the male to which they were more closely related (i.e., she "inbred"). Six females failed to make a choice.

nition develops through learning of both genetic and environmental cues, and discrimination is induced largely by the perception of chemical cues. We are only now learning the fitness benefits accrued by larval amphibians in discriminating between kin and nonkin, but clearly multiple benefits are possible and each reinforces selection for kin recognition.

Our recent work suggests that adults also have the ability and the opportunity to recognize their close relatives. Indeed, given the documented costs of close inbreeding, kin discrimination in the context of inbreeding avoidance may confer fitness benefits greater than those attained by discriminating larvae. Although vocal cues appear sufficient, much work remains to be done to fully understand how females recognize their kin. Given that frequency-dependent selection, such as that generated by mate choice, appears necessary to sustain variation in genetically based recognition cues (Crozier 1987), our findings begin to resolve the paradox of how amphibian kin-recognition systems are maintained.

In recent years evolutionary explanations for mate choice have focused on processes of sexual selection (Ryan 1997). Yet in *Bufo americanus* mate choice appears to be more influenced by selection for genetic compatibility than by selection for "good genes" or runaway secondary sexual characters. Certainly sexual selection plays an important role in mate choice in many amphibians (Sullivan et al. 1995). Nonetheless workers studying mating systems need to consider genetic compatibility as a factor influencing mate choice. Heterozygosity at specific loci, such as the major histocompatibility complex, may confer enhanced disease resistance (Penn and Potts 1999), or heterozygous advantage may result from complementary alleles throughout the genome (Waldman and Tocher 1998). As human activities continue to encroach on natural habitat and disrupt metapopulations, the dynamics of mate choice may be increasingly driven by selection to maintain genetic diversity.

References

Avise J. C., B. W. Bowen, and T. Lamb. 1989. DNA fingerprints from hypervariable mitochondrial genotypes. Molecular Biology and Evolution 6:258–269.

Bateson, P. 1983. Optimal outbreeding. Pp. 257–277. *In* P. Bateson (ed.), Mate Choice. Cambridge University Press, Cambridge.

Bermingham, E., T. Lamb, and J. C. Avise. 1986. Size polymorphism and heteroplasmy in the mitochondrial DNA of lower vertebrates. Journal of Heredity 77:249–252.

Berven, K. A., and T. A. Grudzien. 1990. Dispersal in the wood frog (*Rana sylvatica*): Implications for genetic population structure. Evolution 44:2047–2056.

Blaustein, A. R. 1983. Kin recognition mechanisms: Phenotypic matching or recognition alleles? American Naturalist 121:749–754.

Blaustein, A. R., M. Bekoff, J. A. Byers, and T. J. Daniels. 1991. Kin recognition in vertebrates: What do we really know about adaptive value? Animal Behaviour 41:1079–1083.

Blaustein, A. R., and R. K. O'Hara. 1981. Genetic control for sibling recognition? Nature 290:246–248.

Blaustein, A. R., and R. K. O'Hara. 1982a. Kin recognition cues in *Rana cascadae* tadpoles. Behavioral and Neural Biology 36:77–87.

Blaustein, A. R., and R. K. O'Hara. 1982b. Kin recognition in *Rana cascadae* tadpoles: Maternal and paternal effects. Animal Behaviour 30:1151–1157.

Blaustein, A. R., and R. K. O'Hara. 1986. An investigation of kin recognition in red-legged frog (*Rana aurora*) tadpoles. Journal of Zoology, London A 209:347–353.

Blaustein, A. R., R. K. O'Hara, and D. H. Olson. 1984. Kin preference behaviour is present after metamorphosis in *Rana cascadae* frogs. Animal Behaviour 32:445–450.

Blaustein, A. R., and B. Waldman. 1992. Kin recognition in anuran amphibians. Animal Behaviour 44:207–221.

Blaustein, A. R., and S. C. Walls. 1995. Aggregation and kin recognition. Pp. 568–602. *In* H. Heatwole and B. K. Sullivan (eds.), Amphibian Biology, Vol. 2. Surrey Beatty and Sons, Chipping Norton, New South Wales, Australia.

Blaustein, A. R., T. Yoshikawa, K. Asoh, and S. C. Walls. 1993. Ontogenetic shifts in tadpole kin recognition: Loss of signal and perception. Animal Behaviour 46:525–538.

Bogert, C. M. 1947. A field study of homing in the Carolina toad. American Museum Novitates 1355:1–24.

Bonini, L., G. Bogliana, and F. Barbieri. 1994. Mancanza di associazione preferenziale fra consanguinei nei raggruppamenti di girini di alcuni Anuri italiana. Studi Trentini di Scienze Naturali Acta Biologica 71:157–161.

Breden, F. 1987. The effect of post-metamorphic dispersal on the population genetic structure of Fowler's toad, *Bufo woodhousei fowleri*. Copeia 1987:386–395.

Buckley, D., B. Arano, P. Herrero, and G. Llorente. 1996. Population structure of Moroccan water frogs: Genetic cohesion despite a fragmented distribution. Journal of Zoological Systematics and Evolutionary Research 34:173–179.

Carreno, C. A., T. J. Vess, and R. N. Harris. 1996. An investigation of kin recognition abilities in larval four-toed salamanders, *Hemidactylium scutatum* (Caudata: Plethodontidae). Herpetologica 52:293–300.

Cheney, D. L., and R. M. Seyfarth. 1990. How Monkeys See the World. University of Chicago Press, Chicago.

Christein, D., and D. H. Taylor. 1978. Population dynamics in breeding aggregations of the American toad, *Bufo americanus* (Amphibia, Anura, Bufonidae). Journal of Herpetology 12:17–24.

Cornell, T. J., K. A. Berven, and G. J. Gamboa. 1989. Kin recognition by tadpoles and froglets of the wood frog *Rana sylvatica*. Oecologia 78:312–316.

Crozier, R. H. 1987. Genetic aspects of kin recognition: Concepts, models, and synthesis. Pp. 55–73. *In* D. J. C. Fletcher and C. D. Michener (eds.), Kin Recognition in Animals. John Wiley and Sons, Chichester.

Davies, N. B., and T. R. Halliday. 1977. Optimal mate selection in the toad *Bufo bufo*. Nature 269:56–58.

Densmore, L. D., J. W. Wright, and W M. Brown. 1985. Length variation and heteroplasmy are frequent in mitochondrial DNA from parthenogenetic and bisexual lizards (genus *Cnemidophorus*). Genetics 110:689–707.

Dole, J. W. 1972. Homing and orientation of displaced toads, *Bufo americanus*, to their home sites. Copeia 1972:151–157.

D'Orgeix, C. A., and B. J. Turner. 1995. Multiple paternity in the red-eyed treefrog *Agalychnis callidryas* (Cope). Molecular Ecology 4:505–508.

Driscoll, D. A. 1997. Mobility and metapopulation structure of *Geocrinia alba* and *Geocrinia vitellina*, two endangered frog species from southwestern Australia. Australian Journal of Ecology 22:185–195.

Driscoll, D. A. 1998a. Genetic structure, metapopulation processes and evolution influence the conservation strategies for two endangered frog species. Biological Conservation 83:43–54.

Driscoll, D. A. 1998b. Genetic structure of the frogs *Geocrinia lutea* and *Geocrinia rosea* reflects extreme population divergence and range changes, not dispersal barriers. Evolution 52:1147–1157.

Edenhamn, P. 1996. Spatial dynamics of the European tree frog (*Hyla arborea* L.) in a heterogeneous landscape. Ph.D. thesis. Swedish University of Agricultural Sciences, Uppsala.

Fairchild, L. 1984. Male reproductive tactics in an explosive breeding toad population. American Zoologist 24:407–418.

Falconer, D. S. 1989. Introduction to Quantitative Genetics. 3rd ed. Longman, Burnt Mill, Harlow.

Fisher, R. A. 1958. The Genetical Theory of Natural Selection. Dover, New York.

Fishwild, T. G., R. A. Schemidt, K. M. Jankens, K. A. Berven, G. J. Gamboa, and C. M. Richards. 1990. Sibling recognition by larval frogs (*Rana pipiens, Rana sylvatica* and *Pseudacris crucifer*). Journal of Herpetology 24:40–44.

Forester, D. C., and K. J. Thompson. 1998. Gauntlet behaviour as a male sexual tactic in the American toad (Amphibia: Bufonidae). Behaviour 135:99–119.

Gabor, C. R. 1996. Differential kin discrimination by red-spotted newts (*Notophthalmus viridescens*) and smooth newts (*Triturus vulgaris*). Ethology 102:649–659.

Gamboa, G. J., K. A. Berven, R. A. Schemidt, T. G. Fishwild, and K. M. Jankens. 1991. Kin recognition by larval wood frogs (*Rana sylvatica*): Effects of diet and prior exposure to conspecifics. Oecologia 86:319–324.

Gamboa, G. J., H. K. Reeve, and W. G. Holmes. 1991. Conceptual issues and methodology in kin-recognition research: A critical discussion. Ethology 88:109–127.

Gamboa, G. J., H. K. Reeve, and D. W. Pfennig. 1986. The evolution and ontogeny of nestmate recognition in social wasps. Annual Review of Entomology 31:431–454.

Gatz, A. J., Jr. 1981. Non-random mating by size in American toads, *Bufo americanus*. Animal Behaviour 29:1004–1012.

Getz, W. M. 1981. Genetically based kin recognition systems. Journal of Theoretical Biology 92:209–226.

Grafen, A. 1990. Do animals really recognize kin? Animal Behaviour 39:42–54.

Greenberg, L. 1979. Genetic component of bee odor in kin recognition. Science 206:1095–1097.

Greenberg, L. 1988. Kin recognition in the sweat bee, *Lasioglossum zephyrum*. Behavior Genetics 18:425–438.

Hall, J. A., J. H. Larsen, D. E. Miller, and R. E. Fitzner. 1995. Discrimination of kin-based and diet-based cues by larval spadefoot toads, *Scaphiopus intermontanus* (Anura, Pelobatidae), under laboratory conditions. Journal of Herpetology 29:233–243.

Harrison, R. G. 1989. Animal mitochondrial DNA as a genetic marker in population and evolutionary biology. Trends in Ecology and Evolution 4:6–11.

Hepper, P. G., and B. Waldman. 1992. Embryonic olfactory learning in frogs. Quarterly Journal of Experimental Psychology B 44:179–197.

Heusser, H. 1969. Die Lebensweise der Erdkröte, *Bufo bufo* (L.): Das Orientierungsproblem. Revue Suisse de Zoologie 76:443–518.

Hews, D. K. 1988. Alarm response in larval western toads, *Bufo boreas:* Release of larval chemicals by a natural predator and its effect on predator capture efficiency. Animal Behaviour 36:125–133.

Hews, D. K., and A. R. Blaustein. 1985. An investigation of the alarm response in *Bufo boreas* and *Rana cascadae* tadpoles. Behavioral and Neural Biology 43:47–57.

Hitchings, S. P., and T. J. C. Beebee. 1997. Genetic substructuring as a result of barriers to gene flow in urban *Rana temporaria* (common frog) populations: Implications for biodiversity conservation. Heredity 79:117–127.

Hitchings, S. P., and T. J. C. Beebee. 1998. Loss of genetic diversity and fitness in common toad (*Bufo bufo*) populations isolated by inimical habitat. Journal of Evolutionary Biology 11:269–283.

Hokit, D. G., and A. R. Blaustein. 1994. The effects of kinship on growth and development in tadpoles of *Rana cascadae*. Evolution 48:1383–1388.

Hokit, D. G., and A. R. Blaustein. 1997. The effects of kinship on interactions between tadpoles of *Rana cascadae*. Ecology 78:1722–1735.

Howard, R. D. 1988. Sexual selection on male body size and mating behaviour in American toads, *Bufo americanus*. Animal Behaviour 36:1796–1808.

Howard, R. D., and J. G. Palmer. 1995. Female choice in *Bufo americanus:* Effects of dominant frequency and call order. Copeia 1995:212–217.

Howard, R. D., H. H. Whiteman, and T. I. Schueller. 1994. Sexual selection in American toads: A test of a good-genes hypothesis. Evolution 48:1286–1300.

Howard, R. D., and J. R. Young. 1998. Individual variation in male vocal traits and female mating preferences in *Bufo americanus*. Animal Behaviour 55:1165–1179.

Jaeger, R. G. 1986. Pheromonal markers as territorial advertisement by terrestrial salamanders. Pp. 191–203. *In* D. Duvall, D. Müller-Schwarze, and R. M. Silverstein (eds.), Chemical Signals in Vertebrates 4. Ecology, Evolution, and Comparative Biology. Plenum Press, New York.

Jameson, D. L. 1957. Population structure and homing responses in the Pacific tree frog. Copeia 1957:221–228.

Jasienski, M. 1988. Kinship ecology of competition: Size hierarchies in kin and nonkin laboratory cohorts of tadpoles. Oecologia 77:407–413.

Jeffreys, A. J. 1987. Highly variable minisatellites and DNA fingerprints. Biochemical Society Transactions 15:309–317.

Kimberling, D. N., A. R. Ferreira, S. M. Shuster, and P. Keim. 1996. RAPD marker estimation of genetic structure among isolated northern leopard frog populations in the southwestern USA. Molecular Ecology 5:521–529.

Kruse, K. C. 1981. Mating success, fertilization potential, and male body size in the American toad (*Bufo americanus*). Herpetologica 37:228–233.

Lacy, R. C., and P. W. Sherman. 1983. Kin recognition by phenotype matching. American Naturalist 121:489–512.

Laurila, A. and P. Seppa. 1998. Multiple paternity in the common frog (*Rana temporaria*): Genetic evidence from tadpole kin groups. Biological Journal of the Linnean Society 63:221–232.

Licht, L. E. 1976. Sexual selection in toads (*Bufo americanus*). Canadian Journal of Zoology 54:1277–1284.

Littlejohn, M. J. 1981. Reproductive isolation: A critical review. Pp. 298–334. *In* W. R. Atchley and D. S. Woodruff (eds.), Evolution and Speciation. Cambridge University Press, Cambridge.

Madison, D. M. 1975. Intraspecific odor preferences between salamanders of the same sex: Dependence on season and proximity of residence. Canadian Journal of Zoology 53:1356–1361.

Madsen, T., B. Stille, and R. Shine. 1996. Inbreeding depression in an isolated population of adders *Vipera berus*. Biological Conservation 75:113–118.

Mantel, N. 1967. The detection of disease clustering and a generalized regression approach. Cancer Research 27:209–220.

Masters, B. S., and D.C. Forester. 1995. Kin recognition in a brooding salamander. Proceedings of the Royal Society of London B 261:43–48.

McAlpine, S. 1993. Genetic heterozygosity and reproductive success in the green treefrog, *Hyla cinerea*. Heredity 70:553–558.

McAlpine, S., and M. H. Smith. 1995. Genetic correlates of fitness in the green treefrog, *Hyla cinerea*. Herpetologica 51:393–400.

Mitchell, S. L. 1990. The mating system genetically affects offspring performance in Woodhouse's toad (*Bufo woodhousei*). Evolution 44:502–519.

Mitton, J. B., C. Carey, and T. D. Kocher. 1986. The relation of enzyme heterozygosity to standard and active oxygen consumption and body size of tiger salamanders, *Ambystoma tigrinum*. Physiological Zoology 59:574–582.

Naylor, A. F. 1962. Mating systems which could increase heterozygosity for a pair of alleles. American Naturalist 96:51–60.

O'Hara, R. K., and A. R. Blaustein. 1981. An investigation of sibling recognition in *Rana cascadae* tadpoles. Animal Behaviour 29:1121–1126.

O'Hara, R. K., and A. R. Blaustein. 1982. Kin preference behavior in *Bufo boreas* tadpoles. Behavioral Ecology and Sociobiology 11:43–49.

O'Hara, R. K., and A. R. Blaustein. 1985. *Rana cascadae* tadpoles aggregate with siblings: An experimental field study. Oecologia 67:44–51.

O'Hara, R. K., and A. R. Blaustein. 1988. *Hyla regilla* and *Rana pretiosa* tadpoles fail to display kin recognition behaviour. Animal Behavior 36:946–948.

Pärt, T. 1996. Problems with testing inbreeding avoidance: The case of the collared flycatcher. Evolution 50:1625–1630.

Partridge, L. 1983. Non-random mating and offspring fitness. Pp. 227–255. *In* P. Bateson (ed.), Mate Choice. Cambridge University Press, Cambridge.

Penn, D. J., and W. K. Potts. 1999. The evolution of mating preferences and major histocompatibility complex genes. American Naturalist 153:145–164.

Pfennig, D. W. 1990. "Kin recognition" among spadefoot toad tadpoles: A side-effect of habitat selection? Evolution 44:785–798.

Pfennig, D. W. 1997. Kinship and cannibalism. BioScience 47:667–675.

Pfennig, D. W. 1999. Cannibalistic tadpoles that pose the greatest threat to kin are most likely to discriminate kin. Proceedings of the Royal Society of London B 266:57–61.

Pfennig, D. W., H. K. Reeve, and P. W. Sherman. 1993. Kin recognition and cannibalism in spadefoot toad tadpoles. Animal Behaviour 46:87–94.

Pfennig, D. W., P. W. Sherman, and J. P. Collins. 1994. Kin recognition and cannibalism in polyphenic salamanders. Behavioral Ecology 5:225–232.

Pierce, B. A., and J. B. Mitton. 1982. Allozyme heterozygosity and growth in the tiger salamander, *Ambystoma tigrinum*. Journal of Heredity 73:250–253.

Porter, R. H., S. A. McFadyen-Ketchum, and G. A. King. 1989. Underlying bases of recognition signatures in spiny mice, *Acomys cahirinus*. Animal Behaviour 37:638–644.

Reading, C. J., J. Loman, and T. Madsen. 1991. Breeding pond fidelity in the common toad, *Bufo bufo*. Journal of Zoology, London 225:201–211.

Reeve, H. K. 1989. The evolution of conspecific acceptance thresholds. American Naturalist 133:407–435.

Reh, W., and A. Seitz. 1990. The influence of land use on the genetic structure of populations of the common frog *Rana temporaria*. Biological Conservation 54:239–249.

Ryan, M. J. 1997. Sexual selection and mate choice. Pp. 179–202. *In* J. R. Krebs and N. B. Davies (eds.), Behavioural Ecology. An Evolutionary Approach. Fourth edition. Blackwell, Oxford.

Samallow, B. P., and M. E. Soulé. 1983. A case of stress related heterozygote superiority in nature. Evolution 37:646–649.

Schwarzkopf, L., and R. A. Alford. 2001. Nomadic movement in tropical toads. Oikos (in press).

Sherman, P. W., and W. G. Holmes. 1985. Kin recognition: Issues and evidence. Pp. 437–460. *In* B. Hölldobler and M. Lindauer (eds.), Experimental Behavioral Ecology and Sociobiology. Sinauer, Sunderland, MA.

Shields, W. M. 1982. Philopatry, Inbreeding, and the Evolution of Sex. State University of New York Press, Albany, NY.

Sinsch, U. 1990. Migration and orientation in anuran amphibians. Ethology Ecology and Evolution 2:65–79.

Sinsch, U. 1992. Structure and dynamic of a natterjack toad metapopulation (*Bufo calamita*). Oecologia 90:489–499.

Sinsch, U. 1997. Postmetamorphic dispersal and recruitment of first breeders in a *Bufo calamita* metapopulation. Oecologia 112:42–47.

Sjögren, P. 1991a. Extinction and isolation gradients in metapopulations: The case of the pool frog (*Rana lessonae*). Biological Journal of the Linnean Society 42:135–147.

Sjögren, P. 1991b. Genetic variation in relation to demography of peripheral pool frog populations (*Rana lessonae*). Evolutionary Ecology 5:248–271.

Sjögren Gulve, P. 1994. Distribution and extinction patterns within a northern metapopulation of the pool frog, *Rana lessonae*. Ecology 75:1357–1367.

Smith, D.C. 1990. Population structure and competition among kin in the chorus frog (*Pseudacris triseriata*). Evolution 44:1529–1541.

Sullivan, B. K. 1992. Sexual selection and calling behavior in the American toad (*Bufo americanus*). Copeia 1992:1–7.

Sullivan, B. K., M. J. Ryan, and P. A. Verrell. 1995. Female choice and mating system structure. Pp. 469–517. *In* H. Heatwole and B. K. Sullivan (eds.), Amphibian Biology, Vol. 2. Surrey Beatty and Sons, Chipping Norton, New South Wales, Australia.

Twitty, V. C. 1966. Of Scientists and Salamanders. W. H. Freeman, San Francisco.

Vrijenhoek, R. C. 1994. Genetic diversity and fitness in small populations. Pp. 37–53, *In* V. Loeschcke, J. Tomiuk, and S. K. Jain (eds.), Conservation Genetics. Birkhauser Verlag, Basel.

Waldman, B. 1981. Sibling recognition in toad tadpoles: The role of experience. Zeitschrift für Tierpsychologie 56:341–358.

Waldman, B. 1982. Sibling association among schooling toad tadpoles: Field evidence and implications. Animal Behaviour 30:700–713.

Waldman, B. 1984. Kin recognition and sibling association among wood frog (*Rana sylvatica*) tadpoles. Behavioral Ecology and Sociobiology 14:171–180.

Waldman, B. 1985. Olfactory basis of kin recognition in toad tadpoles. Journal of Comparative Physiology A 156:565–577.

Waldman, B. 1986. Chemical ecology of kin recognition in anuran amphibians. Pp. 225–242. *In* D. Duvall, D. Müller-Schwarze, and D. M. Silverstein (eds.), Chemical Signals in Vertebrates 4. Ecology, Evolution, and Comparative Biology. Plenum Press, New York.

Waldman, B. 1987. Mechanisms of kin recognition. Journal of Theoretical Biology 128:159–185.

Waldman, B. 1988. The ecology of kin recognition. Annual Review of Ecology and Systematics 19:543–571.

Waldman, B. 1989. Do anuran larvae retain kin recognition abilities following metamorphosis? Animal Behaviour 37:1055–1058.

Waldman, B. 1991. Kin recognition in amphibians. Pp. 162–219. *In* P. G. Hepper (ed.), Kin Recognition. Cambridge University Press, Cambridge.

Waldman, B. 1997. Kinship, sexual selection, and female choice in toads. Advances in Ethology 32:200.

Waldman, B., and K. Adler. 1979. Toad tadpoles associate preferentially with siblings. Nature 282:611–613.

Waldman, B., and J. S. McKinnon. 1993. Inbreeding and outbreeding in fishes, amphibians, and reptiles. Pp. 250–282. *In* N. W. Thornhill (ed.), The Natural History of Inbreeding and Outbreeding: Theoretical and Empirical Perspectives. University of Chicago Press, Chicago.

Waldman, B., J. E. Rice, and R. L. Honeycutt. 1992. Kin recognition and incest avoidance in toads. American Zoologist 32:18–30.

Waldman, B., and M. Tocher. 1998. Behavioral ecology, genetic diversity, and declining amphibian populations. Pp. 394–443. *In* T. Caro (ed.), Behavioral Ecology and Conservation Biology. Oxford University Press, New York.

Walls, S. C. 1991. Ontogenetic shifts in the recognition of siblings and neighbours by juvenile salamanders. Animal Behaviour 42:423–434.

Walls, S. C., and A. R. Blaustein. 1995. Larval marbled salamanders, *Ambystoma opacum,* eat their kin. Animal Behaviour 50:537–545.

Walls, S. C., C. S. Conrad, M. L. Murillo, and A. R. Blaustein. 1996. Agonistic behaviour in larvae of the northwestern salamander (*Ambystoma gracile*): The effects of kinship, familiarity and population source. Behaviour 133:965–984.

Walls, S. C., and R. E. Roudebush. 1991. Reduced aggression toward siblings as evidence of kin recognition in cannibalistic salamanders. American Naturalist 138:1027–1038.

Wassersug, R. J. 1973. Aspects of social behavior in anuran larvae. Pp. 273–297. *In* J. L. Vial (ed.), Evolutionary Biology of the Anurans. University of Missouri Press, Columbia, MO.

Welch, A. M., R. D. Semlitsch, and H. C. Gerhardt. 1998. Call duration as an indicator of genetic quality in male gray tree frogs. Science 280:1928–1930.

Wells, K. D. 1977. The social behaviour of anuran amphibians. Animal Behaviour 25:666–693.

Wells, K. D., and T. L. Taigen. 1984. Reproductive behavior and aerobic capacities of male American toads (*Bufo americanus*): Is behavior constrained by physiology? Herpetologica 40:292–298.

Wetton, J. H., R. E. Carter, D. T. Parkin, and D. Walters. 1987. Demographic study of a wild house sparrow population by DNA fingerprinting. Nature 327:147–149.

Wilbur, H. M., D. I. Rubenstein, and L. Fairchild. 1978. Sexual selection in toads: The roles of female choice and male body size. Evolution 32:264–270.

Woodward, B. D. 1986. Paternal effects on juvenile growth in *Scaphiopus multiplicatus.* American Naturalist 128:58–65.

Woodward, B. D. 1987. Paternal effects on offspring traits in *Scaphiopus couchi* (Anura: Pelobatidae). Oecologia 73:626–629.

Wright, M. F., and S. I. Guttman. 1995. Lack of an association between heterozygosity and growth rate in the wood frog, *Rana sylvatica.* Canadian Journal of Zoology 73:569–575.

Addresses of Contributors

Todd Alder
Department of Biology
University of Utah
Salt Lake City, UT 84112
USA

Adolfo Amézquita
Departmento de Ciencias Biológicas
Universidad de los Andes
A.A. 4976
Bogotá
Colombia

Jaime Bosch
Museo Nacional de Ciencias Naturales
José Gutiérez Abascal 2
28006 Madrid
Spain

Eliot A. Brenowitz
Department of Psychology
Box 351525
University of Washington
Seattle, WA 98195-1525
USA

Sergio Castellano
Dipartimento di Biologia Animale
Università di Torino
Via Accademia Albertina, 17
10123 Torino
Italy

Joanne Chu
Department of Zoology
Oregon State University
Cordley Hall
Corvallis, OR 97331
USA

Sharon B. Emerson
Department of Biology
University of Utah
Salt Lake City, UT 84112
USA

H. Carl Gerhardt
Division of Biological Sciences
University of Missouri
Columbia, MO 65201
USA

Cristina Giacoma
Dipartimento di Biologia Animale
Università di Torino
Via Accademia Albertina, 17
10123 Torino
Italy

Sam Horng
Department of Biological Sciences
Sherman Fairchild Center for the Life Sciences
Mail Code 2432
Columbia University
New York, NY 10027
USA

Walter Hödl
Institute of Zoology
University of Vienna
Althanstrasse 14
A-1090 Wien
Austria

Darcy B. Kelley
Department of Biological Sciences
Sherman Fairchild Center for the Life Sciences
Mail Code 2432
Columbia University
New York, NY 10027
USA

Murray J. Littlejohn
Department of Zoology
University of Melbourne
Parkville, Victoria 3052
Australia

Rafael Márquez
Centro de Biologia Ambiental
Faculdade de Ciências
Universidade de Lisboa
P-1700 Lisboa
Portugal

Peter M. Narins
Department of Physiological Science
621 Circle Drive South
Box 951527
University of California
Los Angeles, CA 90095–1527
USA

Steve Phelps
Center for Behavioral Neuroscience
Emory University
Atlanta, GA 30329
USA

Gary J. Rose
Department of Biology
University of Utah
Salt Lake City, UT 84112
USA

A. Stanley Rand
Smithsonian Tropical Research Institute
Apartado 2072
Balboa
Panama

Michael J. Ryan
Section of Integrative Biology C0930
School of Biological Sciences
University of Texas
Austin, TX 78712
USA

Joshua J. Schwartz
Division of Biological Sciences
University of Missouri
Columbia, MO 65201
USA

Martha L. Tobias
Department of Biological Sciences
Sherman Fairchild Center for the Life Sciences
Mail Code 2432
Columbia University
New York, NY 10027
USA

Bruce Waldman
Department of Zoology
University of Canterbury
Private Bag 4800
Christchurch
New Zealand

Kentwood D. Wells
Department of Ecology and Evolutionary Biology
University of Connecticut
Storrs, CT 06269–3043
USA

Walter Wilczynski
Department of Psychology
University of Texas
Austin, TX 78712
USA

Mary Jane West-Eberhard
Smithsonian Tropical Research Institute
Apartado 2072
Balboa
Panama

Index